AF509459

Bioactive Lipids and Lipidomics 2018

Bioactive Lipids and Lipidomics 2018

Special Issue Editors

Mario Ollero
David Touboul

MDPI • Basel • Beijing • Wuhan • Barcelona • Belgrade • Manchester • Tokyo • Cluj • Tianjin

Special Issue Editors

Mario Ollero
Université Paris Est Créteil
France

David Touboul
Université Paris-Saclay
France

Editorial Office
MDPI
St. Alban-Anlage 66
4052 Basel, Switzerland

This is a reprint of articles from the Special Issue published online in the open access journal *International Journal of Molecular Sciences* (ISSN 1422-0067) (available at: https://www.mdpi.com/journal/ijms/special_issues/lipids_lipidomics).

For citation purposes, cite each article independently as indicated on the article page online and as indicated below:

LastName, A.A.; LastName, B.B.; LastName, C.C. Article Title. *Journal Name* **Year**, *Article Number*, Page Range.

ISBN 978-3-03936-276-9 (Hbk)
ISBN 978-3-03936-277-6 (PDF)

Cover image courtesy of Iwona Pranke.

Contents

About the Special Issue Editors

Mario Ollero, DVM, PhD. Currently full professor at Université Paris Est Créteil, he has developed transversal research activity from fatty acid biochemistry and lipid-related oxidative stress to membrane microdomain dynamics and cell signaling in the fields of gamete biology, cystic fibrosis and glomerular diseases. His previous appointments include research positions and professorships at Universidad de Zaragoza (Spain), Harvard Medical School (USA), Institut National de la Santé et la Recherche Médicale (France) and Centre National de la Recheche Scientifique (France).

David Touboul, PhD. is Research Assistant at CNRS and is leading the Mass Spectrometry research group at CNRS-ICSN. He is a specialist in the structural characterization of natural products by mass spectrometry including lipidomic approaches and the development of new chemo-informatics tools such as MetGem for molecular networks. He received the Bronze Medal of CNRS in 2014.

International Journal of
Molecular Sciences

Editorial

Lipidomics Conquers a Niche, Consolidates Growth

David Touboul [1] and Mario Ollero [2,3,*]

[1]   Institut de Chimie des Substances Naturelles, CNRS UPR 2301, Univ. Paris-Sud, Université Paris-Saclay, 8 Avenue de la Terrasse, 91198 Gif-sur-Yvette, France

[2]   Institut Mondor de Recherche Biomédicale, INSERM, U955 EQ21, 8, rue du Général Sarrail, 94010 Créteil, France

[3]   Université Paris Est Créteil, 61, avenue du Général de Gaulle, 94010 Créteil, France

*   Correspondence: Mario.ollero@inserm.fr; Tel.: +33-149813667

Received: 25 June 2019; Accepted: 26 June 2019; Published: 29 June 2019

Sixteen years after the first published article in which the term "lipidomics" was stated [1], one of the latecomers to the omics revolution has consolidated its position in the evolution of analytical approaches in experimental biology and has conquered a specific niche in science. More than 3000 publications since the pioneering work by Han and colls. (colleagues) confirm that widespread lipidomics research has become a reality, even if constituting a modest production as compared to that of proteomics and genomics. As shown in Figure 1, the number of articles published on genomics and proteomics seems to have plateaued since 2017, according to PubMed. Meanwhile, the "minor" omics disciplines (transcriptomics, metabolomics, glycomics, and lipidomics) continue to grow exponentially (Figure 1a). Lipidomics has undergone an explosion of publications since 2013 and continues to experience a sustained and impressive growth (Figure 1b). Most strikingly, since the publication of the last Special Issue on "Bioactive Lipids and Lipidomics", in 2015, more articles have been published on lipidomics than in the preceding twelve years (2003–2012) (Figure 1c).

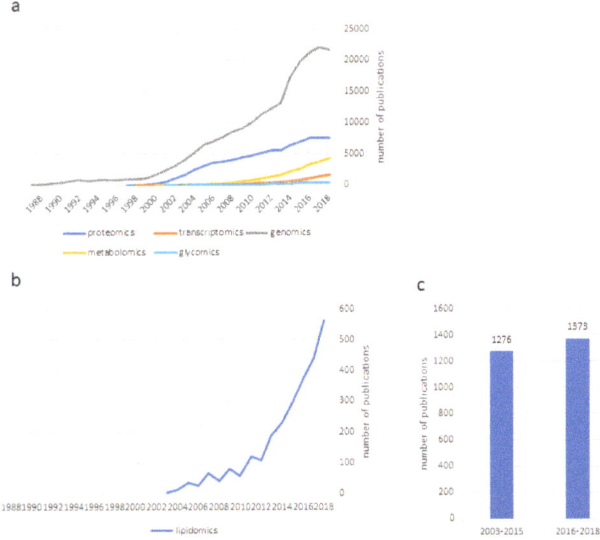

Figure 1. (**a**) Number of listed publications in PubMed over time using the corresponding "omic" discipline name as the keyword. (**b**) Number of listed publications in PubMed over time using "lipidomics" as the keyword. (**c**) Comparison of the number of listed publications in PubMed using "lipidomics" as the keyword during two periods of time, namely 2003–2015 and 2016–2018.

The current special issue provides an instantaneous picture of the situation and witnesses the place of lipidomics in terms of technological advances and fields of application, and hints about the directions research may follow in the near future.

1. Technology

As a technology-driven discipline, the evolution of lipidomics is directly linked to that of the associated technologies for the separation, detection, and identification of compounds. While analytical approaches are consolidated and can be globally applicable to most omics, the specific technological development presently resides in data processing and molecular networking analysis.

A particular challenge is still the identification and characterization of molecules. The combination of mass spectrometry data with other technologies, such as optical rotation analysis and NMR [2], provides functional and structural insight. Differential scanning calorimetry is utilized in the quality control of lipid biomaterials [3].

The integration of different omics analyses has been one of the main challenges of global analysis strategies. Stuani and colls. successfully combined proteomics and lipidomics to address fatty acid metabolism in combination with stable isotope labeling [4].

Supercritical fluid chromatography, addressed in our 2015 special issue [5], is represented in this new edition by a study in which it is applied to glycosphingolipid analysis [6]. In the same study, this novel approach is combined with isotope labelling and high-performance thin layer chromatography to scrutinize metabolic flux, showing the relevance of classic strategies. This is also the case of GC-MS for fatty acid profiling (4). An alternative to the latter is the selective ion monitoring-tandem mass spectrometry (SIM-MS/MS), used in this case for fatty acid analysis in extracellular vesicles [7] and tissues [3].

Lipid imaging remains a promising field, and a cluster TOF-SIMS strategy is presented by Abbas and colls. to detect lipid ions at the kidney glomerular scale [8].

As emerging technologies allow for increasing sensitivity of measurement, studies have to deal with a larger number of variables, thus requiring a profound network analysis. This is being performed with the help of software tools like MZmine [9], IDEOM [10], or online open source platforms, like Cytoscape (https://cytoscape.org/). Databases like the one provided by the Dictionary of Natural Products are of great help.

2. Applications

The expansion of lipidomics leads the discipline to conquering new areas. This includes non-global studies targeting novel bioactive lipids.

The analysis of natural molecules is gaining importance. The need for new perspectives in therapeutics or plague control is directly linked to technological advances. Thorough analyses of new natural sources of bioactive molecules, such as endophyte products (as published by Barthélemy and colls [9]), are highly valuable. The antimicrobial and insecticidal effects of lipoamino acids from entomopathogens have also been characterized by Touré and colls [2]. Abnormal lipid metabolism due to a frequent mutation in acute myeloid leukemia cells has been revealed by Stuani and colls [4].

The roles of lipid-containing supra-structures, like extracellular vesicles, are revolutionizing concepts in cell biology and are also the foci of lipidomic scrutiny. Sagini and colls. analyzed the fatty acid profiling of extracellular vesicles released by senescent cells and found that they are selectively enriched in polyunsaturated and saturated chains, thus prompting intriguing hypotheses [7]. A similar concept, but arising from prokaryote cells, is that of outer membrane vesicles, which play a role in the pathogenic mechanisms of bacterial infections. Their lipidomic analysis has been addressed by Jasim and colls. in *Klebsiella pneumoniae*, providing valuable information towards unveiling the mechanisms of bacterial resistance to Polymyxin B [10].

Out of the pure lipidomic profiling, regulation of lipid metabolism is still an open field. Tian and colls. addressed the molecular mechanisms governing lipid accumulation in trophoblast cells via cell biology and molecular biology approaches [11]. The clinical side of sphingolipid metabolism has been addressed in the article by Malekkou and colls. in which the activity of non-lysosomal glucosylceramidase is evaluated in patients presenting mutations in the gene encoding the enzyme.

Novel synthetic lipids can modulate cell signaling. This is the case reported by Su and colls. who described the effects of ursodeoxycholyl lysophosphatidylethanolamide on integrin signaling and endocytic pathway [12].

The search for new biomaterials of therapeutic use is another goal of lipid-related studies. One example is the successful development of a matrix for the oral administration of hydrophobic compounds by Fratter and colls [3].

The six reviews included in this issue represent some of the main concerns of the community and can provide clues towards current needs as well as future directions. In all cases, the point towards biomedical topics. Public health issues, cancer, and cardiovascular disease are the foci of three of the articles. Two of these reviews address the role of polyunsaturated-derived mediators in hematologic malignancies [13], and the last review is related to the involvement of lipid metabolism modifications in the pathogenesis of viral infection-induced cancers [14]. Solati and colls. lecture on the impact of oxidative stress, in particular that of oxidized lipids on acute coronary disease [15]. The relevance of lipidomics analysis in glomerulopathies, a group of rare kidney diseases, the usefulness of lipid imaging, and the need for improved sensitivity and resolution is addressed by Abbas and colls [8]. A defective resolution of inflammation contributes to the pathogenesis of cystic fibrosis, another rare disease. Philippe and Urbach indicate the state of the art of lipid mediators of resolution in the context of this pathology [16]. Finally, Krivoi and Petrov review the functional role of cholesterol metabolism in neuromuscular junction, a key aspect to understanding the physiology of synaptogenesis and neural transmission, which has implications in motor disorders [17].

3. Future

Biomedical applications and clinical studies will benefit from analytical advances. Sensitivity improvements must be reflected by a subtler research, evolving from total cell to subcellular approaches. This may lead to the growth of "sublipidomics" as a branch of this discipline. Although the pathophysiology of human diseases represents the mainstream and the bulk of research on bioactive lipids and lipidomics, the search for new lipid base biomaterials and the roles and applications of natural substances should gain momentum in the following years. The main technical challenge is still the need to ensure accuracy in the identification of isomers, while the number of lipid molecules keeps expanding. Biomarker research and clinical lipidomics will be highly impacted by these developments. Also, MS imaging must progress in sensitivity and resolution to keep the pace and consolidate as a complementary approach to high resolution microscopy. Finally, studies integrating different omics disciplines are paving the path to a more global view and a better understanding of biological processes.

Conflicts of Interest: The authors have no conflict of interest to declare.

References

1. Han, X.; Gross, R.W. Global analyses of cellular lipidomes directly from crude extracts of biological samples by ESI mass spectrometry: A bridge to lipidomics. *J. Lipid Res.* **2003**, *44*, 1071–1107. [CrossRef] [PubMed]
2. Touré, S.; Desrat, S.; Pellissier, L.; Allard, P.M.; Wolfender, J.L.; Dusfour, I.; Stien, D.; Eparvier, V. Characterization, diversity, and structure-activity relationship study of lipoamino acids from pantoea sp. and synthetic analogues. *Int. J. Mol. Sci.* **2019**, *20*, 1083. [CrossRef] [PubMed]
3. Fratter, A.; Mason, V.; Pellizzato, M.; Valier, S.; Cicero, A.F.G.; Tedesco, E.; Meneghetti, E.; Benetti, F. Lipomatrix: A novel ascorbyl palmitate-based lipid matrix to enhancing enteric absorption of serenoa repens oil. *Int. J. Mol. Sci.* **2019**, *20*, 669. [CrossRef] [PubMed]
4. Stuani, L.; Riols, F.; Millard, P.; Sabatier, M.; Batut, A.; Saland, E.; Viars, F.; Tonini, L.; Zaghdoudi, S.; Linares, L.K.; et al. Stable isotope labeling highlights enhanced fatty acid and lipid metabolism in human acute myeloid leukemia. *Int. J. Mol. Sci.* **2018**, *19*, 3325. [CrossRef] [PubMed]

5. Laboureur, L.; Ollero, M.; Touboul, D. Lipidomics by supercritical fluid chromatography. *Int. J. Mol. Sci.* **2015**, *16*, 13868–13884. [CrossRef] [PubMed]

6. Malekkou, A.; Samarani, M.; Drousiotou, A.; Votsi, C.; Sonnino, S.; Pantzaris, M.; Chiricozzi, E.; Zamba-Papanicolaou, E.; Aureli, M.; Loberto, N.; et al. Biochemical characterization of the GBA2 c.1780G>C missense mutation in lymphoblastoid cells from patients with spastic ataxia. *Int. J. Mol. Sci.* **2018**, *19*, 3099. [CrossRef] [PubMed]

7. Sagini, K.; Urbanelli, L.; Costanzi, E.; Mitro, N.; Caruso, D.; Emiliani, C.; Buratta, S. Oncogenic H-Ras expression induces fatty acid profile changes in human fibroblasts and extracellular vesicles. *Int. J. Mol. Sci.* **2018**, *19*, 3515. [CrossRef] [PubMed]

8. Abbas, I.; Noun, M.; Touboul, D.; Sahali, D.; Brunelle, A.; Ollero, M. Kidney lipidomics by mass spectrometry imaging: A focus on the glomerulus. *Int. J. Mol. Sci.* **2019**, *20*, 1623. [CrossRef] [PubMed]

9. Barthélémy, M.; Elie, N.; Pellissier, L.; Wolfender, J.L.; Stien, D.; Touboul, D.; Eparvier, V. Structural identification of antibacterial lipids from amazonian palm tree endophytes through the molecular network approach. *Int. J. Mol. Sci.* **2019**, *20*, 2006. [CrossRef] [PubMed]

10. Jasim, R.; Han, M.L.; Zhu, Y.; Hu, X.; Hussein, M.H.; Lin, Y.W.; Zhou, Q.; Dong, C.Y.D.; Li, J.; Velkov, T. Lipidomic analysis of the outer membrane vesicles from paired polymyxin-susceptible and -resistant klebsiella pneumoniae clinical isolates. *Int. J. Mol. Sci.* **2018**, *19*, 2356. [CrossRef] [PubMed]

11. Tian, L.; Wen, A.; Dong, S.; Yan, P. Molecular characterization of microtubule affinity-regulating kinase4 from sus scrofa and promotion of lipogenesis in primary porcine placental trophoblasts. *Int. J. Mol. Sci.* **2019**, *20*, 1206. [CrossRef] [PubMed]

12. Su, J.; Gan-Schreier, H.; Goeppert, B.; Chamulitrat, W.; Stremmel, W.; Pathil, A. Bivalent ligand UDCA-LPE inhibits pro-fibrogenic integrin signalling by inducing lipid raft-mediated internalization. *Int. J. Mol. Sci.* **2018**, *19*, 3254. [CrossRef] [PubMed]

13. Loew, A.; Kohnke, T.; Rehbeil, E.; Pietzner, A.; Weylandt, K.H. A role for lipid mediators in acute myeloid leukemia. *Int. J. Mol. Sci.* **2019**, *20*, 2425. [CrossRef] [PubMed]

14. Dutta, A.; Sharma-Walia, N. Curbing lipids: Impacts ON cancer and viral infection. *Int. J. Mol. Sci.* **2019**, *20*, 644. [CrossRef] [PubMed]

15. Solati, Z.; Ravandi, A. Lipidomics of bioactive lipids in acute coronary syndromes. *Int. J. Mol. Sci.* **2019**, *20*, 1051. [CrossRef] [PubMed]

16. Philippe, R.; Urbach, V. Specialized pro-resolving lipid mediators in cystic fibrosis. *Int. J. Mol. Sci.* **2018**, *19*, 2865. [CrossRef] [PubMed]

17. Krivoi, I.I.; Petrov, A.M. Cholesterol and the safety factor for neuromuscular transmission. *Int. J. Mol. Sci.* **2019**, *20*, 1046. [CrossRef] [PubMed]

International Journal of
Molecular Sciences

Article

Multi-Staged Regulation of Lipid Signaling Mediators during Myogenesis by COX-1/2 Pathways

Chenglin Mo [1,*], Zhiying Wang [1], Lynda Bonewald [2] and Marco Brotto [1,*]

[1]   Bone-Muscle Research Center, College of Nursing and Health Innovation, The University of Texas-Arlington, Arlington, TX 76019, USA
[2]   Indiana Center for Musculoskeletal Health, School of Medicine, Indiana University, Indianapolis, IN 46202, USA
*   Correspondence: chenglin.mo@uta.edu (C.M.); marco.brotto@uta.edu (M.B.)

Received: 19 August 2019; Accepted: 21 August 2019; Published: 4 September 2019

Abstract: Cyclooxygenases (COXs), including COX-1 and -2, are enzymes essential for lipid mediator (LMs) syntheses from arachidonic acid (AA), such as prostaglandins (PGs). Furthermore, COXs could interplay with other enzymes such as lipoxygenases (LOXs) and cytochrome P450s (CYPs) to regulate the signaling of LMs. In this study, to comprehensively analyze the function of COX-1 and -2 in regulating the signaling of bioactive LMs in skeletal muscle, mouse primary myoblasts and C2C12 cells were transfected with specific COX-1 and -2 siRNAs, followed by targeted lipidomic analysis and customized quantitative PCR gene array analysis. Knocking down COXs, particularly COX-1, significantly reduced the release of PGs from muscle cells, especially PGE_2 and $PGF_{2\alpha}$, as well as oleoylethanolamide (OEA) and arachidonoylethanolamine (AEA). Moreover, COXs could interplay with LOXs to regulate the signaling of hydroxyeicosatetraenoic acids (HETEs). The changes in LMs are associated with the expression of genes, such as *Itrp1* (calcium signaling) and *Myh7* (myogenic differentiation), in skeletal muscle. In conclusion, both COX-1 and -2 contribute to LMs production during myogenesis in vitro, and COXs could interact with LOXs during this process. These interactions and the fine-tuning of the levels of these LMs are most likely important for skeletal muscle myogenesis, and potentially, muscle repair and regeneration.

Keywords: Cyclooxygenase; skeletal muscle; myogenic differentiation; lipidomics

---

1. Introduction

Skeletal muscle myogenesis, such as muscle regeneration after injury, is a biological process critical for maintaining a functional musculoskeletal system. Myogenesis generally consists of several consecutive stages, including activation of satellite cells, proliferation of myoblasts, myogenic differentiation, and fusion into multinucleated myocytes that can later become fully mature and long, differentiated muscle cells, sometimes referred to as muscle fibers [1]. This process is highly coordinated, and many factors have been shown to be involved in the regulation of myogenesis [2].

Prostaglandins (PGs) are a group of lipid mediators (LMs) playing important roles in various physiological and pharmacological processes, such as fever, inflammation, reproductive function, tissue regeneration, and myogenesis [3–6]. In skeletal muscle, PGE_2 and $PGF_{2\alpha}$, are the two most important PGs. PGE_2 has been shown to enhance myoblast proliferation and differentiation [4,7], and $PGF_{2\alpha}$ is able to promote muscle cell survival and fusion [8,9].

PGs are derived from arachidonic acid (AA) through the activities of a series of specific enzymes. Cyclooxygenases (COXs), including COX-1 and -2, are the rate-limiting enzymes during this process. Generally, COX-1 is constitutively expressed in most cells, while COX-2 is inducible in a variety of pathological situations, such as inflammation and cancer development [10,11]. In skeletal muscle, the knowledge of COXs derives mostly from the studies of COX-2. In muscle repair or regeneration models, COX-2 knockout mice had delayed recovery from muscle injury, suggesting that COX-2 and the downstream PGs from this pathway could be important in regenerative myogenesis, especially in the early inflammatory phase of muscle regeneration for activation of neutrophils, macrophages, and satellite cells [12,13]. However, in this model, the roles of COX-1 and -2 in myogenic processes after the inflammation phase have not been defined. Moreover, in the hind limb suspension mouse model, the induction of COX-2 is essential for muscle recovery from the atrophy caused by unloading [14].

In addition to COXs, AA is also the substrate for lipoxygenases (LOXs) [15] and cytochromes P450 (CYPs) [16]. The metabolites via these two pathways include leukotrienes and hydroxy eicosatetraenoic acids, which are biological activators of intracellular signaling [17,18]. To our knowledge, the interactions between COXs, LOXs, and CYPs have not been studied in skeletal muscle. The changes in the functionalities of COXs would cause indirect effects resulting in modified activities of LOX and/or CYPs.

In this study, we investigated the functional relevance of COX-1 and -2 in myogenesis from myoblasts to the development of multi-nucleated myotubes in both C2C12 cells and mouse primary myoblasts. Since selective inhibitors of COX could reduce the production of PG through COX-independent pathways and could be unselective under certain conditions [19], in the present studies, specific siRNAs for COX-1 and -2 were used to evaluate the effects of COX-1 and -2 on myogenic differentiation. We employed our novel liquid chromatography-mass spectrometry/mass spectrometry (LC-MS/MS) method and a AA-targeted lipidomics method package, which is able to detect 87 compounds derived from AA, 18 eicosapentaenoic acid (EPA)-derived compounds, 16 docosahexaenoic acid (DHA)-derived compounds, and 11 ethanolamides for evaluating the changes in lipid profiling after knocking down COX-1 and -2 during myogenesis. In addition, based on the morphological changes induced by siRNA treatments, a customized skeletal muscle-targeted gene array [4] was used to identify genetic components regulated by COXs and LMs. We further linked these studies with functional measurements of intracellular calcium levels in myotubes, which is an essential surrogate for a host of skeletal muscle functions. Our results demonstrate that knocking down COXs has a significant effect on the synthesis of PGs in skeletal muscle cells. However, they function in a complex LM network not limited to PGs and have significant impacts on the levels of other LMs, such as oleoylethanolamide (OEA) and arachidonoylethanolamine (AEA), which are potentially new factors released from muscle for systemic metabolic regulation. Moreover, COXs play an important role in the regulation of gene expression of contractile apparatus and Ca^{2+} signaling, such as *Myh7*, *Cacna1s*, and *Itrp1*, which can be reflected in the changes observed in morphological and functional tests.

2. Results

2.1. Transfection with Specific siRNAs Targeting COX-1 or -2 Significantly Reduce the Expression Levels of COXs

Two siRNAs specific for each COX were transfected into primary myoblasts. Forty-eight hours after transfection, the total RNA was collected for quantitative RT-PCR to determine the changes in COX expression level. For each gene, both siRNAs efficiently decreased gene expression (Figure 1A,B). Since the siRNA-2 of COX-1 and -2 had higher levels of knockdown efficiencies, resulting in 97.2% and 79.3% downregulation of COX-1 and -2, respectively, compared with negative control (NC), they were used for all remaining experiments. In addition, the protein levels of COX-1 and -2 were shown around 55% reduction at 48 h post transfection with COX-1 or -2 siRNA (Figure 1C–F). Completed Western blot images are shown in supplementary Figure S1.

After 48 h transfection with COX-1 or -2 siRNA, significant morphological changes were observed in myotubes (Figure 1G). Quantified myogenic differentiation data showed that fusion index was reduced from 79.6% (NC) to 49% (COX-1 siRNA) and 45.4% (COX-2 siRNA), respectively (Figure 1H).

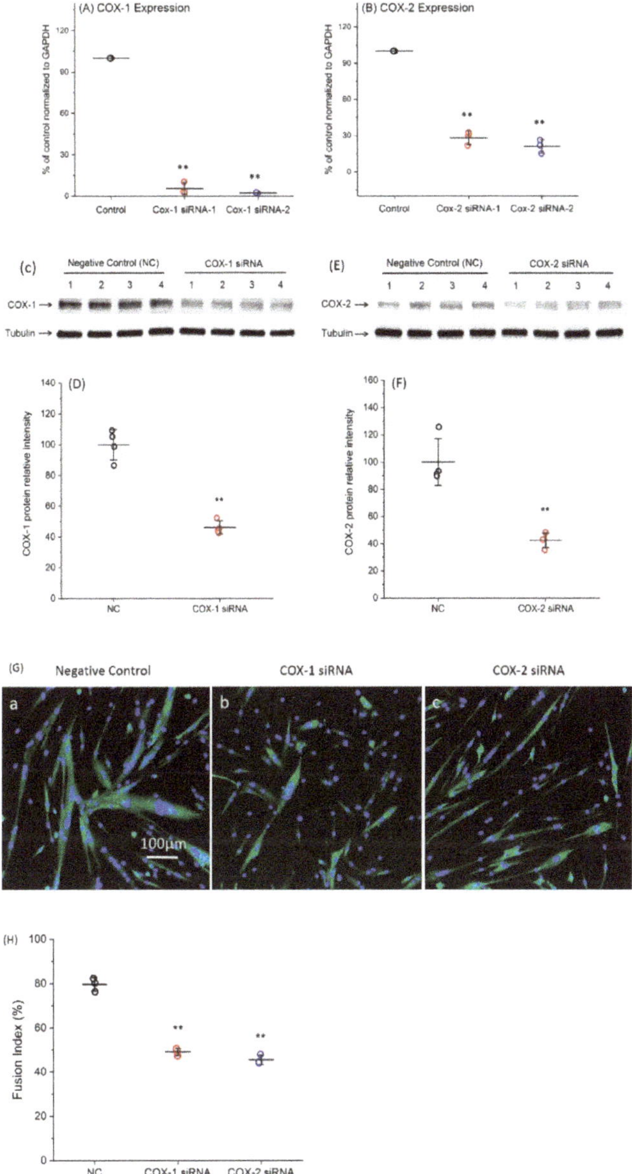

Figure 1. Verification of the high efficiency of COX-1 and COX-2 siRNA knockdown. (**A**) Knockdown efficiency of siRNAs targeting COX-1; (**B**) knockdown efficiency of siRNAs targeting COX-2; (**C**) COX-1 Western blot results after siRNA transfection for 48 h; (**D**) quantification of COX-1 Western blot results using ImageJ; (**E**) COX-2 Western blot results after siRNA transfection for 48 h; (**F**) quantification of COX-2 Western blot results using ImageJ; and (**G**) both COX-1 and COX-2 siRNA transfections inhibit primary myoblast myogenic differentiation. Morphological phenotypes observed after transfections with siRNAs. a: Negative control; b: COX-1 siRNA; and c: COX-2 siRNA. (**H**) Treatments with siRNAs significantly reduces fusion index. $n = 3$–4, ** $p < 0.01$ compared with NC.

2.2. The Changes in Levels of Lipid Mediators after Knocking Down COX-1 or -2 Are not Limited to PGs and Thromboxane B_2 (TXB$_2$)

To investigate the mechanisms responsible for the effect of COXs in skeletal muscle myogenesis, we first used our new lipidomics method to directly quantify 14 LMs selected from our preliminary studies, mostly AA metabolites through COX and other enzymes in cell differentiation medium (DM). Compared with blank medium, after differentiation for 72 h, the levels of PGE$_2$, PGF$_{2\alpha}$, 6-keto-PGF$_{1\alpha}$ (stable metabolite of PGI$_2$), DHA, and OEA in the medium increased significantly, suggesting that these LMs were released from myocytes/myotubes. We then further analyzed the effect of COXs on their production. Knocking down COX-1 using siRNA significantly reduced the levels of PGE$_2$ and PGF$_{2\alpha}$ compared with NC, but had no significant effect on the levels of 6-keto-PGF$_{1\alpha}$. At the same time, knocking down COX-2 also showed a similar impact on PGE$_2$ levels, but the effect was significantly less than knocking down COX-1. In addition to changes in LMs in AA pathway, COX-1 knockdown significantly reduced the concentration of DHA and OEA in the DM after 72 h (Figure 2). These data demonstrate that the functions of COXs are not limited to regulating the production of PGs from AA. The whole list of LMs identified in these experiments, including LMs with lower levels after 72 h differentiation compared with blank medium (LMs could be consumed by myocytes/myotubes during differentiation), is summarized in supplementary Figure S2.

(A)

	Amount of LMs (ng/mg protein/mL) released in differentiation medium at 72 h post-transfection				
	PGE$_2$	PGF$_{2\alpha}$	6-keto-PGF$_{1\alpha}$	DHA	OEA
NC	1.18 ± 0.04	1.42 ± 0.41	0.82 ± 0.22	47.72 ± 8.18	16.70 ± 1.50
COX-1 siRNA	0.63 ± 0.12	0.37 ± 0.01	0.43 ± 0.14	23.92 ± 6.49	10.70 ± 2.50
COX-2 siRNA	0.89 ± 0.07	0.82 ± 0.19	0.48 ± 0.14	31.98 ± 5.81	13.16 ± 1.43

(B)

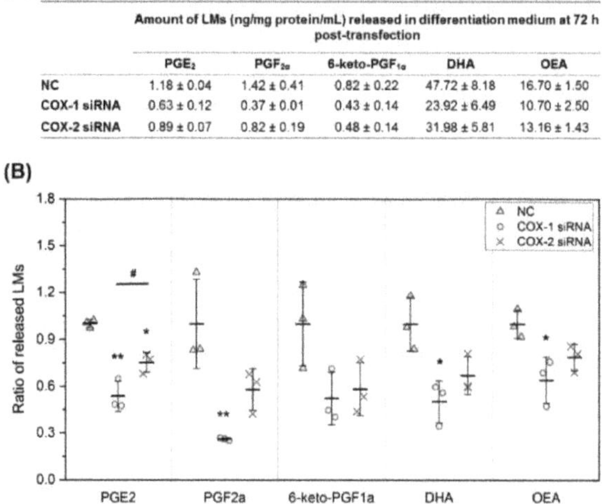

Figure 2. COX-1 and -2 knockdown reduces the levels of key lipid mediators released by primary muscle cells. (**A**) Absolute quantification of lipid mediators (LMs) released in differentiation medium (DM) from primary mouse myocytes/myotubes during differentiation; (**B**) ratio of LMs released in DM at 72 h post-transfection comparing COX-1 siRNA or COX-2 siRNA treatment with NC transfection. $n = 3$, * $p < 0.05$ and ** $p < 0.01$ compared with NC; # $p < 0.05$ compared with COX-1 siRNA.

2.3. COXs could Interact with LOXs to Regulate the Levels of Lipid Mediators

In addition to direct quantification for lipid mediators, lipidomic profiling of 158 lipid mediators in DM also was performed. Our results indicate that the levels of 12-Hydroxyeicosatetraenoic acid (12-HETE), a lipid mediator derived from the 12-LOX pathway, and 15-HETE, a lipid mediator derived from the 15-LOX pathway, significantly decreased after siRNA transfection targeting both COX-1 and -2. In contrast, the levels of 5-HETE, a lipid mediator derived from the 5-LOX pathway was not affected (Figure 3).

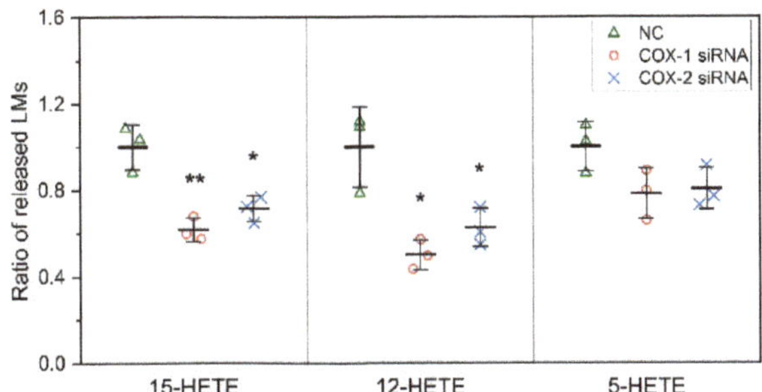

Figure 3. Knockdown of COXs reduces the levels of hydroxyeicosatetraenoic acids (HETEs) released by primary muscle cells. The levels of 12-HETE and 15-HETE, but not 5-HETE are significantly affected by the downregulation of gene expression of both COX-1 and COX-2. $n = 3$, * $p < 0.05$ and ** $p < 0.01$ compared with NC.

2.4. Supplement with LMs Improves Defective Myogenic Differentiation of Primary Myoblast Caused by Knocking Down COX-1 or -2

Based on the results of lipidomic analysis, to confirm that the effects on myogenic differentiation after knocking down COX-1 and -2 were through decreasing the production of LMs, three LMs, including PGE_2, 12-HETE, and 15-HETE, were selected to determine whether replenishment with these LMs could improve defective myogenesis following transfection with siRNAs. Our results indicated that co-treatment with PGE_2 or 15-HETE, but not 12-HETE, partially recovered the inhibition of both siRNAs used against COX-1 or -2 on myogenic differentiation. The fusion indexes increased significantly from 49% to 56.1% and 58.3% in culture treated with COX-1 siRNA, and from 45.4% to 59.8% and 62.3% in the COX-2 siRNA treated group, respectively. However, neither PGE_2 nor 15-HETE brought the fusion index back to normal (negative control) level (Figure 4).

2.5. Results of Lipidomic Analysis of C2C12 Cells Show Similar Patterns as Primary Myoblasts

Following the studies of primary myoblasts, lipidomic analysis was performed in C2C12 cells. Since it is relatively easy to reach cell numbers high enough for reliable lipidomic analysis in C2C12 cell culture, we performed lipidomic studies in both cell culture media and cells.

In C2C12 cell culture media, similar to the results obtained in mouse primary myoblast cultures, PGs from the AA pathway, including PGE_2, $PGF_{2\alpha}$, and 6-keto-$PGF_{1\alpha}$ (PGI_2), were released from cells into media. In addition, AEA and OEA also were identified as LMs released by myocytes/myotubes during differentiation. Knocking down COXs significantly lowered the concentrations of PGE_2, 6-keto-$PGF_{1\alpha}$, AEA, and OEA in media. COX-1 was more effective in modulating the concentrations of PGE_2 and 6-keto-$PGF_{1\alpha}$, but COX-2 knockdown had more impact on the release of $PGF_{2\alpha}$. DHA was not a lipid mediator released by C2C12 cells during differentiation (Figure 5).

In C2C12 cells, for LMs from AA pathway, downregulation of COXs significantly reduced the levels of PGE_2, but had no effect on the levels of $PGF_{2\alpha}$ or 6-keto-$PGF_{1\alpha}$. Moreover, knocking down COX-1, but not COX-2, significantly lowered the concentration of PGD_2. TXB_2 was not detectable in C2C12 cells. Interestingly, knocking down COXs significantly increased the level of AEA in C2C12 cells, but had no effect on OEA levels (Figure 6). These results further confirm that the functional change in COXs affects a more complex network of LMs than just PGs and TXA_2. The whole list of LMs identified in these studies using C2C12 cells is summarized in supplementary Figure S3 for cell culture medium and supplementary Figure S4 for C2C12 cells.

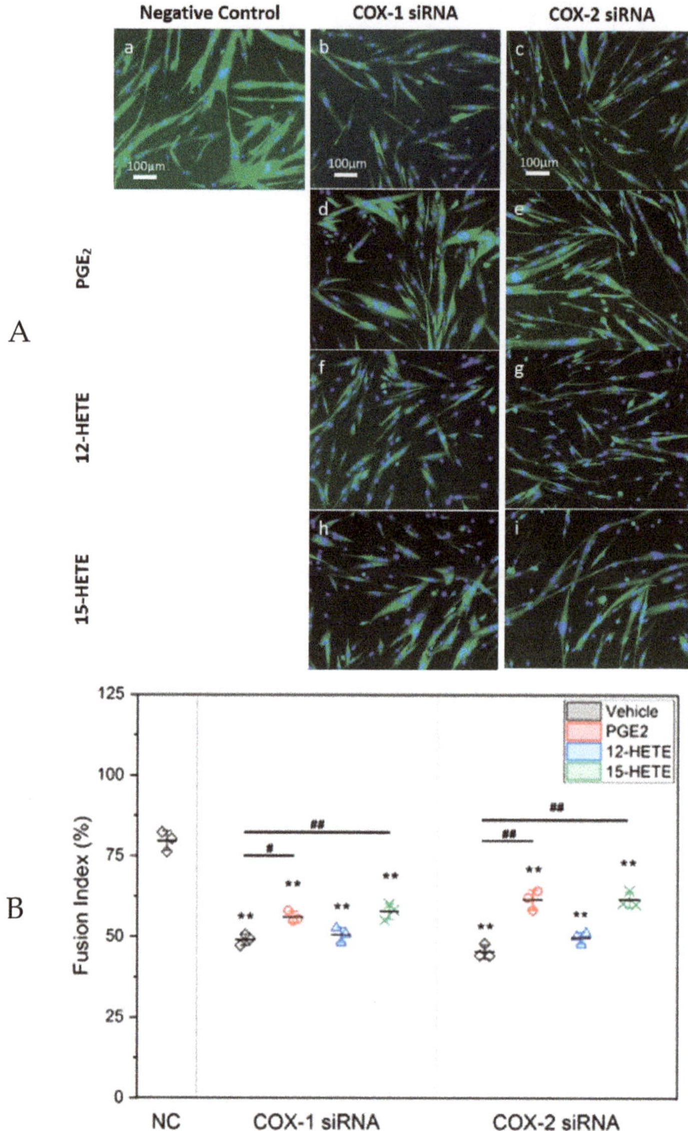

Figure 4. Treatment with PGE$_2$ or 15-HETE partially recovers the impaired myogenesis induced by COX-1 or -2 knockdown. Panel (**A**): Representative fluorescence images of morphological changes of myotubes after siRNA transfection and supplement with LMs. Blue: DAPI (4′,6-diamidino-2-phenylindole) staining; green: MHC (myosin heavy chain) staining. Panel (**B**): Pretreatment with PGE$_2$ and 15-HETE partially but significantly improved Fusion Index. $n = 3$, ** $p < 0.01$ compared with NC; # $p < 0.05$ and ## $p < 0.01$ compared with COX-1 or -2 siRNA.

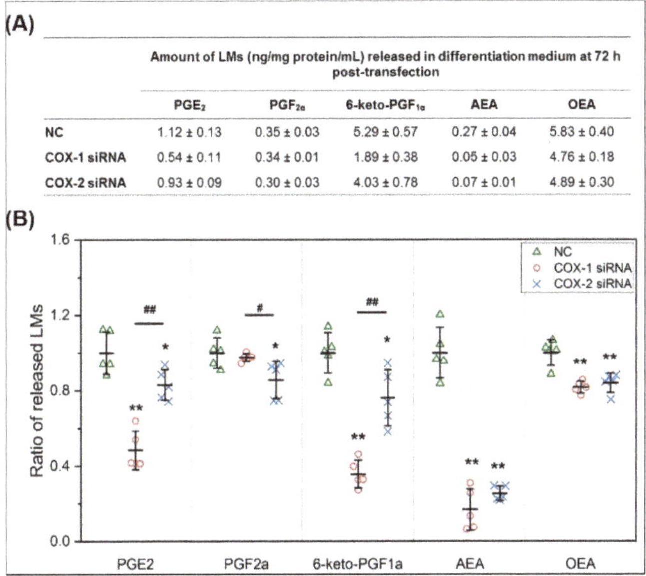

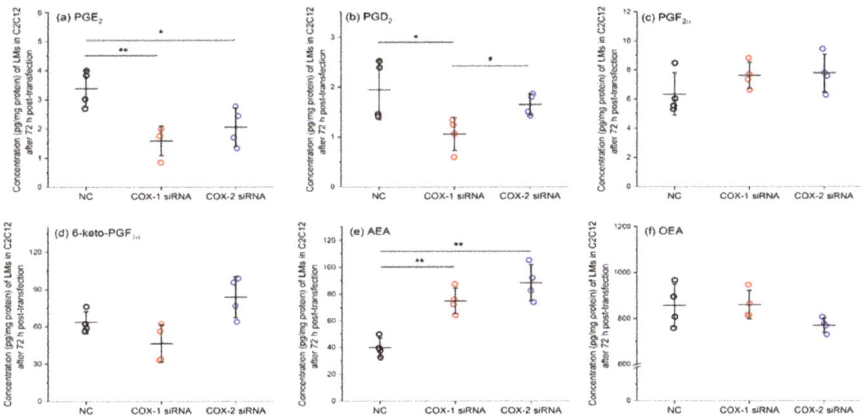

Figure 5. COX-1 or -2 knockdown reduces the levels of key lipid mediators released by C2C12 muscle cells. (**A**) Absolute quantification of LMs released in DM of C2C12; (**B**) ratio of LMs released in DM at 72 h post transfection comparing COX-1 siRNA or COX-2 siRNA treatment with NC transfection. $n = 5$, * $p < 0.05$ and ** $p < 0.01$ compared with NC; # $p < 0.05$ and ## $p < 0.01$ compared with COX-1 siRNA.

Figure 6. COX-1 or -2 knockdown alters the levels of key lipid mediators in C2C12 muscle cells. $n = 4$, * $p < 0.05$ and ** $p < 0.01$ compared with NC; # $p < 0.05$ compared with COX-1 siRNA.

2.6. Changes in Gene Expression Profile after siRNA Transfection Targeting at COX-1 or -2

Next, to study the genetic mechanism(s) related to the changes in lipid mediators after knocking down COX-1 or -2, a customized quantitative RT-PCR gene array, which includes 91 genes associated with cell myogenic differentiation, cell survival, Ca^{2+} signaling and homeostasis, cell metabolism, oxidative stress, and cell growth was performed [4]. After transfection with siRNAs, genes encoding components of contractile apparatus and Ca^{2+} signaling were significantly affected (Figure 7). *Myh7*, *Acta1*, *Ttn*, *Myh1*, and *Myh6* were downregulated by knocking down at least one of the COX isoforms. In contrast, the expression of *ITPR1* gene, which encodes the inositol 1,4,5-triphosphate (IP3) receptor

1, an important regulator of intracellular calcium signaling, was increased. However, knocking down COX-1 significantly reduced the expression levels of *Cacna1c* and *Cacna1s*, which are genes encoding subunits of voltage-sensitive, L-type calcium channel, and *Jph2*. The impact of COX-2 on calcium signaling is more complex, in addition to *Itpr1*, transfection with COX-2 siRNA also upregulated the expression of *Cacna1c*, *Ryr2*, and *Stim2*, but downregulated the expression of *Sypl2*, *Mtmr14*, *Tmem38a*, and *Itpr2*.

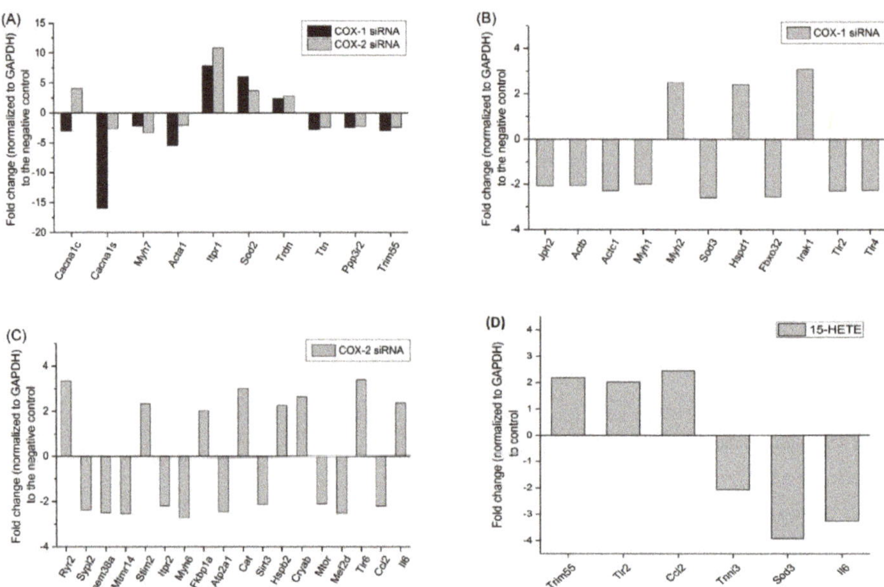

Figure 7. Knocking down COX-1 or -2 affects the expression of genes related with muscle structure and functions. (**A**) Genes affected by both COX-1 and COX-2 siRNA transfection; (**B**) genes affected by COX-1 siRNA transfection only; and (**C**) genes affected by COX-2 siRNA transfection only. (**D**) Changes in gene expression after treatment with 15-HETE for 48 h. Only genes with two-fold or greater changes, which are considered as significant changes, are listed.

In addition, the changes in antioxidative genes (*Sod2*, *Sod3*, and *Cat*) and the upregulation of genes of heat shock protein family (*Hspd1*, *Hspb2*, and *Cryab*) suggest that the cells were under stress after transfection of the siRNAs.

The changes in gene expression after COXs downregulation could be directly related with the decreased levels of lipid mediators. We previously reported the effect of PGE_2 on gene expression in muscle cells using our customized gene array [4]. In this study, using the same method, the changes in gene expression in primary mouse myoblast after 48 h of treatment with 15-HETE were determined. Genes encoding tripartite motif-containing protein 55 (Trim55), Toll-like receptor 2 (TLR2), and CC-chemokine ligand 2 (CCL2) were significantly upregulated after treatment with 15-HETE. *Trim55* is one of genes downregulated after transfection with either COX-1 or -2 siRNA, and the gene expression of *TLR2* and *CCL2* were significantly reduced by knocking down COX-1 and -2, respectively (Figure 7). These results support, at the genetic level, the partial recovery effect of myogenesis induced 15-HETE treatments shown in Figure 4.

2.7. Intracellular Calcium Homeostasis Measurement

Since there are significant changes in gene expression in the contractile apparatus and Ca^{2+} machinery, the measurement of intracellular calcium homeostasis was performed to identify functional changes in myotubes after siRNA transfection.

Both COX-1 and -2 siRNA transfection significantly altered the profile of intracellular calcium homeostasis in response to caffeine stimulation, but there was some difference between COX-1 and -2 knockdown. COX-1 siRNA treated myotubes demonstrated spontaneous cyclical transition in baseline fluorescence and a weaker response to caffeine stimulation compared to the negative control group. While COX-2 siRNA treated myotubes do not show cyclical oscillation in intracellular Ca^{2+} measurement, the amplitude of their responses to caffeine stimulation were further attenuated (Figure 8).

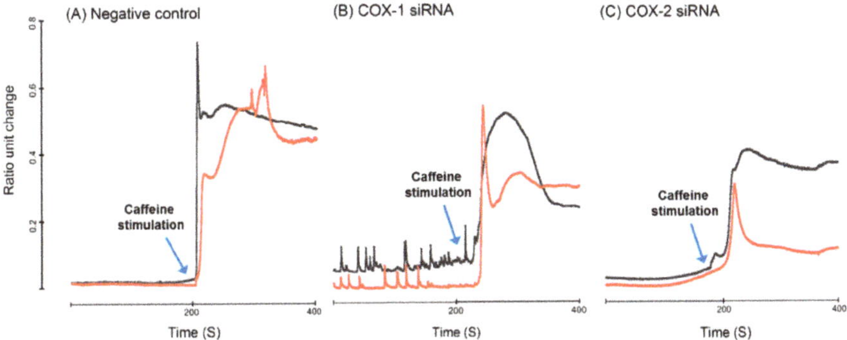

Figure 8. Representative Ca^{2+} transient of mouse primary myotubes loaded with Fura-2/AM in response to 20 mM caffeine (arrows). Treatment with COX-1 siRNA induced spontaneous Ca^{2+} oscillation with reduced response to caffeine stimulation. While Ca^{2+} oscillation was not observed in myotubes treated with COX-2 siRNA, their response to caffeine stimulation was further reduced. (**A**) Negative control; (**B**) COX-1 siRNA knockdown; and (**C**) COX-2 siRNA knockdown.

3. Discussion

COX-1 and -2 are the two most important enzymes in the synthesis of PGs and TXA_2 from AA. Due to the comprehensive functions of PGs and TXA_2 in physiological and pathological processes, COX-1 and -2 have been considered as important targets for the development of new therapeutics for disease [20,21]. In skeletal muscle, previous studies have shown that COXs, through the regulation of their AA metabolites, play important roles in muscle development, regeneration, and diseases [13,22].

To date, most studies concerned with the role of COXs in skeletal muscle have been focused on COX-2, the inducible form of COX. COX-2 increases during muscle regeneration after injury and during recovery from muscle atrophy [14]. Moreover, under normal conditions, the protein levels of both COX-1 and -2 are detectable in rat extensor digitorum longus (EDL) and soleus muscle [23]. Inhibition of COX-2 results in attenuated muscle growth during regeneration after injuries and reduced muscle hypertrophy in animal models [24]. At least part of the effects of COXs are thought to be mediated by the functions of their AA metabolites, which include PGE_2, $PGF_{2\alpha}$, PGI_2, PGD_2, and TXA_2. In skeletal muscle, due to their important functions in the regulation of myoblast proliferation and differentiation, and the function of inflammatory cells, PGE_2 and $PGF_{2\alpha}$ have been considered major mediators of the effects of COXs [7,9]. PGI_2 plays an important role in regulating the migration and fusion of muscle cells [25]. In contrast, treatment with PGD_2 inhibited C2C12 myogenesis in vitro [26].

COX-2 induction during muscle regeneration occurs in the early acute inflammatory phase, which is involved in the recruitment of inflammatory cells, such as macrophages, activation of satellite cells, and myoblast proliferation [13]. However, administration of COX-2 inhibitor after acute inflammatory phase did not affect muscle regeneration and had no noticeable effect in undamaged muscles. These data suggest that COX-1, the constitutive isoform, may compensate for COX-2 and also plays a role in muscle regeneration and in the maintenance of normal muscle functions. By downregulating COX-1 and -2 expression in mouse primary myoblast using siRNAs, we compared the functionalities of these enzymes in myogenic differentiation. Our results demonstrated that knocking down both COX-1 and -2 significantly inhibited myogenesis. However, these two enzymes may have different functions in myogenic differentiation, based on the morphological phenotypes after the transfection of siRNAs.

To our knowledge, there is currently no systematic study comparing the functionalities of COX-1 and -2 during myogenesis in terms of the production of AA metabolites and other aspects important for myoblast migration, proliferation, differentiation, and fusion, such as lipid profiling and intracellular calcium homeostasis.

Our data provide evidence supporting previous findings that PGE_2 and $PGF_{2\alpha}$ could be two major mediators from the AA/COX pathway in skeletal muscle. Moreover, the changes in PGs and TXB_2 clearly indicate that COX-1 plays a critical role in the stages from myoblast to fusion in myogenesis. In addition to affecting the production of PGs and TXs, reduced COXs functionalities also affected the levels of AA metabolites through LOX pathways. 5- and 12/15 LOXs are the LOX isoforms utilizing AA as substrate to generate 5-, 12-, and 15-HETE. Our results demonstrated that reduced COX-1 or -2 expression significantly decreased the levels of 12- and 15-HETE, but had no effect on 5-HETE. These results suggest that COXs could interact with LOX to regulate the production of lipid mediators from AA. 12/15 LOX shares some function with COXs, such as the regulation of inflammatory cytokines. In animal studies, deletion of 12/15 LOX prevents the early onset of inflammation caused by a high-fat diet [27] and denervation-induced muscle atrophy [28]. On the other hand, the same genetic manipulation resulted in exaggerated inflammation and tissue damage in arthritis, and disruption of the translocation of glucose transporter type 4 in cardiac and skeletal muscle. Our results suggest that COX-1 and -2 could function indirectly on LMs by altering the metabolism of AA by LOXs. This could be the first evidence of the interaction between COXs and LOXs in skeletal muscle.

Recently, skeletal muscle has been recognized as an endocrine tissue. Factors released from muscles, such as β-aminoisobutyric acid (BAIBA), a muscle metabolite, can act as endocrine factors to crosstalk with bone, adipose tissue, and other tissues or organs [29,30]. In our study, besides PGs, OEA was also identified as a factor released by skeletal muscle, a metabolite derived from omega-9 fatty acid, oleic acid. BAIBA, via activation of peroxisome proliferator-activated receptor α (PPARα), transient receptor potential vanilloid type-1 (TRPV1), and G protein coupled receptor GPR119 regulates fat catabolism, food intake, and glucose homeostasis [31–33]. In soleus muscle, OEA enhanced the oxidation of fatty acid, but had no significant effect on glucose metabolism [34]. Currently, feeding status and enzymes directly responsible for OEA synthesis or degradation, such as N-acyl transferase and fatty acid amide hydrolase [35], are major factors affecting the OEA level. Our results demonstrated that COXs in skeletal muscle could be an important factor regulating the OEA level. AEA is another candidate lipid mediator acting as a myokine, because it has important functions in metabolic regulation and anti-inflammatory effects through activating TRPV1 and cannabinoid receptors, respectively [36], and in our studies is regulated by the activities of COXs in skeletal muscle. These data could help to expand the pool of myokines and provide new insight for explaining the beneficial effect of exercise.

The regulatory function of skeletal muscle on metabolism is closely related with its status, especially functionality status. After transfection with siRNAs targeting COXs, the development of myotubes is inhibited. Corresponding with this phenotype, genes encoding components of contractile apparatus and cytoskeleton, including *Myh2*, *Myh7*, *Acta1*, *Actb*, and *Actc1*, were also significantly affected. Appropriate cytoskeletal remodeling, which also is related to the assembly of the contractile apparatus, is critical for migration, cell-to-cell recognition, and fusion of myoblasts/myocytes [37]. The changes in gene expression of the contractile apparatus and cellular structural components suggest that COXs are important for assembly of contractile apparatus and cytoskeleton. Moreover, after knocking down COX-1 or -2, functional tests using the measurement of intracellular Ca^{2+} homeostasis in myotubes was performed. Our results indicate that intracellular Ca^{2+} signaling was defective after downregulation of COXs. COX-1 siRNA treated myotubes demonstrated spontaneous cyclical transition in baseline fluorescence and a weaker response to caffeine stimulation. These phenomena could have been resulted from the changes in gene expression of Ca^{2+} machinery. *Cacna1c* and *Cacna1s* are genes encoding subunits of the voltage-sensitive, L-type Ca^{2+} channel, which plays a critical role in gating intracellular Ca^{2+} movement [38]. Significant downregulation of these two genes after knocking down COX-1 could lead to the dysfunction of voltage-sensitive, L-type Ca^{2+} channels, which could be the major reason for the detectable spontaneous Ca^{2+} transients in myotubes. While COX-2 siRNA treated myotubes did not show similar changes in intracellular Ca^{2+} measurement, the amplitude of their responses to caffeine stimulation were further attenuated. Gene expression of *Sypl2* (*Mg29*) and *Mtmr14* significantly decreased after COX-2 knockdown. Previous findings from our group have confirmed that knocking down these genes causes defective Ca^{2+} signaling in skeletal muscle [39,40]. These changes, along with downregulation of *Tmem38a*, a gene encoding trimeric intracellular cation channel type A, which is important for maintenance of rapid intracellular calcium release [41], could contribute to the attenuated response upon caffeine stimulation.

The changes in gene expression after COX-1 and -2 siRNA transfection could be modulated through decreasing levels of 15-HETE. Treatment with 15-HETE significantly increased the expression of *Trim55*, *TLR2*, and *CCL2*. *Trim55*, also called muscle-specific RING finger protein 2 (*MuRF2*), was downregulated after knocking down COX-1 or -2. This gene has been shown to be important for the organization of cytoskeleton and contractile machinery in muscle. A reduced *Trim55* expression level led to delayed myoblast fusion, defective contractile function, and deformation of Z- and M-bands, suggesting that Trim55 is an adaptor for tubulin, titin, and myosin, which has an important impact on structural and functional aspects in muscle [42,43]. *TLR2* and *CCL2* were genes downregulated by knocking down COX-1 and -2, respectively. They are important components in inflammatory responses, which play essential roles in immune responses, muscle regeneration after injuries and muscle atrophy [44,45]. During endurance training, TLR2 signaling mediates the activation of mitogen-activated protein kinase (MAPK) and nuclear factor κB (NF-κB) induced by extracellular nonesterified fatty acids [46]. One the other hand, muscle atrophy after immobilization is closely related with oxidative stress and inflammation through the activation of TLR2 [47]. *CCL2* might be one of the targets of TLR2 signaling in skeletal muscles. Peptidoglycan, an agonist of TLR1 and TLR2, significantly induced *CCL2* expression in C2C12 myotubes [48]. Polymorphisms of *CCL2* are associated with muscle adaption and muscle damage response caused by exercise [49,50]. Research concerned with TLR2 and CCL2 in muscle has been focused on their functions in recruiting immune cells, such as monocytes, during muscle recovery from injury, which involves cell migration and cell adhesion [51,52]. Myoblast migration and adhesion are important steps for differentiation and fusion. Our results imply that COXs-15-HETE signaling could be important for pre-fusion events in myogenesis. Another interesting finding is that transfection with COX-2 siRNA significantly increased the expression of *interleukin-6 (IL-6)*, which was reversed when myoblasts were treated with 15-HETE. IL-6 is a multi-functional factor in skeletal muscle. It can stimulate satellite cell proliferation [53], but chronic exposure to IL-6 led to muscle atrophy [54], which is supported by the previous report that inhibition of IL-6 signaling attenuated muscle atrophy in tail suspension model through the downregulation of atrophy-related genes, such as atrogin-1 [55].

Collectively, these studies provide new insights into the regulation of LMs in skeletal muscle and their crucial function for muscle cell homeostasis.

4. Materials and Methods

4.1. Cell Culture

4.1.1. Myoblast Isolation and Culture

Isolation of primary myoblasts was performed as previously described [7]. Primary myoblasts were isolated from hind limb muscles of 5 months old C57BL/6 mice. Collected muscles were minced and digested using 0.1% pronase (EMD Millipore, Temecula, CA, USA). Isolated cells (fibroblasts and myoblasts) were maintained and expanded in collagen-I (Corning, Corning, NY, USA) coated T-75 flask in growth medium (GM) consisted of Ham's F-10 (Corning), 20% fetal bovine serum (FBS, Atlanta Biologicals, Flowery Branch, GA, USA), 5 ng/mL basic recombinant human fibroblast growth factor (Promega, Fitchburg, WI, USA), 100 µg/mL streptomycin (Thermo Scientific, Rockford, IL, USA), and 100 U/mL penicillin G (Thermo Fisher Scientific, Waltham, MA, USA) for 3 to 4 weeks for purification. For differentiation, purified myoblasts were plated on E-C-L (Millipore)-coated 6-well plates at ~200,000 cells/well and differentiated in DM for 48 or 72 h.

4.1.2. C2C12 Cells

C2C12 cells were cultured as previously described [56]. Briefly, cells [American Type Culture Collection (ATCC), Manassas, VA, USA] were cultured in complete growth medium [CGM, high-glucose Dulbecco's Modified Eagle Medium (DMEM, Corning) with 10% fetal bovine serum, plus 100 U/mL penicillin and 100 µg/mL streptomycin (Thermo Fisher Scientific)], at 37 °C and 5% CO_2. C2C12 myoblasts were maintained at 70–80% confluence and passaged one or two times before being used in experiments.

To initiate differentiation, CGM was replaced by differentiation medium (DM) containing high-glucose DMEM, 2.5% horse serum (Hyclone Laboratories Inc, Logan, UT, USA), 100 U/mL penicillin, and 100 µg/mL streptomycin.

4.2. siRNA Transfection

For primary mouse myoblasts, cells were seeded at ~200,000 cells/well in 6-well plates in primary GM, then differentiated overnight before being transfected with 10nM siRNAs, including negative control siRNA and siRNAs targeting COX-1 or -2 [Integrated DNA Technologies (IDT), Coralville, IA, USA]. Lipofectamine RNAiMAX (Thermo Fisher Scientific) was used as a transfectant following the instructions from the manufacturer.

For recovery experiments with LM supplements, including PGE_2, 12-HETE, and 15-HETE, primary myoblasts were treated with 50 nM of each LM for 2 h in fresh DM before being transfected with siRNAs.

For C2C12 cells, cells grew in CGM until 80–90% confluence in 6-well plates, then differentiated overnight before being treated with siRNAs, as described in primary myoblast experiments.

4.3. Quantitative Real-Time PCR (qRT-PCR)

Total RNA was extracted from primary myoblasts using Direct-zol RNA MiniPrep (Zymo Research, Irvine, CA, USA) according to the manufacturer's instruction, and was quantified in a Nanodrop 1000 spectrophotometer (Thermo Fisher Scientific). An aliquot of RNA sample (0.5–1 µg) with the A260/280 nm absorbance ratio of 1.8 or above was reverse transcribed in a 20 µL reaction volume using a protoscript II first strand cDNA synthesis kit (New England Biolabs, Ipswich, MA, USA).

The RT-PCR reaction mixture contained 2 µL cDNA, 12.5 µL of the RT^2 SYBR Green/Rox PCR master mix (Qiagen, Germantown, MD, USA), 0.4 µL of primer pairs (10 µM) and 10.1 µL of RNase free water to a complete reaction volume of 25 µL. qRT-PCR was performed using Step-One Plus TM RT-PCR System (Thermo Fisher Scientific), and results were normalized to the reference gene GAPDH. Primers used in the experiments include: 1) COX-1: Forward: 5′-TGCCCATGGAGACCAGAAGAAGTT-3′; Reverse: 5′-ATGGGTGTGGAGAAATGGCTCAGT-3′; 2) COX-2: Forward: 5′-ATGACTGGCTGGT GCATCTCATCT-3′; Reverse: 5′-ACTTGCCCTCACGGACAATGTAGT-3′; 3) GAPDH: Forward: 5′-T GCGATGGGTGTGAACCACGAGAA-3′; Reverse: 5′- GAGCCCTTCCACAATGCCAAAGTT-3′.

The customized gene array was previously developed by our laboratory in collaboration with Qiagen and is now commercially available from Qiagen (Item No.: CAPM09345C, Germantown, MD, USA) [4]. Experiments were performed according to the instructions from the manufacturer. Data were uploaded and analyzed by specific software from Qiagen. Changes in gene expression were considered significant when change was two-fold or greater.

4.4. Protein Sample Preparation and Western Blotting

Muscle cells cultured in 6-well plates were washed 3 times with ice-cold Dulbecco's phosphate buffered saline (PBS) before being lysed by RIPA buffer [1× Tris-buffered saline (TBS), 1% Nonidet P-40, 0.5% sodium deoxycholate, 0.1% SDS, 0.004% sodium azide] (Sigma-Aldrich, St. Louis, MO, USA) with 1% cocktail of proteinase and phosphatase inhibitors (Sigma-Aldrich). Lysates were then collected and incubated in ice for 30 min, followed by centrifugation at 16,000× g for 20 min at 4 °C. Supernatants were collected for protein assay.

Protein assay was performed using Micro BCA Protein Assay Kit (Thermo Fisher Scientific) according to the manufacturer's instructions. Protein samples then were mixed with 4× Western blot loading buffer (Bio-Rad, Plano, TX, USA) and denatured at 100 °C for 5 min.

For Western blots, ~30 µg of total proteins were fractionated by 4–15% Mini Protean TGX gels (Bio-Rad) and transferred to polyvinylidene difluoride (PVDF) membranes (Bio-Rad). Membranes were blocked in 5% non-fat dry milk in 1× TBS with 0.1% Tween 20 (TBST) for 1 h at room temperature (RT), followed by incubation with antibodies COX-1 (1:1000, Cell Signaling Technology, Inc, Danvers, MA, USA) and β-tubulin (1:1000, Cell Signaling Technology, Inc, Danvers, MA, USA) in 5% bovine serum in TBST or COX-2 antibody (1 µg/mL, R&D systems, Minneapolis, MN, USA) in 5% non-fat dry milk at 4 °C overnight. HRP-conjugated goat anti-rabbit (For COX-1 and β-tubulin, 1:10,000, Jackson ImmunoResearch, West Grove, PA, USA) or HRP-conjugated rabbit anti-goat (For COX-2, 1:5000, Thermo Fisher Scientific) secondary antibodies were then applied to membranes for 1 h at RT. After five 5-min washes in TBST, Clarity Max ECL Western blotting substrates (Bio-Rad) or Super Signal West Femto substrate (Thermo Fisher Scientific) were used to detect the signal by ChemiDoc MP imaging system (Bio-Rad).

4.5. Immunohistochemistry

After differentiation, cells in 6-well plates were fixed in 10% neutral buffered formalin solution (NBF, Sigma-Aldrich) for 15 min. After removal of NBF, cells were washed 4 times with PBS, followed by permeabilization with 0.1% Triton X-100 in PBS for 15 min. Cells were then incubated with myosin heavy chain (MHC) fluorescein-conjugated antibody (1:100, R&D Systems) overnight at 4 °C. After 3 washes with PBS, DAPI (1:1000, Sigma-Aldrich) was added for 10 min incubation at room temperature. Images were taken with Olympus IX50 system using software cellSens Dimension 1.15 (Olympus Corp., New Orleans, LA, USA).

4.6. LC-MS/MS

4.6.1. Sample Preparation for Lipidomics Analysis

Briefly, cells from four wells of 6-well plates were harvested after experiments and transferred into 1.0 mL of ice-cold 80% methanol in water (*v/v*) to perform homogenization using the TissueLyser II homogenizer (Qiagen) at the frequency of 30/sec, in 6 × 30 s bursts, and 20 s in between to avoid high temperature. An aliquot (20 µL) of cell homogenates was saved separately for future protein content measurement by BCA (bicinchoninic acid) assay (Thermo Scientific, Rockford, IL, USA). The remaining homogenate from each cell sample was added to 10 µL of IS mixture stock solution (5 µg/mL for AA-d$_8$, 2 µg/mL for DHA-d$_5$ and EPA-d$_5$, and 0.5 µg/mL), then agitated on ice in the dark for 1–2 h. For cell culture media, 1 mL of culture media sample was mixed with 1.5 mL of ice-cold methanol and 10 µL of IS mixture stock solution, then agitated on ice in dark for 15 min. After incubating the homogenate or culture media sample on ice, samples were centrifuged at 6000× g at 4 °C for 10 min to remove any precipitated proteins. All LM standards and isotope-labelled LM internal standards were purchased from Cayman Chemical Co (Ann Arbor, MI, USA). Formic acid (reagent grade, ≥95%) was obtained from Sigma-Aldrich. HPLC-MS grade acetonitrile, water, methanol, and ethanol were purchased from J.T. Baker (Phillipsburg, NY, USA).

Both cell and culture media samples were cleaned and concentrated by Solid Phase Extraction (SPE) before being injected into the LCMS. Before loading samples to the preconditioned SPE cartridges (Strata-X 33 μm polymeric reversed phase, Phenomenex, Torrance, CA, USA), 4 or 6 mL of ice-cold 0.1% formic acid was added in the supernatant from cell or medium sample, respectively, to fully protonate the LM species. Once the sample had been totally loaded, cartridges were washed with 1 mL of 0.1% formic acid followed by 1 mL of 15% (*v/v*) ethanol in water to remove excess salts. Then the LMs from the SPE sorbent bed were eluted by methanol. Solvents were removed using an Eppendorf® 5301 concentrator centrifugal evaporator (Eppendorf, Hauppauge, NY, USA), and the dried extracts stored at −80 °C for future LC-MS/MS analysis.

4.6.2. LC-MS/MS Conditions

All components of LC-MS/MS system are from Shimadzu Scientific Instruments, Inc. (Columbia, MD, USA). LC system was equipped with four pumps (Pump A/B: LC-30AD, Pump C/D: LC-20AD XR), a SIL-30AC autosampler (AS), and a CTO-30A column oven containing a 2-channel six-port switching valve. The LC separation was conducted on a C8 column (Ultra C8, 150 × 2.1 mm, 3 μm, RESTEK, Manchaca, TX, USA) along with a Halo guard column (Optimize Technologies, Oregon City, OR, USA). The MS/MS analysis was performed on Shimadzu LCMS-8050 triple quadrupole mass spectrometer. The instrument was operated and optimized under both positive and negative electrospray and multiple reaction monitoring modes (+/− ESI MRM). The settings of flow rate and gradient program for the LC system as well as MS/MS conditions are recommended by a software method package for 158 lipid mediators (Shimadzu Scientific Instruments, Inc., Columbia, MD, USA) and further optimized following our previously published quantification method [57]. Briefly, the optimized conditions are as follows: Interface voltage, 4.0 kV; interface temperature, 275 °C; DL temperature, 275 °C; heating block temperature, 400 °C; drying gas (N_2), 10 L/min; nebulizing gas (N_2), 3 L/min; heating gas (Air), and 10 L/min; CID gas (Ar), 230 kPa. The acquisition was divided into multiple segments. The *m/z* transitions and their tuning voltages were selected based on the best MRM responses from the instrumental method optimization software. All analyses and data processing were completed on Shimadzu LabSolutions V5.91 software (Shimadzu Scientific Instruments, Inc., Columbia, MD, USA).

4.7. Intracellular Ca^{2+} Measurements

Intracellular Ca^{2+} measurements were performed as previously described [58]. A Photon Technology International (PTI) imaging system was used to measure intracellular Ca^{2+} homeostasis. Myotubes imaged were loaded with 2 μM Fura-2 AM (Thermo Fisher Scientific) for 30 min at 37 °C, followed by RT de-esterification for 15 min. Only cells that had an initial ratio below 1.0, indicating healthy and not Ca^{2+}-overloaded myotubes, were selected for application of 20 mM caffeine (Thermo Fisher Scientific) with a perfusion system (Bioscience Tools, San Diego, CA, USA). Ratiometric analysis (350/375 nm excitation ratio; 510 nm emission) was performed using software PTI EasyRatioPro 2 (HORIBA, Edison, NJ, USA). These experiments were repeated 5 times and at least 6 myotubes were tested on each experiment.

4.8. Statistical Analysis

Three to five independent replicates were performed for each experiment, except for the customized gene array study. One-way ANOVA with post hoc Tukey's test was performed for data analysis. Results were expressed as mean ± SD. Differences were considered significant at $p < 0.05$.

5. Conclusions

In conclusion, by using lipidomic analysis, our data have provided new insights regarding the functions of COXs in skeletal muscle. COX-1 may play a major role in the step from myoblast to fusion in myogenic differentiation. Its effect could also be related with the alteration in AA/LOX pathway. Further studies on the lipid mediators from AA/COX pathway, and the interactions between COXs

Int. J. Mol. Sci. **2019**, *20*, 4326

and LOXs will advance the knowledge of COX and related lipid signaling in skeletal muscle and other tissues, which could benefit the development of new treatments for inflammation related diseases in skeletal muscle and other tissues.

Supplementary Materials: Supplementary materials can be found at http://www.mdpi.com/1422-0067/20/18/4326/s1.

Author Contributions: For Conceptualization, C.M. and M.B.; methodology, C.M., Z.W., and M.B.; validation, C.M., Z.W. and M.B.; formal analysis, C.M., Z.W. and M.B.; investigation, C.M. and Z.W.; resources, L.B. and M.B.; data curation, C.M. and Z.W.; writing—original draft preparation, C.M. and Z.W.; writing—review and editing, L.B. and M.B.; supervision, M.B.; project administration, M.B.; funding acquisition, L.B. and M.B.

Funding: This work was supported by NIH-National Institutes of Aging PO1 AG039355 (L.B., M.B.), and the George W. and Hazel M. Jay and Evanston Research Endowments (M.B.). C.M. and Z.W. were partially supported by grants from National Institutes of Aging R01AG056504 and R01 AG060341 (M.B.) and the National Institutes of Diabetes, Digestive, and Kidney Diseases (M.B.).

Acknowledgments: We thank Yating Du for her assistance in the experiments, and we are grateful for instrumentation support from Shimadzu Scientific Instruments, Inc.

Conflicts of Interest: The authors declare no conflict of interest.

References

1. Karalaki, M.; Fili, S.; Philippou, A.; Koutsilieris, M. Muscle regeneration: Cellular and molecular events. *In Vivo* **2009**, *23*, 779–796. [PubMed]
2. Charge, S.B.; Rudnicki, M.A. Cellular and molecular regulation of muscle regeneration. *Physiol. Rev.* **2004**, *84*, 209–238. [CrossRef] [PubMed]
3. Sugimoto, Y.; Inazumi, T.; Tsuchiya, S. Roles of prostaglandin receptors in female reproduction. *J. Biochem.* **2015**, *157*, 73–80. [CrossRef] [PubMed]
4. Mo, C.; Romero-Suarez, S.; Bonewald, L.; Johnson, M.; Brotto, M. Prostaglandin E2: From clinical applications to its potential role in bone—Muscle crosstalk and myogenic differentiation. *Recent Pat. Biotechnol.* **2012**, *6*, 223–229. [CrossRef] [PubMed]
5. Nakanishi, M.; Rosenberg, D.W. Multifaceted roles of PGE2 in inflammation and cancer. *Semin. Immunopathol.* **2013**, *35*, 123–137. [CrossRef]
6. Kalinski, P. Regulation of immune responses by prostaglandin E2. *J. Immunol.* **2012**, *188*, 21–28. [CrossRef] [PubMed]
7. Mo, C.; Zhao, R.; Vallejo, J.; Igwe, O.; Bonewald, L.; Wetmore, L.; Brotto, M. Prostaglandin E2 promotes proliferation of skeletal muscle myoblasts via EP4 receptor activation. *Cell Cycle* **2015**, *14*, 1507–1516. [CrossRef]
8. Jansen, K.M.; Pavlath, G.K. Prostaglandin F2alpha promotes muscle cell survival and growth through upregulation of the inhibitor of apoptosis protein BRUCE. *Cell Death Differ.* **2008**, *15*, 1619–1628. [CrossRef]
9. Horsley, V.; Pavlath, G.K. Prostaglandin F2(α) stimulates growth of skeletal muscle cells via an NFATC2-dependent pathway. *J. Cell Biol.* **2003**, *161*, 111–118. [CrossRef]
10. Smith, W.L.; DeWitt, D.L.; Garavito, R.M. Cyclooxygenases: Structural, cellular, and molecular biology. *Annu. Rev. Biochem.* **2000**, *69*, 145–182. [CrossRef]
11. Rouzer, C.A.; Marnett, L.J. Cyclooxygenases: Structural and functional insights. *J. Lipid. Res.* **2009**, *50*, S29–S34. [CrossRef] [PubMed]
12. Shen, W.; Prisk, V.; Li, Y.; Foster, W.; Huard, J. Inhibited skeletal muscle healing in cyclooxygenase-2 gene-deficient mice: The role of PGE2 and PGF2alpha. *J. Appl. Physiol.* **2006**, *101*, 1215–1221. [CrossRef] [PubMed]
13. Bondesen, B.A.; Mills, S.T.; Kegley, K.M.; Pavlath, G.K. The COX-2 pathway is essential during early stages of skeletal muscle regeneration. *Am. J. Physiol. Cell Physiol.* **2004**, *287*, C475–C483. [CrossRef] [PubMed]
14. Bondesen, B.A.; Mills, S.T.; Pavlath, G.K. The COX-2 pathway regulates growth of atrophied muscle via multiple mechanisms. *Am. J. Physiol. Cell Physiol.* **2006**, *290*, C1651–C1659. [CrossRef] [PubMed]
15. Ding, X.Z.; Hennig, R.; Adrian, T.E. Lipoxygenase and cyclooxygenase metabolism: New insights in treatment and chemoprevention of pancreatic cancer. *Mol. Cancer* **2003**, *2*, 10. [CrossRef]

16. Kroetz, D.L.; Zeldin, D.C. Cytochrome P450 pathways of arachidonic acid metabolism. *Curr. Opin. Lipidol.* **2002**, *13*, 273–283. [CrossRef] [PubMed]

17. Singh, R.K.; Tandon, R.; Dastidar, S.G.; Ray, A. A review on leukotrienes and their receptors with reference to asthma. *J. Asthma* **2013**, *50*, 922–931. [CrossRef]

18. Moreno, J.J. New aspects of the role of hydroxyeicosatetraenoic acids in cell growth and cancer development. *Biochem. Pharmcol.* **2009**, *77*, 1–10. [CrossRef]

19. Brenneis, C.; Maier, T.J.; Schmidt, R.; Hofacker, A.; Zulauf, L.; Jakobsson, P.J.; Scholich, K.; Geisslinger, G. Inhibition of prostaglandin E_2 synthesis by SC-560 is independent of cyclooxygenase 1 inhibition. *FASEB J.* **2006**, *20*, 1352–1360. [CrossRef]

20. Paiotti, A.P.; Marchi, P.; Miszputen, S.J.; Oshima, C.T.; Franco, M.; Ribeiro, D.A. The role of nonsteroidal antiinflammatory drugs and cyclooxygenase-2 inhibitors on experimental colitis. *In Vivo* **2012**, *26*, 381–393.

21. Daikoku, T.; Wang, D.; Tranguch, S.; Morrow, J.D.; Orsulic, S.; DuBois, R.N.; Dey, S.K. Cyclooxygenase-1 is a potential target for prevention and treatment of ovarian epithelial cancer. *Cancer Res.* **2005**, *65*, 3735–3744. [CrossRef] [PubMed]

22. Otis, J.S.; Burkholder, T.J.; Pavlath, G.K. Stretch-induced myoblast proliferation is dependent on the COX2 pathway. *Exp. Cell Res.* **2005**, *310*, 417–425. [CrossRef] [PubMed]

23. Testa, M.; Rocca, B.; Spath, L.; Ranelletti, F.O.; Petrucci, G.; Ciabattoni, G.; Naro, F.; Schiaffino, S.; Volpe, M.; Reggiani, C. Expression and activity of cyclooxygenase isoforms in skeletal muscles and myocardium of humans and rodents. *J. Appl. Physiol.* **2007**, *103*, 1412–1418. [CrossRef] [PubMed]

24. Novak, M.L.; Billich, W.; Smith, S.M.; Sukhija, K.B.; McLoughlin, T.J.; Hornberger, T.A.; Koh, T.J. COX-2 inhibitor reduces skeletal muscle hypertrophy in mice. *Am. J. Physiol. Regul. Integr. Comp. Physiol.* **2009**, *296*, R1132–R1139. [CrossRef] [PubMed]

25. Bondesen, B.A.; Jones, K.A.; Glasgow, W.C.; Pavlath, G.K. Inhibition of myoblast migration by prostacyclin is associated with enhanced cell fusion. *FASEB J.* **2007**, *21*, 3338–3345. [CrossRef] [PubMed]

26. Velica, P.; Khanim, F.L.; Bunce, C.M. Prostaglandin D2 inhibits C2C12 myogenesis. *Mol. Cell. Endocrinol.* **2010**, *319*, 71–78. [CrossRef]

27. Sears, D.D.; Miles, P.D.; Chapman, J.; Ofrecio, J.M.; Almazan, F.; Thapar, D.; Miller, Y.I. 12/15-lipoxygenase is required for the early onset of high fat diet-induced adipose tissue inflammation and insulin resistance in mice. *PLoS ONE* **2009**, *4*, e7250. [CrossRef] [PubMed]

28. Bhattacharya, A.; Hamilton, R.; Jernigan, A.; Zhang, Y.; Sabia, M.; Rahman, M.M.; Li, Y.; Wei, R.; Chaudhuri, A.; Van Remmen, H. Genetic ablation of 12/15-lipoxygenase but not 5-lipoxygenase protects against denervation-induced muscle atrophy. *Free Radic. Biol. Med.* **2014**, *67*, 30–40. [CrossRef] [PubMed]

29. Kitase, Y.; Vallejo, J.A.; Gutheil, W.; Vemula, H.; Jahn, K.; Yi, J.; Zhou, J.; Brotto, M.; Bonewald, L.F. β-aminoisobutyric Acid, l-BAIBA, Is a Muscle-Derived Osteocyte Survival Factor. *Cell Rep.* **2018**, *22*, 1531–1544. [CrossRef]

30. Roberts, L.D.; Bostrom, P.; O'Sullivan, J.F.; Schinzel, R.T.; Lewis, G.D.; Dejam, A.; Lee, Y.K.; Palma, M.J.; Calhoun, S.; Georgiadi, A.; et al. β-Aminoisobutyric acid induces browning of white fat and hepatic β-oxidation and is inversely correlated with cardiometabolic risk factors. *Cell Metab.* **2014**, *19*, 96–108. [CrossRef]

31. Lauffer, L.M.; Iakoubov, R.; Brubaker, P.L. GPR119 is essential for oleoylethanolamide-induced glucagon-like peptide-1 secretion from the intestinal enteroendocrine L-cell. *Diabetes* **2009**, *58*, 1058–1066. [CrossRef] [PubMed]

32. Wang, X.; Miyares, R.L.; Ahern, G.P. Oleoylethanolamide excites vagal sensory neurones, induces visceral pain and reduces short-term food intake in mice via capsaicin receptor TRPV1. *J. Physiol.* **2005**, *564 Pt 2*, 541–547. [CrossRef]

33. Fu, J.; Gaetani, S.; Oveisi, F.; Lo Verme, J.; Serrano, A.; Rodriguez De Fonseca, F.; Rosengarth, A.; Luecke, H.; Di Giacomo, B.; Tarzia, G.; et al. Oleylethanolamide regulates feeding and body weight through activation of the nuclear receptor PPAR-α. *Nature* **2003**, *425*, 90–93. [CrossRef] [PubMed]

34. Guzman, M.; Lo Verme, J.; Fu, J.; Oveisi, F.; Blazquez, C.; Piomelli, D. Oleylethanolamide stimulates lipolysis by activating the nuclear receptor peroxisome proliferator-activated receptor alpha (PPAR-α). *J. Biol. Chem.* **2004**, *279*, 27849–27854. [CrossRef] [PubMed]

35. Bowen, K.J.; Kris-Etherton, P.M.; Shearer, G.C.; West, S.G.; Reddivari, L.; Jones, P.J.H. Oleic acid-derived oleoylethanolamide: A nutritional science perspective. *Prog. Lipid Res.* **2017**, *67*, 1–15. [CrossRef]

36. Pillarisetti, S.; Alexander, C.W.; Khanna, I. Pain and beyond: Fatty acid amides and fatty acid amide hydrolase inhibitors in cardiovascular and metabolic diseases. *Drug Discov. Today* **2009**, *14*, 1098–1111. [CrossRef] [PubMed]

37. Martinez-Huenchullan, S.; McLennan, S.V.; Verhoeven, A.; Twigg, S.M.; Tam, C.S. The emerging role of skeletal muscle extracellular matrix remodelling in obesity and exercise. *Obes. Rev.* **2017**, *18*, 776–790. [CrossRef] [PubMed]

38. Gurkoff, G.; Shahlaie, K.; Lyeth, B.; Berman, R. Voltage-gated calcium channel antagonists and traumatic brain injury. *Pharmaceuticals (Basel)* **2013**, *6*, 788–812. [CrossRef]

39. Shen, J.; Yu, W.M.; Brotto, M.; Scherman, J.A.; Guo, C.; Stoddard, C.; Nosek, T.M.; Valdivia, H.H.; Qu, C.K. Deficiency of MIP/MTMR14 phosphatase induces a muscle disorder by disrupting Ca^{2+} homeostasis. *Nat. Cell Biol.* **2009**, *11*, 769–776. [CrossRef]

40. Weisleder, N.; Brotto, M.; Komazaki, S.; Pan, Z.; Zhao, X.; Nosek, T.; Parness, J.; Takeshima, H.; Ma, J. Muscle aging is associated with compromised Ca^{2+} spark signaling and segregated intracellular Ca^{2+} release. *J. Cell Biol.* **2006**, *174*, 639–645. [CrossRef]

41. Yazawa, M.; Ferrante, C.; Feng, J.; Mio, K.; Ogura, T.; Zhang, M.; Lin, P.H.; Pan, Z.; Komazaki, S.; Kato, K.; et al. TRIC channels are essential for Ca^{2+} handling in intracellular stores. *Nature* **2007**, *448*, 78–82. [CrossRef] [PubMed]

42. Perera, S.; Holt, M.R.; Mankoo, B.S.; Gautel, M. Developmental regulation of MURF ubiquitin ligases and autophagy proteins nbr1, p62/SQSTM1 and LC3 during cardiac myofibril assembly and turnover. *Dev. Biol.* **2011**, *351*, 46–61. [CrossRef] [PubMed]

43. McElhinny, A.S.; Perry, C.N.; Witt, C.C.; Labeit, S.; Gregorio, C.C. Muscle-specific RING finger-2 (MURF-2) is important for microtubule, intermediate filament and sarcomeric M-line maintenance in striated muscle development. *J. Cell Sci.* **2004**, *117 Pt 15*, 3175–3188. [CrossRef]

44. Oliveira-Nascimento, L.; Massari, P.; Wetzler, L.M. The Role of TLR2 in Infection and Immunity. *Front. Immunol.* **2012**, *3*, 79. [CrossRef] [PubMed]

45. Mojumdar, K.; Giordano, C.; Lemaire, C.; Liang, F.; Divangahi, M.; Qureshi, S.T.; Petrof, B.J. Divergent impact of Toll-like receptor 2 deficiency on repair mechanisms in healthy muscle versus Duchenne muscular dystrophy. *J. Pathol.* **2016**, *239*, 10–22. [CrossRef] [PubMed]

46. Zbinden-Foncea, H.; Raymackers, J.M.; Deldicque, L.; Renard, P.; Francaux, M. TLR2 and TLR4 activate p38 MAPK and JNK during endurance exercise in skeletal muscle. *Med. Sci Sports Exerc.* **2012**, *44*, 1463–1472. [CrossRef] [PubMed]

47. Kim, D.S.; Cha, H.N.; Jo, H.J.; Song, I.H.; Baek, S.H.; Dan, J.M.; Kim, Y.W.; Kim, J.Y.; Lee, I.K.; Seo, J.S.; et al. TLR2 deficiency attenuates skeletal muscle atrophy in mice. *Biochem. Biophys. Res. Commun.* **2015**, *459*, 534–540. [CrossRef] [PubMed]

48. Boyd, J.H.; Divangahi, M.; Yahiaoui, L.; Gvozdic, D.; Qureshi, S.; Petrof, B.J. Toll-like receptors differentially regulate CC and CXC chemokines in skeletal muscle via NF-κB and calcineurin. *Infect. Immun.* **2006**, *74*, 6829–6838. [CrossRef] [PubMed]

49. Hubal, M.J.; Devaney, J.M.; Hoffman, E.P.; Zambraski, E.J.; Gordish-Dressman, H.; Kearns, A.K.; Larkin, J.S.; Adham, K.; Patel, R.R.; Clarkson, P.M. CCL2 and CCR2 polymorphisms are associated with markers of exercise-induced skeletal muscle damage. *J. Appl. Physiol.* **2010**, *108*, 1651–1658. [CrossRef]

50. Harmon, B.T.; Orkunoglu-Suer, E.F.; Adham, K.; Larkin, J.S.; Gordish-Dressman, H.; Clarkson, P.M.; Thompson, P.D.; Angelopoulos, T.J.; Gordon, P.M.; Moyna, N.M.; et al. CCL2 and CCR2 variants are associated with skeletal muscle strength and change in strength with resistance training. *J. Appl. Physiol.* **2010**, *109*, 1779–1785. [CrossRef]

51. Lu, H.; Huang, D.; Ransohoff, R.M.; Zhou, L. Acute skeletal muscle injury: CCL2 expression by both monocytes and injured muscle is required for repair. *FASEB J.* **2011**, *25*, 3344–3355. [CrossRef] [PubMed]

52. Hindi, S.M.; Kumar, A. Toll-like receptor signalling in regenerative myogenesis: Friend and foe. *J. Pathol.* **2016**, *239*, 125–128. [CrossRef]

53. Cantini, M.; Massimino, M.L.; Rapizzi, E.; Rossini, K.; Catani, C.; Dalla Libera, L.; Carraro, U. Human satellite cell proliferation in vitro is regulated by autocrine secretion of IL-6 stimulated by a soluble factor(s) released by activated monocytes. *Biochem. Biophys. Res. Commun.* **1995**, *216*, 49–53. [CrossRef] [PubMed]

54. Haddad, F.; Zaldivar, F.; Cooper, D.M.; Adams, G.R. IL-6-induced skeletal muscle atrophy. *J. Appl. Physiol.* **2005**, *98*, 911–917. [CrossRef] [PubMed]

55. Yakabe, M.; Ogawa, S.; Ota, H.; Iijima, K.; Eto, M.; Ouchi, Y.; Akishita, M. Inhibition of interleukin-6 decreases atrogene expression and ameliorates tail suspension-induced skeletal muscle atrophy. *PLoS ONE* **2018**, *13*, e0191318. [CrossRef] [PubMed]

56. Huang, J.; Hsu, Y.H.; Mo, C.; Abreu, E.; Kiel, D.P.; Bonewald, L.F.; Brotto, M.; Karasik, D. METTL21C is a potential pleiotropic gene for osteoporosis and sarcopenia acting through the modulation of the NF-κB signaling pathway. *J. Bone Min. Res.* **2014**, *29*, 1531–1540. [CrossRef] [PubMed]

57. Wang, Z.; Bian, L.; Mo, C.; Kukula, M.; Schug, K.A.; Brotto, M. Targeted quantification of lipid mediators in skeletal muscles using restricted access media-based trap-and-elute liquid chromatography-mass spectrometry. *Anal. Chim. Acta* **2017**, *984*, 151–161. [CrossRef]

58. Huang, J.; Romero-Suarez, S.; Lara, N.; Mo, C.; Kaja, S.; Brotto, L.; Dallas, S.L.; Johnson, M.L.; Jähn, K.; Bonewald, L.F.; et al. Crosstalk between MLO-Y4 osteocytes and C2C12 muscle cells is mediated by the Wnt/β-catenin pathway. *JBMR Plus* **2017**, *1*, 86–100. [CrossRef]

International Journal of
Molecular Sciences

Article

Bivalent Ligand UDCA-LPE Inhibits Pro-Fibrogenic Integrin Signalling by Inducing Lipid Raft-Mediated Internalization

Jie Su [1,2], Hongying Gan-Schreier [1], Benjamin Goeppert [3], Walee Chamulitrat [1], Wolfgang Stremmel [1] and Anita Pathil [1,*]

[1]   Department of Internal Medicine IV, Gastroenterology and Hepatology, University of Heidelberg, Im Neuenheimer Feld 410, 69120 Heidelberg, Germany; suj@sustc.edu.cn (J.S.); Hongying.Gan-Schreier@med.uni-heidelberg.de (H.G.-S.); Walee.Chamulitrat@med.uni-heidelberg.de (W.C.); Wolfgang.Stremmel@med.uni-heidelberg.de (W.S.)
[2]   Department of Biology, Southern University of Science and Technology, Shenzhen 518055, China
[3]   Institute of Pathology, University Hospital Heidelberg, Im Neuenheimer Feld 224, 69120 Heidelberg, Germany; Benjamin.Goeppert@med.uni-heidelberg.de
*   Correspondence: anita.pathil-warth@med.uni-heidelberg.de; Tel.: +49-6221-56-38102; Fax: +49-6221-56-5687

Received: 5 September 2018; Accepted: 10 October 2018; Published: 20 October 2018

Abstract: Ursodeoxycholyl lysophosphatidylethanolamide (UDCA-LPE) is a synthetic bile acid-phospholipid conjugate with profound hepatoprotective and anti-fibrogenic functions in vitro and in vivo. Herein, we aimed to demonstrate the inhibitory effects of UDCA-LPE on pro-fibrogenic integrin signalling. UDCA-LPE treatment of human embryonic liver cell line CL48 and primary human hepatic stellate cells induced a non-classical internalization of integrin β1 resulting in dephosphorylation and inhibition of SRC and focal adhesion kinase (FAK). Signalling analyses suggested that UDCA-LPE may act as a heterobivalent ligand for integrins and lysophospholipid receptor1 (LPAR1) and co-immunoprecipitation demonstrated the bridging effect of UDCA-LPE on integrin β1 and LPAR1. The disruption of either the UDCA-moiety binding to integrins by RGD-containing peptide GRGDSP or the LPE-moiety binding to LPAR1 by LPAR1 antagonist Ki16425 reversed inhibitory functions of UDCA-LPE. The lack of inhibitory functions of UDCA-PE and UDCA-LPE derivatives (14:0 and 12:0, LPE-moiety containing shorter fatty acid chain) as well as the consistency of the translocation of UDCA-LPE and integrins, which co-fractionated with LPE but not UDCA, suggested that the observed UDCA-LPE-induced translocation of integrins was mediated by LPE endocytic transport pathway.

Keywords: integrin signalling; lipid raft-mediated internalization; hepatic fibrosis

---

1. Introduction

Liver fibrosis is characterized by pathological accumulation of extracellular matrix (ECM). ECM is a collection of molecules which are secreted by cells and distributed in all organs and tissues consisting of collagens, proteoglycans, glycoproteins and glycosaminoglycans [1]. Although some other cell types in the liver can also contribute to fibrosis, it is generally accepted that activated hepatic stellate cells (HSC) are the main source of excessive ECM. Integrins are a family of heterodimeric transmembrane receptors composed of an α and a β subunit, which are involved in cell-cell and cell-matrix interactions. Binding of integrins to ECM components mediates the recruitment and activation of signalling proteins such as focal adhesion kinase (FAK) and SRC kinase, which play a central role in the transduction of intracellular integrin signalling events. Furthermore, integrins have been reported to be involved in HSC activation and migration [2] and found to be upregulated during liver fibrosis [3].

UDCA-LPE is a synthetic bile acid-phospholipid conjugate, which has exhibited profound hepatoprotective and anti-fibrogenic functions in vitro and in vivo [4,5]. The conjugate contains an ursodeoxycholic acid (UDCA) moiety, which by itself also exhibits protective effects against hydrophobic bile-acid-induced hepatocellular apoptosis in cholestatic liver disease [6] and has been approved for the treatment of primary biliary cirrhosis [7].

Notably, our former studies revealed that protective functions of UDCA-LPE are critically dependent on the conjugation between the bile acid and the phospholipid whereas the individual compounds UDCA or LPE showed only little efficacy in different in vitro [4] and in vivo models [8]. These results imply that the conjugation due to its hydrophobicity is decisive in order to facilitate the interaction of UDCA-LPE with lipid membranes [9,10] rendering it a promising drug candidate for membrane lipid therapy [11].

Herein, we demonstrate the interaction of UDCA-LPE with integrins leading to integrin internalization via lipid rafts and subsequent inhibition of fibrogenic signalling. These events represent a novel mechanism of UDCA-LPE in support of its potent anti-fibrogenic effects previously observed in experimental mouse models of liver disease [5].

2. Results

2.1. UDCA-LPE Induces Translocation of Integrins

The interaction with ECM leads to an autophosphorylation of FAK at Tyr397 with subsequent binding of FAK to SRC, which in turn activates SRC leading to phosphorylation of FAK at Tyr576/577 and Tyr925, which is known to be essential for its kinase activity [12]. After phosphorylation in response to integrin engagement, FAK and SRC trigger pro-fibrogenic signalling both in vivo and in vitro [13,14]. As non-kinase receptors, integrins activate FAK by a conformational change. Thus, the co-localization and interaction between integrins and FAK/SRC are considered to be essential for proper signalling. In the absence of UDCA-LPE, integrin β1 and SRC were found to be localized predominantly at cell-to-cell contacts of CL48 liver cells (Figure 1A). Upon addition of UDCA-LPE for 30 min, most of integrin β1 migrated away from plasma membrane while SRC localization was not affected (Figure 1A). After 2 h treatment, integrin β1 accumulated more pronouncedly at the nuclear envelope (Figure 1A). Despite of this integrin β1 translocation, the localization of active FAK (pFAK Tyr397) at the focal adhesions of CL48 cells was not affected by UDCA-LPE (Figure S1). UDCA-LPE-induced internalization of integrin β1 was also observed in HHStec cells (Figure S2A). The co-localization of integrin β1 and the endoplasmic reticulum (ER) marker calnexin in CL48 cells (Figure 1B) and HHStec cells (Figure S2B) upon UDCA-LPE treatment suggests the localization of endocytosed integrin β1 at the ER. Notably, treatment of CL48 cells with UDCA, LPE or UDCA + LPE had no effect on integrin β1 localization (Figure S3). Besides integrin β1, UDCA-LPE similarly induced the translocation of other integrins including integrin α2, α3, α5, αv, β4 and β5 (Figure S4). In the absence of UDCA-LPE, these integrins displayed some differences in terms of localization, that is, integrins α2, α3 and α5 localized at plasma membrane, integrin αv at the cytoplasm and integrin β5 at focal adhesions (Figure S5). However, the internalization of these integrins by UDCA-LPE was similar to that of integrin β1.

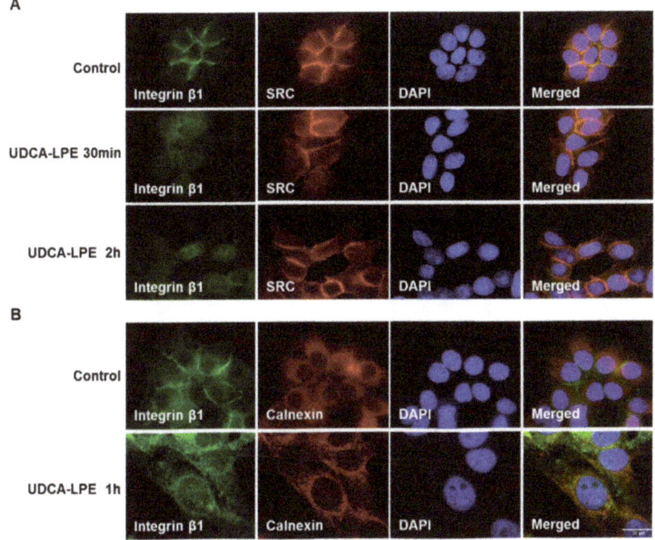

Figure 1. UDCA-LPE modulates the localization of integrin β1. Representative fluorescence microscopy images of CL48 cells after treatment with 90 μM UDCA-LPE for (**A**) 30 min or 2 h and (**B**) 1 h. Immunofluorescence showed the staining of (**A**) integrin β1 (green), SRC (red) and DAPI (blue) and (**B**) integrin β1 (green), calnexin (red) and DAPI (blue). DAPI was used for nuclear staining.

2.2. Integrin Translocation by UDCA-LPE Suppresses FAK and SRC Phosphorylation

Translocation of integrin β1 with subsequent loss of its co-localization with SRC (Figure 1A) was associated with decreased phosphorylation of FAK (Tyr925 and Tyr576/577) and SRC (Tyr416) upon UDCA-LPE treatment of CL48 (Figure 2A) and HHStec cells (Figure 2B) in a time-dependent manner from 15 min to 2 h. In CL48 cells, phosphorylation of c-Jun N-terminal kinases (JNK) which is a downstream target protein of FAK was also decreased by UDCA-LPE treatment (Figure 2A). Thus, UDCA-LPE inhibited integrin signalling after induction of integrin internalization via an inhibition of FAK and SRC phosphorylation.

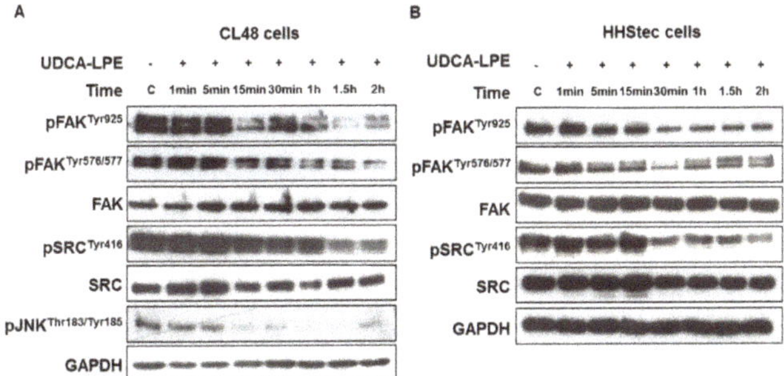

Figure 2. UDCA-LPE inhibits phosphorylation of SRC and FAK. (**A**) CL48 cells and (**B**) HHStec cells were treated with 90 μM UDCA-LPE for 1 min to 2 h. Cell lysates were probed with antibodies against phospho-FAK (Tyr925), phospho-FAK (Tyr576/577), FAK, phospho-SRC (Tyr416), SRC and phospho-JNK (Thr183/Tyr185). GAPDH was used as control for equal protein loading.

2.3. RGD-Containing Peptide GRGDSP Inhibits UDCA-LPE-Induced Translocation of Integrins

The most prevalent integrin recognition site present in ECM contains a tripeptide motif composed of L-arginine, glycine and L-aspartic acid (RGD). RGD-containing peptides, which bind to the RGD-recognition site of integrin, inhibit their binding to the ECM. Although UDCA-LPE mediated the translocation of multiple integrins, it did not induce the translocation of integrin α1 as observed in CL48 cells (Figure S5). Integrin α1 can uniquely form a α1β1 heterodimer, which unlike most other integrins does not recognize RGD motif in ECM [15]. The lack of integrin α1 translocation implies that UDCA-LPE-induced translocation of integrins may solely depend on the RGD-recognition motif. GRGDSP peptide which blocks the RGD-recognition motif in integrins was therefore used to disrupt the binding of integrins to the RGD motif in ECM. GRGDSP alone had no effect on integrin localization (Figure 3). However, pre-treatment with GRGDSP markedly blocked UDCA-LPE-induced translocation of integrin β1 (Figure 3).

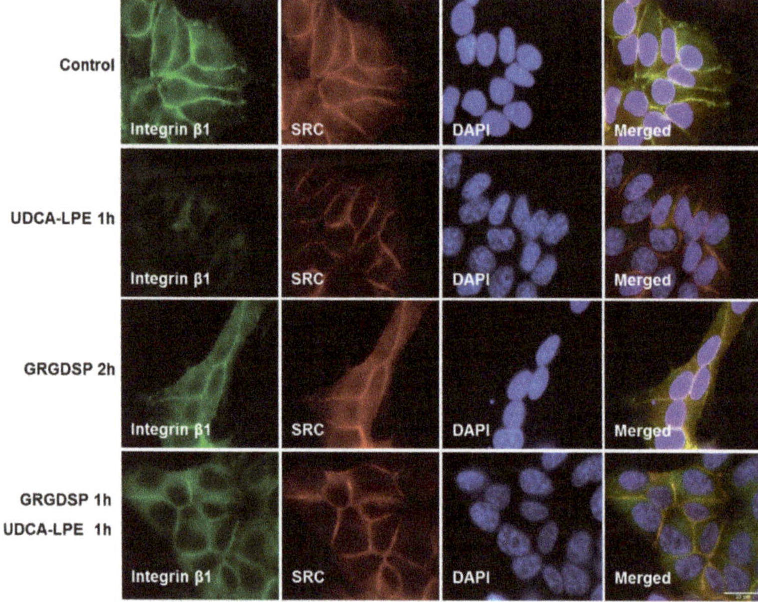

Figure 3. RGD-containing peptide GRGDSP inhibits UDCA-LPE–induced translocation of integrins. Representative fluorescence microscopy images of CL48 cells after treatment with 200 μg/mL RGD-containing peptide GRGDSP for 1 h and 90 μM UDCA-LPE for additional 1 h. IF staining of integrin β1 (green), SRC (red) and DAPI (blue). DAPI was used for nuclear staining.

2.4. UDCA-LPE Binds to Integrins with Its UDCA-Moiety

It is known that activation of integrin signalling involves autophosphorylation of FAK at Tyr397, which leads to an interaction of FAK with SRC [12]. With a short incubation time of 1–5 min, UDCA-LPE stimulated the phosphorylation of FAK (Tyr397) as well as the downstream targets c-Raf (p-Ser338) and ERK (p-Thr202/Tyr204) (Figure 4A). This activation was inhibited by GRGDSP pre-treatment. Phosphorylation of FAK at Tyr397, c-Raf and ERK was also observed by UDCA treatment (Figure 4B). Similar to GRGDSP, pre-treatment with FAK inhibitor 1,2,4,5-benzenetetraamine tetrahydrochloride (Y15) significantly blocked UDCA-LPE-induced phosphorylation of FAK (Tyr397), c-Raf and ERK (Figure 4C), suggesting a FAK-dependent mechanism. We found that RGD peptide alone also induced the phosphorylation of c-Raf and ERK after 1-5 min treatment time (Figure 4D) in a similar manner as UDCA-LPE (Figure 4D) and UDCA (Figure 4B). This suggested that these compounds triggered integrin signalling in a similar manner like RGD peptide. Taken together, our results suggest that an interaction of UDCA-LPE with integrins may employ the UDCA-moiety of the molecule. Further binding experiments have to prove this hypothesis.

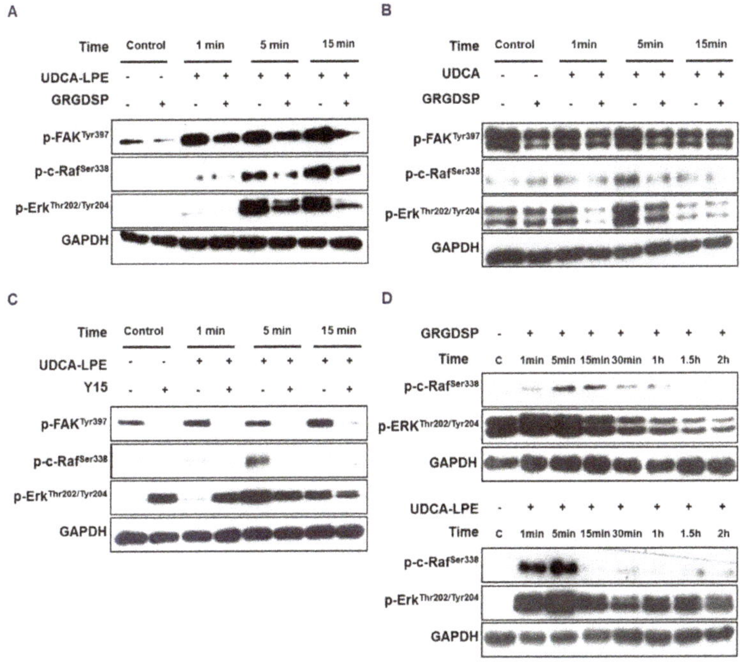

Figure 4. UDCA-LPE and UDCA induce integrin-dependent phosphorylation of c-Raf and ERK. (**A–C**) CL48 cells were treated with (**A,B**) 100 µg/mL RGD containing peptide GRGDSP or (**C**) 100 µM FAK inhibitor 1,2,4,5-benzenetetraamine tetrahydrochloride (Y15) for 30 min and (**A,C**) 90 µM UDCA-LPE or (**B**) 90 µM UDCA for 1 to 15 min. Lysates were probed with antibodies against phospho-FAK (Tyr397), phospho-c-Raf (Ser338) and phospho-ERK 1/2 (Thr202/Tyr204). (**D**) CL48 cells were treated with 100 µg/mL RGD peptide or 90 µM UDCA-LPE for 1 min to 2 h. Lysates were probed with antibodies against phospho-c-Raf (Ser338) and phospho-ERK 1/2 (Thr202/Tyr204). GAPDH was used as control for equal protein loading.

2.5. UDCA-LPE Serves as a Bivalent Ligand Bridging Between Integrin β1 and LPAR1

We found that treatment of CL48 cells with UDCA-LPE or LPE was able to induce phosphorylation of b-Raf at Ser445 in the first 15 min (Figure 5A). It is known that LPE interacts with a G protein-coupled receptor LPAR1 [16] and that LPAR activation induces the activation of PKA [17,18]. We used anti-PKA substrates (RRXS*/T*) antibody to determine the activity of PKA. UDCA-LPE was able to rapidly induce phosphorylation of PKA substrates maximizing at 15 min (Figure 5B). The UDCA-LPE-induced activation of b-Raf (but not c-Raf) and ERK was inhibited by pre-treatment with PKA antagonist Rp-cAMP (Figure 5C). These data showed the ability of UDCA-LPE to trigger LPE/LPAR1 signalling via PKA/b-Raf/ERK pathways. We further performed immunoprecipitation of integrin β1 followed by immunoblotting with an anti-LPAR1 antibody. LPAR1 was nearly undetectable in the pull-downs of control cells whereas LPAR1 protein levels were markedly elevated in those of UDCA-LPE-treated cells (Figure 5D). Our results suggest that UDCA-LPE may act as a bivalent ligand bridging between integrins and LPAR1 to form a tri-component complex.

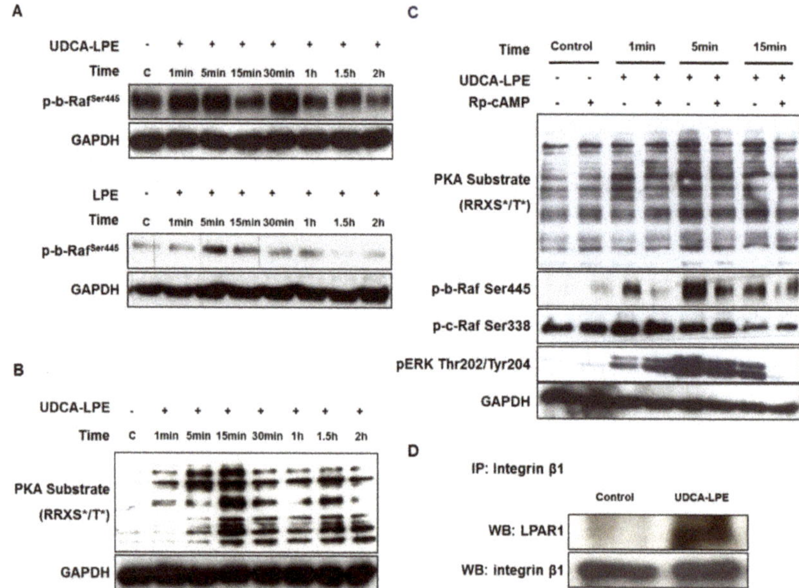

Figure 5. UDCA-LPE induces the LPE signalling and complex formation between LPAR1 and integrin β1. (**A,B**) CL48 cells were treated with (**A,B**) 90 μM UDCA-LPE or (**A**) 90 μM LPE for 1 min to 2 h. Lysates were probed with antibodies against (**A**) phospho-b-Raf (Ser445) and (**B**) PKA substrates (RRXS*/T*). (**C**) CL48 cells were treated with 200 μM Rp-cAMP for 30 min and 90 μM UDCA-LPE for 1 min to 15 min. Lysates were probed with antibodies against PKA substrate (RRXS */T *), phospho-b-Raf (Ser445), phospho-c-Raf (Ser338) and phospho-ERK (Thr202/Tyr204). (**D**) CL48 cells were treated with 90 μM UDCA-LPE for 1 h. Integrin β1-containing proteins were immunoprecipitated with a polyclonal anti-integrin β1 antibody and immunoblotted using anti-LPAR1 or anti-integrin β1 antibody.

2.6. LPE-Moiety is Necessary for UDCA-LPE-Induced Translocation of Integrin β1 and Suppressed FAK and SRC Phosphorylation

To dissect the role of LPAR1, we further utilized an LPAR antagonist Ki16425, which was reported to disrupt the binding of LPE to LPAR1 [16]. We found that Ki16425 pre-treatment significantly blocked UDCA-LPE-induced translocation of integrin β1 in a concentration-dependent manner (Figure 6A). Additionally, UDCA-LPE-induced inhibition of phosphorylation of FAK (Tyr576/577 and Tyr925) (Figure 6B) and SRC (Tyr416) (Figure 6C) was nearly completely abolished by pre-incubation with Ki16425. It has been reported that the activity of lysophosphatidic acids to bind and activate LPAR decreases with a shorter fatty-acid chain length [19,20]. Thus, we treated CL48 cells with UDCA-PE (a conjugate of UDCA and 18:1, 18:1 PE), UDCA-LPE (12:0) (UDCA conjugated with 12:0 LPE) or UDCA-LPE (14:0) (UDCA conjugated with 14:0 LPE). Unlike UDCA-LPE (UDCA conjugated with 18:1 LPE), UDCA-PE, UDCA-LPE (12:0) or UDCA-LPE (14:0) did not decrease but rather slightly increase the phosphorylation of FAK (Tyr 925 and Tyr576/577) and SRC (Tyr416), which was found to be similar to UDCA and tauro-UDCA (TUDCA) (Figure 6D). Taken together, our results suggested that the LPE-moiety and its association with LPAR1 were essential for UDCA-LPE-induced translocation of integrin β1 and inhibition of SRC and FAK phosphorylation.

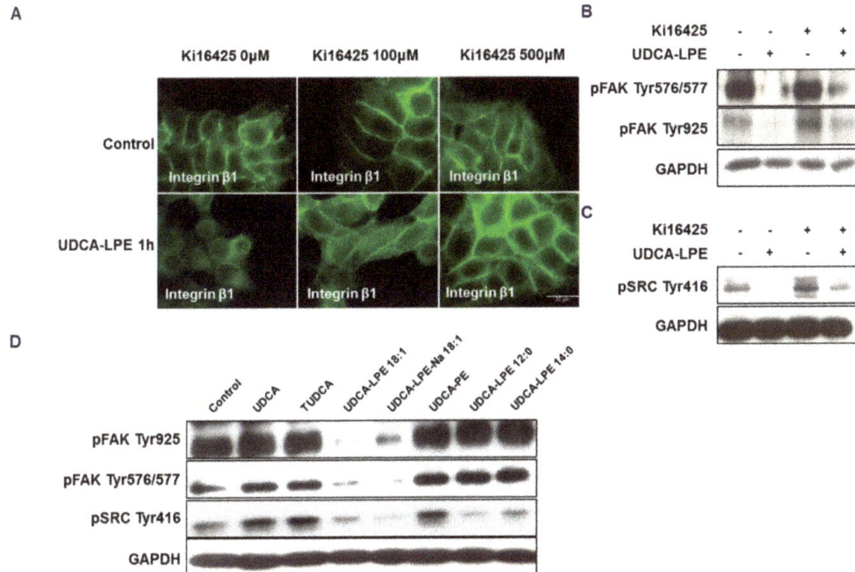

Figure 6. Translocation of integrin β1 and dephosphorylation of SRC and FAK is dependent on the LPE moiety of UDCA-LPE. (**A**) Representative fluorescence microscopy images of CL48 cells after treatment with LPAR antagonist Ki16425 at 100 µM or 500 µM for 1 h and 90 µM UDCA-LPE for additional 1 h. IF of anti-integrin β1 (green). (**B**) CL48 cells were treated with 50 µM Ki16425 for 1 h, followed with 90 µM UDCA-LPE for additional 1h. Lysates were probed with antibodies against phospho-FAK (Tyr925) and phospho-FAK (Tyr576/577). (**C**) CL48 cells were treated with 1mM Ki16425 for 1 h, followed with 90 µM UDCA-LPE for additional 1 h. Lysates were probed with antibodies against phospho-SRC (Tyr416). (**D**) CL48 cells were treated with 90 µM UDCA, TUDCA, UDCA-LPE, UDCA-LPE-Na, UDCA-PE, UDCA-LPE 12:0 or UDCA-LPE 14:0 for 2 h. Lysates were probed with antibodies against phospho-FAK (Tyr925) and phospho-FAK (Tyr576/577). GAPDH was used as control for equal protein loading.

2.7. UDCA-LPE Mediates the Compartmentalization of Integrins into Lipid Rafts

Cell lysates were subjected to lipid fractionation and the levels of various integrins in 12 fractions were analysed by western blotting. A marker for lipid rafts caveolin-1 was mostly detected in fractions 2–4 of control CL48 cell lysates and in fractions 1–4 in cells treated with UDCA-LPE for 30 min (Figure 7A). This indicated that the integrity of lipid rafts was not disturbed by UDCA-LPE and that lipid rafts were maintained in lower density fractions 1–4. UDCA-LPE treatment did not alter SRC protein concentrations in any of lipid fractions (Figure 7H) but markedly increased concentrations of integrin α2, α3, α5, αv, β1 and β4 in lipid-raft fractions 1–4 concomitant with decreased concentrations in fractions 5–8 (Figure 7B–G). Moreover, co-incubation with GRGDSP inhibited UDCA-LPE-induced translocation of these integrins to lipid-raft fractions (Figure 7B–G).

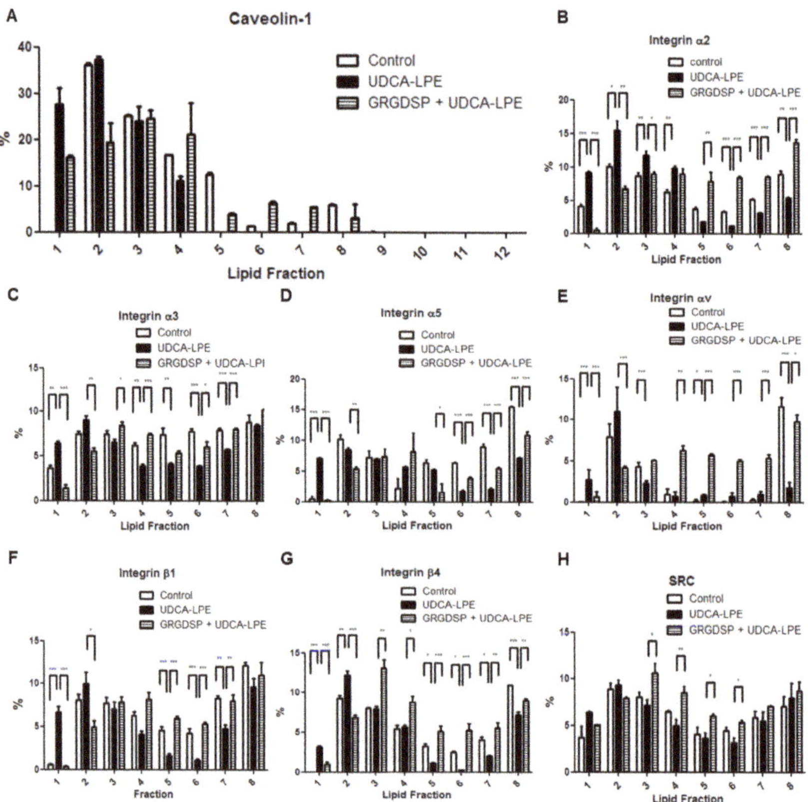

Figure 7. UDCA-LPE mediates compartmentalization of integrins into lipid rafts. (**A–C**) Lipid fractionation of CL48 cells after treatment with 200 μg/mL RGD-containing peptide GRGDSP for 1 h and 90 μM UDCA-LPE for 30 min. Separated fractions were immunoblotted with antibodies against (**A**) caveolin-1, (**B**) integrin α2, (**C**) integrin α3, (**D**) integrin α5, (**E**) integrin αv, (**F**) integrin β1, (**G**) integrin β4 or (**H**) SRC respectively. The protein of interest was normalized to the amount of all proteins in 12 fractions as 100%. Data are means ± the standard deviation of three independent experiments. *** $p < 0.001$, ** $p < 0.01$, * $p < 0.05$.

2.8. Integrin-Bound UDCA-LPE Translocated into Lipid Rafts, Which Co-Fractionated with LPE but Not UDCA

The intracellular transport of a heterobivalent ligand could be determined by one of its receptors [21]. To investigate which receptor determined the localization of UDCA-LPE, we treated CL48 cells with UDCA, LPE or UDCA-LPE for 30 min and cell lysates were subjected to lipid-raft fractionation and the concentrations of UDCA, LPE or UDCA-LPE in 12 fractions were respectively determined by high-performance liquid chromatography-tandem mass spectrometry. UDCA was localized only in non-raft fractions, whereas LPE was present in both raft- and non-raft fractions (Figure 8A), suggesting that UDCA receptors were localized only in non-raft fractions whereas LPE receptors were present in both fractions. UDCA-LPE displayed an integrated localization of both UDCA and LPE and the proportion of UDCA-LPE in raft fractions 1–4 was in parallel to that of LPE (Figure 8A, Inset) suggesting that the initial localization of UDCA-LPE was determined by both UDCA- and LPE-receptors. GRGDSP, which inhibited the binding of UDCA-LPE to integrins, decreased the proportion of UDCA-LPE in non-raft fractions and increased the proportion in lipid-raft fractions, indicating that integrin-bound UDCA-LPE was initially localized in non-raft fractions. After incubation with UDCA-LPE for 2 h or overnight an increased proportion of UDCA-LPE was detected in lipid rafts in a time-dependent manner (Figure 8B), suggesting a translocation of integrin-bound UDCA-LPE to lipid rafts at a longer incubation time. These data were consistent with the translocation of integrins into lipid rafts by UDCA-LPE treatment (Figure 7), suggesting a co-translocation of integrins with UDCA-LPE. Taken together, the co-translocation of integrins and UDCA-LPE was determined by LPE receptors.

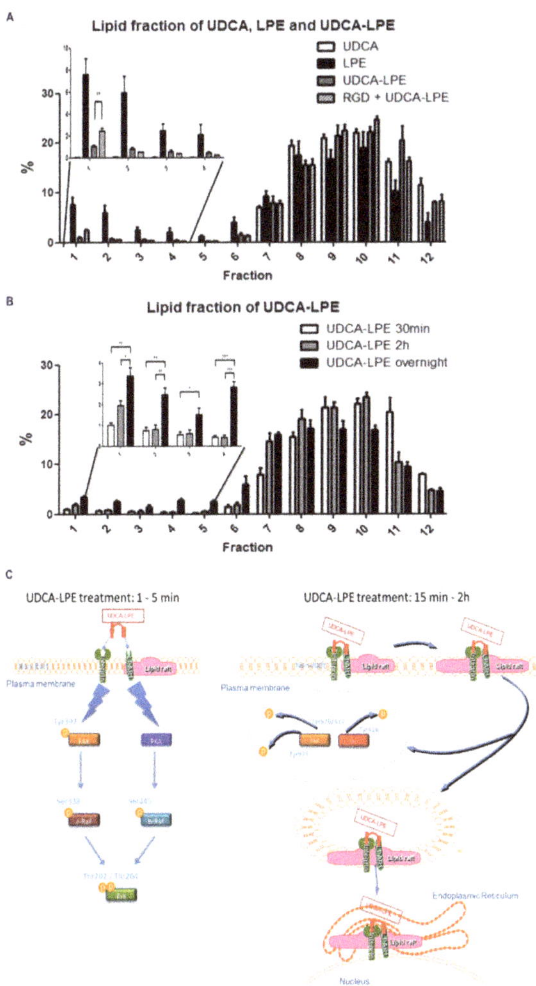

Figure 8. Distribution of UDCA, LPE and UDCA-LPE in lipid fractions. (**A**,**B**) Lipid-raft fractionation of CL48 cells after treatment with (**A**) 90 µM UDCA, 90 µM LPE, 90 µM UDCA-LPE for 30 min or 200 µg/mL GRGDSP for 1 h and 90 µM UDCA-LPE for additional 30 min or (**B**) 90 µM UDCA-LPE for 30 min, 2 h or overnight. Separated fractions were subjected to liquid-chromatography mass spectrometry for quantification of UDCA, LPE or UDCA-LPE in UDCA, LPE or UDCA-LPE-treated cells, respectively. In each treated group, the total levels of UDCA, LPE or UDCA-LPE in 12 fractions was normalized as 100%. Data are means ± the standard deviation of three independent experiments. *** $p < 0.001$, ** $p < 0.01$, * $p < 0.05$. (**C**) Schematic time-dependent model for anti-fibrogenic effects of UDCA-LPE. The lightning graphic means stimulation: the binding of UDCA-LPE with integrin activates phosphorylation of FAK and SRC. The arrows mean (1) translocation of the UDCA-LPE complex into lipid rafts, which (2) results in the dephosphorylation of FAK and SRC.

3. Discussion

As effective therapeutic options against liver fibrosis are limited to date, the proposal of novel compounds which target pro-fibrogenic pathways is urgently needed. The bile acid-phospholipid conjugate UDCA-LPE has been proven to exhibit potent anti-fibrogenic functions in vitro and in vivo [5]. In this study, we analysed enforced translocation of integrins by UDCA-LPE as a possible mechanism for its anti-fibrogenic effects. We showed that UDCA-LPE can associate to the RGD-recognition motif in integrins and LPAR1 with its UDCA- and LPE-moiety, respectively. The latter binding acts as a transporter of UDCA-LPE into lipid-rafts occurring simultaneously with an internalization of UDCA-LPE-bound integrins to the ER and the nuclear envelope. The subsequent loss of SRC co-localization with integrins decreased phosphorylation levels of SRC and FAK leading to an inhibition of pro-fibrogenic activity.

Recent studies have reported that TUDCA stimulates integrin-dependent phosphorylation of SRC, FAK, ERK and p38MAPK [22,23]. Similar to TUDCA, UDCA and UDCA-LPE stimulated integrin- and FAK-dependent c-Raf and ERK phosphorylation in CL48 cells as well (Figure 4). Interestingly, recent results using a 3D model of integrin $\alpha5\beta1$ have shown the importance of the RGD-recognition motif as a sensor of TUDCA. However, TUDCA has an intracellular effect on integrin $\alpha5\beta1$ rather than at the plasma membrane [24]. As UDCA and TUDCA are known to be located at the interfacial outer surface of plasma membrane [25,26], this may be the case for the binding of UDCA-LPE to the extracellular domain of integrins as we showed that UDCA-LPE was able to induce integrin internalization at plasma membrane (Figure 1) and that GRGDSP could inhibit the translocation of integrins (Figure 2B–D and Figure 3).

The design of heterobivalent ligands to target two different receptors has previously been used for pharmacological purposes [27]. As a novel heterobivalent ligand (Figure 8C), UDCA-LPE was not only able to bind to integrins with its UDCA-moiety but also triggered LPE/LPAR1 signalling through its LPE-moiety (Figure 5A–C). The stimulation of UDCA and LPE signalling occurred in the first 5 min of UDCA-LPE treatment by association of UDCA-LPE with integrins and LPAR1 (Figure 8, left). This process may be equivalent to UDCA + LPE treatment. However, the character of UDCA-LPE to bridge integrins and LPE receptors, which was confirmed by co-immunoprecipitation of integrins and LPAR1 (Figure 5D), rendered UDCA-LPE to have a specific function in pulling integrins into the intracellular transport pathway of LPE. We hypothesize that this results in the translocation of integrins from the plasma membrane to the ER and the nuclear envelop observed at a longer incubation time (Figure 8, right). Additional experiments have to further evaluate the localization and intracellular trafficking of endocytosed integrins. Our results showed that LPE alone had no effect on the localization of integrins and that UDCA-LPE-induced translocation of integrins and the inhibition of integrin signalling were dependent on the LPE-moiety of UDCA-LPE.

It has been reported that LPAR1 is localized partially in lipid rafts [28] and that disruption of lipid rafts impairs the function of LPAR1 [29,30]. As LPAR1 is a receptor of LPE [16], we also found that ~18% of LPE was localized in lipid rafts upon LPE treatment for 30 min (Figure 8A). Although UDCA has been reported to antagonize the deoxycholate-induced cholesterol depletion [31], it has been shown that UDCA owns a much higher affinity to non-raft than to lipid-raft fractions [25]. Consistent with this, almost no UDCA was detectable in raft fractions of UDCA-treated cells (Figure 8A). The localization of integrins in non-rafts and LPE endocytic transport pathway destined in lipid-rafts indeed allowed an opportunity for UDCA-LPE to be the mediator for integrin translocation (Figure 8C). UDCA-LPE was translocated into lipid rafts via LPE/LPAR1 axis (Figure 8) concomitant with its internalization (Figure 7) via UDCA/integrin axis.

The general mechanism for endocytic transport of LPE has not been well understood. LPARs are normally localized in both clathrin and caveolar endocytic microdomains and the latter is thought to respond to LPAR internalization because of LPAR co-localization with caveolin-1 in the nucleus [28]. It has been shown that LPAR-induced gene expression is insensitive to caveolea-disrupting agents filipin and methyl-β-cyclodextrin [28], suggesting that LPAR internalization does not necessarily rely on the structure of caveolea. Our data also supported this notion as filipin or methyl-β-cyclodextrin treatment did not inhibit integrin translocation induced by UDCA-LPE (data not shown). The independency from caveolea was one of the features of UDCA-LPE-induced internalization of integrins which may be different from the previously reported integrin endocytosis/recycling pathway [32].

Integrins cross-talk with crucial pro-fibrogenic pathways such as TGFβ1 and PDGF signalling [33] and are therefore regarded as attractive therapeutic targets for the treatment of fibrotic disease. Most inhibitors of integrins including antibodies and cyclic RGD-containing peptides [34,35] have focused on the inhibition of integrin-induced cell-to-ECM and cell-to-cell interactions. However, the use of RGD peptides for fibrosis treatment is quite limited [36,37] because of their lack of persistent effects [38]. Due to multiple binding sites of integrins for ECM [39], an exclusive blockade of RGD-recognition motif may not completely disrupt the binding of integrins to ECM. Here, we could demonstrate that UDCA-LPE not only occupied the RGD-binding sites in integrins but also induced integrin internalization which completely disrupted the ECM-binding to integrins at the plasma membrane (Figure 8C). Thus, UDCA-LPE emerged as an effective inhibitor of RGD-binding integrins more potent than the typical RGD-containing peptide.

It is well-recognized that integrin-induced signalling plays a crucial role in fibrogenesis and that the downstream proteins FAK and SRC play an essential during pro-fibrotic signalling [40,41]. RGD peptide has been reported to activate integrins [42,43], which may also promote fibrogenic signalling. Our data supported this notion as we found that RGD peptide was able to induce integrin signalling (Figure 4D). Unlike RGD peptide, by removing the activator of FAK and SRC UDCA-LPE treatment led to persistent inhibition of integrin signalling after long incubation of CL48 cells and HHStec cells (Figures 2 and 8C, right) thus displaying a very potent anti-fibrogenic effect.

In present study, we demonstrated a possible novel pharmacological tool for integrin inhibition, where UDCA-LPE did not function as a direct inhibitor of integrins per se but as a heterobivalent ligand bridging between integrins and LPAR1. By the action of LPE/LPAR1 transporters in cells, UDCA-LPE was able to induce the translocation of integrins leading to a loss of co-localization with SRC, which resulted in dephosphorylation of FAK and SRC and inhibition of downstream fibrogenic targets. This elucidated mechanism of action renders UDCA-LPE as a drug candidate for the treatment of liver fibrosis.

4. Materials and Methods

4.1. Reagents and Cell Culture

All reagents as well as the cultures and treatment of human embryonic liver CL48 cell line and Human Hepatic Stellate Cells (HHStec) are shown in Supplementary Materials.

4.2. Western Blotting

Lysates of treated cells were subjected to western blotting analysis (Supplementary Materials).

4.3. Immunofluorescence

Paraformaldehyde-fixed cells were subjected to immunofluorescence (Supplementary Materials).

4.4. Lipid Fractionation

For each treatment group, CL48 cells were cultured in 20×75 cm^2 culture flasks. After UDCA, LPE or UDCA-LPE treatment, cells were rinsed with PBS and scraped into 10 mL buffer containing 2 mM HEPES, 150 mM NaCl, 1 mM EGTA, 5 mM sodium vanadate, 10 mM sodium azide, 10 mM sodium pyrophosphate, 100 µg/mL PMSF, 1 mM sodium orthovanadate and 10 µl/mL protease inhibitor cocktail. Cells were homogenized and the lysates were centrifuged at $800 \times g$ at 4 °C for 10 min. Two mL of supernatants were incubated at 37 °C for 4 min and then incubated with 0.02 g Brij 98 at 37 °C for 5 min. The extracts were adjusted to 4 mL with 2 M Sucrose and cooled down in ice for 1 h. The extracts were gently overlaid with successive decreasing sucrose densities solutions (0.9–0.8–0.75–0.7–0.6–0.5–0.4–0.2 mol/L Sucrose) to prepare a discontinuous sucrose gradient. The gradients were centrifuged at 200,000× *g* in a Beckman SW 41Ti rotor for 22 h at 4 °C. Twelve fractions (1 mL for each fraction) were collected and used for western blotting and liquid-chromatography mass spectrometry (LC/MS-MS) analyses. The concentrations of total targets (proteins or lipids) in 12 fractions were normalized to 100% and the proportion or abundance of each target was reported in %. Cell lysates were subjected to sucrose density-gradient centrifugation for lipid fractionation.

4.5. Quantification of UDCA-LPE, UDCA and LPE

Following lipid fractionation, 500 µL of each fraction were extracted with 3 mL chloroform-methanol 2:1 mixture, 500 µL water and 20 µL internal standard D4-UDCA. Following centrifugation at 2500 rpm for 5 min, the lower chloroform phase was collected. Three mL of 2:1 chloroform-methanol mixture was added to the upper phase, extracted the second time and again centrifuged at 2500 rpm for 5 min. The lower phase was collected, combined with the previous chloroform phase and added to 0.4 mL 50 mM citric acid. Following mixing and centrifugation the lower phase was collected in a glass tube and the solvent was evaporated to dryness. The dried lipids were dissolved in 180 µL methanol. Concentrations of UDCA-LPE, UDCA and LPE in each lipid fraction were quantified using a liquid-chromatography mass spectrometer. The responses were calculated from the ratio of UDCA-LPE, UDCA, or LPE peak and D4-UDCA. Concentrations in nmol/mg protein were calculated from response of UDCA-LPE, UDCA and LPE used in standard curves. LC/MS-MS machine and running conditions are described in our published work [44]. Briefly, the separation was achieved by using a Phenomenex Luna C18 (Phenomenex, Aschaffenburg, Germany) column (100 × 2.0 mm, 3 µm) fitted on a separation module of a Waters 2695 (Waters, Milford, MA, USA). Binary solvents were 80% H2O/MeOH with 8 mM ammonium acetate, pH 8.0 (solvent A) and 95% MeOH/H2O with 8 mM ammonium acetate, pH 8.0 (solvent B). The flow rate was maintained at 0.2 mL/min and the gradient was started with 100% solvent A for 2.5 min, changed to 100% solvent B in 1 min, held for 16.5 min and returned to the initial condition in 3 min. Separated fractions were detected on-line by an electrospray ionization source of the tandem mass spectrometer (Quattro micro API, Micromass Waters, Waters, Milford, MA, USA).

4.6. Immunoprecipitation

Lysates of treated cells were subjected to immunoprecipitation analysis (Supplementary Materials).

4.7. Statistical Analysis

Statistical analysis was performed using Prism Software version 4.0 (GraphPad, La Jolla, San Diego, CA, USA).

Please see Supplementary Materials for detailed information.

5. Conclusions

UDCA-LPE enforces internalization of integrins leading to an inhibition of downstream signalling pathways. As a possible novel mode of integrin inhibition, we described the simultaneous bivalent ligation of integrins and LPAR1 by via the LPE endocytic transport pathway. Thus, UDCA-LPE emerges as drug candidate for treatment of liver fibrosis by inhibiting integrin signalling via its internalization.

Supplementary Materials: Supplementary materials can be found at http://www.mdpi.com/1422-0067/19/10/3254/s1.

Author Contributions: Conceptualization, J.S. and A.P.; Formal analysis, J.S.; Funding acquisition, A.P.; Investigation, J.S.; Methodology, J.S., H.G.-S. and B.G.; Resources, W.S.; Supervision, A.P.; Validation, W.C. and A.P.; Writing–original draft, J.S.; Writing–review & editing, W.C., W.S. and A.P.

Funding: This study was supported by the German Research Foundation (PA 2365/1-1). A.P. was funded by the Olympia Morata Postdoctoral Fellowship of the Medical Faculty of University of Heidelberg.

Acknowledgments: The authors thank Fortunata Jung for excellent technical assistance. We acknowledge financial support by Deutsche Forschungsgemeinschaft within the funding programme Open Access Publishing, by the Baden-Württemberg Ministry of Science, Research and the Arts and by Ruprecht-Karls-Universität Heidelberg.

Conflicts of Interest: The authors who have taken part in this study declare that they do not have anything to disclose regarding funding or conflict of interests. W.S. has a patent on UDCA-LPE (no industrial funding).

Abbreviations

ECM	extracellular matrix
ER	endoplasmic reticulum
ERK	extracellular signal-regulated kinase
FAK	focal adhesion kinase
HHStec	primary human hepatic stellate cells
LPAR1	lysophosphatidic acid receptor 1
LPE	lysophosphatidylethanolamine
RGD	L-arginine, glycine and L-aspartic acid
TUDCA	tauro-ursodeoxycholic acid
UDCA	ursodeoxycholic acid
UDCA-LPE	ursodeoxycholyl lysophosphatidylethanolamide

References

1. Schuppan, D.; Ruehl, M.; Somasundaram, R.; Hahn, E.G. Matrix as a modulator of hepatic fibrogenesis. *Semin. Liver Dis.* **2001**, *21*, 351–372. [CrossRef] [PubMed]
2. Patsenker, E.; Popov, Y.; Wiesner, M.; Goodman, S.L.; Schuppan, D. Pharmacological inhibition of the vitronectin receptor abrogates PDGF-BB-induced hepatic stellate cell migration and activation in vitro. *J. Hepatol.* **2007**, *46*, 878–887. [CrossRef] [PubMed]
3. Margadant, C.; Sonnenberg, A. Integrin-TGF-β crosstalk in fibrosis, cancer and wound healing. *EMBO Rep.* **2010**, *11*, 97–105. [CrossRef] [PubMed]
4. Chamulitrat, W.; Burhenne, J.; Rehlen, T.; Pathil, A.; Stremmel, W. Bile salt-phospholipid conjugate ursodeoxycholyl lysophosphatidylethanolamide as a hepatoprotective agent. *Hepatology* **2009**, *50*, 143–154. [CrossRef] [PubMed]
5. Pathil, A.; Mueller, J.; Ludwig, J.M.; Wang, J.; Warth, A.; Chamulitrat, W.; Stremmel, W. Ursodeoxycholyl lysophosphatidylethanolamide attenuates hepatofibrogenesis by impairment of TGF-β1/SMAD2/3 signalling. *Br. J. Pharmacol.* **2014**, *171*, 5113–5126. [CrossRef] [PubMed]
6. Roma, M.G.; Toledo, F.D.; Boaglio, A.C.; Basiglio, C.L.; Crocenzi, F.A.; Sanchez Pozzi, E.J. Ursodeoxycholic acid in cholestasis: Linking action mechanisms to therapeutic applications. *Clin. Sci.* **2011**, *121*, 523–544. [CrossRef] [PubMed]
7. Poupon, R.E.; Lindor, K.D.; Pares, A.; Chazouilleres, O.; Poupon, R.; Heathcote, E.J. Combined analysis of the effect of treatment with ursodeoxycholic acid on histologic progression in primary biliary cirrhosis. *J. Hepatol.* **2003**, *39*, 12–16. [CrossRef]
8. Pathil, A.; Warth, A.; Chamulitrat, W.; Stremmel, W. The synthetic bile acid-phospholipid conjugate ursodeoxycholyl lysophosphatidylethanolamide suppresses TNFα-induced liver injury. *J. Hepatol.* **2011**, *54*, 674–684. [CrossRef] [PubMed]
9. Esteves, M.; Ferreira, M.J.; Kozica, A.; Fernandes, A.C.; Goncalves da Silva, A.; Saramago, B. Interaction of cytotoxic and cytoprotective bile acids with model membranes: Influence of the membrane composition. *Langmuir* **2015**, *31*, 8901–8910. [CrossRef] [PubMed]
10. Fahey, D.A.; Carey, M.C.; Donovan, J.M. Bile acid/phosphatidylcholine interactions in mixed monomolecular layers: Differences in condensation effects but not interfacial orientation between hydrophobic and hydrophilic bile acid species. *Biochemistry* **1995**, *34*, 10886–10897. [CrossRef] [PubMed]
11. Escriba, P.V.; Busquets, X.; Inokuchi, J.; Balogh, G.; Torok, Z.; Horvath, I.; Harwood, J.L.; Vigh, L. Membrane lipid therapy: Modulation of the cell membrane composition and structure as a molecular base for drug discovery and new disease treatment. *Prog. Lipid Res.* **2015**, *59*, 38–53. [CrossRef] [PubMed]

12. Mitra, S.K.; Schlaepfer, D.D. Integrin-regulated FAK-SRC signaling in normal and cancer cells. *Curr. Opin. Cell Biol.* **2006**, *18*, 516–523. [CrossRef] [PubMed]

13. Parsons, C.J.; Takashima, M.; Rippe, R.A. Molecular mechanisms of hepatic fibrogenesis. *J. Gastroenterol. Hepatol.* **2007**, *22* (Suppl. 1), S79–S84. [CrossRef]

14. Wu, H.J.; Zhang, Z.Q.; Yu, B.; Liu, S.; Qin, K.R.; Zhu, L. Pressure activates Src-dependent FAK-AKT and ERK1/2 signaling pathways in rat hepatic stellate cells. *Cell. Physiol. Biochem.* **2010**, *26*, 273–280. [CrossRef] [PubMed]

15. Eble, J.A.; Golbik, R.; Mann, K.; Kuhn, K. The alpha 1 beta 1 integrin recognition site of the basement membrane collagen molecule [alpha 1(iv)]2 alpha 2(iv). *EMBO J.* **1993**, *12*, 4795–4802. [CrossRef] [PubMed]

16. Park, S.J.; Lee, K.P.; Kang, S.; Chung, H.Y.; Bae, Y.S.; Okajima, F.; Im, D.S. Lysophosphatidylethanolamine utilizes LPA and CD97 in MDA-MB-231 breast cancer cells. *Cell. Signal.* **2013**, *25*, 2147–2154. [CrossRef] [PubMed]

17. Jang, I.S.; Rhim, J.H.; Park, S.C.; Yeo, E.J. Downstream molecular events in the altered profiles of lysophosphatidic acid-induced camp in senescent human diploid fibroblasts. *Exp. Mol. Med.* **2006**, *38*, 134–143. [CrossRef] [PubMed]

18. Spohr, T.C.; Dezonne, R.S.; Rehen, S.K.; Gomes, F.C. LPA-primed astrocytes induce axonal outgrowth of cortical progenitors by activating PKA signaling pathways and modulating extracellular matrix proteins. *Front. Cell. Neurosci.* **2014**, *8*, 296. [CrossRef] [PubMed]

19. Jalink, K.; Hengeveld, T.; Mulder, S.; Postma, F.R.; Simon, M.F.; Chap, H.; van der Marel, G.A.; van Boom, J.H.; van Blitterswijk, W.J.; Moolenaar, W.H. Lysophosphatidic acid-induced Ca^{2+} mobilization in human A431 cells: Structure-activity analysis. *Biochem. J.* **1995**, *307*, 609–616. [CrossRef] [PubMed]

20. Bandoh, K.; Aoki, J.; Taira, A.; Tsujimoto, M.; Arai, H.; Inoue, K. Lysophosphatidic acid (LPA) receptors of the EDG family are differentially activated by LPA species. Structure activity relationship of cloned LPA receptors. *FEBS Lett.* **2000**, *478*, 159–165. [CrossRef]

21. Harikumar, K.G.; Akgun, E.; Portoghese, P.S.; Miller, L.J. Modulation of cell surface expression of nonactivated cholecystokinin receptors using bivalent ligand-induced internalization. *J. Med. Chem.* **2010**, *53*, 2836–2842. [CrossRef] [PubMed]

22. Schliess, F.; Kurz, A.K.; vom Dahl, S.; Haussinger, D. Mitogen-activated protein kinases mediate the stimulation of bile acid secretion by tauroursodeoxycholate in rat liver. *Gastroenterology* **1997**, *113*, 1306–1314. [CrossRef] [PubMed]

23. Haussinger, D.; Kurz, A.K.; Wettstein, M.; Graf, D.; Vom Dahl, S.; Schliess, F. Involvement of integrins and src in tauroursodeoxycholate-induced and swelling-induced choleresis. *Gastroenterology* **2003**, *124*, 1476–1487. [CrossRef]

24. Gohlke, H.; Schmitz, B.; Sommerfeld, A.; Reinehr, R.; Haussinger, D. Alpha5 beta1-integrins are sensors for tauroursodeoxycholic acid in hepatocytes. *Hepatology* **2013**, *57*, 1117–1129. [CrossRef] [PubMed]

25. Mello-Vieira, J.; Sousa, T.; Coutinho, A.; Fedorov, A.; Lucas, S.D.; Moreira, R.; Castro, R.E.; Rodrigues, C.M.; Prieto, M.; Fernandes, F. Cytotoxic bile acids, but not cytoprotective species, inhibit the ordering effect of cholesterol in model membranes at physiologically active concentrations. *Biochim. Biophys. Acta* **2013**, *1828*, 2152–2163. [CrossRef] [PubMed]

26. Ben Mouaz, A.; Lindheimer, M.; Montet, J.C.; Zajac, J.; Lagerge, S. A study of the adsorption of bile salts onto model lecithin membranes. *Colloids Surf. B Biointerfaces* **2001**, *20*, 119–127. [CrossRef]

27. Hiller, C.; Kuhhorn, J.; Gmeiner, P. Class a G-protein-coupled receptor (GPCR) dimers and bivalent ligands. *J. Med. Chem.* **2013**, *56*, 6542–6559. [CrossRef] [PubMed]

28. Gobeil, F., Jr.; Bernier, S.G.; Vazquez-Tello, A.; Brault, S.; Beauchamp, M.H.; Quiniou, C.; Marrache, A.M.; Checchin, D.; Sennlaub, F.; Hou, X.; et al. Modulation of pro-inflammatory gene expression by nuclear lysophosphatidic acid receptor type-1. *J. Biol. Chem.* **2003**, *278*, 38875–38883. [CrossRef] [PubMed]

29. Zhao, J.; He, D.; Su, Y.; Berdyshev, E.; Chun, J.; Natarajan, V.; Zhao, Y. Lysophosphatidic acid receptor 1 modulates lipopolysaccharide-induced inflammation in alveolar epithelial cells and murine lungs. *Am. J. Physiol. Lung Cell. Mol. Physiol.* **2011**, *301*, L547–L556. [CrossRef] [PubMed]

30. Peres, C.; Yart, A.; Perret, B.; Salles, J.P.; Raynal, P. Modulation of phosphoinositide 3-kinase activation by cholesterol level suggests a novel positive role for lipid rafts in lysophosphatidic acid signalling. *FEBS Lett.* **2003**, *534*, 164–168. [CrossRef]

31. Heuman, D.M.; Bajaj, R. Ursodeoxycholate conjugates protect against disruption of cholesterol-rich membranes by bile salts. *Gastroenterology* **1994**, *106*, 1333–1341. [CrossRef]

32. Shi, F.; Sottile, J. Caveolin-1-dependent beta1 integrin endocytosis is a critical regulator of fibronectin turnover. *J. Cell Sci.* **2008**, *121*, 2360–2371. [CrossRef] [PubMed]

33. Ivaska, J.; Heino, J. Cooperation between integrins and growth factor receptors in signaling and endocytosis. *Annu. Rev. Cell Dev. Biol.* **2011**, *27*, 291–320. [CrossRef] [PubMed]

34. Hedin, U.L.; Daum, G.; Clowes, A.W. Disruption of integrin alpha 5 beta 1 signaling does not impair PDGF-BB-mediated stimulation of the extracellular signal-regulated kinase pathway in smooth muscle cells. *J. Cell. Physiol.* **1997**, *172*, 109–116. [CrossRef]

35. Horton, M.A.; Taylor, M.L.; Arnett, T.R.; Helfrich, M.H. Arg-Gly-Asp (RGD) peptides and the anti-vitronectin receptor antibody 23C6 inhibit dentine resorption and cell spreading by osteoclasts. *Exp. Cell Res.* **1991**, *195*, 368–375. [CrossRef]

36. Ylanne, J. Rgd peptides may only temporarily inhibit cell adhesion to fibronectin. *FEBS Lett.* **1990**, *267*, 43–45. [CrossRef]

37. Katow, H.; Yazawa, S.; Sofuku, S. A fibronectin-related synthetic peptide, Pro-Ala-Ser-Ser, inhibits fibronectin binding to the cell surface, fibronectin-promoted cell migration in vitro, and cell migration in vivo. *Exp. Cell Res.* **1990**, *190*, 17–24. [CrossRef]

38. Staubli, U.; Chun, D.; Lynch, G. Time-dependent reversal of long-term potentiation by an integrin antagonist. *J. Neurosci.* **1998**, *18*, 3460–3469. [CrossRef] [PubMed]

39. Plow, E.F.; Haas, T.A.; Zhang, L.; Loftus, J.; Smith, J.W. Ligand binding to integrins. *J. Biol. Chem.* **2000**, *275*, 21785–21788. [CrossRef] [PubMed]

40. Leask, A. Focal adhesion kinase: A key mediator of transforming growth factor beta signaling in fibroblasts. *Adv. Wound Care* **2013**, *2*, 247–249. [CrossRef] [PubMed]

41. Grimminger, F.; Gunther, A.; Vancheri, C. The role of tyrosine kinases in the pathogenesis of idiopathic pulmonary fibrosis. *Eur. Respir. J.* **2015**, *45*, 1426–1433. [CrossRef] [PubMed]

42. Du, X.P.; Plow, E.F.; Frelinger, A.L.; O'Toole, T.E.; Loftus, J.C.; Ginsberg, M.H. Ligands "activate" integrin alpha IIb beta 3 (platelet GPIIb-IIIa). *Cell* **1991**, *65*, 409–416. [CrossRef]

43. Mayo, K.H.; Fan, F.; Beavers, M.P.; Eckardt, A.; Keane, P.; Hoekstra, W.J.; Andrade-Gordon, P. Rgd induces conformational transition in purified platelet integrin GPIIb/IIIa-SDS system yielding multiple binding states for fibrinogen gamma-chain C-terminal peptide. *FEBS Lett.* **1996**, *378*, 79–82. [CrossRef]

44. Jiao, L.; Gan-Schreier, H.; Tuma-Kellner, S.; Stremmel, W.; Chamulitrat, W. Sensitization to autoimmune hepatitis in group VIA calcium-independent phospholipase A2-null mice led to duodenal villous atrophy with apoptosis, goblet cell hyperplasia and leaked bile acids. *Biochim. Biophys. Acta* **2015**, *1852*, 1646–1657. [CrossRef] [PubMed]

International Journal of
Molecular Sciences

Review

Cholesterol and the Safety Factor for Neuromuscular Transmission

Igor I. Krivoi [1] and Alexey M. Petrov [2,3,*]

[1] Department of General Physiology, St. Petersburg State University, St. Petersburg 199034, Russia;
    iikrivoi@gmail.com
[2] Institute of Neuroscience, Kazan State Medical University, Butlerova st. 49, Kazan 420012, Russia
[3] Laboratory of Biophysics of Synaptic Processes, Kazan Institute of Biochemistry and Biophysics,
    Federal Research Center "Kazan Scientific Center of RAS", P. O. Box 30, Lobachevsky Str., 2/31,
    Kazan 420111, Russia
* Correspondence: fysio@rambler.ru

Received: 30 January 2019; Accepted: 24 February 2019; Published: 28 February 2019

Abstract: A present review is devoted to the analysis of literature data and results of own research. Skeletal muscle neuromuscular junction is specialized to trigger the striated muscle fiber contraction in response to motor neuron activity. The safety factor at the neuromuscular junction strongly depends on a variety of pre- and postsynaptic factors. The review focuses on the crucial role of membrane cholesterol to maintain a high efficiency of neuromuscular transmission. Cholesterol metabolism in the neuromuscular junction, its role in the synaptic vesicle cycle and neurotransmitter release, endplate electrogenesis, as well as contribution of cholesterol to the synaptogenesis, synaptic integrity, and motor disorders are discussed.

Keywords: skeletal muscle; neuromuscular transmission; safety factor; cholesterol and lipid rafts; oxysterols; synaptic vesicle cycle; quantal release; Na,K-ATPase; nicotinic acetylcholine receptor

1. Introduction

Reliable transmission of brain commands through motor nerve to skeletal muscle and physical activity are required to maintain both motor function and muscle mass and, thus, important to ensure a healthy life. Skeletal muscle neuromuscular junction (NMJ) represents specialized region where the axon of a motor neuron establishes synaptic contact with a striated muscle fiber. The efficiency of the neuromuscular transmission strongly depends on a variety of pre- and postsynaptic factors [1,2].

Proteins that regulate quantal neurotransmitter release are clustered at specialized regions of the presynaptic membrane, called active zones (AZs). AZs represent highly specific compartments that contain synaptic vesicles (SVs) and serve as a molecular platform for precise spatial and temporal control of vesicle fusion and quantal neurotransmitter release [3]. Following release from the motor nerve terminals, neurotransmitter acetylcholine (ACh) diffuses across the synaptic cleft; during this travel ACh molecules are partially hydrolyzed by the enzyme acetylcholinesterase (AChE). The remaining ACh molecules interact with the postsynaptic nicotinic acetylcholine receptors (nAChRs), which are distributed at very high densities ($\sim$10,000/μm^2) at the crests of the primary postsynaptic folds of the adult NMJ [4]. Molecules of ACh released from one SV (a 'quantum' of transmitter) activate approximately two thousands of nAChRs, leading to opening cationic channels of the nAChRs and a net influx of positive ions. The latter produces a local transient (within milliseconds) small ($\sim$1 mV) depolarization, called the miniature endplate potential (MEPP). MEPPs occur spontaneously without presynaptic action potential (AP); the frequency of MEPPs reflects spontaneous neurotransmitter release. Motor neuron activity is translated to firing of APs, which, upon arriving to nerve terminals, trigger release of the multiple SVs. As a result, greater levels of neurotransmitter in synaptic cleft produces more depolarization than MEPP called the endplate potential (EPP). This local depolarization extends towards the depths of the secondary synaptic folds where, in turn, the voltage-gated sodium channels are located. The sodium channel's opening triggers the AP generation, thereby transforming the local EPP into propagating AP of the muscle fiber.

The fidelity of the neuromuscular transmission will depend on whether the EPP amplitude exceeds the threshold for muscle AP generation. The EPP amplitude depends on the amount of ACh quanta released (quantal content) and individual quanta size, AChE activity, the density of the nAChRs packaging, and the resting membrane potential (RMP). The AP generation threshold (membrane excitability) also depends on the RMP [5]. The term 'safety factor' refers to the ability of neuromuscular transmission to be effective and is quantified by the ratio of the EPP amplitude to AP generation threshold [1,2]. In different muscles, depending on their specialization, the value of the safety factor ranges from 2 to 5, however, it can decrease under various physiological and pathophysiological conditions such as fatigue, injury or diseases [1,6,7].

A normal segregation of plasma membrane lipid phases is vital to the maintenance of membrane fluidity, curvature, ion channel, transporter functions, as well as compartmentalization. Importantly, membrane proteins that encompass about 50% of the cell membrane are often cocrystalized with membrane cholesterol [8]. Cholesterol's presence in membranes regulates the function of many proteins, including transporters, receptors, and ion channels. There are two main ways of how cholesterol can affect protein properties. First mechanism is through specific binding of cholesterol with high affinity sites. This may contribute to control of protein conformation and function. Second mechanism is dependent on the nonspecific physicochemical influence of cholesterol on membrane fluidity and thicknesses that consequently affects numerous protein-dependent processes. Distinguishing between mechanisms of cholesterol action can be challenging [8,9].

The basic principles of lipid-protein interactions have still not been completely elucidated. Cholesterol-associated changes impact membrane fluidity and stiffness, lateral lipid diffusion, protein mobility, excitability, and a variety of other key functional membrane properties including fusion and fission. In addition, membrane cholesterol modulates interactions between the lipid bilayer and ions environment as well as the interplay between plasma membrane and underlying cytoskeleton suggesting the involvement in mechanotransduction. Cholesterol is also a key molecule in the formation of lipid rafts known as molecular platform involved in numerous cellular processes like apoptosis, signaling, and cell differentiation. Little is known, however, about the precise consequences of changes in membrane cholesterol levels of different cell type. Notably, responses to membrane cholesterol enrichment or depletion vary depending on cell type and can be very complex and unpredictable [10–14].

While the essential role of cholesterol for many vital functions is well documented, much less is known about its role at the NMJ, in particular, in maintaining the safety factor for neuromuscular transmission. The review focuses on the crucial role of cholesterol to maintain a reliability of neuromuscular transmission in health and diseases.

2. Cholesterol Production

Cholesterol production is energetically expensive and requires more than 30 enzymatic reactions. Different cells in nerve system have a different capacity to produce cholesterol. Oligodendrocytes are characterized by a higher capacity than astrocytes which in turn, had at least 2–3-times higher ability to cholesterol synthesis than neurons [15]. During nerve system development the intensity of cholesterol synthesis by different cells undergoes significant changes. Cholesterol production by oligodendrocytes is responsible for the great increase in cholesterol content at period of active myelinization with its peak during the first postnatal weeks. After myelinization, synthesis occurs at low rate in nerve system of adult mammalians [16]. Before glial cell differentiation, a neuron could cover its own cholesterol requirements by producing cholesterol. In the postnatal period, neuronal ability to cholesterol production is strongly reduced and neurons rely on the cholesterol mainly synthesized in glia [17,18].

Neuronal requirements for cholesterol are extremely high, as it is indispensable for formation of synaptic membranes [17]. Studies in central and neuromuscular synapses suggest that all steps of SV cycle (exocytosis, endocytosis and vesicular traffic) could be dependent on cholesterol content in presynaptic membrane and membrane of SVs [19,20]. The presynaptic part of NMJs is far from the motor neuron soma, where the main biosynthetic apparatus is located, and this part is adapted to translate high frequency patterns of motor neuron activity to muscle. To prevent the loss of connectivity, pre- and postsynaptic compartment are tightly stickled by extracellular matrix and presynaptic part is covered by Schwann cell. Accordingly, mechanism of cholesterol delivery to neurons and maintenance of cholesterol steady state level in synapse should be crucial for neurotransmission, especially in presynaptic compartment of NMJs. The best candidate to provide NMJs with cholesterol is terminal Schwann cell, which can produce apolipoprotein (particularly, apolipoprotein E) and cholesterol containing particles. Consistent with this, stimulation of apolipoprotein receptors are important for the formation, maintenance and regeneration of peripheral NMJs [21,22].

Skeletal muscles comprise approximately 50% of total body weight and contain T-tubule membranes, enriched with cholesterol. Cholesterol sources for muscle are both local biosynthesis and uptake from the circulation. The later has a predominant relevance and high-rich diet could lead to enrichment of muscle fiber membranes with cholesterol [23,24]. However, certain basal level of local cholesterol production in the skeletal muscle is probably essential as inhibition of cholesterol synthesis (with statins) caused muscle fatigue and weakness. This may be linked with both the high requirements of cholesterol for muscle membrane remodeling, and the importance of cholesterol biosynthesis intermediates (e.g., ubiquinone, dolichol, farnesyl, and geranyl pyrophosphates) [25]. Additionally, skeletal muscle can store cholesterol in the form of cholesterol esters in lipid droplets, and their accumulation could be accompanied by myopathy [26]. Cholesterol elimination from muscle membranes theoretically may be linked with muscle activity. Membrane cholesterol is susceptible to oxidation by reactive oxygen species (ROS) [27] and ROS produced during muscle loading could oxidize membrane cholesterol, leading to cholesterol elimination in the form of oxysterols. Oxysterols can activate nuclear liver X receptor in skeletal muscle [28,29] and, thus, trigger increased cholesterol synthesis and (or) uptake to compensate cholesterol elimination. Therefore, skeletal muscle cholesterol metabolism probably relies on balance between cholesterol synthesis, consumption, storage, and elimination.

3. Cholesterol and Quantal Neurotransmission Release

Motor nerve endings are specialized on fast neurotransmitter release in response to arriving AP. Neurotransmitter molecules are packed into SVs which distributed into cytoplasm near sites of exocytosis. Small population of SVs is docked at the presynaptic membrane in specialized regions, AZs. Some of these SVs, consisting of ready releasable pool, are able to fuse with short delay after Ca^{2+} influx through voltage-gated Ca^{2+} channels located within membrane of the AZ. After exocytosis, lipid and protein components of vesicular membrane are internalized by endocytotic mechanism. Newly formed endocytotic vesicles may be repeatedly used for neurotransmission after refilling with neurotransmitter. Thus, synaptic activity is accompanied by SV exo-endocytotic cycles (recycling) [19,30]. These cycles put several challenges for presynaptic machinery. First, fast and intensive exocytosis and compensatory endocytosis require changes in membrane curvature and rigidity; second, SV and AZ membranes have specific protein–lipid composition, which should be maintained, despite fusion of the SVs into the presynaptic membrane; third, AZ, and surrounding presynaptic membranes (peri-AZ) should have different properties to allow fast exocytosis and relatively fast endocytosis; forth, during innervation and reinnervation membrane properties should be different to easily adapt shape of distal axon to target postsynaptic compartment. Of course, protein-based mechanisms are primary responsible for resolving these challenges, but cholesterol as one of the main component of presynaptic vesicle membranes are involved in maintenance of presynaptic function [31].

Initially, in electron microscopic study it was found that sterol-binding antibiotic filipin did not form complex with cholesterol in AZs in the frog nerve terminals. At the same time, filipin–sterol complexes were detected in most axonal areas, including regions of nerve terminals lacking AZ, and membranes of Schwann cell membranes [32]. Moreover, during degeneration or regeneration, despite AZ disorganization or presence in primitive form (without normal double row organization) filipin–sterol complexes were excluded from these membrane sites, while Schwann cells occupied synaptic gutters displayed high abundance with filipin–sterol complexes [33]. At the same time, more intensive filipin staining was detected in axons of crayfish, frog, mouse, and rat, suggesting cholesterol abundance in nerve terminals [34–36]. This cholesterol could organize membrane microdomains, lipid rafts. Indeed, axonal synaptic membranes of frog, mouse and rat showed a high intensity of staining with lipid raft markers [35,37–39]. Moreover, acute cholesterol depletion with methyl-β-cyclodextrin (MβCD) led to decrease a lipid raft labeling [37,39]. All together, these results suggest that the membrane regions of AZ can have a relatively low cholesterol content, but other regions of the presynaptic membranes are cholesterol rich and contain lipid rafts, which could organize fence around AZ. Thus, presynaptic membrane cholesterol could have functional rather than structural meaning for SVs exocytosis.

It is conceivable that cholesterol within and in close vicinity of exocytotic sites could have a regulatory role and be kept lower for maintenance of regulation range. In a pioneer study by Zamir and Charlton [34] it was found that cholesterol depletion blocked evoked neurotransmitter release, without marked postsynaptic effects. Suppression of evoked neurotransmitter release was linked with hyperpolarization of presynaptic axon and lack of AP propagation, whereas Ca^{2+}-dependent exocytosis functioned and spontaneous exocytosis was even enhanced. All these effects were reversed by cholesterol supplementation. This study points to importance of cholesterol level for presynaptic transmitter release. Cholesterol depletion suppressed evoked neurotransmitter release, but increased spontaneous exocytosis in frog and mice NMJs as well as central synapses [40–45]. Thus, cholesterol level could determine the ratio between spontaneous and evoked neurotransmission. In context of NMJs, it means that lower cholesterol content could lead to decrease safety factor due to suppressed AP-evoked exocytosis and increased energy expenditure and desensitization of postsynaptic receptors as a result of upregulated spontaneous neurotransmitter release. Notably, that enhanced spontaneous exocytosis could facilitate uptake of extracellular macromolecules, as was shown for botulin neurotoxin A [46]. Speculatively, it may be a part of homeostatic response to cholesterol depletion, because compensatory endocytosis following by exocytosis could internalize of cholesterol-containing lipoprotein particles, which are likely main source of cholesterol for motor nerve terminals.

Several signaling mechanisms could be responsible for increased spontaneous exocytosis after cholesterol depletion. In frog NMJs, decreased cholesterol content led to increase in NADPH oxidase-dependent ROS production, which stimulated an increase in cytosolic Ca^{2+} (probably via TRPV channels) and, subsequently, calcineurin (phosphatase PP2B) activation, which, in turn, facilitates the involvement of SVs to spontaneous exocytosis. Under these conditions, increased ROS production led to marked lipid peroxidation of synaptic membranes [41]. Moreover, activation of protein kinase C in response to cholesterol depletion can support SV exocytosis in full mode, whereas inhibition of protein kinase C (but not phospholipase C) converts exocytosis to kiss-and-run mode (when transmitter release occurs through transient fusion pore) in frog NMJs [42]. Involvement of protein kinases A and C in the effect of cholesterol depletion on spontaneous neurotransmitter release was also observed in central synapses [47]. Notably, cholesterol depletion may provoke spontaneous exocytosis from separate SV pool, which reluctantly participates in evoked neurotransmitter release in frog NMJs [44]. Thus, presynaptic cholesterol level acting via signaling mechanism can limit both spontaneous neurotransmitter release and overactivation of protein kinases, phosphatase PP2B as well as production of ROS and increase in cytosolic Ca^{2+}. This housekeeping role of membrane cholesterol in NMJs could maintain neuromuscular transmission and signaling at steady state level.

Molecular study indicates that SV membrane contains very high amount of cholesterol [48]. Decreased cholesterol content in membranes of SVs in Drosophila, frog and rat NMJs as well as central synapses led to inhibition of SV endocytosis and recycling [36,40,49,50]. Furthermore, in Drosophila NMJs, cholesterol extraction from SV membranes caused dispersion of SV proteins (synaptotagmin, VGLUT, and CSP) throughout presynaptic membrane and disturbance of actin polymerization, suggesting requirement of vesicular cholesterol for maintenance of protein composition of SVs and regulation of cytoskeleton dynamic [51]. Therefore, vesicular cholesterol is essential for SV endocytosis and could serve as glue for keeping SV proteins and lipids in one membrane microdomains (lipid rafts), which are selectively trapped by endocytotic mechanism [35,52]. Along with this, antiganglioside GM1 antibodies (hallmark of Guillain–Barré syndrome leading to acute motor axonal neuropathy), which recognize components of lipid rafts, are intensively uptaken by motor nerve terminals, counteracting damage of the nerve terminals [53].

SV exocytosis is a major mechanism of neurotransmitter secretion. However, nonvesicular (nonquantal) secretion of neurotransmitter also occurs in NMJ and central synapses. The functional role of this type of ACh secretion is elusive. In rat NMJs, depletion of cholesterol from SV membranes led to an increase in nonquantal ACh release due to the incorporation of neurotransmitter transporters (in particular, vesicular ACh transporter) into the presynaptic membrane and enhancement of ACh/H^+ exchange [36]. Thus, cholesterol level in SVs could limit neurotransmitter leakage via nonquantal release.

4. Effects of Cholesterol Derivatives on Neurotransmitter Release and Synaptic Vesicle Cycle

Cholesterol is oxidized by ROS and specific enzymes to oxysterols, which could modulate multiple processes, linked with inflammatory reactions and neurodegeneration [19,54,55]. Synaptic transmission in NMJs of frog and mouse was a high sensitive to different oxysterols and enzymatic cholesterol oxidation [37–39,56,57]. 24S-hydroxycholesterol (24HC) is a product of cholesterol metabolism, which is generated specifically in brain and passes across a blood–brain barrier into the circulation. The levels of 24HC in the circulation are significantly changed at early and (or) advanced stage of neurodegenerative disease [15,19,54]. 24HC (0.4–4 µM; 1/3 h-application) enhanced synaptic transmission in mice NMJs due to increase in a rate of SVs recycling. The mechanism of 24HC action was linked with decreased NO production, likely by endothelial NO synthase [57]. Activity of this enzyme is dependent on its association with lipid rafts, partially, via caveolin-dependent mechanism.

In an amyotrophic lateral sclerosis (ALS) mouse model carrying a mutant superoxide dismutase 1 (SOD1^{G93A}) 24HC also decreased NO production, but led to an opposite effect on neuromuscular transmission, because the role of NO is likely reversed in ALS mice as compared to wild type mice. Increased NO synthesis promoted SV exocytosis during intense activity in SOD1^{G93A} mice, but it suppressed the exocytosis in wild type mice [39]. Probably, attenuation of NO production was mediated by increased lipid raft integrity in response to 24HC. Indeed, disturbance of lipid rafts prevented the effect of 24HC on NO synthesis [39]. However, it is unclear why NO had opposite actions in the ALS versus wild type mice. It is conceivable that increased mitochondrial ROS production [58] could invert the effect of NO at the NMJs of SOD1^{G93A} mice. For example, peroxynitrite formed from the reaction between NO and ROS can promote exocytosis probably due to direct modification of AZ proteins SNAP-25 and Munc18 [59]; while NO, acting as retrograde messenger, may inhibit neurotransmitter release due to activation of guanylyl cyclase/cGMP pathway [60] or direct S-nitrosylation of SV fusion-clamp protein complexin [61].

Another oxysterol, 5α-cholestan-3-one (5Ch3), is produced as an intermediate sterol in the biosynthetic pathway for cholestanol. The levels of 5Ch3 were increased in *Cerebrotendinous xanthomatosis* patients [62], having muscle dysfunction. 5Ch3 (0.2 µM) decreased the amount of SVs which were actively involved in neurotransmitter release in mouse NMJs. The effect of 5Ch3 was linked with decrease in lipid raft integrity and was dependent on membrane cholesterol content [37]. Similarly, 5Ch3 reduced lipid raft integrity and the number of SVs participating in exo- and endocytosis during synaptic transmission in frog NMJs [38]. In contrast, structurally similar oxysterol olesoxime (cholest-4-en-3-one, oxime; TRO19622) increased evoked ACh release as well as the number of SVs involved in exo-endocytosis and the rate of SV recycling. Moreover, olesoxime was able to increase lipid raft integrity in frog NMJs [38]. Note that olesoxime is potential neuroprotective compound in models of ALS, multiple sclerosis, Parkinson's, and Huntington's disease [63–66]. These data show that these oxysterols induce marked different changes in neuromuscular transmission which are related with the alteration in SV cycle and lipid raft behavior.

Similarly, oxidation of endogenous cholesterol by cholesterol oxidase significantly impaired lipid raft integrity as well as affected mode of SV exocytosis (toward to kiss-and-run mechanism) and disturbed SV clusterization [56]. The effects of cholesterol oxidase on SV cycle were different from cholesterol depletion [40], suggesting that oxidative cholesterol derivative (cholest-4-en-3-one) could mediate action of cholesterol oxidase. Taken together, oxidized cholesterol metabolites could present a new class of presynaptic neurotransmitter release modulators, which may contribute to adaptation of muscle activity to current physiological status of organism.

5. Cholesterol and Proteins Involved in Synaptic Vesicle Cycle

Cholesterol-interacting proteins could serve as transducer of changes in local cholesterol level to presynaptic processes. Cholesterol microdomain can clusterize Ca^{2+} channels (e.g., N-, L-, and P/Q types) in the presynaptic membrane of neuronal cells, affecting distance from the channels to the site of exocytosis and, thus, neurotransmitter release [67–69]. Also, a main Ca^{2+} sensor—synaptotagmin 1—triggers SV exocytosis and is a lipid raft resident [70]. Studies with cholesterol depletion suggest that neurotransmitter transporter distribution and (or) their activity in presynaptic terminals could be dependent on cholesterol availability [36,71–73]. Also a vesicular H^+ pump, which creates a proton gradient for neurotransmitter flux into SV, was also found in cholesterol microdomains and cholesterol depletion attenuated the H^+-ATPase activity [74]. Cholesterol depletion could also suppress SV swelling mediated by coordinated activity of H^+ pump and water channel aquaporin-6 [75]. Several studies suggested that clusterization of syntaxin, an essential component of exocytotic machinery, is affected by membrane cholesterol [76] and depolarization of synaptosomal membrane increases redistribution of syntaxin into lipid raft fraction [77]. Furthermore, cholesterol may be a part of the fusion pore, connecting lumen of SV with extracellular space, and increasing cholesterol content favors fusion pore opening [78,79]. This is in agreement with extremely high cholesterol content (40 mol%) in SVs [48]. Interaction of most abundant SV protein, synaptophysin, with cholesterol could be important for SV endocytosis [52]. Interestingly, a mutation in DJ-1 (a genetic factor for early-onset autosomal recessive Parkinson's disease) impaired SV endocytosis, without inducing structural alteration in synapses, via a reduction in cholesterol level [80]. In addition, the main SV clustering protein synapsin can affect cholesterol content in microdomains, promoting lipid raft formation [81].

Thus, changes in cholesterol levels can affect triggering exocytosis by Ca^{2+} (via Ca^{2+} channel and synaptotagmin), SV fusion (syntaxin) and endocytosis (synaptophysin), vesicle refilling with neurotransmitter (neurotransmitter transporters, H^+ pump), and clusterization of SV (synapsin). Of course, changes in intracellular signaling molecules (e.g., phospholipases, protein kinases, and small GTPases) could mediate effects of cholesterol on synaptic transmission. Putative cholesterol-dependent steps in presynaptic vesicular cycle and cholesterol-sensitive proteins are shown in Figure 1 and Table 1.

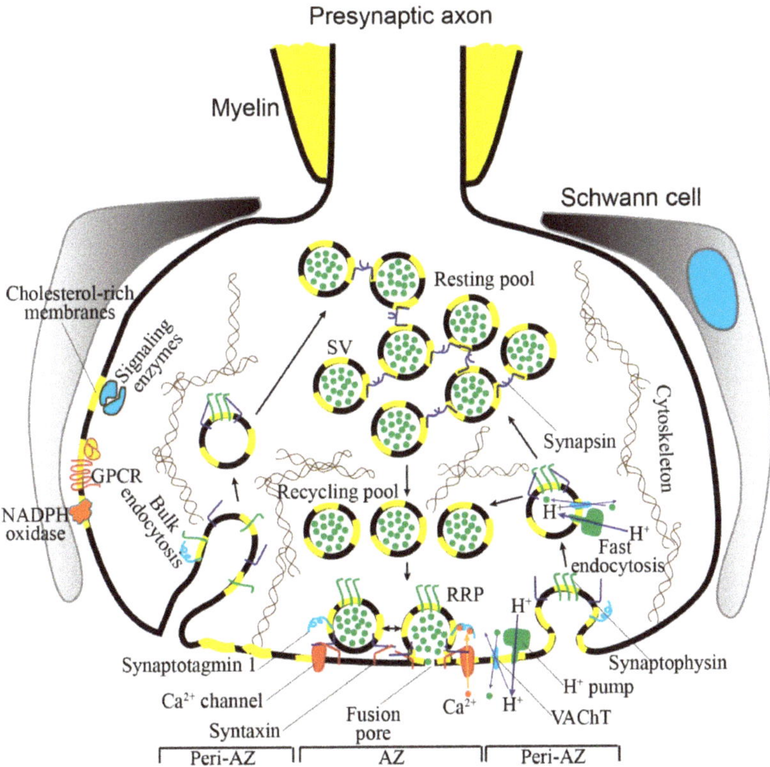

Figure 1. Putative role of cholesterol in presynaptic processes. Cholesterol organizes microdomains in presynaptic membrane and SVs. Several proteins, essential for presynaptic function, reside in these microdomains, and/or directly bind with cholesterol. These interactions are involved in control of multiple aspects of SV cycle that guarantees the maintenance of neurotransmitter release. Additionally, numerous presynaptic G-protein coupled receptors (GPCRs) and signaling enzymes (e.g., protein kinases and small GTPases), as well as a ROS-generating enzyme (NADPH oxidase), which regulates the steps of the SV cycle, could be located in cholesterol-rich microdomains. SV exocytosis occurs due to fusion of SVs from ready-releasable pool (RRP) with presynaptic membrane in AZ region. Under condition of moderate motor nerve activity, replenishment of RRP is mediated by delivery of SVs from recycling pool. After exocytosis, these SVs are able to rapidly recover through fast endocytotic mechanism, refill with ACh (green circles), and repeatedly participate in neurotransmitter release. Intense activation of motor neuron can led to mobilization of SVs from resting (reserve) pool to the exocytotic sites. After massive exocytosis, the resting pool is replenished by mainly slow endocytotic route (bulk endocytosis) associated with formation of endosome-like structures in NMJs. The cytoskeleton can contribute to spatial organization of both SV pools and routes for SV traffic. Key events, where the presence of cholesterol is required, are SV exocytosis (fusion pore formation, Ca²⁺ triggering step), endocytosis (vesicular protein holding in cluster), SV refilling with ACh (formation of proton gradient, exchange ACh and proton), SV clusterization (SV interconnections by synapsins), limitation of ACh leakage via spontaneous (clamping of signaling enzymes activity and ROS production), and nonquantal (decrease in activity of vesicular ACh transporter and H⁺-pump in presynaptic membrane) secretion. Please, for details, see Sections 3 and 5.

AChE mainly resides in synaptic cleft, but the pool of AChE is located in association with lipid rafts through a GPI anchor [82]. Close distribution of raft-linked AChE to plasma membrane enriched with the receptors (e.g., M3 muscarinic receptors) could be important for control local levels of ACh near the receptors [83]. Accordingly, presence of cholesterol-rich microdomains may control sensitivity of feedback mechanisms which regulate ACh release from nerve terminals. Moreover, cholesterol-rich microdomains seem to be an important for development of depressant effect of ATP (cotransmitter of ACh) on neurotransmitter release at the frog NMJ. This effect is realized via P2Y12 receptors and ROS production by NADPH oxidase and it is reduced and largely delayed by cholesterol depletion [84]. In Drosophila NMJs, lipid rafts are required for presynaptic growth promoting effect of mannosyl glucosylceramide, which facilitates presynaptic Wnt1/Wingless signaling [85]. Thus, cholesterol is important component for coordination of receptor-dependent signaling, which regulates both neurotransmitter release and presynaptic bouton formation.

Table 1. Some potential interactions between cholesterol and proteins involved in regulation of neurotransmitter release from motor nerve terminals.

Protein	Role in Neuromuscular Transmission	Potential Role of Interaction with Cholesterol	Ref.
P2Y12 receptor	Inhibition of ACh release	Acceleration of the downstream receptor signaling	[84]
Ca^{2+} channels (N, L, P/Q types)	Triggering SV exocytosis in response to AP	Clusterization of channels near exocytotic sites, thereby facilitating exocytosis	[67–69]
NADPH oxidase (ROS generating enzyme)	Regulation of AP-evoked and spontaneous ACh release	Limitation of background activity	[41,84]
Proton pump	Formation of H^+ gradient necessary for SV filling with ACh; regulation of SV size	Regulation of precise location and potentiation of H^+ transport function	[74,75]
Signaling enzymes (protein kinases A/C)	Control of neurotransmitter release	Limitation of background activity of the protein kinases	[42,47]
Synapsin	Clusterization of SVs in pools	Lipid raft organization in SV membranes	[81]
Synaptophysin	Regulation of exo- and endocytosis	Induction of SV curvature during endocytosis	[52]
Synaptotagmin 1	A major Ca^{2+} sensor for neurotransmitter release	Location in lipid rafts and precise distribution in presynaptic membrane	[51,70]
Syntaxin	Component of SNARE complex mediated SV fusion	Clusterization in membrane; activity-dependent redistribution in lipid rafts	[76,77]
Vesicular ACh transporter	Uptake of ACh into SV, nonvesicular ACh release	Precise location in SV membranes and regulation of activity	[36]

6. Cholesterol–Na,K-ATPase Interactions

Maintaining a steady state level of the RMP of muscle fibers is essential for the normal functioning of skeletal muscle. Membrane depolarization leads to a decrease in EPP amplitude, as well as inactivation of sodium channels and suppression of membrane excitability [1,6]. A smaller EPP amplitude and higher the AP generation threshold reduce safety factor at the NMJ. Among different mechanisms involved in maintaining skeletal muscle electrogenesis and contractile function, the activity of Na,K-ATPase plays a crucial role [86,87]. This transport system, discovered by Dr. J. Skou [88] (Nobel Prize in Chemistry, 1997), is an ubiquitous transmembrane protein that functions as a Na,K pump. Na,K-ATPase catalyzes the active transport of K^+ into and Na^+ out of the cell, thereby maintaining the steep Na^+ and K^+ gradients that provide electrical excitability and the driving force for many other transport processes [86,87,89–91]. Na,K-ATPase is critically important for excitability, electrogenesis, and contractility of skeletal muscles, which contain the main pool of the whole body Na,K-ATPase. The density of the distribution of Na,K-ATPase molecules in the sarcolemma is extremely high and ranges from 1000 to 3350/μm^2 [86].

The influence of cholesterol on transmembrane protein (receptors, ion channels, transporters) function and properties is well established [8,9,92–94]. The function of Na,K-ATPase is also regulated by the lipid environment and strongly depends on membrane cholesterol content [93]. This regulation is realized via direct protein–lipid interactions or by influencing physical properties of the lipid bilayer [8,93,95].

Na,K-ATPase is composed of α catalytic and β glycoprotein subunits. Four isoforms of the α subunit are known to exist in tissues of vertebrates. It is generally accepted that the ubiquitous $\alpha 1$ isoform plays the main housekeeping role while the other isoforms expressing in a cell- and tissue-specific manner possess additional regulatory functions [89,90,96,97]. In skeletal muscles, the $\alpha 1$ and $\alpha 2$ isoforms of α subunit are coexpressed and the $\alpha 2$ Na,K-ATPase isozyme is predominant in adult skeletal muscle [98,99]. Distinct isoform-specific functions of Na,K-ATPase $\alpha 1$ and $\alpha 2$ isozyme in skeletal muscle are proposed [100–104]. The $\alpha 2$ Na,K-ATPase isozyme is specifically regulated by muscle use and enables working muscles to maintain excitability, contraction and resistance to fatigue [100–103]. Two main pools of the $\alpha 2$ Na,K-ATPase exist: the majority of $\alpha 2$ isozyme is expressed in the interior transverse tubule membranes [102], which are highly enriched in sphingomyelin and cholesterol compared to the surface sarcolemma [24]. The smaller $\alpha 2$ Na,K-ATPase isozyme pool is localized to the junctional (endplate) membrane [100,103], where cholesterol and lipid rafts serve as a potential signaling platform for nAChRs clustering [105,106]. Notably, these distinct $\alpha 2$ Na,K-ATPase membrane pools are regulated differently [103]. Both these $\alpha 2$ Na,K-ATPase membrane pools display the loss of electrogenic activity in response to hindlimb suspension (HS). However, the extrajunctional pool depends strongly on muscle disuse, and even the increased protein and mRNA content as well as enhanced $\alpha 2$ Na,K-ATPase membrane abundance after 12 h of HS cannot counteract this sustained inhibition. In contrast, additional factors (possibly circulating factors related to HS) may regulate the junctional $\alpha 2$ Na,K-ATPase pool that is able to recover during HS. Notably, acute, low intensity muscle workload restores functioning of both pools [103].

Subcellular compartmentalization is one of the basic principles of cellular organization [107,108]. A certain pool of Na,K-ATPase is localized in specialized lipid microdomains of the membrane—caveolae—where it forms regulatory multimolecular complexes and performs new functions, in particular, signal transduction [109–113]. As regulators of protein functions in caveolae and planar rafts may be the lipids themselves, including cholesterol. It was shown that cholesterol interacts with caveolins, principal components of caveolar membranes. Presumably, the $\alpha 1$ isoform of Na,K-ATPase contains binding sites for caveolin-1 located near the M1 and M10 transmembrane domains and interacted with the N-terminus of caveolin-1 [112,114,115]. Notably, caveolin-1 is involved in regulation of intracellular cholesterol traffic [116]. In experiments on cell lines, it has been shown that $\alpha 1$ Na,K-ATPase, through interaction with caveolin-1, participates in the distribution of cholesterol between intracellular membranes and the plasma membrane. On the one hand, the impaired expression of $\alpha 1$ Na,K-ATPase affects the formation of caveolae, cholesterol synthesis, and its traffic [117]. On the other hand, cholesterol itself is involved in the regulation of Na,K-ATPase. Thus, the reduction of cholesterol level in the plasma membrane stimulates endocytosis and degradation of $\alpha 1$ Na,K-ATPase via Src- and ubiquitin-dependent regulatory pathways [118]. The data obtained on the cell lines suggest the possibility of mutual regulation between cholesterol and $\alpha 1$ Na,K-ATPase which is carried out with the participation of caveolin-1.

In skeletal muscle, MβCD-induced partial membrane cholesterol removal selectively decreases the $\alpha 2$ Na,K-ATPase isozyme electrogenic activity without changing the $\alpha 1$ isozyme activity [119]. A similar specific dysfunction of the $\alpha 2$ Na,K-ATPase isozyme is also observed under the conditions of partial loss of membrane cholesterol in response to skeletal muscle motor unloading (disuse) [120]. Conversely, specific inhibition of the $\alpha 2$ Na,K-ATPase activity by ouabain induces a disturbance of lipid rafts due to partial cholesterol loss [120]. Collectively, these findings, obtained in alive skeletal muscles, suggest the reciprocal interactions between membrane cholesterol and the α2 Na,K-ATPase activity. It is important to note that the listed changes were most pronounced in the junctional membrane region [119,120].

While the molecular mechanisms of Na,K-ATPase regulation by surrounding lipids is being studied [93,95,121], little is known on how α2 Na,K-ATPase may regulate lipid raft stability. For example, this regulation may be mediated by α2 Na,K-ATPase ouabain receptor site. Notably, ouabain inhibits Na,K-ATPase transport by stabilizing the enzyme in an E2 conformation. Three specific lipid/Na,K-ATPase interactions are proposed that either stabilize the protein or stimulate/inhibit Na,K-ATPase activity, with separate binding sites and distinct kinetic mechanisms. Both stimulatory and inhibitory lipid interactions poise the conformational equilibrium toward the E2 state [121]. Altogether, these findings lead us to propose that reciprocal interactions between cholesterol and α2 Na,K-ATPase is more favored in the E2 enzyme conformation.

Regulation of α2 Na,K-ATPase is also specifically determined by its less stable integration into the lipid membrane compared with other α subunit isoforms. When studying Na,K-ATPase using *Pichia pastoris* yeast as an expression system, lipids, including cholesterol, have been shown to play an important role in stabilizing the Na,K-ATPase in the plasma membrane, and in an isoform-specific manner. Thus, a number of altering factors (heating, the use of detergents) predominantly affect the stability of insertion of the α2 Na,K-ATPase into the membrane compared to the α1 and α3 isoforms [122]. It is assumed that the relatively lower stability of the α2 isoform is due to the peculiarities of its transmembrane domains M8, M9, and M10, which are responsible for interaction with phospholipids, as well as weaker association with protein FXYD1 (phospholemman, an auxiliary regulatory subunit of the Na,K-ATPase) [123].

In sum, in skeletal muscle, the α1 Na,K-ATPase isozyme is functionally more stable compared with the α2 isozyme, whose adaptive plasticity is determined by specific localization and regulation of different enzyme pools and functional interactions with molecular environment including cholesterol.

7. Cholesterol–nAChR Interactions

Localization of neurotransmitter receptors in lipid microdomains and their interaction with cholesterol play an important role in the synaptic function and its plasticity, as well as in the development of a number of neurodegenerative diseases [54,124–126]. One of the important effects of the lipid environment is a change in receptor kinetics and their affinity for specific ligands. These changes can be explained both by the modulating effect of cholesterol due to its direct interaction with receptors, and by changing the properties of the membrane lipid bilayer [92,127–129].

nAChRs, as well as Na,K-ATPase, are integral membrane proteins that play key roles in membrane excitation. Role of cholesterol in the endplate membrane integrity is poorly understood due to deficit of ex vivo and in vivo studies. Both cholesterol and the nAChRs are localized at the junctional region (Figure 2A) and lipid rafts serve as a signaling platform for nAChRs clustering [105,106]. Previous investigations have demonstrated a direct molecular interaction between cholesterol and the nAChRs [92,130–132]. Surrounding lipids influence nAChRs kinetic mechanisms and cholesterol is critical for the channel function [129].

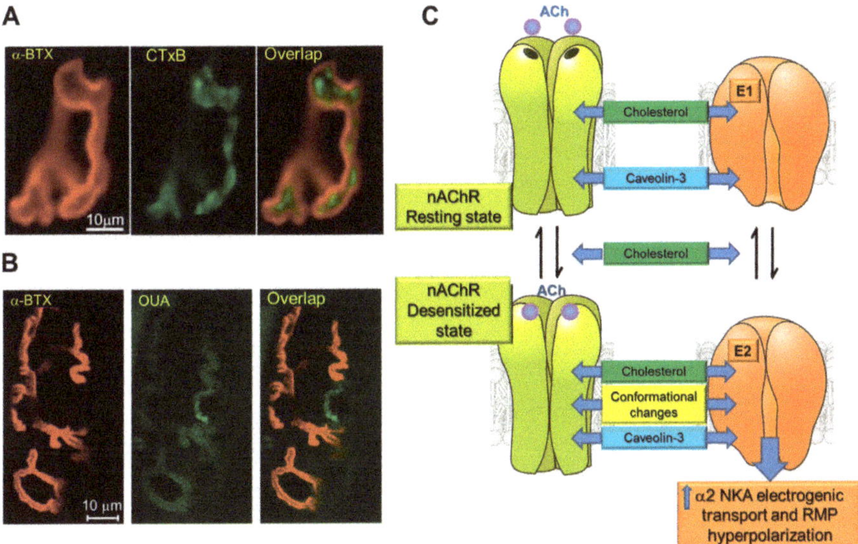

Figure 2. Cholesterol as a part of nAChR/α2Na,K-ATPase multimolecular regulatory complex. (**A**) The nAChRs and lipid rafts reside at the endplate region. A single endplate of rat soleus muscle was dual-labeled with α-BTX (nAChRs, red channel) and fluorescent cholera toxin B subunit to stain lipid rafts (CTxB, green channel). Overlap (orange channel). (**B**) The nAChR and the α2 Na,K-ATPase colocalization at the muscle endplate. A single endplate of mouse extensor digitorum longus muscle was dual-labeled with α-BTX (nAChRs, red channel) and BODIPY-conjugated ouabain (α2 Na,K-ATPase, green channel). Overlap (orange channel). Scale bars, 10 μm. (**C**) Hypothetical scheme of nAChR/α2Na,K-ATPase/Caveolin-3/Cholesterol interactions stimulating electrogenic active transport to hyperpolarize the resting membrane potential (RMP). Modified from Petrov et al. (2017) ref. [120] (A) and Heiny et al. (2010) ref. [100] (B).

Cell studies suggest that peripheral membrane proteins, like rapsyn, contribute to organization of nAChRs cluster (prototype of postsynaptic nAChRs cluster) and cholesterol rich-microdomains are required for intracellular sorting and targeting of the nAChRs and rapsyn into the plasma membranes [133,134]. Also, lipid rafts and cholesterol content could be essential in the initial clustering and later stabilization of nAChRs clusters and organization signaling complex, including nAChRs, rapsyn, MuSK, and Src-family kinases [106,135,136]. Lipid rafts may regulate nAChRs clustering by facilitating the agrin/MuSK signaling and the interaction between the nAChRs and rapsyn; conversely, membrane cholesterol depletion inhibits nAChRs cluster formation [105]. Also, membrane cholesterol depletion disturbs the sarcolemma distribution of β-dystroglycan and its interaction with dystrophin [137], a key protein of the endplate integrity. Cholesterol stabilized nAChRs cluster in denervated muscle and could trigger maturation of nerve sprout-elicited nAChRs cluster into a "pretzel" shape [90]. An actin polymerization-dependent mechanism could facilitate lipid raft coalescence and thereby the formation of large nAChRs clusters [138].

Prolonged nerve activity and other conditions such as myopathy led to membrane depolarization. This depolarization causes inactivation of the sodium channels (localized at the bottom of postsynaptic folds), which plays an essential role in the loss of muscle fiber excitability [5]. The reduction of the excitability is most severe for the junctional region of the sarcolemma, where the local EPP is transformed into propagating AP.

Membrane depolarization increases the threshold of muscle AP generation and reduces the safety factor of neuromuscular transmission; hyperpolarization has the opposite effects [2]. In this regard, the fact that the junctional membrane of mammals is hyperpolarized by 2–4 mV compared to the extrajunctional region of the sarcolemma is especially important. Presumably, this local hyperpolarization resulted from the activation of Na,K-ATPase by nonhydrolyzed ACh, which is constantly present in the synaptic cleft in nanomolar concentration even with active AChE [139,140]. This residual ACh remains in the synaptic cleft for some time following quantal transmitter release, and also appears due to nonquantal ACh release. The physiological consequence of a local junctional membrane hyperpolarization is expected to be more effective muscle excitation and neuromuscular transmission.

A regulatory mechanism whereby nAChR and α2 Na,K-ATPase functionally and molecularly interact to modulate the RMP of skeletal muscle was identified [96,100,141]. The nAChR/Na,K-ATPase reciprocal interactions were demonstrated in a purified membrane preparation from *Torpedo californica*, enriched by nAChRs and Na,K-ATPase [100,141]. In this preparation, specific ligand binding to the nAChRs modulates specific ligand binding to the Na,K-ATPase and vice versa, suggesting the direct molecular interaction between these two proteins. Notably, similar reciprocal interaction between nAChR of neuronal type and Na,K-ATPase was further confirmed in an insect nervous system [142].

Moreover, the α2 Na,K-ATPase isozyme is enriched in junctional membrane region where it colocalizes with the nAChRs (Figure 2B). These proteins coimmunoprecipitate with each other, as well as with phospholemman (FXYD1 protein) and caveolin-3 [100]. Caveolin-3 is enriched at the junctional region where it co-localized with the nAChRs and promotes their clustering in the endplate membrane. The α subunit of the nAChR has a putative caveolin-binding motif and a lack of caveolin-3 inhibits nAChR clustering [143]. Also, the caveolin/Na,K-ATPase interactions are well-documented [112,114]. Since caveolin-3 is associated with caveolae in fully differentiated skeletal muscles [144], the nAChR/α2 Na,K-ATPase interaction likely takes place within caveolae [100].

In this interaction, the binding of ACh at nanomolar concentrations to the nAChRs stimulates electrogenic transport by the α2 Na,K-ATPase isozyme, causing a local junctional membrane hyperpolarization. Notably, the nAChRs oscillate between resting (micromolar affinity for ACh), open or desensitized (nonconducting state with nanomolar apparent affinity for ACh) conformations [145,146]. Micromolar concentrations of ACh promote channel opening following by spontaneous transitions into the desensitized state. Desensitization of the nAChRs can also occur without channel opening and is favored by prolonged exposure to low concentrations of agonist. Moreover, a number of facts suggest that it is the conformational change to desensitized state of the nAChRs that is responsible for the interaction with the α2 Na,K-ATPase [96,100]. Cholesterol and other lipids influence the rates of transitions between different nAChRs conformational states and a "conformational selection" model proposed that cholesterol modulate the equilibrium between resting and desensitized states [129]. In addition, sarcolemma cholesterol specifically contributes to maintaining endplate electrogenesis and cholesterol depletion by MβCD selectively decreases the α2 Na,K-ATPase isozyme electrogenic activity and eliminates local hyperpolarization [119] suggesting the involvement of cholesterol in formation and function of the nAChR/α2 Na,K-ATPase molecular complex.

Collectively, these findings suggest a mechanism by which the nAChRs in a nonconducting, desensitized state, with high apparent affinity for ACh, directly interacts with the α2 Na,K-ATPase to stimulate electrogenic active transport. The interaction utilizes a membrane-associated regulatory complex that includes the nAChR, the α2 Na,K-ATPase, FXYD1, caveolin-3 and cholesterol and is responsible for maintaining RMP and muscle excitability during intensive muscle use (Figure 2C).

8. Cholesterol and Motor Dysfunction

Physical exercise is extremely important to ensure a healthy life and, particularly, is essential for lipids metabolism [147]. While the impact of cholesterol and lipid rafts in pathogenesis of neurodegenerative disorders is well documented [54,124–126] much less is known about the role of cholesterol in neuromuscular disorders.

ALS is known to have a severe dysfunction of NMJs. Hence, ALS mouse model is an extensively studied model for NMJ diseases. The progression of ALS could be attributed to the alterations in cholesterol metabolism. High levels of plasma cholesterol and low density lipoprotein have neuroprotective effects in ALS patients, whereas inhibition of cholesterol synthesis with statins aggravates ALS progression [148–150]. Excess cholesterol could facilitate production of oxysterols which are ligands for liver X receptors (LXR). These receptors are expressed in skeletal muscles and the number of NMJs reduces in LXRβ-deficient mice [151]. Moreover, ablation of LXRβ led to an ALS-like pathology [152,153] and polymorphism of LXRs significantly affects ALS phenotype [154]. Stimulation of LXRs by oxysterols may decrease inflammation, increase antioxidant defense and affect lipid metabolism, thereby reducing the muscle damage [29]. However, certain oxysterols can have toxic effects. Notably, the plasma levels of 25HC were markedly increased in ALS patients and correlated with the disease severity [155]. One of the main sources of 25HC is mast cells and macrophages that may come into contact with degenerating NMJs, thereby accelerating axonopathy in the SOD1^{G93A} rats [156]. Lipid raft alterations can contribute to ALS progression. Decrease in lipid raft scaffold protein caveolin-1 was found in skeletal muscle of SOD1^{G93A} mice [157]. Additionally, omics analysis suggested that changes in abundance of numerous protein-residents of lipid rafts occur in spinal cord of ALS model mice [158]. In diaphragm, staining with lipid raft marker showed the presence of two populations of NMJs in SOD1^{G93A} mice containing less and more lipid rafts [39]. This may reflect the co-occurrence of opposite processes: a progressive NMJ dysfunction and a compensatory enhancement of synaptic function in the remaining NMJs.

The latest data indicate the leading role of motor activity in maintaining the level of membrane cholesterol. Recently, it has been shown that increased plasma lipid levels exacerbate muscle pathology in the mdx mouse model of Duchenne muscular dystrophy [159]. Another neuromuscular disorder is dysferlinopathy, induced by a deficiency of dysferlin protein. Dysferlin plays a key role in the multimolecular complex responsible for the repair of the integrity of sarcolemma during contractile activity. Dysferlin-deficient Bla/J mice (one of the models of dysferlinopathy) are also characterized by impaired lipid metabolism [160]. Notably, both mdx and Bla/J mice models are characterized by membrane depolarization resulting from lowered Na,K-ATPase electrogenic activity [6,161]. Dysferlin is a membrane-associated protein implicated in vesicle fusion, trafficking, and membrane resealing. Loss of dysferlin has a wide range of implications, such as decrease in membrane integrity, disturbances of the dynamics of membrane-associated molecules, Ca^{2+} dysregulation, Ca^{2+}-induced proteolysis, and oxidative stress [162,163]. Together, these processes contribute to increased levels of necrosis and inflammation and result in the loss of motility. Furthermore, dysferlin-deficient muscles demonstrated an impaired glucose and lipid metabolism. Progressive adipocyte replacement and accumulation of lipid droplets within dysferlin-deficient myofibers [160] as well as extramyocellular lipid deposition [164,165] were observed. New evidence suggests that plasma membrane cholesterol accumulation in mice fed on a western high-fat diet may contribute by damaging the cortical F-actin structure that is essential for insulin-regulated GLUT4 translocation and glucose transport. It was interesting to note that exercise training has a preventive effect [166]. Also, elevated plasma cholesterol levels correlated with muscular pathology and can aggravate muscle fiber damage in dysferlinopathies [167]. In sum, such lipid abnormalities accompanied by fundamental metabolic disturbance suggest additional progressive decline of muscle function in dysferlinopathies.

Another muscular protein whose function could be linked with cholesterol availability is a member of the synaptophysin family, Mitsugumin 29, which contains cholesterol-binding MARVEL domain. Mitsugumin 29 is specifically located in the triad junction of muscle fibers and important for T-tubule formation, efficient excitation–contraction coupling, Ca^{2+} signaling, and resistance to fatigue [168–170]. Decreased expression of Mitsugumin 29 which leads to disturbance of triad junction structure and Ca^{2+} signaling could be partially linked with contractile dysfunction during muscle aging [171]. Similarly, cholesterol depletion affecting T-tubules reduced L-type Ca^{2+} current in freshly isolated fetal skeletal muscle cells [172]. Interestingly, a low-frequency coding variant in the gene encoding Mitsugumin 29 was associated with morbid obesity, suggesting a link between Mitsugumin 29 and lipid metabolism [173].

NMJ ultrastructure undergoes continual morphological remodeling in response to changes in the pattern of motor activity. Reduced activity (denervation, injury, bed rest and other form of disuse, microgravity at spaceflight, muscle diseases, and aging) triggers alterations in NMJ stability and integrity [174–176]. Even acute disuse (6–12 h of HS) disrupts plasma membrane lipid-ordered rafts in sarcolemma of rat soleus muscle [120]. This disturbance is accompanied by membrane depolarization due to loss of the α2 Na,K-ATPase electrogenic activity [103,177]. The greatest disturbances are observed at the junctional membrane region [103]. Moreover, specific inhibition of α2 Na,K-ATPase by 1 μM ouabain induces a disturbance of lipid rafts similar to that induced by disuse alone [120]. In both cases, lipid rafts were able to recover with cholesterol supplementation, suggesting that disturbance results from cholesterol loss. Repetitive nerve stimulation also restored both α2 Na,K-ATPase electrogenic activity and lipid rafts integrity. These findings suggest that the stability of lipid rafts is subject to regulation by skeletal muscle motor activity [120]. Intriguingly, important role of sphingolipids (including ceramide) in regulation of skeletal muscle function in health and disease is known [178] and it was shown that acute HS-induced cholesterol loss is accompanied by ceramide accumulation [179]. These reciprocal ceramide/cholesterol changes may reflect the ability of an excess of ceramide to displace cholesterol from sarcolemma [180].

Notably, adenosine monophosphate-activated protein kinase (AMPK) is an energy-sensing enzyme that is known to be a key regulator of glucose, lipid, and protein metabolism in skeletal muscle [181]. Particularly, AMPK has been implicated in control of sarcolemma cholesterol levels [166,182]. In addition, AMPK and AMPK-activated autophagy have been implicated in NMJ remodeling [183–185]. Lowered contractile activity should provide an accumulation of phosphorylated high-energy phosphates. Consistent with this, decreased phosphorylation of AMPK during 6–24 h of HS was demonstrated in rat soleus muscle [186–188].

These results provide evidence to suggest that the ordering of lipid rafts strongly depends on motor nerve input and may involve interactions with the α2 Na,K-ATPase and AMPK. Lipid raft disturbance, accompanied by the loss of α2 Na,K-ATPase activity and decreased phosphorylation of AMPK, are among the earliest remodeling events induced by skeletal muscle disuse. However, we should note that the effects of prolonged immobilization or unloading regarding skeletal muscle membrane remain poorly understood. Continuous membrane stretch can modulate the activity of channels and hence the dynamics of a variety of molecules incorporated into the lipid bilayer. Accordingly, stretch-induced lipid stress may trigger membrane remodeling including changes in cholesterol distribution. Thus, the question of whether membrane cholesterol disturbance results from muscle inactivity per se, or mediated by sarcolemma stretching, remains open.

9. Conclusions

Cholesterol, as important component of cell membranes, plays a crucial role in segregation of plasma membrane lipid phases and is essential to maintain membrane fluidity, curvature, ion channel and transporter functions, compartmentalization, and signaling. Cholesterol is specifically important for synaptic function, and a link between impaired cholesterol metabolism and neurodegenerative disorders is well recognized. However, much less is known about the role of cholesterol in peripheral neuromuscular transmission and corresponding disorders. Our review is the first attempt to summarize known experimental data on the role of cholesterol and its derivatives in the NMJ in health and some motor dysfunctions. Accumulated novel date indicates that cholesterol is included in all key pre- and postsynaptic processes on which the safety factor at the NMJ depends. Moreover, an imbalance of cholesterol homeostasis accompanied by membrane cholesterol changes can modulates these steps and disruption of lipid rafts is an early event in the disuse-induced muscle atrophy. Possible cholesterol-dependent steps, essential for maintaining safety factor for the neuromuscular transmission, are summarized in Figure 3. Future studies will be required to identify the precise molecular interactions between cholesterol and skeletal muscle motor activity.

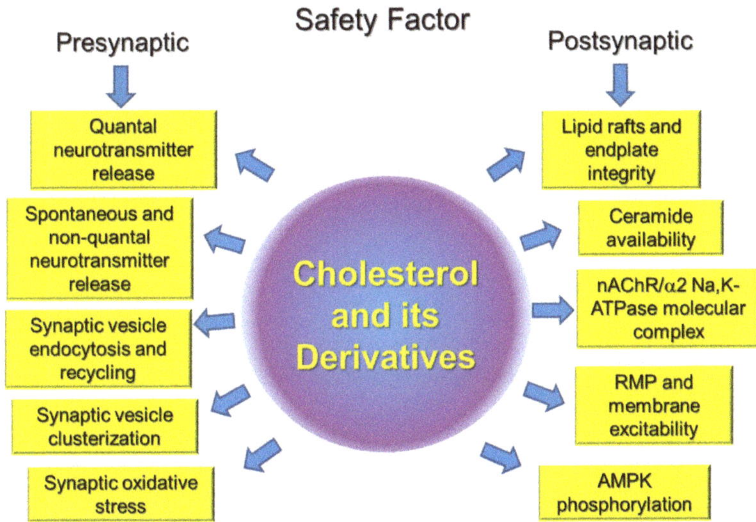

Figure 3. Key pre- and postsynaptic points that cross-talked with plasma membrane cholesterol and are responsible for the safety factor at the neuromuscular junction (NMJ).

Author Contributions: The authors contributed equally.

Funding: This review was funded by the Russian Science Foundation (RSF) Grant No. 18-15-00043 (for IIK, Sections 1 and 6–9) and Russian Foundation for Basic Research Grant No. 17-04-00046 (for AMP, Sections 2–5).

Acknowledgments: We are very grateful to our colleagues A.L. Zefirov and V.V. Kravtsova for their seminal contribution in our long-standing collaboration.

Conflicts of Interest: The authors declare no conflicts of interest.

Abbreviations

α-BTX	rhodamine-conjugated α-bungarotoxin
ACh	acetylcholine
AChE	acetylcholinesterase
ALS	amyotrophic lateral sclerosis
AMPK	$5'$ adenosine monophosphate-activated protein kinase
AP	action potential
ATP	adenosine triphosphate
AZs	active zones
5Ch3	5α-cholestan-3-one
CSP	cysteine string protein
CTxB	cholera toxin B subunit
EPP	endplate potential
GPCRs	G-protein coupled receptors
24HC	24S-hydroxycholesterol
HS	hindlimb suspension
LXR	liver X receptor
MβCD	methyl-β-cyclodextrin
MEPP	miniature endplate potential

nAChR	nicotinic acetylcholine receptor
NADPH	nicotinamide adenine dinucleotide phosphate
NMJ	neuromuscular junction
RMP	resting membrane potential
ROS	reactive oxygen species
SOD1^{G93A}	superoxide dismutase 1
SV	synaptic vesicle
VGLUT	vesicular glutamate transporter

References

1. Wood, S.J.; Slater, C.R. Safety factor at the neuromuscular junction. *Prog. Neurobiol.* **2001**, *64*, 393–429. [CrossRef]
2. Ruff, R.L. Endplate contributions to the safety factor for neuromuscular transmission. *Muscle Nerve* **2011**, *44*, 854–861. [CrossRef] [PubMed]
3. Ackermann, F.; Waites, C.L.; Garner, C.C. Presynaptic active zones in invertebrates and vertebrates. *EMBO Rep.* **2015**, *16*, 923–938. [CrossRef] [PubMed]
4. Sanes, J.S.; Lichtman, J.W. Development of the vertebrate neuromuscular junction. *Annu. Rev. Neurosci.* **1999**, *22*, 389–442. [CrossRef] [PubMed]
5. Filatov, G.N.; Pinter, M.J.; Rich, M.M. Resting Potential-dependent Regulation of the Voltage Sensitivity of Sodium Channel Gating in Rat Skeletal Muscle In Vivo. *J. Gen. Physiol.* **2005**, *126*, 161–172. [CrossRef] [PubMed]
6. Miles, M.T.; Cottey, E.; Cottey, A.; Stefanski, C.; Carlson, C.G. Reduced resting potentials in dystrophic (mdx) muscle fibers are secondary to NF-κB-dependent negative modulation of ouabain sensitive Na$^+$-K$^+$ pump activity. *J. Neurosci.* **2011**, *303*, 53–60. [CrossRef] [PubMed]
7. Serra, A.; Ruff, R.L.; Leigh, R.J. Neuromuscular transmission failure in myasthenia gravis: Decrement of safety factor and susceptibility of extraocular muscles. *Ann. N. Y. Acad. Sci.* **2012**, *1275*, 129–135. [CrossRef] [PubMed]
8. Grouleff, J.; Irudayam, S.J.; Skeby, K.K.; Schiøtt, B. The influence of cholesterol on membrane protein structure, function, and dynamics studied by molecular dynamics simulations. *Biochim. Biophys. Acta* **2015**, *1848*, 1783–1795. [CrossRef] [PubMed]
9. Belani, J.D. Chirality Effect on Cholesterol Modulation of Protein Function. *Adv. Exp. Med. Biol.* **2019**, *1115*, 3–19. [CrossRef] [PubMed]
10. Oh, H.; Mohler, E.R., 3rd; Tian, A.; Baumgart, T.; Diamond, S.L. Membrane cholesterol is a biomechanical regulator of neutrophil adhesion. *Arterioscler. Thromb. Vasc. Biol.* **2009**, *29*, 1290–1297. [CrossRef] [PubMed]
11. Chun, Y.S.; Oh, H.G.; Park, M.K.; Kim, T.W.; Chung, S. Increasing Membrane Cholesterol Level Increases the Amyloidogenic Peptide by Enhancing the Expression of Phospholipase C. *J. Neurodegener. Dis.* **2013**, *2013*, 407903. [CrossRef] [PubMed]
12. Magarkar, A.; Dhawan, V.; Kallinteri, P.; Viitala, T.; Elmowafy, M.; Róg, T.; Bunker, A. Cholesterol level affects surface charge of lipid membranes in saline solution. *Sci. Rep.* **2014**, *4*, 5005. [CrossRef] [PubMed]
13. Amsalem, M.; Poilbout, C.; Ferracci, G.; Delmas, P.; Padilla, F. Membrane cholesterol depletion as a trigger of Nav1.9 channel-mediated inflammatory pain. *EMBO J.* **2018**, *37*, e97349. [CrossRef] [PubMed]
14. Meleleo, D.; Sblano, C. Influence of cholesterol on human calcitonin channel formation. Possible role of sterol as molecular chaperone. *AIMS Biophys.* **2019**, *6*, 23–38. [CrossRef]
15. Arenas, F.; Garcia-Ruiz, C.; Fernandez-Checa, J.C. Intracellular Cholesterol Trafficking and Impact in Neurodegeneration. *Front. Mol. Neurosci.* **2017**, *10*, 382. [CrossRef] [PubMed]
16. Saher, G.; Stumpf, S.K. Cholesterol in myelin biogenesis and hypomyelinating disorders. *Biochim. Biophys. Acta* **2015**, *1851*, 1083–1094. [CrossRef] [PubMed]
17. Pfrieger, F.W.; Ungerer, N. Cholesterol metabolism in neurons and astrocytes. *Prog. Lipid Res.* **2011**, *50*, 357–371. [CrossRef] [PubMed]
18. Funfschilling, U.; Jockusch, W.J.; Sivakumar, N.; Mobius, W.; Corthals, K.; Li, S.; Quintes, S.; Kim, Y.; Schaap, I.A.; Rhee, J.S.; et al. Critical time window of neuronal cholesterol synthesis during neurite outgrowth. *J. Neurosci.* **2012**, *32*, 7632–7645. [CrossRef] [PubMed]

19. Petrov, A.M.; Kasimov, M.R.; Zefirov, A.L. Brain Cholesterol Metabolism and Its Defects: Linkage to Neurodegenerative Diseases and Synaptic Dysfunction. *Acta Nat.* **2016**, *8*, 58–73.
20. Egawa, J.; Pearn, M.L.; Lemkuil, B.P.; Patel, P.M.; Head, B.P. Membrane lipid rafts and neurobiology: Age-related changes in membrane lipids and loss of neuronal function. *J. Physiol.* **2016**, *594*, 4565–4579. [CrossRef] [PubMed]
21. Comley, L.H.; Fuller, H.R.; Wishart, T.M.; Mutsaers, C.A.; Thomson, D.; Wright, A.K.; Ribchester, R.R.; Morris, G.E.; Parson, S.H.; Horsburgh, K.; et al. ApoE isoform-specific regulation of regeneration in the peripheral nervous system. *Hum. Mol. Genet.* **2011**, *20*, 2406–2421. [CrossRef] [PubMed]
22. Choi, H.Y.; Liu, Y.; Tennert, C.; Sugiura, Y.; Karakatsani, A.; Kröger, S.; Johnson, E.B.; Hammer, R.E.; Lin, W.; Herz, J. APP interacts with LRP4 and agrin to coordinate the development of the neuromuscular junction in mice. *eLife* **2013**, *2*, e00220. [CrossRef] [PubMed]
23. Yokoyama, M.; Seo, T.; Park, T.; Yagyu, H.; Hu, Y.; Son, N.H.; Augustus, A.S.; Vikramadithyan, R.K.; Ramakrishnan, R.; Pulawa, L.K.; et al. Effects of lipoprotein lipase and statins on cholesterol uptake into heart and skeletal muscle. *J. Lipid Res.* **2007**, *48*, 646–655. [CrossRef] [PubMed]
24. Barrientos, G.; Sánchez-Aguilera, P.; Jaimovich, E.; Hidalgo, C.; Llanos, P. Membrane Cholesterol in Skeletal Muscle: A Novel Player in Excitation-Contraction Coupling and Insulin Resistance. *J. Diabetes Res.* **2017**, *2017*, 3941898. [CrossRef] [PubMed]
25. Norata, G.D.; Tibolla, G.; Catapano, A.L. Statins and skeletal muscles toxicity: From clinical trials to everyday practice. *Pharmacol. Res.* **2014**, *88*, 107–113. [CrossRef] [PubMed]
26. Greenberg, A.S.; Coleman, R.A.; Kraemer, F.B.; McManaman, J.L.; Obin, M.S.; Puri, V.; Yan, Q.W.; Miyoshi, H.; Mashek, D.G. The role of lipid droplets in metabolic disease in rodents and humans. *J. Clin. Investig.* **2011**, *121*, 2102–2110. [CrossRef] [PubMed]
27. Murphy, R.C.; Johnson, K.M. Cholesterol, reactive oxygen species, and the formation of biologically active mediators. *J. Biol. Chem.* **2008**, *283*, 15521–15525. [CrossRef] [PubMed]
28. Archer, A.; Laurencikiene, J.; Ahmed, O.; Steffensen, K.R.; Parini, P.; Gustafsson, J.A.; Korach-André, M. Skeletal muscle as a target of LXR agonist after long-term treatment: Focus on lipid homeostasis. *Am. J. Physiol. Endocrinol. Metab.* **2014**, *306*, E494–E502. [CrossRef] [PubMed]
29. Webb, R.; Hughes, M.G.; Thomas, A.W.; Morris, K. The Ability of Exercise-Associated Oxidative Stress to Trigger Redox-Sensitive Signalling Responses. *Antioxidants* **2017**, *6*, 63. [CrossRef] [PubMed]
30. Rizzoli, S.O. Synaptic vesicle recycling: Steps and principles. *EMBO J.* **2014**, *33*, 788–822. [CrossRef] [PubMed]
31. Rizzoli, S.O.; Betz, W.J. Synaptic vesicle pools. *Nat. Rev. Neurosci.* **2005**, *6*, 57–69. [CrossRef] [PubMed]
32. Nakajima, Y.; Bridgman, P.C. Absence of filipin-sterol complexes from the membranes of active zones and acetylcholine receptor aggregates at frog neuromuscular junctions. *J. Cell. Biol.* **1981**, *88*, 453–458. [CrossRef] [PubMed]
33. Ko, C.P.; Propst, J.W. Absence of sterol-specific complexes at active zones of degenerating and regenerating frog neuromuscular junctions. *J. Neurocytol.* **1986**, *15*, 231–240. [CrossRef] [PubMed]
34. Zamir, O.; Charlton, M.P. Cholesterol and synaptic transmitter release at crayfish neuromuscular junctions. *J. Physiol.* **2006**, *571*, 83–99. [CrossRef] [PubMed]
35. Petrov, A.M.; Kudryashova, K.E.; Odnoshivkina, Y.G.; Zefirov, A.L. Cholesterol and lipid rafts in the plasma membrane of nerve terminal and membrane of synaptic vesicles. *Neurochem. J.* **2011**, *5*, 13–19. [CrossRef]
36. Petrov, A.M.; Naumenko, N.V.; Uzinskaya, K.V.; Giniatullin, A.R.; Urazaev, A.K.; Zefirov, A.L. Increased non-quantal release of acetylcholine after inhibition of endocytosis by methyl-β-cyclodextrin: The role of vesicular acetylcholine transporter. *Neuroscience* **2011**, *186*, 1–12. [CrossRef] [PubMed]
37. Kasimov, M.R.; Giniatullin, A.R.; Zefirov, A.L.; Petrov, A.M. Effects of 5α-cholestan-3-one on the synaptic vesicle cycle at the mouse neuromuscular junction. *Biochim. Biophys. Acta* **2015**, *1851*, 674–685. [CrossRef] [PubMed]
38. Kasimov, M.R.; Zakyrjanova, G.F.; Giniatullin, A.R.; Zefirov, A.L.; Petrov, A.M. Similar oxysterols may lead to opposite effects on synaptic transmission: Olesoxime versus 5 α-cholestan-3-one at the frog neuromuscular junction. *Biochim. Biophys. Acta* **2016**, *1861*, 606–616. [CrossRef] [PubMed]
39. Mukhutdinova, K.A.; Kasimov, M.R.; Giniatullin, A.R.; Zakyrjanova, G.F.; Petrov, A.M. 24S-hydroxycholesterol suppresses neuromuscular transmission in SOD1(G93A) mice: A possible role of NO and lipid rafts. *Mol. Cell. Neurosci.* **2018**, *88*, 308–318. [CrossRef] [PubMed]

40. Petrov, A.M.; Kasimov, M.R.; Giniatullin, A.R.; Tarakanova, O.I.; Zefirov, A.L. The role of cholesterol in the exo- and endocytosis of synaptic vesicles in frog motor nerve endings. *Neurosci. Behav. Physiol.* **2010**, *40*, 894–901. [CrossRef] [PubMed]

41. Petrov, A.M.; Yakovleva, A.A.; Zefirov, A.L. Role of membrane cholesterol in spontaneous exocytosis at frog neuromuscular synapses: Reactive oxygen species-calcium interplay. *J. Physiol.* **2014**, *592*, 4995–5009. [CrossRef] [PubMed]

42. Petrov, A.M.; Zakyrjanova, G.F.; Yakovleva, A.A.; Zefirov, A.L. Inhibition of protein kinase C affects on mode of synaptic vesicle exocytosis due to cholesterol depletion. *Biochem. Biophys. Res. Commun.* **2015**, *456*, 145–150. [CrossRef] [PubMed]

43. Tarakanova, O.I.; Petrov, A.M.; Zefirov, A.L. The role of membrane cholesterol in neurotransmitter release from motor nerve terminals. *Dokl. Biol. Sci.* **2011**, *438*, 138–140. [CrossRef] [PubMed]

44. Rodrigues, H.A.; Lima, R.F.; Fonseca Mde, C.; Amaral, E.A.; Martinelli, P.M.; Naves, L.A.; Gomez, M.V.; Kushmerick, C.; Prado, M.A.; Guatimosim, C. Membrane cholesterol regulates different modes of synaptic vesicle release and retrieval at the frog neuromuscular junction. *Eur. J. Neurosci.* **2013**, *38*, 2978–2987. [CrossRef] [PubMed]

45. Teixeira, G.; Vieira, L.B.; Gomez, M.V.; Guatimosim, C. Cholesterol as a key player in the balance of evoked and spontaneous glutamate release in rat brain cortical synaptosomes. *Neurochem. Int.* **2012**, *61*, 1151–1159. [CrossRef] [PubMed]

46. Thyagarajan, B.; Potian, J.G.; McArdle, J.J.; Baskaran, P. Perturbation to Cholesterol at the Neuromuscular Junction Confers Botulinum Neurotoxin A Sensitivity to Neonatal Mice. *Toxicol. Sci.* **2017**, *159*, 179–188. [CrossRef] [PubMed]

47. Smith, A.J.; Sugita, S.; Charlton, M.P. Cholesterol-dependent kinase activity regulates transmitter release from cerebellar synapses. *J. Neurosci.* **2010**, *30*, 6116–6121. [CrossRef] [PubMed]

48. Takamori, S.; Holt, M.; Stenius, K.; Lemke, E.A.; Grønborg, M.; Riedel, D.; Urlaub, H.; Schenck, S.; Brügger, B.; Ringler, P.; et al. Molecular anatomy of a trafficking organelle. *Cell* **2006**, *127*, 831–846. [CrossRef] [PubMed]

49. Dason, J.S.; Smith, A.J.; Marin, L.; Charlton, M.P. Vesicular sterols are essential for synaptic vesicle cycling. *J. Neurosci.* **2010**, *30*, 15856–15865. [CrossRef] [PubMed]

50. Yue, H.Y.; Xu, J. Cholesterol regulates multiple forms of vesicle endocytosis at a mammalian central synapse. *J. Neurochem.* **2015**, *134*, 247–260. [CrossRef] [PubMed]

51. Dason, J.S.; Smith, A.J.; Marin, L.; Charlton, M.P. Cholesterol and F-actin are required for clustering of recycling synaptic vesicle proteins in the presynaptic plasma membrane. *J. Physiol.* **2014**, *592*, 621–633. [CrossRef] [PubMed]

52. Thiele, C.; Hannah, M.J.; Fahrenholz, F.; Huttner, W.B. Cholesterol binds to synaptophysin and is required for biogenesis of synaptic vesicles. *Nat. Cell. Biol.* **2000**, *2*, 42–49. [CrossRef] [PubMed]

53. Fewou, S.N.; Rupp, A.; Nickolay, L.E.; Carrick, K.; Greenshields, K.N.; Pediani, J.; Plomp, J.J.; Willison, H.J. Anti-ganglioside antibody internalization attenuates motor nerve terminal injury in a mouse model of acute motor axonal neuropathy. *J. Clin. Investig.* **2012**, *122*, 1037–1051. [CrossRef] [PubMed]

54. Petrov, A.M.; Kasimov, M.R.; Zefirov, A.L. Cholesterol in the Pathogenesis of Alzheimer's, Parkinson's Diseases and Autism: Link to Synaptic Dysfunction. *Acta Nat.* **2017**, *9*, 26–37.

55. Testa, G.; Rossin, D.; Poli, G.; Biasi, F.; Leonarduzzi, G. Implication of oxysterols in chronic inflammatory human diseases. *Biochimie* **2018**, *153*, 220–231. [CrossRef] [PubMed]

56. Petrov, A.M.; Kasimov, M.R.; Giniatullin, A.R.; Zefirov, A.L. Effects of Oxidation of Membrane Cholesterol on the Vesicle Cycle in Motor Nerve Terminals in the Frog Rana Ridibunda. *Neurosci. Behav. Physiol.* **2014**, *44*, 1020–1030. [CrossRef]

57. Kasimov, M.R.; Fatkhrakhmanova, M.R.; Mukhutdinova, K.A.; Petrov, A.M. 24S-Hydroxycholesterol enhances synaptic vesicle cycling in the mouse neuromuscular junction: Implication of glutamate NMDA receptors and nitric oxide. *Neuropharmacology* **2017**, *117*, 61–73. [CrossRef] [PubMed]

58. Xiao, Y.; Karam, C.; Yi, J.; Zhang, L.; Li, X.; Yoon, D.; Wang, H.; Dhakal, K.; Ramlow, P.; Yu, T.; et al. ROS-related mitochondrial dysfunction in skeletal muscle of an ALS mouse model during the disease progression. *Pharmacol. Res.* **2018**, *138*, 25–36. [CrossRef] [PubMed]

59. Di Stasi, A.M.; Mallozzi, C.; Macchia, G.; Maura, G.; Petrucci, T.C.; Minetti, M. Peroxynitrite affects exocytosis and SNARE complex formation and induces tyrosine nitration of synaptic proteins. *J. Neurochem.* **2002**, *82*, 420–429. [CrossRef] [PubMed]

60. Thomas, S.; Robitaille, R. Differential frequency-dependent regulation of transmitter release by endogenous nitric oxide at the amphibian neuromuscular synapse. *J. Neurosci.* **2001**, *21*, 1087–1095. [CrossRef] [PubMed]

61. Robinson, S.W.; Bourgognon, J.M.; Spiers, J.G.; Breda, C.; Campesan, S.; Butcher, A.; Mallucci, G.R.; Dinsdale, D.; Morone, N.; Mistry, R.; et al. Nitric oxide-mediated posttranslational modifications control neurotransmitter release by modulating complexin farnesylation and enhancing its clamping ability. *PLoS Biol.* **2018**, *16*, e2003611. [CrossRef] [PubMed]

62. DeBarber, A.E.; Sandlers, Y.; Pappu, A.S.; Merkens, L.S.; Duell, P.B.; Lear, S.R.; Erickson, S.K.; Steiner, R.D. Profiling sterols in cerebrotendinous xanthomatosis: Utility of Girard derivatization and high resolution exact mass LC-ESI-MS(n) analysis. *J. Chromatogr. B Anal. Technol. Biomed. Life Sci.* **2011**, *879*, 1384–1392. [CrossRef] [PubMed]

63. Martin, L.J. Olesoxime, a cholesterol-like neuroprotectant for the potential treatment of amyotrophic lateral sclerosis. *IDrugs* **2010**, *13*, 568–580. [PubMed]

64. Gouarné, C.; Tracz, J.; Paoli, M.G.; Deluca, V.; Seimandi, M.; Tardif, G.; Xilouri, M.; Stefanis, L.; Bordet, T.; Pruss, R.M. Protective role of olesoxime against wild-type α-synuclein-induced toxicity in human neuronally differentiated SHSY-5Y cells. *Br. J. Pharmacol.* **2015**, *172*, 235–245. [CrossRef] [PubMed]

65. Magalon, K.; Le Grand, M.; El Waly, B.; Moulis, M.; Pruss, R.; Bordet, T.; Cayre, M.; Belenguer, P.; Carré, M.; Durbec, P. Olesoxime favors oligodendrocyte differentiation through a functional interplay between mitochondria and microtubules. *Neuropharmacology* **2016**, *111*, 293–303. [CrossRef] [PubMed]

66. Weber, J.J.; Ortiz Rios, M.M.; Riess, O.; Clemens, L.E.; Nguyen, H.P. The calpain-suppressing effects of olesoxime in Huntington's disease. *Rare Dis.* **2016**, *4*, e1153778. [CrossRef] [PubMed]

67. Taverna, E.; Saba, E.; Rowe, J.; Francolini, M.; Clementi, F.; Rosa, P. Role of lipid microdomains in P/Q-type calcium channel (Cav2.1) clustering and function in presynaptic membranes. *J. Biol. Chem.* **2004**, *279*, 5127–5134. [CrossRef] [PubMed]

68. Thoreson, W.B.; Mercer, A.J.; Cork, K.M.; Szalewski, R.J. Lateral mobility of L-type calcium channels in synaptic terminals of retinal bipolar cells. *Mol. Vis.* **2013**, *19*, 16–24. [PubMed]

69. Ronzitti, G.; Bucci, G.; Emanuele, M.; Leo, D.; Sotnikova, T.D.; Mus, L.V.; Soubrane, C.H.; Dallas, M.L.; Thalhammer, A.; Cingolani, L.A.; et al. Exogenous α-synuclein decreases raft partitioning of Cav2.2 channels inducing dopamine release. *J. Neurosci.* **2014**, *34*, 10603–10615. [CrossRef] [PubMed]

70. Lv, J.H.; He, L.; Sui, S.F. Lipid rafts association of synaptotagmin I on synaptic vesicles. *Biochemistry* **2008**, *73*, 283–288. [CrossRef] [PubMed]

71. Matthies, H.J.; Han, Q.; Shields, A.; Wright, J.; Moore, J.L.; Winder, D.G.; Galli, A.; Blakely, R.D. Subcellular localization of the antidepressant-sensitive norepinephrine transporter. *BMC Neurosci.* **2009**, *10*, 65. [CrossRef] [PubMed]

72. De Juan-Sanz, J.; Núñez, E.; Zafra, F.; Berrocal, M.; Corbacho, I.; Ibáñez, I.; Arribas-González, E.; Marcos, D.; López-Corcuera, B.; Mata, A.M.; et al. Presynaptic control of glycine transporter 2 (GlyT2) by physical and functional association with plasma membrane Ca^{2+}-ATPase (PMCA) and Na^+-Ca^{2+} exchanger (NCX). *J. Biol. Chem.* **2014**, *289*, 34308–34324. [CrossRef] [PubMed]

73. Rahbek-Clemmensen, T.; Lycas, M.D.; Erlendsson, S.; Eriksen, J.; Apuschkin, M.; Vilhardt, F.; Jørgensen, T.N.; Hansen, F.H.; Gether, U. Super-resolution microscopy reveals functional organization of dopamine transporters into cholesterol and neuronal activity-dependent nanodomains. *Nat. Commun.* **2017**, *8*, 740. [CrossRef] [PubMed]

74. Yoshinaka, K.; Kumanogoh, H.; Nakamura, S.; Maekawa, S. Identification of V-ATPase as a major component in the raft fraction prepared from the synaptic plasma membrane and the synaptic vesicle of rat brain. *Neurosci. Lett.* **2004**, *363*, 168–172. [CrossRef] [PubMed]

75. Lee, J.S.; Cho, W.J.; Shin, L.; Jena, B.P. Involvement of cholesterol in synaptic vesicle swelling. *Exp. Biol. Med.* **2010**, *235*, 470–477. [CrossRef] [PubMed]

76. Murray, D.H.; Tamm, L.K. Molecular mechanism of cholesterol- and polyphosphoinositide-mediated syntaxin clustering. *Biochemistry* **2011**, *50*, 9014–9022. [CrossRef] [PubMed]

77. Gil, C.; Cubí, R.; Blasi, J.; Aguilera, J. Synaptic proteins associate with a sub-set of lipid rafts when isolated from nerve endings at physiological temperature. *Biochem. Biophys. Res. Commun.* **2006**, *348*, 1334–1342. [CrossRef] [PubMed]

78. Cho, W.J.; Jeremic, A.; Jin, H.; Ren, G.; Jena, B.P. Neuronal fusion pore assembly requires membrane cholesterol. *Cell Biol. Int.* **2007**, *31*, 1301–1308. [CrossRef] [PubMed]

79. Kreutzberger, A.J.; Kiessling, V.; Tamm, L.K. High cholesterol obviates a prolonged hemifusion intermediate in fast SNARE-mediated membrane fusion. *Biophys. J.* **2015**, *109*, 319–329. [CrossRef] [PubMed]

80. Kyung, J.W.; Kim, J.M.; Lee, W.; Ha, T.Y.; Cha, S.H.; Chung, K.H.; Choi, D.J.; Jou, I.; Song, W.K.; Joe, E.H.; et al. DJ-1 deficiency impairs synaptic vesicle endocytosis and reavailability at nerve terminals. *Proc. Natl. Acad. Sci. USA* **2018**, *115*, 1629–1634. [CrossRef] [PubMed]

81. Kao, H.T.; Ryoo, K.; Lin, A.; Janoschka, S.R.; Augustine, G.J.; Porton, B. Synapsins regulate brain-derived neurotrophic factor-mediated synaptic potentiation and axon elongation by acting on membrane rafts. *Eur. J. Neurosci.* **2017**, *45*, 1085–1101. [CrossRef] [PubMed]

82. Moral-Naranjo, M.T.; Montenegro, M.F.; Muñoz-Delgado, E.; Campoy, F.J.; Vidal, C.J. The levels of both lipid rafts and raft-located acetylcholinesterase dimers increase in muscle of mice with muscular dystrophy by merosin deficiency. *Biochim. Biophys. Acta* **2010**, *1802*, 754–764. [CrossRef] [PubMed]

83. Montenegro, M.F.; Cabezas-Herrera, J.; Campoy, F.J.; Muñoz-Delgado, E.; Vidal, C.J. Lipid rafts of mouse liver contain nonextended and extended acetylcholinesterase variants along with M3 muscarinic receptors. *FASEB J.* **2017**, *31*, 544–555. [CrossRef] [PubMed]

84. Giniatullin, A.; Petrov, A.; Giniatullin, R. The involvement of P2Y12 receptors, NADPH oxidase, and lipid rafts in the action of extracellular ATP on synaptic transmission at the frog neuromuscular junction. *Neuroscience* **2015**, *285*, 324–332. [CrossRef] [PubMed]

85. Huang, Y.; Huang, S.; Di Scala, C.; Wang, Q.; Wandall, H.H.; Fantini, J.; Zhang, Y.Q. The glycosphingolipid MacCer promotes synaptic bouton formation in *Drosophila* by interacting with Wnt. *eLife* **2018**, *7*, e38183. [CrossRef] [PubMed]

86. Clausen, T. Na$^+$-K$^+$ pump regulation and skeletal muscle contractility. *Physiol. Rev.* **2003**, *83*, 1269–1324. [CrossRef] [PubMed]

87. Clausen, T. Quantification of Na$^+$,K$^+$ pumps and their transport rate in skeletal muscle: Functional significance. *J. Gen. Physiol.* **2013**, *142*, 327–345. [CrossRef] [PubMed]

88. Skou, J.C. The influence of some cations on an adenosine triphosphatase from peripheral nerves. *Biochim. Biophys. Acta* **1957**, *23*, 394–401. [CrossRef]

89. Blanco, G.; Mercer, R.W. Isozymes of the Na-K-ATPase: Heterogeneity in structure, diversity in function. *Am. J. Physiol.* **1998**, *275*, F633–F655. [CrossRef] [PubMed]

90. Mobasheri, A.; Avila, J.; Cózar-Castellano, I.; Brownleader, M.D.; Trevan, M.; Francis, M.J.; Lamb, J.F.; Martín-Vasallo, P. Na$^+$,K$^+$-ATPase isozyme diversity; comparative biochemistry and physiological implications of novel functional interactions. *Biosci. Rep.* **2000**, *20*, 51–91. [CrossRef] [PubMed]

91. Sejersted, O.M.; Sjogaard, G. Dynamics and consequences of potassium shifts in skeletal muscle and heart during exercise. *Physiol. Rev.* **2000**, *80*, 1411–1481. [CrossRef] [PubMed]

92. Levitan, I.; Singh, D.K.; Rosenhouse-Dantsker, A. Cholesterol binding to ion channels. *Front. Physiol.* **2014**, *5*, 65. [CrossRef] [PubMed]

93. Cornelius, F.; Habeck, M.; Kanai, R.; Toyoshima, C.; Karlish, S.J. General and specific lipid-protein interactions in Na,K-ATPase. *Biochim. Biophys. Acta* **2015**, *1848*, 1729–1743. [CrossRef] [PubMed]

94. Sibarov, D.A.; Poguzhelskaya, E.E.; Antonov, S.M. Downregulation of calcium-dependent NMDA receptor desensitization by sodium-calcium exchangers: A role of membrane cholesterol. *BMC Neurosci.* **2018**, *19*, 73. [CrossRef] [PubMed]

95. Haviv, H.; Habeck, M.; Kanai, R.; Toyoshima, C.; Karlish, S.J. Neutral phospholipids stimulate Na,K-ATPase activity: A specific lipid-protein interaction. *J. Biol. Chem.* **2013**, *288*, 10073–10081. [CrossRef] [PubMed]

96. Matchkov, V.V.; Krivoi, I.I. Specialized functional diversity and interactions of the Na,K-ATPase. *Front. Physiol.* **2016**, *7*, 179. [CrossRef] [PubMed]

97. Clausen, M.V.; Hilbers, F.; Poulsen, H. The Structure and Function of the Na,K-ATPase Isoforms in Health and Disease. *Front. Physiol.* **2017**, *8*, 371. [CrossRef] [PubMed]

98. Orlowski, J.; Lingrel, J.B. Tissue-specific and developmental regulation of rat Na,K-ATPase catalytic α isoform and β subunit mRNAs. *J. Biol. Chem.* **1988**, *263*, 10436–10442. [PubMed]

99. He, S.; Shelly, D.A.; Moseley, A.E.; James, P.F.; James, J.H.; Paul, R.J.; Lingrel, J.B. The α1- and α2-isoforms of Na-K-ATPase play different roles in skeletal muscle contractility. *Am. J. Physiol. Regul. Integr. Comp. Physiol.* **2001**, *281*, R917–R925. [CrossRef] [PubMed]

100. Heiny, J.A.; Kravtsova, V.V.; Mandel, F.; Radzyukevich, T.L.; Benziane, B.; Prokofiev, A.V.; Pedersen, S.E.; Chibalin, A.V.; Krivoi, I.I. The nicotinic acetylcholine receptor and the Na,K-ATPase α2 isoform interact to regulate membrane electrogenesis in skeletal muscle. *J. Biol. Chem.* **2010**, *285*, 28614–28626. [CrossRef] [PubMed]

101. Radzyukevich, T.L.; Neumann, J.C.; Rindler, T.N.; Oshiro, N.; Goldhamer, D.J.; Lingrel, J.B.; Heiny, J.A. Tissue-specific role of the Na,K-ATPase α2 isozyme in skeletal muscle. *J. Biol. Chem.* **2013**, *288*, 1226–1237. [CrossRef] [PubMed]

102. DiFranco, M.; Hakimjavadi, H.; Lingrel, J.B.; Heiny, J.A. Na,K-ATPase α2 activity in mammalian skeletal muscle T-tubules is acutely stimulated by extracellular K$^+$. *J. Gen. Physiol.* **2015**, *146*, 281–294. [CrossRef] [PubMed]

103. Kravtsova, V.V.; Petrov, A.M.; Matchkov, V.V.; Bouzinova, E.V.; Vasiliev, A.N.; Benziane, B.; Zefirov, A.L.; Chibalin, A.V.; Heiny, J.A.; Krivoi, I.I. Distinct α2 Na,K-ATPase membrane pools are differently involved in early skeletal muscle remodeling during disuse. *J. Gen. Physiol.* **2016**, *147*, 175–188. [CrossRef] [PubMed]

104. Kutz, L.C.; Mukherji, S.T.; Wang, X.; Bryant, A.; Larre, I.; Heiny, J.A.; Lingrel, J.B.; Pierre, S.V.; Xie, Z. Isoform-specific role of Na/K-ATPase α1 in skeletal muscle. *Am. J. Physiol. Endocrinol. Metab.* **2018**, *314*, E620–E629. [CrossRef] [PubMed]

105. Zhu, D.; Xiong, W.C.; Mei, L. Lipid rafts serve as a signaling platform for nicotinic acetylcholine receptor clustering. *J. Neurosci.* **2006**, *26*, 4841–4851. [CrossRef] [PubMed]

106. Willmann, R.; Pun, S.; Stallmach, L.; Sadasivam, G.; Santos, A.F.; Caroni, P.; Fuhrer, C. Cholesterol and lipid microdomains stabilize the postsynapse at the neuromuscular junction. *EMBO J.* **2006**, *25*, 4050–4060. [CrossRef] [PubMed]

107. Saks, V.; Monge, C.; Guzun, R. Philosophical basis and some historical aspects of systems biology: From Hegel to Noble—Applications for bioenergetic research. *Int. J. Mol. Sci.* **2009**, *10*, 1161–1192. [CrossRef] [PubMed]

108. Lingwood, D.; Simons, K. Lipid rafts as a membrane-organizing principle. *Science* **2010**, *327*, 46–50. [CrossRef] [PubMed]

109. Schoner, W.; Scheiner-Bobis, G. Endogenous and exogenous cardiac glycosides and their mechanisms of action. *Am. J. Cardiovasc. Drugs* **2007**, *7*, 173–189. [CrossRef] [PubMed]

110. Tian, J.; Xie, Z. The Na-K-ATPase and calcium-signaling microdomains. *Physiology* **2008**, *23*, 205–211. [CrossRef] [PubMed]

111. Lingrel, J.B. The physiological significance of the cardiotonic steroid/ouabain-binding site of the Na,K-ATPase. *Annu. Rev. Physiol.* **2010**, *72*, 395–412. [CrossRef] [PubMed]

112. Morrill, G.A.; Kostellow, A.B.; Askari, A. Caveolin-Na/K-ATPase interactions: Role of transmembrane topology in non-genomic steroid signal transduction. *Steroids* **2012**, *77*, 1160–1168. [CrossRef] [PubMed]

113. Cui, X.; Xie, Z. Protein Interaction and Na/K-ATPase-Mediated Signal Transduction. *Molecules* **2017**, *22*, 990. [CrossRef]

114. Wang, H.; Haas, M.; Liang, M.; Cai, T.; Tian, J.; Li, S.; Xie, Z. Ouabain assembles signaling cascades through the caveolar Na$^+$/K$^+$-ATPase. *J. Biol. Chem.* **2004**, *279*, 17250–17259. [CrossRef] [PubMed]

115. Cai, T.; Wang, H.; Chen, Y.; Liu, L.; Gunning, W.T.; Quintas, L.E.; Xie, Z.J. Regulation of caveolin-1 membrane trafficking by the Na/K-ATPase. *J. Cell Biol.* **2008**, *182*, 1153–1169. [CrossRef] [PubMed]

116. Razani, B.; Woodman, S.E.; Lisanti, M.P. Caveolae: From Cell Biology to Animal Physiology. *Pharmacol. Rev.* **2002**, *54*, 431–467. [CrossRef] [PubMed]

117. Chen, Y.; Cai, T.; Wang, H.; Li, Z.; Loreaux, E.; Lingrel, J.B.; Xie, Z. Regulation of intracellular cholesterol distribution by Na/K-ATPase. *J. Biol. Chem.* **2009**, *284*, 14881–14890. [CrossRef] [PubMed]

118. Chen, Y.; Li, X.; Ye, Q.; Tian, J.; Jing, R.; Xie, Z. Regulation of α1 Na/K-ATPase expression by cholesterol. *J. Biol. Chem.* **2011**, *286*, 15517–15524. [CrossRef] [PubMed]

119. Kravtsova, V.V.; Petrov, A.M.; Vasiliev, A.N.; Zefirov, A.L.; Krivoi, I.I. Role of cholesterol in the maintenance of endplate electrogenesis in rat diaphragm. *Bull. Exp. Biol. Med.* **2015**, *158*, 298–300. [CrossRef] [PubMed]

120. Petrov, A.M.; Kravtsova, V.V.; Matchkov, V.V.; Vasiliev, A.N.; Zefirov, A.L.; Chibalin, A.V.; Heiny, J.A.; Krivoi, I.I. Membrane lipid rafts are disturbed in the response of rat skeletal muscle to short-term disuse. *Am. J. Physiol. Cell Physiol.* **2017**, *312*, C627–C637. [CrossRef] [PubMed]

121. Habeck, M.; Haviv, H.; Katz, A.; Kapri-Pardes, E.; Ayciriex, S.; Shevchenko, A.; Ogawa, H.; Toyoshima, C.; Karlish, S.J. Stimulation, inhibition, or stabilization of Na,K-ATPase caused by specific lipid interactions at distinct sites. *J. Biol. Chem.* **2015**, *290*, 4829–4842. [CrossRef] [PubMed]

122. Lifshitz, Y.; Petrovich, E.; Haviv, H.; Goldshleger, R.; Tal, D.M.; Garty, H.; Karlish, S.J.D. Purification of the human α2 isoform of Na,K-ATPase expressed in *Pichia pastoris*. Stabilization by lipids and FXYD1. *Biochemistry* **2007**, *46*, 14937–14950. [CrossRef] [PubMed]

123. Kapri-Pardes, E.; Katz, A.; Haviv, H.; Mahmmoud, Y.; Ilan, M.; Khalfin-Penigel, I.; Carmeli, S.; Yarden, O.; Karlish, S.J.D. Stabilization of the α2 isoform of Na,K-ATPase by mutations in a phospholipid binding pocket. *J. Biol. Chem.* **2011**, *286*, 42888–42899. [CrossRef] [PubMed]

124. Allen, J.A.; Halverson-Tamboli, R.A.; Rasenick, M.M. Lipid raft microdomains and neurotransmitter signalling. *Nat. Rev. Neurosci.* **2007**, *8*, 128–140. [CrossRef] [PubMed]

125. Hicks, D.A.; Nalivaeva, N.N.; Turner, A.J. Lipid rafts and Alzheimer's disease: Protein-lipid interactions and perturbation of signaling. *Front. Physiol.* **2012**, *3*, 189. [CrossRef] [PubMed]

126. Sebastiao, A.M.; Colino-Oliveira, M.; Assaife-Lopes, N.; Dias, R.B.; Ribeiro, J.A. Lipid rafts, synaptic transmission and plasticity: Impact in age-related neurodegenerative diseases. *Neuropharmacology* **2013**, *64*, 97–107. [CrossRef] [PubMed]

127. Sooksawate, T.; Simmonds, M.A. Effects of membrane cholesterol on the sensitivity of the GABAA receptor to GABA in acutely dissociated rat hippocampal neurones. *Neuropharmacology* **2001**, *40*, 178–184. [CrossRef]

128. Eroglu, C.; Brugger, B.; Wieland, F.; Sinning, I. Glutamate-binding affinity of *Drosophila* metabotropic glutamate receptor is modulated by association with lipid rafts. *Proc. Natl. Acad. Sci. USA* **2003**, *100*, 10219–10224. [CrossRef] [PubMed]

129. Baenziger, J.E.; Domville, J.A.; Therien, J.P.D. The Role of Cholesterol in the Activation of Nicotinic Acetylcholine Receptors. *Curr. Top. Membr.* **2017**, *80*, 95–137. [CrossRef] [PubMed]

130. Burger, K.; Gimpl, G.; Fahrenholz, F. Regulation of receptor function by cholesterol. *Cell. Mol. Life Sci.* **2000**, *57*, 1577–1592. [CrossRef] [PubMed]

131. Brannigan, G.; Hénin, J.; Law, R.; Eckenhoff, R.; Klein, M.L. Embedded cholesterol in the nicotinic acetylcholine receptor. *Proc. Natl. Acad. Sci. USA* **2008**, *105*, 14418–14423. [CrossRef] [PubMed]

132. Barrantes, F.J. Cholesterol and nicotinic acetylcholine receptor: An intimate nanometer-scale spatial relationship spanning the billion year time-scale. *Biomed. Spectrosc. Imaging* **2016**, *5*, S67–S86. [CrossRef]

133. Scher, M.G.; Bloch, R.J. The lipid bilayer of acetylcholine receptor clusters of cultured rat myotubes is organized into morphologically distinct domains. *Exp. Cell. Res.* **1991**, *195*, 79–91. [CrossRef]

134. Marchand, S.; Devillers-Thiéry, A.; Pons, S.; Changeux, J.P.; Cartaud, J. Rapsyn escorts the nicotinic acetylcholine receptor along the exocytic pathway via association with lipid rafts. *J. Neurosci.* **2002**, *22*, 8891–8901. [CrossRef] [PubMed]

135. Campagna, J.A.; Fallon, J. Lipid rafts are involved in C95 (4,8) agrin fragment-induced acetylcholine receptor clustering. *Neuroscience* **2006**, *138*, 123–132. [CrossRef] [PubMed]

136. Stetzkowski-Marden, F.; Gaus, K.; Recouvreur, M.; Cartaud, A.; Cartaud, J. Agrin elicits membrane lipid condensation at sites of acetylcholine receptor clusters in C2C12 myotubes. *J. Lipid Res.* **2006**, *47*, 2121–2133. [CrossRef] [PubMed]

137. Vega-Moreno, J.; Tirado-Cortes, A.; Álvarez, R.; Irles, C.; Mas-Oliva, J.; Ortega, A. Cholesterol depletion uncouples β-dystroglycans from discrete sarcolemmal domains, reducing the mechanical activity of skeletal muscle. *Cell Physiol. Biochem.* **2012**, *29*, 905–918. [CrossRef] [PubMed]

138. Pato, C.; Stetzkowski-Marden, F.; Gaus, K.; Recouvreur, M.; Cartaud, A.; Cartaud, J. Role of lipid rafts in agrin-elicited acetylcholine receptor clustering. *Chem. Biol. Interact.* **2008**, *175*, 64–67. [CrossRef] [PubMed]

139. Nikolsky, E.E.; Zemkova, H.; Voronin, V.A.; Vyskocil, F. Role of non-quantal acetylcholine release in surplus polarization of mouse diaphragm fibres at the endplate zone. *J. Physiol.* **1994**, *477*, 497–502. [CrossRef] [PubMed]

140. Vyskocil, F.; Malomouzh, A.I.; Nikolsky, E.E. Non-quantal acetylcholine release at the neuromuscular junction. *Physiol. Res.* **2009**, *58*, 763–784. [PubMed]

141. Krivoi, I.I.; Drabkina, T.M.; Kravtsova, V.V.; Vasiliev, A.N.; Eaton, M.J.; Skatchkov, S.N.; Mandel, F. On the functional interaction between nicotinic acetylcholine receptor and Na$^+$,K$^+$-ATPase. *Pflugers Arch.* **2006**, *452*, 756–765. [CrossRef] [PubMed]

142. Bao, H.; Sun, H.; Xiao, Y.; Zhang, Y.; Wang, X.; Xu, X.; Liu, Z.; Fang, J.; Li, Z. Functional interaction of nicotinic acetylcholine receptors and Na$^+$/K$^+$ ATPase from *Locusta migratoria manilensis* (Meyen). *Sci. Rep.* **2015**, *5*, 8849. [CrossRef] [PubMed]

143. Hezel, M.; de Groat, W.C.; Galbiati, F. Caveolin-3 promotes nicotinic acetylcholine receptor clustering and regulates neuromuscular junction activity. *Mol. Biol. Cell* **2010**, *21*, 302–310. [CrossRef] [PubMed]

144. Galbiati, F.; Razani, B.; Lisanti, M.P. Caveolae and caveolin-3 in muscular dystrophy. *Trends Mol. Med.* **2001**, *7*, 435–441. [CrossRef]

145. Prince, R.J.; Sine, S.M. Acetylcholine and epibatidine binding to muscle acetylcholine receptors distinguish between concerted and uncoupled models. *J. Biol. Chem.* **1999**, *274*, 19623–19629. [CrossRef] [PubMed]

146. Mourot, A.; Rodrigo, J.; Kotzyba-Hibert, F.; Bertrand, S.; Bertrand, D.; Goeldner, M. Probing the reorganization of the nicotinic acetylcholine receptor during desensitization by time-resolved covalent labeling using [^{3}H]AC5, a photoactivatable agonist. *Mol. Pharmacol.* **2006**, *69*, 452–461. [CrossRef] [PubMed]

147. Radak, Z.; Suzuki, K.; Higuchi, M.; Balogh, L.; Boldogh, I.; Koltai, E. Physical exercise, reactive oxygen species and neuroprotection. *Free Radic. Biol. Med.* **2016**, *98*, 187–196. [CrossRef] [PubMed]

148. Dupuis, L.; Corcia, P.; Fergani, A.; Gonzalez De Aguilar, J.L.; Bonnefont-Rousselot, D.; Bittar, R.; Seilhean, D.; Hauw, J.J.; Lacomblez, L.; Loeffler, J.P.; et al. Dyslipidemia is a protective factor in amyotrophic lateral sclerosis. *Neurology* **2008**, *70*, 1004–1009. [CrossRef] [PubMed]

149. Dorst, J.; Kuhnlein, P.; Hendrich, C.; Kassubek, J.; Sperfeld, A.D.; Ludolph, A.C. Patients with elevated triglyceride and cholesterol serum levels have a prolonged survival in amyotrophic lateral sclerosis. *J. Neurol.* **2011**, *258*, 613–617. [CrossRef] [PubMed]

150. Zheng, Z.; Sheng, L.; Shang, H. Statins and amyotrophic lateral sclerosis: A systematic review and meta-analysis. *Amyotroph. Lateral. Scler. Frontotemporal. Degener.* **2013**, *14*, 241–245. [CrossRef] [PubMed]

151. Bigini, P.; Steffensen, K.R.; Ferrario, A.; Diomede, L.; Ferrara, G.; Barbera, S.; Salzano, S.; Fumagalli, E.; Ghezzi, P.; Mennini, T.; et al. Neuropathologic and biochemical changes during disease progression in liver X receptor beta$^{-/-}$ mice, a model of adult neuron disease. *J. Neuropathol. Exp. Neurol.* **2010**, *69*, 593–605. [CrossRef] [PubMed]

152. Andersson, S.; Gustafsson, N.; Warner, M.; Gustafsson, J.A. Inactivation of liver X receptor beta leads to adultonset motor neuron degeneration in male mice. *Proc. Natl. Acad. Sci. USA* **2005**, *102*, 3857–3862. [CrossRef] [PubMed]

153. Kim, H.J.; Fan, X.; Gabbi, C.; Yakimchuk, K.; Parini, P.; Warner, M.; Gustafsson, J.A. Liver X receptor beta (LXRbeta): A link between beta-sitosterol and amyotrophic lateral sclerosis-Parkinson's dementia. *Proc. Natl. Acad. Sci. USA* **2008**, *105*, 2094–2099. [CrossRef] [PubMed]

154. Mouzat, K.; Molinari, N.; Kantar, J.; Polge, A.; Corcia, P.; Couratier, P.; Clavelou, P.; Juntas-Morales, R.; Pageot, N.; Lobaccaro, J.-A.; et al. Liver X Receptor Genes Variants Modulate ALS Phenotype. *Mol. Neurobiol.* **2018**, *55*, 1959–1965. [CrossRef] [PubMed]

155. Kim, S.M.; Noh, M.Y.; Kim, H.; Cheon, S.Y.; Lee, K.M.; Lee, J.; Cha, E.; Park, K.S.; Lee, K.W.; Sung, J.J.; et al. 25-Hydroxycholesterol is involved in the pathogenesis of amyotrophic lateral sclerosis. *Oncotarget* **2017**, *8*, 11855–11867. [CrossRef] [PubMed]

156. Trias, E.; Ibarburu, S.; Barreto-Núñez, R.; Varela, V.; Moura, I.C.; Dubreuil, P.; Hermine, O.; Beckman, J.S.; Barbeito, L. Evidence for mast cells contributing to neuromuscular pathology in an inherited model of ALS. *JCI Insight* **2017**, *2*, e95934. [CrossRef] [PubMed]

157. Flis, D.J.; Dzik, K.; Kaczor, J.J.; Halon-Golabek, M.; Antosiewicz, J.; Wieckowski, M.R.; Ziolkowski, W. Swim Training Modulates Skeletal Muscle Energy Metabolism, Oxidative Stress, and Mitochondrial Cholesterol Content in Amyotrophic Lateral Sclerosis Mice. *Oxid. Med. Cell. Longev.* **2018**, *2018*, 5940748. [CrossRef] [PubMed]

158. Zhai, J.; Ström, A.L.; Kilty, R.; Venkatakrishnan, P.; White, J.; Everson, W.V.; Smart, E.J.; Zhu, H. Proteomic characterization of lipid raft proteins in amyotrophic lateral sclerosis mouse spinal cord. *FEBS J.* **2009**, *276*, 3308–3323. [CrossRef] [PubMed]

159. Milad, N.; White, Z.; Tehrani, A.Y.; Sellers, S.; Rossi, F.M.V.; Bernatchez, P. Increased plasma lipid levels exacerbate muscle pathology in the mdx mouse model of Duchenne muscular dystrophy. *Skelet. Muscle* **2017**, *7*, 19. [CrossRef] [PubMed]

160. Grounds, M.D.; Terrill, J.R.; Radley-Crabb, H.G.; Robertson, T.; Papadimitriou, J.; Spuler, S.; Shavlakadze, T. Lipid accumulation in dysferlin-deficient muscles. *Am. J. Pathol.* **2014**, *184*, 1668–1676. [CrossRef] [PubMed]

161. Kravtsova, V.V.; Timonina, N.A.; Zakir'yanova, G.F.; Sokolova, A.V.; Mikhailov, V.M.; Zefirov, A.L.; Krivoi, I.I. The Structural and Functional Characteristics of the Motor End Plates of Dysferlin-Deficient Mice. *Neurochem. J.* **2018**, *12*, 305–310. [CrossRef]

162. Kerr, J.P.; Ward, C.W.; Bloch, R.J. Dysferlin at transverse tubules regulates Ca^{2+} homeostasis in skeletal muscle. *Front. Physiol.* **2014**, *5*, 89. [CrossRef] [PubMed]

163. Demonbreun, A.R.; Allen, M.V.; Warner, J.L.; Barefield, D.Y.; Krishnan, S.; Swanson, K.E.; Earley, J.U.; McNally, E.M. Enhanced Muscular Dystrophy from Loss of Dysferlin Is Accompanied by Impaired Annexin A6 Translocation after Sarcolemmal Disruption. *Am. J. Pathol.* **2016**, *186*, 1610–1622. [CrossRef] [PubMed]

164. Nagy, N.; Nonneman, R.J.; Llanga, T.; Dial, C.F.; Riddick, N.V.; Hampton, T.; Moy, S.S.; Lehtimäki, K.K.; Ahtoniemi, T.; Puoliväli, J.; et al. Hip region muscular dystrophy and emergence of motor deficits in dysferlin-deficient Bla/J mice. *Physiol. Rep.* **2017**, *5*, e13173. [CrossRef] [PubMed]

165. Llanga, T.; Nagy, N.; Conatser, L.; Dial, C.; Sutton, R.B.; Hirsch, M.L. Structure-Based Designed Nano-Dysferlin Significantly Improves Dysferlinopathy in BLA/J Mice. *Mol. Ther.* **2017**, *25*, 2150–2162. [CrossRef] [PubMed]

166. Ambery, A.G.; Tackett, L.; Penque, B.A.; Brozinick, J.T.; Elmendorf, J.S. Exercise training prevents skeletal muscle plasma membrane cholesterol accumulation, cortical actin filament loss, and insulin resistance in C57BL/6J mice fed a western-style high-fat diet. *Physiol. Rep.* **2017**, *5*, e13363. [CrossRef] [PubMed]

167. Sellers, S.L.; Milad, N.; White, Z.; Pascoe, C.; Chan, R.; Payne, G.W.; Seow, C.; Rossi, F.; Seidman, M.A.; Bernatchez, P. Increased nonHDL cholesterol levels cause muscle wasting and ambulatory dysfunction in the mouse model of LGMD2B. *J. Lipid Res.* **2018**, *59*, 261–272. [CrossRef] [PubMed]

168. Nagaraj, R.Y.; Nosek, C.M.; Brotto, M.A.; Nishi, M.; Takeshima, H.; Nosek, T.M.; Ma, J. Increased susceptibility to fatigue of slow- and fast-twitch muscles from mice lacking the MG29 gene. *Physiol. Genom.* **2000**, *4*, 43–49. [CrossRef] [PubMed]

169. Brandt, N.R.; Franklin, G.; Brunschwig, J.P.; Caswell, A.H. The role of mitsugumin 29 in transverse tubules of rabbit skeletal muscle. *Arch. Biochem. Biophys.* **2001**, *385*, 406–409. [CrossRef] [PubMed]

170. Brotto, M.A.; Nagaraj, R.Y.; Brotto, L.S.; Takeshima, H.; Ma, J.J.; Nosek, T.M. Defective maintenance of intracellular Ca^{2+} homeostasis is linked to increased muscle fatigability in the MG29 null mice. *Cell Res.* **2004**, *14*, 373–378. [CrossRef] [PubMed]

171. Weisleder, N.; Brotto, M.; Komazaki, S.; Pan, Z.; Zhao, X.; Nosek, T.; Parness, J.; Takeshima, H.; Ma, J. Muscle aging is associated with compromised Ca^{2+} spark signaling and segregated intracellular Ca^{2+} release. *J. Cell Biol.* **2006**, *174*, 639–645. [CrossRef] [PubMed]

172. Pouvreau, S.; Berthier, C.; Blaineau, S.; Amsellem, J.; Coronado, R.; Strube, C. Membrane cholesterol modulates dihydropyridine receptor function in mice fetal skeletal muscle cells. *J. Physiol.* **2004**, *555*, 365–381. [CrossRef] [PubMed]

173. Jiao, H.; Arner, P.; Gerdhem, P.; Strawbridge, R.J.; Näslund, E.; Thorell, A.; Hamsten, A.; Kere, J.; Dahlman, I. Exome sequencing followed by genotyping suggests SYPL2 as a susceptibility gene for morbid obesity. *Eur. J. Hum. Genet.* **2015**, *23*, 1216–1222. [CrossRef] [PubMed]

174. Rudolf, R.; Khan, M.M.; Labeit, S.; Deschenes, M.R. Degeneration of neuromuscular junction in age and dystrophy. *Front. Aging Neurosci.* **2014**, *6*, 99. [CrossRef] [PubMed]

175. Tintignac, L.A.; Brenner, H.R.; Rüegg, M.A. Mechanisms regulating neuromuscular junction development and function and causes of muscle wasting. *Physiol. Rev.* **2015**, *95*, 809–852. [CrossRef] [PubMed]

176. Willadt, S.; Nash, M.; Slater, C.R. Age-related fragmentation of the motor endplate is not associated with impaired neuromuscular transmission in the mouse diaphragm. *Sci. Rep.* **2016**, *6*, 24849. [CrossRef] [PubMed]

177. Kravtsova, V.V.; Matchkov, V.V.; Bouzinova, E.V.; Vasiliev, A.N.; Razgovorova, I.A.; Heiny, J.A.; Krivoi, I.I. Isoform-specific Na,K-ATPase alterations precede disuse-induced atrophy of rat soleus muscle. *BioMed Res. Int.* **2015**, *2015*, 720172. [CrossRef] [PubMed]

178. Nikolova-Karakashian, M.N.; Reid, M.B. Sphingolipid metabolism, oxidant signaling, and contractile function of skeletal muscle. *Antioxid. Redox Signal.* **2011**, *15*, 2501–2517. [CrossRef] [PubMed]

179. Bryndina, I.G.; Shalagina, M.N.; Sekunov, A.V.; Zefirov, A.L.; Petrov, A.M. Clomipramine counteracts lipid raft disturbance due to short-term muscle disuse. *Neurosci. Lett.* **2018**, *664*, 1–6. [CrossRef] [PubMed]

Int. J. Mol. Sci. **2019**, *20*, 1046

180. Yu, C.; Alterman, M.; Dobrowsky, R.T. Ceramide displaces cholesterol from lipid rafts and decreases the association of the cholesterol binding protein caveolin-1. *J. Lipid Res.* **2005**, *46*, 1678–1691. [CrossRef] [PubMed]
181. Hardie, D.G.; Schaffer, B.E.; Brunet, A. AMPK: An energy-sensing pathway with multiple inputs and outputs. *Trends Cell Biol.* **2016**, *26*, 190–201. [CrossRef] [PubMed]
182. Habegger, K.M.; Hoffman, N.J.; Ridenour, C.M.; Brozinick, J.T.; Elmendorf, J.S. AMPK enhances insulin-stimulated GLUT4 regulation via lowering membrane cholesterol. *Endocrinology* **2012**, *153*, 2130–2141. [CrossRef] [PubMed]
183. Carnio, S.; LoVerso, F.; Baraibar, M.A.; Longa, E.; Khan, M.M.; Maffei, M.; Reischl, M.; Canepari, M.; Loefler, S.; Kern, H.; et al. Autophagy impairment in muscle induces neuromuscular junction degeneration and precocious aging. *Cell Rep.* **2014**, *8*, 1509–1521. [CrossRef] [PubMed]
184. Cerveró, C.; Montull, N.; Tarabal, O.; Piedrafita, L.; Esquerda, J.E.; Calderó, J. Chronic treatment with the AMPK agonist AICAR prevents skeletal muscle pathology but fails to improve clinical outcome in a mouse model of severe spinal muscular atrophy. *Neurotherapeutics* **2016**, *13*, 198–216. [CrossRef] [PubMed]
185. Khan, M.M.; Strack, S.; Wild, F.; Hanashima, A.; Gasch, A.; Brohm, K.; Reischl, M.; Carnio, S.; Labeit, D.; Sandri, M.; et al. Role of autophagy, SQSTM1, SH3GLB1, and TRIM63 in the turnover of nicotinic acetylcholine receptors. *Autophagy* **2014**, *10*, 123–136. [CrossRef] [PubMed]
186. Chibalin, A.V.; Benziane, B.; Zakyrjanova, G.F.; Kravtsova, V.V.; Krivoi, I.I. Early endplate remodeling and skeletal muscle signaling events following rat hindlimb suspension. *J. Cell. Physiol.* **2018**, *233*, 6329–6336. [CrossRef] [PubMed]
187. Vilchinskaya, N.A.; Mochalova, E.P.; Nemirovskaya, T.L.; Mirzoev, T.M.; Turtikova, O.V.; Shenkman, B.S. Rapid decline in MyHC I(β) mRNA expression in rat soleus during hindlimb unloading is associated with AMPK dephosphorylation. *J. Physiol.* **2017**, *595*, 7123–7134. [CrossRef] [PubMed]
188. Vilchinskaya, N.A.; Krivoi, I.I.; Shenkman, B.S. AMP-Activated Protein Kinase as a Key Trigger for the Disuse-Induced Skeletal Muscle Remodeling. *Int. J. Mol. Sci.* **2018**, *19*, 3558. [CrossRef] [PubMed]

International Journal of
Molecular Sciences

Article

Molecular Characterization of Microtubule Affinity-Regulating Kinase4 from *Sus scrofa* and Promotion of Lipogenesis in Primary Porcine Placental Trophoblasts

Liang Tian [1,*], Aiyou Wen [2], Shusheng Dong [1] and Peishi Yan [1]

[1]   College of Animal Science and Technology, Nanjing Agricultural University, Nanjing 210095, China;
      2014105052@njau.edu.cn (S.D.); yanps@njau.edu.cn (P.Y.)
[2]   College of Animal Science, Anhui Science and Technology University, Fengyang 233100, China;
      aywen2008@126.com
*   Correspondence: tianliang2013@njau.edu.cn

Received: 31 January 2019; Accepted: 5 March 2019; Published: 9 March 2019

Abstract: This study aimed to characterize the full-length cDNA of MARK4 in *Sus scrofa*, and evaluated its potential role in the regulation of lipid accumulation in pig placental trophoblasts and analyzed signaling pathways involved, thereby providing insights into mechanisms for placental lipotoxicity induced by excessive back-fat during pregnancy of sows. The cDNA obtained with 5′ and 3′ RACE amplification covered 3216 bp with an open reading frame of 2259 bp encoding 752 amino acids. Multiple alignments and phylogenetic analysis revealed MARK4 protein of *Sus scrofa* had a high homology (95%–99%) to that of other higher vertebrates. After transfection, enhanced MARK4 significantly promoted lipogenesis in pig trophoblasts, as evidenced by accelerated lipid accumulation and consistently increased mRNA expressions of lipogenic genes DGAT1, LPIN1, LPIN3, LPL, PPARδ and SREBP-1c. Meanwhile, PPARγ remarkably inhibited the stimulating effect of MARK4 on non-receptor-mediated lipid accumulation in trophoblasts. Further analyses revealed WNT signaling enhanced lipid accumulation and activation of MARK4 in pig trophoblast cells. Finally, we demonstrated that WNT/β-catenin signal pathway is involved in MARK4 activated lipogenesis. These results suggest that MARK4 promotes lipid accumulation in porcine placental trophoblasts and can be considered as a potential regulator of lipotoxicity associated with maternal obesity in the pig placenta.

Keywords: MARK4; pig; lipogenesis; placenta; WNT; molecular cloning; PPARγ

---

1. Introduction

Obese pregnancy has been demonstrated to provoke an adverse intrauterine milieu, and as a result, poor pregnancy outcomes in human beings and some animal species, such as pig [1,2]. Although the connection between fetal development and maternal obesity is confirmed, the underlying mechanisms connecting adverse maternal environment to the fetus remain elusive. As the interface between the fetus and maternal environment, the placenta has become an important source of pathogenic factors affecting fetal metabolism and development [3,4]. Recently, several studies suggested that maternal obesity during pregnancy is associated with elevated maternal circulating levels of fatty acids and inflammatory cytokines, resulting in a lipotoxic milieu within the placenta characterized by increased placental lipid, inflammation and oxidative stress [5–7]. Lipotoxicity has been demonstrated to induce placental dysfunction evidenced by maternal obesity associated dysregulation of lipid transport and metabolism in the human or pig full-term placenta [8,9]. Recent evidences further revealed that maternal obesity contributes to decreased placental efficiency (a ratio of fetal weight to placental weight) and excessive placental fat accumulation through an aberrant activation of WNT signaling and PPARδ in placenta from an obesity-prone rat model [10], thus leading to compromised fetal development. Furthermore, our studies showed that WNT signaling and inflammatory NF-κB and JNK signaling are activated in term placenta from sows with excessive back-fat [11], suggesting that maternal obesity may induce lipotoxicity in the full-term porcine placenta. However, the precise cellular and molecular mechanisms responsible for maternal obesity associated lipid accumulation in the pig placenta are still barely understood.

Microtubule affinity-regulating kinase 4 (MARK4) is a member of the AMP-activated protein kinase (AMPK)-related family of kinases, which has been reported to expressed in multiple tissues [12]. As the mammalian homologs of nematode Par-1, microtubule affinity regulatory kinases (MARKs) family contains four members, MARK1(Par-1c), MARK2(Par-1b/EMK1), MARK3(Par-1a/C-TAK1) and MARK4 (Par-1d/MARKL-1), and they share a highly conserved structure consisting of three distinct domains: a catalytic kinase domain, a ubiquitin-associated domain and a kinase associated domain [13]. Studies have implicated Mark4 in diverse physiological processes, including regulation of programmed cell death [14], cell proliferation [15], and glucose homeostasis and energy metabolism [16]. Recent evidences demonstrate that MARK4 promotes adipogenesis and triggers adipocytes apoptosis along with increased adipose inflammation and oxidative stress [17,18]. Moreover, our findings indicated that excessive back-fat is associated with increased activation of MARK4 in pig term placenta, suggesting a potential mechanism for increased activation of JNK mediated mitochondrial apoptotic pathway [19]. All these findings suggest that MARK4 is a versatile protein involved in large number of metabolic processes. However, the regulatory role of MARK4 on placental lipid accumulation, especially in maternal obese condition, is still unknown in porcine. To date, MARK4 gene has been characterized molecularly in several vertebrate species, including pigs [13,20], while the knowledge of molecular structure of MARK4 in *Sus scrofa* (Pig) is still limited, as warrants further studies.

Given the regulatory role played by MARK4 in adipogenesis and energy metabolism, we aimed to evaluate whether MARK4 expression is correlated with lipid accumulation in pig placental trophoblast cells *in vitro*. In addition, we cloned the full-length cDNA of the MARK4 gene from the placenta of porcine using 5′ and 3′ RACE amplification and employed bioinformatics analysis to identify the molecular characterization and structure of MARK4 from *Sus scrofa*. In this study, we demonstrated that, through activating the WNT/β-catenin and inhibiting the PPARγ pathways, MARK4 promoted lipogenesis in pig placental trophoblasts, implicating MARK4 as a potential regulator of lipid accumulation associated with maternal obesity in the pig placenta.

2. Results

2.1. Molecular Characterization of MARK4 Gene

After performing core fragment amplification and 5′ and 3′ RACE, the full-length cDNA of MARK4 gene (GenBank accession number: MH926032) from *Sus scrofa* was obtained (Figure S1). The full-length cDNA covered 3216 bp with an ORF of 2259 bp encoding 752 amino acids. The MARK4 protein had a calculated molecular weight (Mw) of 82535.70 Da and isoelectric point (PI) of 9.70. This amino acid (AA) sequence contained several conserved functional sites, including one proton acceptor (Asp181), one protein kinase ATP-binding region signature (IIe65-Lys88), one serine/threonine protein kinase active-site signature (IIe177-Leu189) and one protein kinase domain (Tyr59-IIe310). Based on the results predicted by the online SABLE program, the secondary structure of this MARK4 protein consisted of 13 α-helices, 13 β-strands and 26 coils (Figure S2).

Additionally, conserved motifs were identified in the amino acid sequence of the MARK4 protein, including the activation loop, the catalytic kinase domain (KD), the ubiquitin-associated domain (UBA), the kinase associated domain1 (KA1) and three conserved functional sites (lysine 88 ATP binding site, aspartic 181 active site and threonine 214 phosphorylation site; Figure 1). This MARK4 protein sequence had a high similarity, and showed similar structural features to the MARK4 protein of other species (Figure S3).

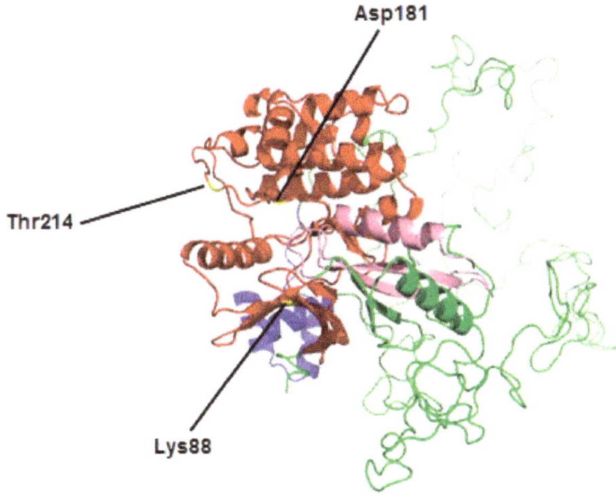

Figure 1. The tertiary protein structures of MARK4 protein in Pig (*Sus scrofa*) modeled by the ProModII program. Kinase domain (KD) colored red, ubiquitin-associated domain (UBA) is blue and kinase associated domain1 (KA1) in pink.

2.2. Phylogenetic Analysis

The phylogenetic tree among 11 species based on the amino acid (AA) sequences of MARK4 protein was presented in Figure S4. MARK4 of pig (*Sus scrofa*) showed a close phylogenetic relationship with that of human (*Homo sapiens*) and chimpanzee (*Pan troglodytes*). Conservation of MARK4 was also evident from similarity comparisons in NCBI, as the MARK4 protein of *Sus scrofa* showed a high identity (95%–99%) to that of David's myotis (*Myotis davidii*), Chimpanzee (*Pan troglodytes*), American beaver (*Castor canadensis*), Domestic guinea pig (*Cavia porcellus*), Norway rat (*Rattus norvegicus*), House mouse (*Mus musculus*), Dingo (*Canis lupus dingo*), Horse (*Equus caballus*) and Human (*Homo sapiens*).

2.3. MARK4 Increases Lipid Droplet Accumulation in Pig Placental Trophoblast Cells

In this study, we speculated that MARK4 could modulate lipid accumulation in porcine placental trophoblast cells. To validate our hypothesis, we initially tested whether overexpression of MARK4 influences the accumulation of fatty acid in cultured term primary pig trophoblasts exposed to 400 µM FA. The results of Bodipy 493/503 fluorescence staining and TG content assay indicated that lipid droplet accumulation was significantly increased in trophoblasts from the Myc- MARK4 group compared with the sh- MARK4 or vector control groups ($p < 0.05$; control panel in Figure 2A,B).

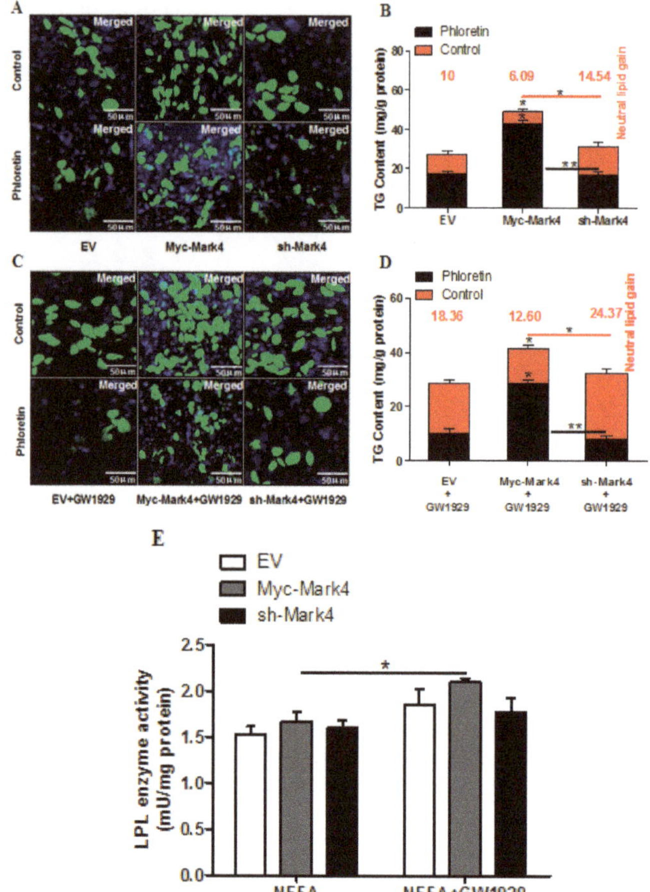

Figure 2. MARK4 promotes lipid accumulation in pig primary trophoblast cells challenged with 400 μM NEFA. (A and C) Representative images (100×) of Bodipy staining after transfection with Myc-MARK4, sh-MARK4 for 48 h in primary (trophoblast cells) isolated from pig placentas. Primary trophoblasts were then incubated with 400 μM NEFA, 2 μM GW1929 or 500 μM phloretin for 24 h ($n = 3$). (**B** and **D**) Quantification of corresponding triglyceride (TG) in (**A**) and (**C**) by ELISA analysis ($n = 3$). The values in red indicate receptor (transport proteins)-mediated fatty acid accumulation by subtracting the values in the presence of phloretin from those in the absence of phloretin. (E) LPL activity (mU/mg protein) after transfection with Myc-MARK4, sh-MARK4 for 48 h in pig primary trophoblasts. Cells were then treated with 400 μM NEFA or 2 μM GW1929 for 24 h ($n = 3$). Values are expressed as mean ± SEM. ** $p < 0.01$; * $p < 0.05$ compared with the control group. Myc-MARK4 group: overexpression of MARK4 group, sh-MARK4 group: knock down of MARK4 group, Control: empty vector (EV) group.

We next examined whether MARK4 affected receptor (transport proteins)-mediated fatty acid accumulation in cultured trophoblast cells. As shown in Figure 2B, sh-MARK4 treatment increased receptor-mediated fatty acid accumulation in trophoblasts compared with Myc-MARK4 group following 24 h exposure to FA (sh-MARK4: 14.54 ± 2.41 mg/g versus Myc-MARK4: 6.09 ± 1.61 mg/g, $p < 0.05$). Previous studies have shown that PPARγ is involved in regulating fatty acid transport and accumulation in primary human placental trophoblasts [21]. We therefore hypothesized that activation of PPARγ might increase the accumulation of fatty acid in cultured pig placental trophoblast cells. To test this hypothesis, we incubated trophoblasts in the presence or absence of PPARγ-specific agonist GW1929. As shown in Figure 2B,D, activation of PPARγ promoted receptor-mediated fatty acid accumulation in sh-MARK4 treatment following 24 h exposure to FA (sh-MARK4+GW1929: 24.37 ± 1.39 mg/g versus sh-MARK4: 14.54 ± 2.41 mg/g, $p < 0.05$), whereas non- receptor-mediated fatty acid accumulation was significantly decreased in Myc-MARK4 group following GW1929 + phloretin treatment (Myc-MARK4+GW1929: 28.75 ± 1.03 mg/g versus Myc-MARK4: 42.87 ± 1.89 mg/g, $p < 0.05$). In accord with increased receptor-mediated fatty acid accumulation in Myc-MARK4+GW1929 group (Myc-MARK4+GW1929: 12.60 ± 1.22 mg/g versus Myc-MARK4: 6.09 ± 1.61 mg/g, $p < 0.05$), the LPL activity in Myc-MARK4 + GW1929 group was markedly higher than that in Myc-MARK4 group ($p < 0.05$; Figure 2E).

2.4. Effect of MARK4 on Key Factors of Lipid Metabolism in Pig Placental Trophoblasts

We first determined the overexpression of MARK4 by testing protein content of MARK4 gene following transfection and FA treatment. As shown in Figure 3A,B, MARK4 protein increased in Myc-MARK4 group, while sh-MARK4 treatment reduced MARK4 protein ($p < 0.05$). Consistent with increased lipid droplet accumulation following FA treatment, the mRNA expression of genes associated with fatty acid uptake and accumulation, including LPL and DGAT1, was significantly increased in Myc-MARK4 group, whereas the mRNA content of lipid metabolism-related genes, including PPARG (PPARγ), ADRP and ACSL1, was reduced in Myc-MARK4 group compared with the sh-MARK4 or vector control groups ($p < 0.05$; Figure 3D). GW1929, the potent and specific agonist of PPARγ (Figure 3C), was used to examine the regulatory role of PPARγ on MARK4-induced increases in lipid accumulation of trophoblasts. As shown in Figure 3D, GW1929 promoted the mRNA expression of PPARG, ADRP and ACSL1 in Myc-MARK4 group, but the mRNA content of LPL and DGAT1 was decreased in Myc-MARK4+ GW1929 treatment ($p < 0.05$). In accordance with elevated receptor-mediated fatty acid accumulation following GW1929 + sh- MARK4 treatment, GW1929 increased the mRNA content of several fatty acid transporters, including FATP1, FATP4, CD36, FABP1 and FABP4, in sh-MARK4 group ($p < 0.05$; Figure 3E).

2.5. WNT Signaling Promotes Lipid Accumulation and Activation of MARK4 in Pig Trophoblasts

Previous experiments in our laboratory and others have shown that an aberrant activation of WNT signaling contributes to significant placental lipid accumulation in obese model of rat or pig [10,11]. In order to further reveal the mechanisms responsible for the increased placental lipid accumulation induced by WNT signaling, we first performed Bodipy fluorescence staining to evaluate lipid droplet accumulation in pig trophoblasts from three groups: Flag-DKK1, sh-DKK1 and Vector control. DKK1 (dickkopf family protein1) is an inhibitor of the canonical WNT signaling pathway [22]. As shown in Figure 4A,B, Flag-DKK1 treatment reduced lipid droplet accumulation in trophoblasts following 24 h exposure to FA ($p < 0.05$), whereas activation of WNT signaling by GSK3β inhibitor LiCL, which was the downstream of DKK1 and blocked the phosphorylation of β-catenin and subsequent proteolytic degradation, significantly increased lipid accumulation in sh-DKK1 treatment ($p < 0.01$). The LPL activity was not affected by Flag-DKK1 or sh-DKK1 treatment in the presence or absence of LiCL (Figure 4C).

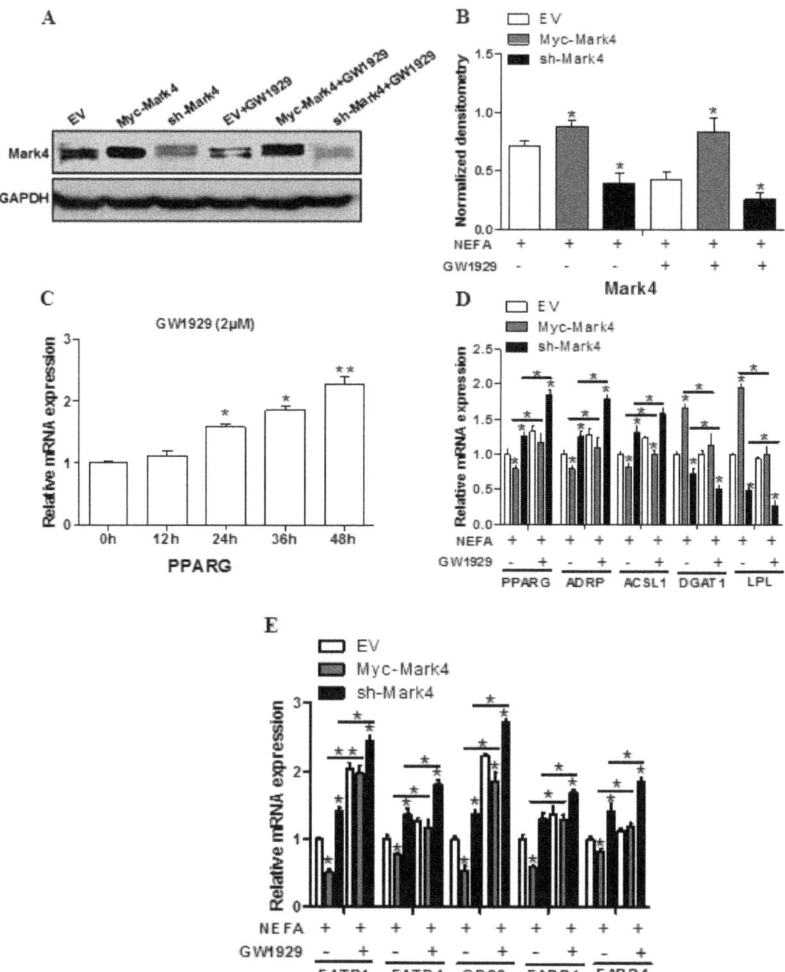

Figure 3. Effects of MARK4 on key molecules of lipid metabolism in pig primary trophoblast cells. (**A–B**) Representative immunoblots and densitometric quantification for MARK4 after transfection with Myc-MARK4, sh-MARK4 for 48 h in primary trophoblast cells isolated from pig placentas. Cells were then incubated with 400 μM NEFA or 2 μM GW1929 for 24 h (n = 3). (**C**) Primary trophoblasts were cultured and incubated for 0 h, 12 h, 24 h, 36 h and 48 h in the presence of 2 μM GW1929. Relative mRNA expression of PPARγ was detected (n = 3). (**D–E**) Relative mRNA expression of lipid metabolism-related genes (**D**) and fatty acid (FA) transporters (**E**) after transfection with Myc-MARK4, sh-MARK4 for 48 h in primary (trophoblast cells). Cells were then treated with 400 μM NEFA or 2 μM GW1929 for 24 h (n = 3). Values are expressed as mean $\pm$ SEM. ** $p < 0.01$; * $p < 0.05$ compared with the control group. Myc-MARK4 group: over expression of MARK4 group, sh-MARK4 group: knock down of MARK4 group, Control: empty vector (EV) group.

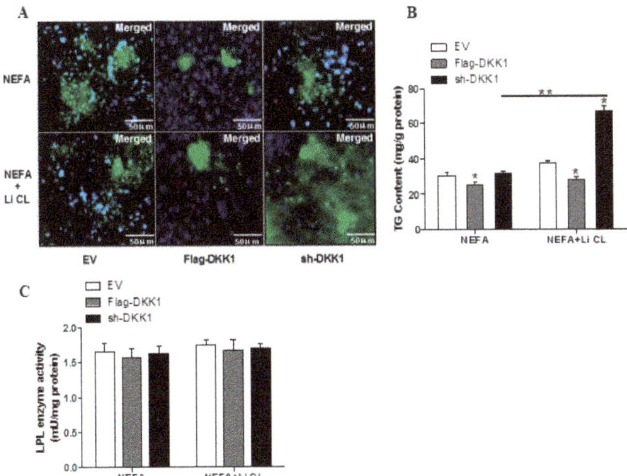

Figure 4. Activation of the Wnt/β-catenin pathway promotes lipid accumulation in pig primary trophoblast cells challenged with 400 μM NEFA. (**A**) Representative images (100×) of Bodipy staining after transfection with Flag-DKK1, sh-DKK1 for 48 h in primary trophoblast cells isolated from pig placentas. Cells were then incubated with 400 μM NEFA or 20 μM Li CL for 24 h (*n* = 3). (**B**) Quantification of corresponding triglyceride (TG) in (**A**) by ELISA analysis (*n* = 3). (**C**) LPL activity (mU/mg protein) after transfection with Flag-DKK1, sh-DKK1for 48 h in pig primary trophoblasts. Cells were then treated with400 μM NEFA or 20 μM Li CL for 24 h (*n* = 3). Values are expressed as mean ± SEM. ** $p < 0.01$; * $p < 0.05$ compared with the control group. Flag-DKK1 group: over expression of DKK1 group, sh-DKK1 group: knock down of DKK1 group, Control: empty vector (EV) group.

We next determined whether inhibition of WNT signaling affected lipid metabolism in pig placental trophoblasts. Not surprisingly, overexpression of DKK1 increased DKK1 protein content ($p < 0.05$; Figure 5A,B) and reduced β-catenin protein expression within the nucleus (Figure 5C). Notably, LiCL treatment prevented DKK1-induced degradation of β-catenin ($p < 0.05$; Figure 5D); this result was also confirmed by immunofluorescence assay for β-catenin ($p < 0.05$; Figure 5E,F). Consistent with elevated lipid accumulation in sh-DKK1 group following exposure to FA + LiCL, the mRNA expression of genes associated with TG synthesis, including DGAT1, LPL, LPIN3 and PPARδ, were higher in sh-DKK1 + LiCL treatment ($p < 0.05$; Figure 5G), while LiCL treatment reduced the mRNA content of fatty acid transport -related genes, including PPARγ, FATP1, FATP4, CD36 and FABP4, in Flag-DKK1 or sh-DKK1 group ($p < 0.05$; Figure 5G,H). Moreover, phos- MARK4(Thr214) was decreased in Flag-DKK1 compared with the sh-DKK1 or vector control groups, but increased activation of Mark4 was observed in Flag-DKK1+ LiCL treatment ($p < 0.05$; Figure 5A,B).

2.6. WNT/β-Catenin Signal is Essential for MARK4 Activated Lipogenesis in Pig Trophoblast Cells

Having determined that WNT signaling enhanced the accumulation of fatty acids and activation of MARK4 in pig placental trophoblast cells, we next addressed whether WNT/β-catenin pathway was involved in Mark4-induced lipid accumulation in pig trophoblasts. To test this hypothesis, we incubated trophoblasts in the presence or absence of WNT signaling pathway specific inhibitor JW74. As shown in Figure 6B,D, inhibition of WNT/β-catenin signaling by JW74 reduced non-receptor-mediated fatty acid accumulation in sh-MARK4 group following 24 h exposure to FA + phloretin (sh-MARK4+JW74: 3.56 ± 0.80 mg/g versus sh-MARK4: 16.47 ± 1.61 mg/g, $p < 0.05$), whereas receptor-mediated fatty acid accumulation was significantly increased in Myc-MARK4 group following JW74 treatment (Myc-MARK4+ JW74: 9.76 ± 0.90 mg/g versus Myc-MARK4: 4.79 ± 1.85 mg/g, $p < 0.05$). No differences were found in the LPL activity among Myc-MARK4, sh-MARK4 and Vector control in the presence or absence of JW74 (Figure 6E).

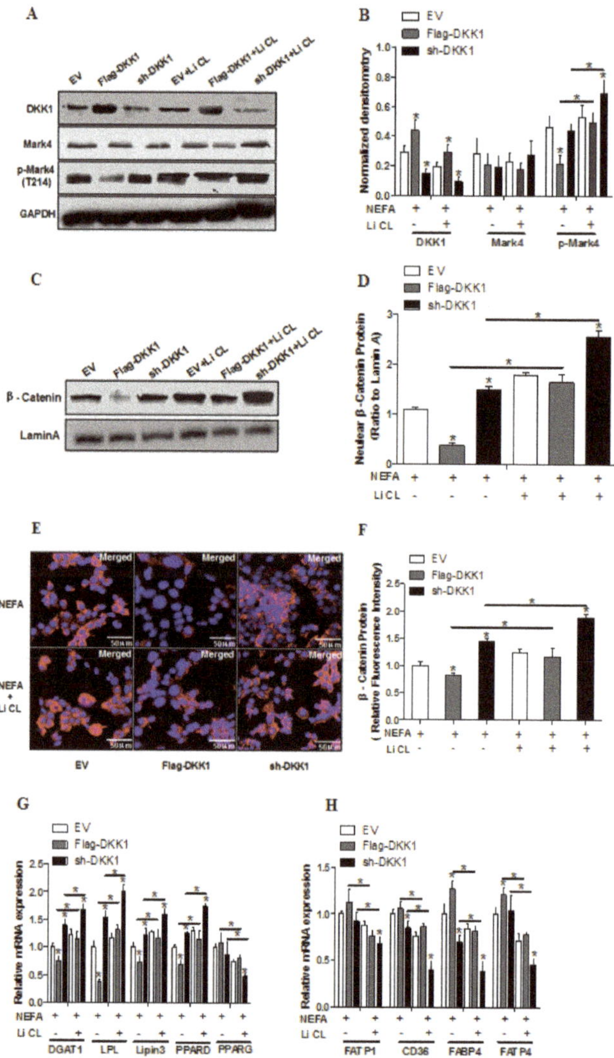

Figure 5. Inhibition of the Wnt/β-catenin pathway blocks key molecules of lipid metabolism and activation of MARK4 in pig primary trophoblast cells. (**A–D**) Representative immunoblots and densitometric quantification for p-MARK4 (T214), DKK1 and β-catenin after transfection with Flag-DKK1, sh-DKK1 for 48 h in primary trophoblast cells isolated from pig placentas. Cells were then incubated with 400 μM NEFA or 20 μM Li CL for 24 h (n = 3). (**E**) Representative images (100×) of β-catenin immunofluorescent staining after transfection with Flag-DKK1, sh-DKK1 for 48 h in pig primary trophoblast cells. Cells were then incubated with 400 μM NEFA or 20 μM Li CL for 24 h (n = 3). (**F**) Quantification of red fluorescence intensity in (**E**) relative to control group (n = 3). (**G–H**) Relative mRNA expression of lipid metabolism-related genes (**G**) and fatty acid (FA) transporters (**H**) after transfection with Flag-DKK1, sh-DKK1 for 48 h in primary trophoblast cells. Cells were then treated with 400 μM NEFA or 20 μM Li CL for 24 h (n = 3). Values are expressed as mean ± SEM. * p < 0.05 compared with the control group. Flag-DKK1 group: overexpression of DKK1 group, sh-DKK1 group: knock down of DKK1 group, Control: empty vector (EV) group.

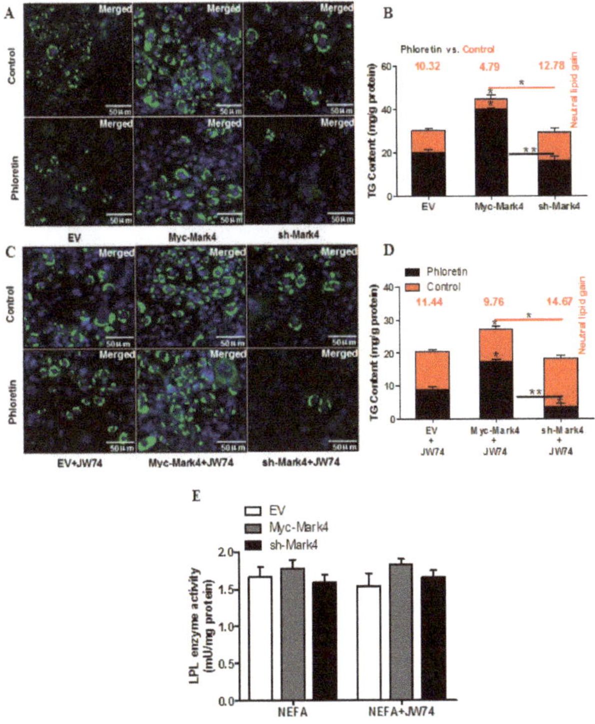

Figure 6. Activation of the WNT/β-catenin pathway by MARK4 promotes lipid accumulation in pig primary trophoblast cells challenged with 400 μM NEFA. (**A** and **C**) Representative images (100×) of Bodipy staining after transfection with Myc-MARK4, sh-MARK4 for 48 h in primary (trophoblast cells) isolated from pig placentas. Cells were then incubated with 400 μM NEFA, 10 μM JW74 or 500 μM phloretin for 24 h (*n* = 3). (**B** and **D**) Quantification of corresponding triglyceride (TG) in (**A**) and (**C**) by ELISA analysis (*n* = 3). The values in red indicate receptor (transport proteins)-mediated fatty acid accumulation by subtracting the values in the presence of phloretin from those in the absence of phloretin. (**E**) LPL activity (mU/mg protein) after transfection with Myc-MARK4, sh-MARK4 for 48 h in pig primary trophoblasts. Cells were then treated with 400 μM NEFA or 10 μM JW74 for 24 h (*n* = 3). Values are expressed as mean ± SEM. ** $p < 0.01$; * $p < 0.05$ compared with the control group. Myc-MARK4 group: overexpression of MARK4 group, sh-MARK4 group: knock down of MARK4 group, Control: empty vector (EV) group.

We further confirmed the role of WNT signaling on MARK4 activated lipogenesis in trophoblasts by Western blot analysis. Specifically, overexpression of MARK4 increased the protein contents of Mark4 and β-catenin ($p < 0.05$; Figure 7A,C), while no changes were noted for DKK1 expression in Myc-MARK4 or sh-MARK4 treatment in the presence or absence of JW74 (Figure 7B). Despite with JW74 treatment, MARK4 still increased the content of β-catenin within the nucleus ($p < 0.05$; Figure 7D). In accordance with increased receptor-mediated fatty acid accumulation in Myc-MARK4 + JW74 treatment, the mRNA expression of genes associated with fatty acid transport, including ACSL1, ADRP, PPARγ, FATP1, FATP4, CD36, FABP1 and FABP4, were up-regulated in Myc- MARK4 group following exposure to JW74($p < 0.05$; Figure 7E,G), whereas the mRNA content of genes associated with TG and lipid droplet synthesis, including ACACA, FASN, DGAT1, LPIN1, LPIN3, LPL, PPARδ and SREBP-1c, were decreased in sh-MARK4 group following JW74 treatment ($p < 0.05$; Figure 7E,F,H), in agreement with reduced non- receptor-mediated fatty acid accumulation in sh-MARK4 + JW74 group following exposure to FA + phloretin.

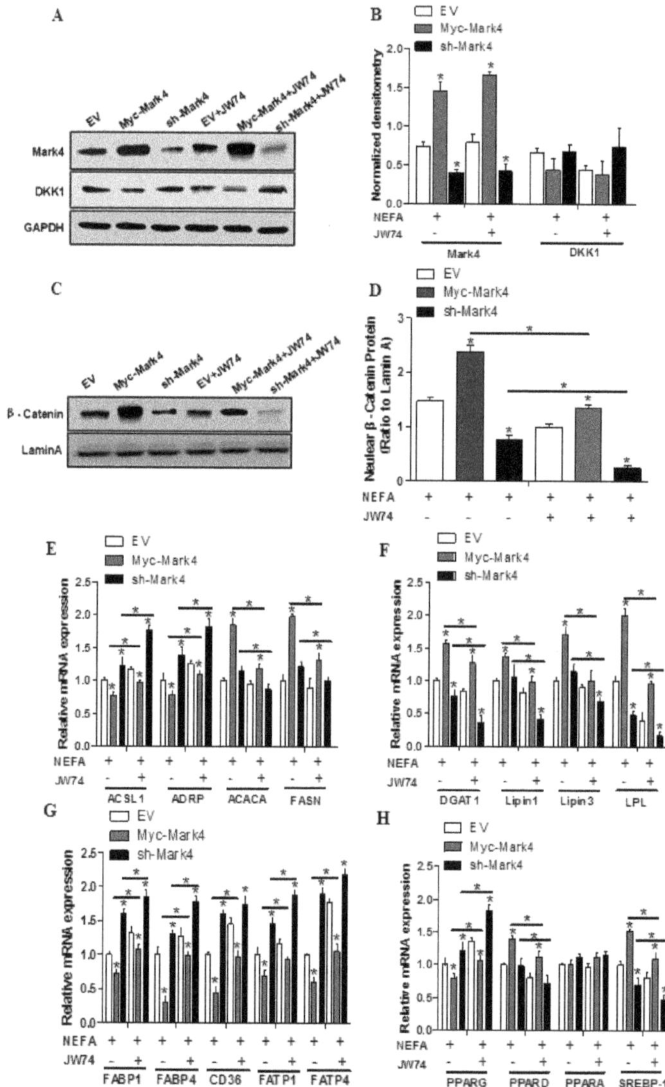

Figure 7. Activation of the WNT/β-catenin pathway by MARK4 promotes lipogenesis in pig primary trophoblast cells challenged with 400 µM NEFA. (**A–D**) Representative immunoblots and densitometric quantification for MARK4, DKK1 and β-catenin after transfection with Myc-MARK4, sh-MARK4 for 48 h in primary trophoblast cells isolated from pig placentas. Cells were then incubated with 400 µM NEFA or 10 µM JW74 for 24 h (*n* = 3). (**E–H**) Relative mRNA expression of lipid metabolism-related genes (**E** and **F**), fatty acid (FA) transporters (**G**) and regulators of lipid metabolism (**H**) after transfection with Myc-MARK4, sh-MARK4 for 48 h in primary trophoblast cells. Cells were then treated with 400 µM NEFA or 10 µM JW74 for 24 h (*n* = 3). Values are expressed as mean ± SEM.* *p* < 0.05 compared with the control group. Myc-MARK4 group: over expression of MARK4 group, sh-MARK4 group: knock down of MARK4 group, Control: empty vector (EV) group.

3. Discussion

At present, the MARK4 gene has been widely explored in mammal species [13]. However, such information is still quite limited in *Sus scrofa* (Pig). In this study, the full-length cDNA of MARK4 was characterized from a lean breed swine (Landrace), including an ORF of 2259 bp nucleotides in length, encoding 752 amino acids (AA) residues, in agreement with the previous study [20]. Sequence alignments and phylogenetic analysis showed that MARK4 is highly conserved between *Sus scrofa* (Pig) and other mammals. In addition, several functional sites were also observed, including a protein kinase ATP-binding region, a serine/threonine protein kinase active-site and a protein kinase domain, which represent the typical characters of the protein kinase superfamily [23]. Meanwhile, the catalytic kinase domain (KD), the ubiquitin-associated domain (UBA), the kinase associated domain1 (KA1) and three conserved functional sites (lysine 88 ATP binding site, aspartic 181 active site and threonine 214 phosphorylation site) were also identified through the multiple alignment analysis, which are regarded as the typical structures of microtubule affinity regulatory kinases family [12,13]. Several studies on mammals indicated that the activation of MARK4 is mediated by the major active site (Asp 181) that is activated by phosphorylation of Thr 214 located in the activation loop (T-loop) on protein kinase domain, whereas phosphorylation of Ser 218 in T-loop inactivates MARK4 [24,25]. Accordingly, compared with other mammals, we found the AA sequences of MARK4 protein in *Sus scrofa* has a conserved T-loop sequence, LDTFCGSPP, including the regulatory phosphorylation sites of Thr 214 and Ser 218. Furthermore, the predicted tertiary protein structure of MARK4 in *Sus scrofa* showed high similarity (AA sequence identity is 99%) with that of human (*Homo sapiens*). This was further confirmed by the observation that the key structural residues (Lys 88, Asp 181 and Thr 214) of human MARK4 protein are all well conserved in that of porcine (See Figure 1).

MARK4, the fourth member of microtubule affinity regulatory kinases (MARKs) family, is implicated in the regulation of dynamic biological functions, including glucose homeostasis and energy metabolism [16]. Recently, MARK4 has been found to promote adipogenesis and trigger apoptosis in 3T3-L1 adipocytes [17], suggesting MARK4 may play an important role in regulating lipid metabolism in adipose tissue. In addition, hyperlipidemia associated with obesity has been suggested to contribute to the ectopic lipid accumulation (lipotoxicity) often seen in highly metabolic tissues, including liver, skeletal muscle and placenta [10,11,26], a process that has been implicated as an important mediator of cellular stress and altered tissue function. Regarding the impact of MARK4 on adipogenesis, we hypothesized that Mark4 may potentially stimulate lipid accumulation in other cell types besides placental trophoblast cells from porcine, and this study was designed to investigate the role of MARK4 in modulating lipid metabolic signaling in pig placental trophoblasts in vitro. We found that in pig trophoblast cells MARK4 significantly increased the expression of lipogenic genes, including FASN, ACACA, DGAT1, LPIN1, LPIN3, LPL and SREBP-1c, suggesting increased TG and lipid droplet synthesis by MARK4 expression, as evidenced by dramatically increased lipid droplet accumulation in trophoblast cells. Thus, our data indicated that Mark4 is involved in regulating lipogenesis of pig placental trophoblasts upon the status of lipotoxic insult.

Studies have suggested that activation of PPARγ stimulates fatty acid uptake and fatty acid accumulation in cultured human trophoblast cells [21,27]. However, our data suggests that the stimulating effect of Mark4 on lipid accumulation of trophoblasts is not mediated by increased activation of PPARγ. Furthermore, the MARK4 effect on fatty acid accumulation is unlikely to be due to an activation of LPL activity since MARK4 did not regulate trophoblast LPL activity in vitro. PPARγ is known to be required for placental development and placental uptake of fatty acids [21,28]. Activation of PPARγ regulates gene expression of several proteins involved in lipid transport, including FA transport proteins (FATPs/SLC27As), intracellular FA binding proteins (FABPs), FA translocase (FAT/CD36), adipose differentiation-related protein (ADRP) and Acyl CoA synthase (ACS) [21,27,29]. Our finding showed that MARK4 inhibited the mRNA expression of PPARγ, ADRP, ACSL1, FATP1, FATP4, CD36, FABP1 and FABP4 in cultured trophoblast cells, suggesting impaired FA uptake by trophoblasts in vitro, as evidenced by significantly decreased receptor (transport proteins)-mediated fatty acid accumulation by MARK4. Recent studies determined that PPARγ and MARK4 play an opposing role in adipose inflammation response and oxidative stress [18]. Consistently, we preliminarily determined that activation of PPARγ by PPARγ-specific agonist GW1929 prevented MARK4 from stimulating lipogenesis and non-receptor-mediated lipid accumulation in cultured pig trophoblasts, suggesting that MARK4 promotes lipid synthesis in pig trophoblast cells by inhibiting the PPARγ pathways. However, the precise mechanism for inhibition of PPARγ by MARK4 in regulating lipogenesis of trophoblasts needs to be further studied.

In this study we determined fatty acid accumulation in trophoblast cells which is dependent upon uptake as well as cellular metabolism. As previously documented, WNT signaling pathway is involved in increased placental lipid accretion in obesity-prone rats or obese women [5,10]. In support of the role of WNT signaling in regulating lipid synthesis, our data showed that inhibition of WNT signaling by DKK1 remarkably reduced the mRNA expression of genes associated with TG and lipid droplet synthesis in pig trophoblasts, including DGAT1, LPL, LPIN3 and PPARδ, which is confirmed by decreased lipid droplet accumulation by DKK1. On the contrary, we found that activation of WNT signaling by GSK3β inhibitor LiCL significantly decreased the expression of PPARγ and several FA transporters, including FATP1, CD36, FABP4 and FATP4, in cultured trophoblast cells. Previous studies have shown that β-catenin (a key target of WNT signaling) and PPARγ functionally interact to negatively regulate each other's activity, and activation of WNT signaling prevents induction of C/EBPα and PPARγ during preadipocyte differentiation [30,31]. Hence, accumulation of fatty acids in pig trophoblast cells in response to lipotoxic insult may be attributed to altered intracellular metabolism of fatty acids rather than changes in cellular uptake.

This study pointed out a significant correlation between MARK4 and WNT signaling. The WNT signal is a cytosolic sensor which activates and promotes β-catenin nuclear translocation and DNA binding [22]. Sun et al. reported that Par-1, the mammalian ortholog of MARKs, is a positive regulator of Wnt/β-catenin pathway in mammalian cells and Drosophila embryos [32]. Consistently, we demonstrated that MARK4 was potent to activate WNT signaling through promoting translocation of β-catenin into the nucleus in cultured pig trophoblasts. It is noticed that activation of WNT signaling pathway by LiCL prevented DKK1 from inhibiting phosphorylation of endogenous Mark4 (Thr214) in trophoblast cells, in agreement with previous studies that WNT signaling stimulates endogenous Par-1 kinase activity [32]. Activation of WNT pathway leads to the phosphorylation of Dishevelled (Dvl) protein, which then inhibits the activity of GSK3β [33]. GSK3β has been shown to inhibit MARK4 protein by phosphorylating the serine residue (Ser218), near the threonine activation site (Thr214) in the activation loop of MARK4 [24]. Therefore, inhibition of GSK3β could be a possible mechanism involved in the activation of MARK4 by WNT signaling. In addition, our experiments employing the WNT specific inhibitor JW74 further confirmed that the WNT pathway is involved in the promotion of lipogenesis via MARK4, suggesting WNT signal is central to MARK4 performing lipid synthesis function in pig trophoblast cells in response to lipotoxic insult.

4. Materials and Methods

4.1. Experimental Animals and Reagents

For the analysis of full-length cDNA cloning of MARK4 gene and isolation of porcine placental trophoblast cells, samples of placenta from *Sus scrofa* (Landrace) were collected at Research Farm of Nan Jing Agricultural University. The collection of porcine full-term placental tissue was specifically approved by the Laboratory Animal Care and Use Committee of Nan Jing Agricultural University. (SYXK2015-0072, 6 September 2015)

For the isolation of porcine placental trophobalst cells, the following reagents were purchased, including Ham's F12/Dulbecco's Modified Eagle Medium(DMEM/F12) (HyClone, Logan, UT, USA), fetal bovine serum (FBS) (HyClone, Logan, UT, USA), Trypsin (Gibco, Grand Island, NY, USA), Phosphate-buffered saline (PBS) (Life Technologies, Grand Island, NY, USA), Bovine serum albumin

(BSA) (Amresco, Solon, OH, USA); Percoll (Pharmacia, London, UK), 100× Penicillin-Streptomycin (10,000 U/mL) (Invitrogen, Carlsbad, CA, USA), 100× Insulin–Transferrin–Selenium (ITS; Sigma, Saint Louis, MO, USA) and epidermal growth factor (EGF; Invitrogen, Carlsbad, CA, USA).

4.2. Full-length cDNA Cloning of the MARK4 Gene

Total RNA was isolated from the placenta of Landrace sows using RNAiso Plus (TaKaRa, Tokyo, Japan) and then was treated with DNase I using Recombinant RNase-free DNase I kit (TaKaRa, Tokyo, Japan) to degrade genomic DNA. 1% agarose gels electrophoresis and spectrophotometric analysis (260/280 ratio) were used to assess the quantity and quality of isolated RNA.

The cDNA was synthesized with PrimeScript 1 st strand cDNA Synthesis kit (TaKaRa, Tokyo, Japan) using total RNA (1 μg) from the placenta as template and Oligo dT18 as primer according to the manufacturer's instructions. Degenerated primer pairs of MARK4F/MARK4R (Table 1) were designed based on highly conserved regions from the available sequences of various vertebrate species. PCR amplification was performed with 1 μL of reverse-transcribed (RT) reactions in a total volume of 50 μL and 1 μL Tks Gflex DNA Polymerase (1.25 U/μL; TaKaRa, Tokyo, Japan). The PCR cycling conditions were one cycle of 94 °C for 1 min, 35 cycles of 98 °C for 10sec, 55 °C for 15sec, and 68 °C for 1 min, followed by one cycle of 72 °C for 5 min. The PCR products were purified with MiniBEST Agarose Gel DNA Extraction Kit Ver.4.0 (TaKaRa, Tokyo, Japan) and sequenced by Takara Biotechnology (Dalian) Co.Ltd (Dalian, China). Sequencing was performed in both forward and reverse directions by using an ABI PRISMTM3730XL DNA Sequencer (Applied Biosystems, Waltham, MA, USA). The forward and reverse sequences were assembled using SeqMan NGen15 software in DNASTAR Lasergene 15.2 (DNASTAR, Madison, WI, USA), through which the core fragment of MARK4 gene was obtained. According to the sequence information of this fragment, gene-specific primers were designed for the 3′RACE and 5′RACE.

Table 1. Primers for the cDNA cloning of MARK4 gene.

Primers	Sequence, 5′-/-3′	Use
MARK4-F	CAACGATCGGAACTCGGACA	Used with MARK4-R for RT-PCR of core fragment
MARK4-R	ATTTGGCAACAGGGACGGGC	Used with MARK4-F
3′RACE Adaptor	CTGATCTAGAGGTACCGGATCC(T)$_{16}$	Used for synthesis of the first- strand cDNA for 3′RACE
MARK4,3-F1	CAAGCGCAGCCCAACCAGCACAG	Used with 3′Outer Primer for first PCR of 3′RACE
3′Outer Primer	TACCGTCGTTCCACTAGTGATTT	Used with MARK4,3-F1
MARK4,3-F2	ACAAGGCAGAGATCCCAGAGCGA	Used with 3′Inner Primer for nested PCR of 3′RACE
3′Inner Primer	CGCGGATCCTCCACTAGTGATTTCA-CTATAGG	Used with MARK4,3-F2
MARK4,5-R1	AGCTTCACAATGTTGGGGTGGTT	Used for synthesis of the first-strand cDNA for 5′RACE
MARK4,5-R2	TGGGGTTCAGCTGGGTTTTGTCG	Used with UPM Primer for first PCR of 5′RACE
UPM Primer	CTAATACGACTCACTATAGGGCAA-GCAGTGGTATCAACGCAGAGT	Used with MARK4,5-R2
MARK4,5-R3	AGGATGTGCCGGGCCAGCTTGAC	Used with UPS Primer for nested PCR of 5′RACE
UPS Primer	CTAATACGACTCACTATAGGGC	Used with MARK4,5-R3

Rapid amplification of the 3′ end was performed using the 3′-full RACE Core Set with PrimeScriptTM RTase (TaKaRa, Tokyo, Japan) following the manufacturer's instructions. The primers used for 3′ RACE are shown in Table 1. Firstly, total RNA (1 μg) from placenta was reverse-transcribed using 3′RACE Adaptor (Table 1) as the primer for the synthesis of first strand cDNA. Then, the cDNA was amplified by a specific forward primer MARK4,3-F1 and 3′ RACE Outer Primer containing the anchor sequence. After the first PCR, 1 μL of Outer PCR reactions was re-amplified using 3′ RACE

Inner Primer and a specific forward primer MARK4,3-F2. The nested PCR product was separated by electrophoresis using 1% agarose gels and sequenced by the methods aforementioned.

Rapid amplification of the 5′ end was conducted with SMARTerTM RACE 5′/3′ cDNA Amplification Kit (TaKaRa, Tokyo, Japan) according to the manufacturer's instructions. Briefly, total RNA (1 μg) was reverse-transcribed with a specific reverse primer MARK4,5-R1(Table 1). After the synthesis of first strand cDNA, 2 μL of RT reactions was amplified by prime pairs MARK4,5-R2 and UPM Primer. Then, 1 μL of the first PCR product was used as a template for the nested PCR, which was performed with UPS Primer and MARK4,5-R3. The nested PCR product was separated by 1% agarose gel test and sequenced following the methods aforementioned.

4.3. Bioinformatics Analysis

The full-length cDNA of MARK4 gene was obtained using SeqMan NGen15 software in DNASTAR Lasergene (version 15.2) to assemble the core fragment, 3′ end and 5′ end sequences. The resulting nucleotide sequence was edited and analyzed by Open Reading Frame (ORF) Finder on NCBI (https://www.ncb-i.nlm.nih.gov/orffinder), and then translated into amino acids (AA) using standard genetic codes. The molecular weight (MW) and isolectric point (PI) of the Mark4 protein were predicted using the compute PI/MW software at https://web.expasy.org/compute_pi. Multiple alignments were generated by the MegAlign 15 program in DNASTAR Lasergene (version 15.2). The secondary and three-dimensional (3D) structures of Mark4 protein were predicted by the SABLE program (http://sable.cchmc.org) and the SWISS-MODEL program (https://swissmodel.expasy.org) as previously described [34], respectively. Illustration of the MARK4 model was performed in PyMOL 2.2 program (https://pymol.org). Phylogeny tree was inferred by the MEGA7 program, and distance analysis was conducted using the Neighbour-Joining (NJ) algorithm. 1000 bootstrap-replications were generated to evaluate the reliability for each code.

4.4. Porcine Placental Trophobalst Cell Isolation and Culture

The isolation and culture of porcine placental trophobalst cells were performed as previously described with some modifications [35]. Briefly, placental villous tissue, obtained from vaginal delivery, were dissected from fetal amnion and rinsed thoroughly in cold PBS containing 100 U/mL penicillin and 100 μg/mL streptomycin, and then cut into 1–3 mm 3 pieces. The tissue fragments were digested with 0.125% (*w/v*) Type I collagenase (trypsin) at 37 °C for 30 min with continuous shaking, followed by filtration through a 70 μm cell strainer. The filtrate was further purified by Percoll gradient centrifugation. Placental trophoblast cells were collected from the appropriate layers between 35% and 45% Percoll density gradient separated layers, and cultured in DMEM/F12 supplemented with 10% FBS, 1% (*v/v*) ITS, 10 ng/mL of EGF, 100 U/mL penicillin and 100 μg/mL streptomycin at 37 °C under 5% CO2 as previously described [35]. The purity of trophoblast cells isolated from full-term placentas was determined by flow cytometry as previously described [36], using FITC fluorescein-labeled antibody against cytokeratin-7 (Santa Cruz Tech, Dallas, CA, USA) as a specific marker of trophoblast cells.

4.5. Cell Transfection and Drug Treatment

DNA constructs including Myc-MARK4 and Flag-DKK1 were made by Generay Biotech Company (Shanghai, China) using pEGFP-N1 expression vector. shRNA sequences against MARK4 or DKK1 were contrived and synthesized by Genepharma Company (Shanghai, China) using pGpU6/GFP/Neo shRNA expression vector. After transfection efficiency detection, the optimal shRNA of MARK4 or DKK1 was chosen and named sh-MARK4 and sh-DKK1. Cells were plated at a concentration of 6×10^5–2×10^6/dish in 60-mm dishes. 2 μg interference or expression plasmids DNA were mixed with X-treme GENE HP Reagent (Roche, Basel, Switzerland) and Opti-MEMI media (Invitrogen, Carlsbad, CA, USA) following the instruction. The transfection mixture was then added into each dish for 48 h to allow the expression of DNA or shRNA constructs described above.

In order to induce lipid accumulation in trophoblast cells in vitro, cells were treated with 400 µM Fatty Acid (FA) Supplement containing 2 mol of linoleic acid and 2 mol of oleic acid per mole of albumin (L9655; Sigma-Aldrich, Saint Louis, MO, USA) in triplicate as previously described [10,37]. The optimal treatment concentration of 400 µM was chosen based on results of concentration gradient studies (Figure S5) indicating that fat accumulation was significantly increased by 50, 100 and 200 µM fatty acids when compared to 0µM, with the most significant increase following the 400 µM treatment. Treatment media without fatty acids was added with bovine serum albumin (FA free) to maintain the same osmolarity. In some experiment, cells were treated with one of the following specific agonists or inhibitors: 2 µM GW1929 (PPARγ-specific agonist; MCE, Shanghai, China), 20 µM LiCL (GSK3β inhibitor; Millipore, Billerica, MA, USA), or 10 µM JW74 (WNT signaling pathway specific inhibitor; MCE, Shanghai, China) for the amount of time specified in the individual figures.

4.6. Oil Red O Staining

After 24 h of FA treatment, cells were fixed in 4% paraformaldehyde for 30 min at room temperature for Oil Red O staining. Each well was then briefly washed in PBS and 60% isopropanol and then stained for 10 min in a 60% working Oil Red O solution (Sigma-Aldrich). For quantification of Oil Red O staining, cells were extracted by 100% isopropanol for colorimetric analysis at an optical density of 490 nm as previously described [10].

4.7. Cell Viability and Reactive Oxygen Species (ROS) Assay

Cell viability was detected using cell counting kit-8 (CKK-8; KeyGen BioTECH, Nanjing, China). The isolated cells were seeded into 96-well plates at a density of 5×10^3 and cultured with 0, 400 and 500 µM fatty acids for the amount of time specified in Figure S5, respectively. 10 µL CKK-8 solution was then added into each well and incubated for 2 h at 37 °C. Absorbance was measured at 450 nm using a Multiskan Go Microplate Spectrophotometer (Thermo Scientific, Waltham, MA, USA).

The intracellular level of ROS test was performed using Oxygen Species Assay Kit (KeyGen BioTECH, Nanjing, China) according to the manufacturer's instructions. Briefly, the dye loading was performed by incubating the cells with 10 µM 2', 7'-dichlorofluorescin diacetate (DCFH-DA) at 37 °C for 1 h. The production of ROS was examined using a Luminescence Spectrophotometer (Promega Corporation, Madison, WI, USA) by measuring the fluorescence intensity of DCF at emission wavelength of 525 nm.

4.8. Lipid Accumulation Assay

The Bodipy 493/503 lipid probes (D-3922; Thermo Scientific) was used to visualize fatty acid accumulation in cultured trophoblast cells as previously described [38]. Briefly, cells were washed in PBS and 4% paraformaldehyde in PBS was added to fix the cells for 30 min at room temperature. After fixation, cells were washed in PBS containing 0.1% Triton X-100 for 5 min. Bodipy dye was diluted in PBS at a concentration of 10 µg/mL and applied to cells for 15 min. For nuclei staining, 10 µg/mL of 4', 6-diamidino-2-phenylindole (DAPI) solution was incubated with each sample for 30 min, and then the samples were examined on confocal laser scanning microscope (Zeiss LSM 700 META, Jena, Germany). For quantification of lipid accumulation, triglyceride (TG) content in cultured trophoblast cells was evaluated with a spectrophotometer (Thermo Scientific) at 510 nm using Tissue Triglyceride Assay Kit (APPLYGEN, Beijing, China) as previously described [17]. Because phloretin blocks receptor (transport proteins)-mediated fatty acid transport and accumulation [39,40] it was used to determine receptor-mediated fatty acid accumulation by subtracting the TG content in the presence of phloretin (500 µM) from those in the absence of phloretin as previously described [21].

4.9. Immunofluorescence Assay

β-catenin immunofluorescence analysis was performed as previously described [10]. Briefly, cells were grown on coverslips, fixed with 4% paraformaldehyde for 30 min and permeabilized with 0.25% Triton X-100 for 10 min. After blocked with 5% BSA-supplemented PBS for 1 h, cells were incubated overnight at 4 °C with rabbit anti-β-Catenin primary antibody (8480, dilution 1:300, Cell Signaling Technology, Danvers, MA, USA), followed by incubation of Goat anti-rabbit Cy3 fluorescein-labeled secondary antibody (BA1032, dilution 1:500, Boster, China). Meanwhile, the cell nuclei were counterstained with 4′, 6-diamidino-2-phenylindole for 10 min, and then the samples were mounted on glass slides and examined on confocal laser scanning microscope (Zeiss LSM 700 META, Jena, Germany). Quantification of the fluorescence intensity from the red channel (β-Catenin) was performed using the Image J software (NIH Image).

4.10. Measurement of LPL Activity

For LPL activity detection, cells were harvested after the medium removed, washed with ice-cold PBS and lysed with cell lysis buffer (20 mM Tris, 150mM NaCl, 1% Triton X-100). The lysate was centrifuged at 10,000× g for 5 min at 4 °C. Then LPL enzyme activity was measured in the supernatant by the enzyme fluorescence method using Biovision LPL Activity assay kit (Biovision Incorporated, Milpitas, CA, USA) according to the manufacturer's instructions as previously described [9]. Results were normalized to the amount of protein (mU per mg of bulk cellular protein). Protein concentration was determined using Pierce BCA Protein Assay Kit (Thermo Scientific, Waltham, MA, USA) according to the manufacturer's instructions.

4.11. Real-time Quantitative PCR Analysis

Total RNA was extracted from cultured cells with the High Pure RNA tissue kit (Omega Bio-Tek, Norcross, GA, USA) and 500 ng of total RNA was reverse transcribed using PrimeScript RT Master Mix Kit (TaKaRa, Tokyo, Japan). Real-time RT-PCR was conducted on the Step One Plus Real-Time PCR System (ABI, Waltham, MA, USA) with the following program: 95 °C for 30 sec, 95 °C for 5 sec, 60 °C for 30 sec, 95 °C for 15 sec, 60 °C for 1 min, and 95 °C for 15 sec, with 40 cycles of steps 2 and 3. Primers were synthesized by Invitrogen (Shanghai, China). Amplication was performed in 25 µL reaction system containing specific primers (Table S1) and SYBR Premix Ex Taq II (TaKaRa, Tokyo, Japan). Relative gene expression was calculated using the comparative Ct method with the formula $2^{-\Delta\Delta Ct}$ [41]. The two reference genes GAPDH and HPRT1 were used. The geometric mean of relative gene expression was calculated and used for further analysis as previously reported [42].

4.12. Protein Extraction and Western Blotting Analysis

Total protein from cultured trophobalst cells was extracted using cell lysis buffer (Beyotime Co, China) by procedures as previously described [38]. Nuclear protein isolation was performed using Nuclear and Cytoplasmic Protein Extraction Kit (KenGEN BioTECH, Nanjing, China) according to the manufacturer's instructions as previously reported [11]. The concentration of protein was quantified using BCA Protein Assay kit (Thermo Scientific, Waltham, MA, USA). Proteins (50 µg) were separated by SDS-PAGE and transferred to PVDF nitrocellulose membrane (Bio-Rad Laboratories, Hercules, CA, USA). After blocking in 5% fat-free milk for 1 h at room temperature, the membranes were incubated with rabbit anti-Mark4(4834, 1:1000 dilution, Cell Signaling Technology, Danvers, MA, USA), β-Catenin (8480, dilution 1:1000, Cell Signaling Technology), GAPDH (2118, dilution 1:1000, Cell Signaling Technology), and Phospho-Mark4 (SAB4504258, 1:500 dilution, Sigma, Saint Louis, MO, USA) antibody, Goat anti-DKK1 (LS-B194, dilution 1:1000, LifeSpan BioSciences, Seattle, WA, USA) antibody, or Mouse anti-LaminA (sc-376248, dilution 1:1000, Santa Cruz Biotechnology, Dallas, TX, USA) antibody overnight at 4 °C, followed by incubation with Donkey anti-goat, Goat anti- mouse or rabbit IgG horseradish peroxidase (HRP)-conjugated secondary antibodies (HAF109, HAF007 and HAF008, dilution 1:2000, RD SYSTEMS, Minneapolis, USA) for 1 h at room temperature. Proteins were visualized using the LumiGLO Reagent and Peroxide system (Cell Signaling Technology, Danvers, USA), and then the blots were quantified using Bio-Rad ChemiDoc imaging system (Bio-Rad Laboratories, Hercules, USA). Band density was normalized according to the GAPDH content.

4.13. Statistical Analysis

All the data were obtained from at least three independent experiments. Statistical analyses were conducted using SPSS Statistics 20.0 software (IBM SPSS, Armonk, NY, USA). Data were analyzed using One-way ANOVA for comparisons among groups, followed by Duncan test. Results were expressed as means $\pm$ SEM. A p-value < 0.05 was considered statistically significant, and very significant was indicated when $p < 0.01$.

5. Conclusions

In summary, our present study demonstrates that MARK4 stimulates fatty acid accumulation in porcine trophoblast cells, which could contribute to a lipotoxic placental milieu in conditions associated with elevated maternal fatty acids such as excessive back-fat during pregnancy of sows. Moreover, WNT/β-catenin signal is essential for MARK4 promoting lipogenesis in pig placental trophoblasts (Figure 8). Thus, our results indicate that MARK4 has potential as a regulator of lipotoxicity associated with maternal obesity in the pig placenta.

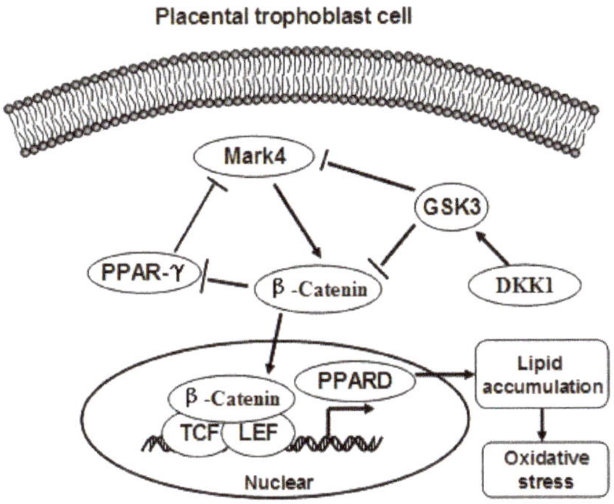

Figure 8. A proposed model for role of MARK4 in regulating lipogenesis in pig placental trophoblast cells. MARK4 promotes lipogenesis by activating WNT/β-catenin signaling pathway. Arrows indicates a positive regulation and bar-headed lines show negative regulation. Interactions depicted are based on studies performed in various tissues (in some cases placenta) and have been previously published.

Supplementary Materials: Supplementary Materials can be found at http://www.mdpi.com/1422-0067/20/5/1206/s1, Figure S1: The full-length cDNA of MARK4 gene in porcine and the deduced amino acid sequence, Figure S2: Clustal W alignment of the MARK4 protein from pig and other organisms, Figure S3: The secondary structure of the MARK4 protein in porcine constructed by the SABLE program, Figure S4: Phylogenetic tree based on MARK4 sequences made by MEGA 7.0 software using the neighbor-joining method, Figure S5: Identification of optimal fatty acid (FA) concentration for induction of FA accumulation in pig primary cytotrophoblasts, Table S1: Primer sets used for real-time PCR.

Author Contributions: All the authors contributed to this manuscript. Planed experiments: L.T. and P.S.; Performed experiments: L.T. and D.S.; Analyzed data: L.T. and A.W.; Contributed reagents/materials/analysis tools: L.T., A.W. and D.S.; Wrote the paper: L.T.

Funding: This research was surported by National Nature Science Foundation of China (No. 31702120), Fundamental Research Funds for the Central Universities of China (No. KJQN201831) and grants from the Nature Science Foundation of Jiangsu Province of China (No. BK20150672).

Acknowledgments: We sincerely acknowledge staffs in Research Farm and National Experimental Teaching Demonstration Center of Animal Science of Nan Jing Agricultural University for their helpful assistance in samples collection and offering technical platform.

Conflicts of Interest: The authors declare no conflict of interest.

Abbreviations

ACACA	Acetyl-CoA carboxylase
ADRP	Adipose differentiation-related protein
ACSL1	Acyl CoA synthase long chain family member 1
cDNA	Complementary DNA
CD36	Fatty acid translocase
DGAT1	Diglyceride acyltransferase 1
DKK1	Dickkopf family protein1
FASN	Fatty acid synthase
FA	Fatty acid
FATP1	Fatty acid transport protein 1
FATP4	Fatty acid transport protein 4
FABP1	Intracellular fatty acid binding protein 1
FABP4	Intracellular fatty acid binding protein 4
FITC	Fluorescein isothiocyanate
GSK3β	Glycogen synthase kinase 3 beta
GAPDH	Glyceraldehyde-3-phosphate dehydrogenase
HPRT1	Hypoxanthine phosphoribosyltransferase 1
LPL	Lipoprotein lipase
LPIN1	Phosphatidic acid phosphatase 1
LPIN3	Phosphatidic acid phosphatase 3
MARK1	Microtubule affinity-regulating kinase 1
MARK2	Microtubule affinity-regulating kinase 2
MARK3	Microtubule affinity-regulating kinase 3
MARK4	Microtubule affinity-regulating kinase 4
ORF	Open reading frame
PPARγ	Peroxisome proliferators-activated receptor gamma
PPARD	Peroxisome proliferators-activated receptor delta
PPARA	Peroxisome proliferators-activated receptor alpha
RACE	Rapid Amplification of cDNA Ends
RT-PCR	Reverse transcription-polymerase chain reaction
SREBP-1c	Sterol regulatory element binding protein 1c
TG	Triglyceride

References

1. Brewer, C.J.; Balen, A.H. The adverse effects of obesity on conception and implantation. *Reproduction* **2010**, *140*, 347–364. [CrossRef] [PubMed]
2. Zhou, Y.; Xu, T.; Cai, A.; Wu, Y.; Wei, H.; Jiang, S.; Peng, J. Excessive backfat of sows at 109 d of gestation induces lipotoxic placental environment and is associated with declining reproductive performance. *J. Anim. Sci.* **2018**, *96*, 250–257. [CrossRef] [PubMed]
3. Myatt, L. Placental adaptive responses and fetal programming. *J. Physiol.* **2006**, *572 Pt 1*, 25–30. [CrossRef]
4. Liguori, A.; D'Armiento, F.P.; Palagiano, A.; Balestrieri, M.L.; Williams-Ignarro, S.; de Nigris, F.; Lerman, L.O.; D'Amora, M.; Rienzo, M.; Fiorito, C.; et al. Effect of gestational hypercholesterolaemia on omental vasoreactivity, placental enzyme activity and transplacental passage of normal and oxidised fatty acids. *BJOG* **2007**, *114*, 1547–1556. [CrossRef] [PubMed]
5. Saben, J.; Lindsey, F.; Zhong, Y.; Thakali, K.; Badger, T.M.; Andres, A.; Gomez-Acevedo, H.; Shankar, K. Maternal obesity is associated with a lipotoxic placental environment. *Placenta* **2014**, *35*, 171–177. [CrossRef] [PubMed]
6. Aye, I.L.; Lager, S.; Ramirez, V.I.; Gaccioli, F.; Dudley, D.J.; Jansson, T.; Powell, T.L. Increasing maternal body mass index is associated with systemic inflammation in the mother and the activation of distinct placental inflammatory pathways. *Biol. Reprod.* **2014**, *90*, 129. [CrossRef] [PubMed]

7. Malti, N.; Merzouk, H.; Merzouk, S.A.; Loukidi, B.; Karaouzene, N.; Malti, A.; Narce, M. Oxidative stress and maternal obesity: Feto-placental unit interaction. *Placenta* **2014**, *35*, 411–416. [CrossRef] [PubMed]

8. Dube, E.; Gravel, A.; Martin, C.; Desparois, G.; Moussa, I.; Ethier-Chiasson, M.; Forest, J.C.; Giguere, Y.; Masse, A.; Lafond, J. Modulation of fatty acid transport and metabolism by maternal obesity in the human full-term placenta. *Biol. Reprod.* **2012**, *87*, 1–11. [CrossRef]

9. Tian, L.; Dong, S.S.; Hu, J.; Yao, J.J.; Yan, P.S. The effect of maternal obesity on fatty acid transporter expression and lipid metabolism in the full-term placenta of lean breed swine. *J. Anim. Physiol. Anim. Nutr.* **2018**, *102*, e242–e253. [CrossRef]

10. Strakovsky, R.S.; Pan, Y.X. A decrease in DKK1, a WNT inhibitor, contributes to placental lipid accumulation in an obesity-prone rat model. *Biol. Reprod.* **2012**, *86*, 81. [CrossRef]

11. Liang, T.; Jinglong, X.; Shusheng, D.; Aiyou, W. Maternal obesity stimulates lipotoxicity and up-regulates inflammatory signaling pathways in the full-term swine placenta. *Anim. Sci. J.* **2018**, *89*, 1310–1322. [CrossRef] [PubMed]

12. Hurov, J.; Piwnica-Worms, H. The Par-1/MARK family of protein kinases: From polarity to metabolism. *Cell Cycle* **2007**, *6*, 1966–1969. [CrossRef] [PubMed]

13. Naz, F.; Anjum, F.; Islam, A.; Ahmad, F.; Hassan, M.I. Microtubule affinity-regulating kinase 4: Structure, function, and regulation. *Cell Biochem. Biophys.* **2013**, *67*, 485–499. [CrossRef] [PubMed]

14. Schneider, A.; Laage, R.; von Ahsen, O.; Fischer, A.; Rossner, M.; Scheek, S.; Grunewald, S.; Kuner, R.; Weber, D.; Kruger, C.; et al. Identification of regulated genes during permanent focal cerebral ischaemia: Characterization of the protein kinase 9b5/MARKL1/MARK4. *J. Neurochem.* **2004**, *88*, 1114–1126. [CrossRef]

15. Beghini, A.; Magnani, I.; Roversi, G.; Piepoli, T.; Di Terlizzi, S.; Moroni, R.F.; Pollo, B.; Fuhrman Conti, A.M.; Cowell, J.K.; Finocchiaro, G.; et al. The neural progenitor-restricted isoform of the MARK4 gene in 19q13.2 is upregulated in human gliomas and overexpressed in a subset of glioblastoma cell lines. *Oncogene* **2003**, *22*, 2581–2591. [CrossRef]

16. Sun, C.; Tian, L.; Nie, J.; Zhang, H.; Han, X.; Shi, Y. Inactivation of MARK4, an AMP-activated protein kinase (AMPK)-related kinase, leads to insulin hypersensitivity and resistance to diet-induced obesity. *J. Biol. Chem.* **2012**, *287*, 38305–38315. [CrossRef]

17. Feng, M.; Tian, L.; Gan, L.; Liu, Z.; Sun, C. Mark4 promotes adipogenesis and triggers apoptosis in 3T3-L1 adipocytes by activating JNK1 and inhibiting p38MAPK pathways. *Biol. Cell* **2014**, *106*, 294–307. [CrossRef]

18. Liu, Z.; Gan, L.; Chen, Y.; Luo, D.; Zhang, Z.; Cao, W.; Zhou, Z.; Lin, X.; Sun, C. Mark4 promotes oxidative stress and inflammation via binding to PPARgamma and activating NF-kappaB pathway in mice adipocytes. *Sci. Rep.* **2016**, *6*, 21382. [CrossRef]

19. Tian, L.; Wen, A.Y.; Dong, S.S.; Xiao, K.Y.; Li, H.; Yan, P.S. Excessive backfat of sows at mating promotes oxidative stress and up-regulates mitochondrial mediated apoptotic pathway in the full-term placenta. *Livest. Sci.* **2019**, *222*, 71–82. [CrossRef]

20. Wang, K.J.; Li, W.T.; Bai, Y.; Yang, W.J.; Ling, Y.; Fang, M.Y. ssc-miR-7134-3p regulates fat accumulation in castrated male pigs by targeting MARK4 gene. *Int. J. Biol. Sci.* **2017**, *13*, 189–197. [CrossRef]

21. Schaiff, W.T.; Bildirici, I.; Cheong, M.; Chern, P.L.; Nelson, D.M.; Sadovsky, Y. Peroxisome proliferator-activated receptor-gamma and retinoid X receptor signaling regulate fatty acid uptake by primary human placental trophoblasts. *J. Clin. Endocrinol. Metab.* **2005**, *90*, 4267–4275. [CrossRef] [PubMed]

22. Kawano, Y.; Kypta, R. Secreted antagonists of the Wnt signalling pathway. *J. Cell Sci.* **2003**, *116 Pt 13*, 2627–2634. [CrossRef]

23. Hanks, S.K.; Quinn, A.M.; Hunter, T. The protein kinase family: Conserved features and deduced phylogeny of the catalytic domains. *Science* **1988**, *241*, 42–52. [CrossRef] [PubMed]

24. Timm, T.; Balusamy, K.; Li, X.; Biernat, J.; Mandelkow, E.; Mandelkow, E.M. Glycogen synthase kinase (GSK) 3beta directly phosphorylates Serine 212 in the regulatory loop and inhibits microtubule affinity-regulating kinase (MARK) 2. *J. Biol. Chem.* **2008**, *283*, 18873–18882. [CrossRef] [PubMed]

25. Timm, T.; Li, X.Y.; Biernat, J.; Jiao, J.; Mandelkow, E.; Vandekerckhove, J.; Mandelkow, E.M. MARKK, a Ste20-like kinase, activates the polarity-inducing kinase MARK/PAR-1. *EMBO J.* **2003**, *22*, 5090–5101. [CrossRef] [PubMed]

26. Gustafson, B.; Gogg, S.; Hedjazifar, S.; Jenndahl, L.; Hammarstedt, A.; Smith, U. Inflammation and impaired adipogenesis in hypertrophic obesity in man. *Am. J. Physiol. Endocrinol. Metab.* **2009**, *297*, E999–E1003. [CrossRef] [PubMed]

27. Duttaroy, A.K. Transport of fatty acids across the human placenta: A review. *Prog. Lipid Res.* **2009**, *48*, 52–61. [CrossRef] [PubMed]

28. Xu, Y.; Wang, Q.; Cook, T.J.; Knipp, G.T. Effect of placental fatty acid metabolism and regulation by peroxisome proliferator activated receptor on pregnancy and fetal outcomes. *J. Pharm. Sci.* **2007**, *96*, 2582–2606. [CrossRef] [PubMed]

29. Bildirici, I.; Roh, C.R.; Schaiff, W.T.; Lewkowski, B.M.; Nelson, D.M.; Sadovsky, Y. The lipid droplet-associated protein adipophilin is expressed in human trophoblasts and is regulated by peroxisomal proliferator-activated receptor-gamma/retinoid X receptor. *J. Clin. Endocrinol. Metab.* **2003**, *88*, 6056–6062. [CrossRef] [PubMed]

30. Liu, J.; Farmer, S.R. Regulating the balance between peroxisome proliferator-activated receptor gamma and beta-catenin signaling during adipogenesis. A glycogen synthase kinase 3beta phosphorylation-defective mutant of beta-catenin inhibits expression of a subset of adipogenic genes. *J. Biol. Chem.* **2004**, *279*, 45020–45027. [PubMed]

31. Liu, J.; Wang, H.; Zuo, Y.; Farmer, S.R. Functional interaction between peroxisome proliferator-activated receptor gamma and beta-catenin. *Mol. Cell. Biol.* **2006**, *26*, 5827–5837. [CrossRef] [PubMed]

32. Sun, T.Q.; Lu, B.; Feng, J.J.; Reinhard, C.; Jan, Y.N.; Fantl, W.J.; Williams, L.T. PAR-1 is a Dishevelled-associated kinase and a positive regulator of Wnt signalling. *Nat. Cell Biol.* **2001**, *3*, 628–636. [CrossRef] [PubMed]

33. Mikels, A.J.; Nusse, R. Purified Wnt5a protein activates or inhibits beta-catenin-TCF signaling depending on receptor context. *PLoS Biol.* **2006**, *4*, e115. [CrossRef] [PubMed]

34. Kiefer, F.; Arnold, K.; Kunzli, M.; Bordoli, L.; Schwede, T. The SWISS-MODEL Repository and associated resources. *Nucleic Acids Res.* **2009**, *37*, D387–D392. [CrossRef] [PubMed]

35. Zhang, H.; Huang, Y.; Wang, L.; Yu, T.; Wang, Z.; Chang, L.; Zhao, X.; Luo, X.; Zhang, L.; Tong, D. Immortalization of porcine placental trophoblast cells through reconstitution of telomerase activity. *Theriogenology* **2016**, *85*, 1446–1456. [CrossRef] [PubMed]

36. Daoud, G.; Simoneau, L.; Masse, A.; Rassart, E.; Lafond, J. Expression of cFABP and PPAR in trophoblast cells: Effect of PPAR ligands on linoleic acid uptake and differentiation. *Biochim. Biophys. Acta* **2005**, *1687*, 181–194. [CrossRef] [PubMed]

37. Magnusson-Olsson, A.L.; Lager, S.; Jacobsson, B.; Jansson, T.; Powell, T.L. Effect of maternal triglycerides and free fatty acids on placental LPL in cultured primary trophoblast cells and in a case of maternal LPL deficiency. *Am. J. Physiol. Endocrinol. Metab.* **2007**, *293*, E24–E30. [CrossRef] [PubMed]

38. Lager, S.; Jansson, N.; Olsson, A.L.; Wennergren, M.; Jansson, T.; Powell, T.L. Effect of IL-6 and TNF-alpha on fatty acid uptake in cultured human primary trophoblast cells. *Placenta* **2011**, *32*, 121–127. [CrossRef]

39. Abumrad, N.; Harmon, C.; Ibrahimi, A. Membrane transport of long-chain fatty acids: Evidence for a facilitated process. *J. Lipid Res.* **1998**, *39*, 2309–2318.

40. Krischer, S.M.; Eisenmann, M.; Bock, A.; Mueller, M.J. Protein-facilitated export of arachidonic acid from pig neutrophils. *J. Biol. Chem.* **1997**, *272*, 10601–10607. [CrossRef]

41. Livak, K.J.; Schmittgen, T.D. Analysis of relative gene expression data using real-time quantitative PCR and the $2^{-\Delta\Delta Ct}$ method. *Methods Cell Sci.* **2001**, *25*, 402–408. [CrossRef] [PubMed]

42. Vandesompele, J.; De Preter, K.; Pattyn, F.; Poppe, B.; Van Roy, N.; De Paepe, A.; Speleman, F. Accurate normalization of real-time quantitative RT-PCR data by geometric averaging of multiple internal control genes. *Genome Biol.* **2002**, *3*, research0034.1. [CrossRef] [PubMed]

International Journal of
Molecular Sciences

Review

Lipidomics of Bioactive Lipids in Acute Coronary Syndromes

Zahra Solati [1,2] and Amir Ravandi [1,2,3,*]

[1]   Institute of Cardiovascular Sciences, St. Boniface Hospital Research Centre, University of Manitoba, Winnipeg, MB R2H 2A6, Canada; solatiz@myumanitoba.ca

[2]   Department of Physiology and Pathophysiology, University of Manitoba, Winnipeg, MB R3E 3P5, Canada

[3]   Section of Cardiology, Department of Internal Medicine, Max Rady College of Medicine, Faculty of Health Sciences, University of Manitoba, 409 Tache Avenue, Winnipeg, MB R2H 2A6, Canada

*   Correspondence: aravandi@sbgh.mb.ca; Fax: +204-237-2023

Received: 1 February 2019; Accepted: 24 February 2019; Published: 28 February 2019

Abstract: Acute coronary syndrome (ACS) refers to ischemic conditions that occur as a result of atherosclerotic plaque rupture and thrombus formation. It has been shown that lipid peroxidation may cause plaque instability by inducing inflammation, apoptosis, and neovascularization. There is some evidence showing that these oxidized lipids may have a prognostic value in ACS. For instance, higher levels of oxidized phospholipids on apo B-100 lipoproteins (OxPL/apoB) predicted cardiovascular events independent of traditional risk factors, C-reactive protein (hsCRP), and the Framingham Risk Score (FRS). A recent cross-sectional study showed that levels of oxylipins, namely 8,9-DiHETrE and 16-HETE, were significantly associated with cardiovascular and cerebrovascular events, respectively. They found that with every 1 nmol/L increase in the concentrations of 8,9-DiHETrE, the odds of ACS increased by 454-fold. As lipid peroxidation makes heterogonous pools of secondary products, therefore, rapid multi-analyte quantification methods are needed for their assessment. Conventional lipid assessment methods such as chemical reagents or immunoassays lack specificity and sensitivity. Lipidomics may provide another layer of a detailed molecular level to lipid assessment, which may eventually lead to exploring novel biomarkers and/or new treatment options. Here, we will briefly review the lipidomics of bioactive lipids in ACS.

Keywords: oxidized phospholipids; oxylipins; bioactive lipids; coronary disease; myocardial infarction; ischemic heart disease; ischemia reperfusion injury; mass spectrometry; lipids

1. Introduction

Acute coronary syndrome (ACS) comprises a set of ischemic conditions including unstable angina (UA), myocardial infarction (MI) (with or without ST-segment elevation), and sudden cardiac death. It is the most common cause of morbidity and mortality worldwide, and accounts for roughly seven million deaths and 129 million loss of disability-adjusted life years (DALYs) annually [1]. The main cause of ischemia is the reduction of blood flow into coronary microcirculation as a result of atherosclerotic plaque rupture and thrombus formation [2]. Complete occlusion of coronary arteries usually presents with ST-segment elevation myocardial infarction (STEMI), which is accompanied by tissue injury and presents with elevated troponin levels. Partially occluded coronary arteries may result in non-STEMI or UA, depending on whether or not myocardial injury occurs [3].

Coronary angiography has shown that the atherosclerosis extent index (including the number of diseased vessels, stenosis and occlusions) is generally lower in ACS patients than in patients with stable angina, suggesting that plaque vulnerability rather than the extent of atherosclerosis may be the determinant of ACS [4]. The mechanisms leading to the progression of an asymptomatic plaque to a vulnerable one are not fully understood. A thin fibrous cap and a large lipid core ($\geq$40% plaque volume), inflammatory cells, and high neovascularity are suggested as factors causing plaque vulnerability [5].

The oxidation of lipoproteins, namely oxidized low-density lipoproteins (Ox-LDLs), has been considered as a key factor in this transition through various mechanisms. Following the infiltration of LDL into the injured endothelium, LDL becomes oxidized to form Ox-LDL. This modified LDL elevates the expression of cell adhesion molecules such as intercellular cell adhesion molecule-1 (ICAM-1) and vascular cell adhesion molecule-1 (VCAM-1), resulting in leukocyte (mainly monocytes and T-lymphocytes) recruitment and migration into the intima. In the intima, monocytes differentiate into macrophages. These lipid laden macrophages, which are called foam cells, along with the migrated T-lymphocytes release a variety of cytokines that promote inflammation and the generation of reactive oxygen species (ROS) [6]. Ox-LDL increases the infiltration of macrophages into the plaque (foam cell formation), up-regulates the expression of matrix metalloproteinase (MMP), and triggers proinflammatory reactions leading to plaque rupture [7].

Several clinical studies have confirmed that Ox-LDL concentrations are significantly higher in MI patients when compared with stable angina or age-matched controls [8–10]. Lipid peroxidation can occur within the LDL membrane through non-enzymatic and/or enzymatic mechanisms, producing diverse secondary products such as 4-hydroxynonenal (4-HNE), malondialdehyde (MDA), oxidized phospholipids (OxPLs), and oxylipins. These oxidized lipids are bioactive and can be bound to proteins, peptides, phospholipids, and nucleic acids, generating structural neo-epitopes called oxidation-specific epitopes (OSEs). Consequently, chronic elevations of OSEs may induce inflammation through the secretion of chemokines and proinflammatory cytokines, leading to plaque instability [11]. Clinical studies have also confirmed higher levels of these bioactive molecules in ACS patients when compared with patients in control groups [12–15].

Previous studies have shown that bioactive lipids can predict ACS occurrence in various populations. For instance, higher levels of OxPLs on Ox-LDL have been found to predict the progression of first or second major coronary events [16,17]. In addition, a recent cross-sectional study showed that levels of oxylipins, namely dihydroxy-eicosatrienoic acid (DiHETrE) and 16- hydroxy-eicosatetraenoic acid (HETE), were significantly associated with cardiovascular and cerebrovascular events, respectively. In this study, levels of 8,9-DiHETrE were significantly higher in patients with ACS (n = 24) compared to those without ACS (n = 74). Univariate and multivariate logistic regression also revealed that 8,9-DiHETrE concentrations were significantly associated with the presence of ACS. Moreover, they found that with every 1 nmol/L increase in the 8,9-DiHETrE concentrations, the odds of ACS increased by 454-fold. In this particular study, 8,9-DiHETrE elevated the odds of ACS by 92-fold [18].

Bioactive lipids have been measured conventionally by the use of chemical reagents, immunoassays, or chromatography [19]; however, these methods have limitations such as the lack of sensitivity and specificity. The main drawback of using conventional methods is that only one analyte can be assessed with one set of analysis. Considering the heterogeneity of pools of oxidized lipids, rapid multi-analyte quantification methods are needed. With the advent of robust mass spectrometric techniques, various groups of compounds can be assessed at the same time in a targeted and non-targeted fashion. By using soft ionization mass spectrometry (MS) such as electrospray ionization (ESI), the identification and quantification of non-volatile and thermolabile samples such as OxPL and oxylipins are feasible. Lipidomics is a powerful tool providing another layer of the detailed molecular levels of lipid assessments that may help to explore novel biomarkers and new treatment options in ACS [20].

In this article, we will briefly review the mechanisms in which bioactive lipids are generated. Then, we will focus on the analytical methods used by previous studies to measure these compounds. Finally, we will review the clinical studies that have assessed the roles of bioactive lipids in ACS patients.

2. Bioactive Lipid Generation

About 700 phospholipid (PL) molecules have been identified on the surface of LDL particles [6]. Phosphatidylcholine (PC) and sphingomyelin (SM) are the main PLs in LDL particles [21]. Most PLs contain polyunsaturated fatty acids (PUFAs), with 14–24 carbons in their sn-2 position, which make them susceptible to oxidation. They can undergo non-enzymatic oxidation mainly by ROS, making heterogeneous pools of oxidized lipids. Hydroperoxides (LOOH) are the first products of PUFA oxidation by ROS. During degradation of LOOH, a large variety of secondary products are produced such as 4-hydroxynonenal (4-HNE), malondialdehyde (MDA), non-fragmented (full length), and fragmented (shorten chain) OxPLs [22] (Figure 1).

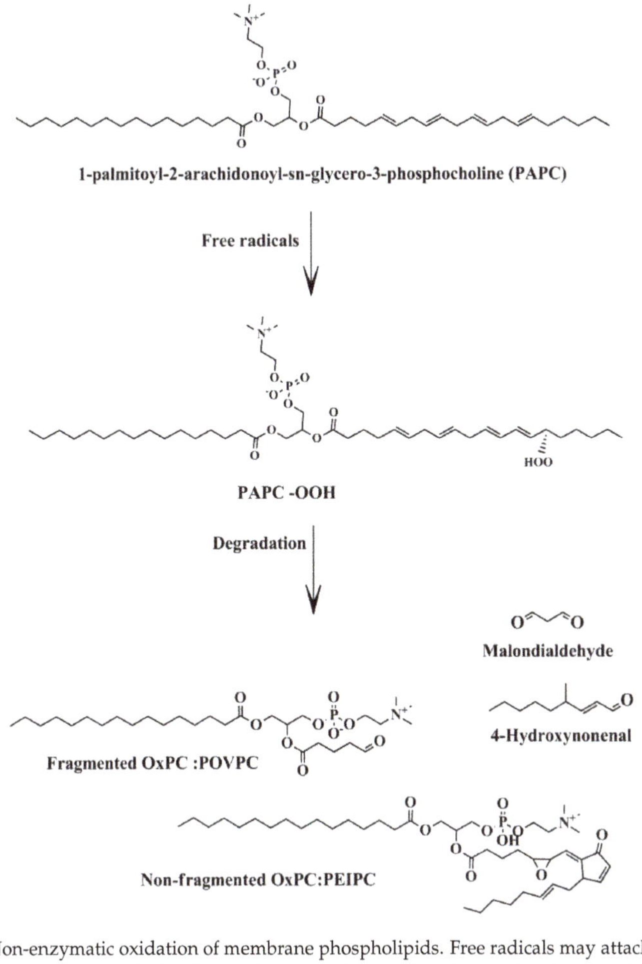

Figure 1. Non-enzymatic oxidation of membrane phospholipids. Free radicals may attack membrane phospholipids such as PAPC, leading to the production of bioactive lipid molecules. Abbreviations: PAPC-OOH, PAPC hydroproxide; OxPC, oxidized phosphatidylcholine; PEIPC, 1-palmitoyl-2-(5,6-epoxyisoprostane E2)-sn-glycero-3-phosphocholine.

4-HNE is a α,β-hydroxyalkenal which is formed through the peroxidation of arachidonic acid (AA) (20-carbon compounds) and linoleic acid (LA) (18-carbon compounds). Its reaction with the histidine, cysteine, or lysine residues of proteins makes Schiff bases or Michael adducts. MDA is a three-carbon aldehyde that is similarly produced through the non-enzymatic oxidation of PUFA. It can also be produced as a side product of thromboxane A2 (TXA2) synthesis. AA and docosahexaenoic acid (DHA) are the main precursors of MDA [23]. Levels of 4-HNE and MDA increase during oxidative stress and have been widely accepted as markers of oxidative stress.

OxPLs can be divided into two groups of non-fragmented (with the same number of carbon with precursor) and fragmented (with shorter chain) OxPLs. Non-fragmented OxPLs are formed following the initial phase of lipid oxidation. Then, they may undergo intramolecular cyclization, rearrangement, and further oxidation and make OxPLs with terminal furans, isoprostanes, and long-chain products with functional groups such as hydroperoxides, hydroxides, keto- and epoxy-groups [24]. Fragmented OxPLs have hydroxide and carbonyl groups in their structures, which are highly bioactive and can rapidly interact with biomolecules causing tissue injury [25] (Figure 2).

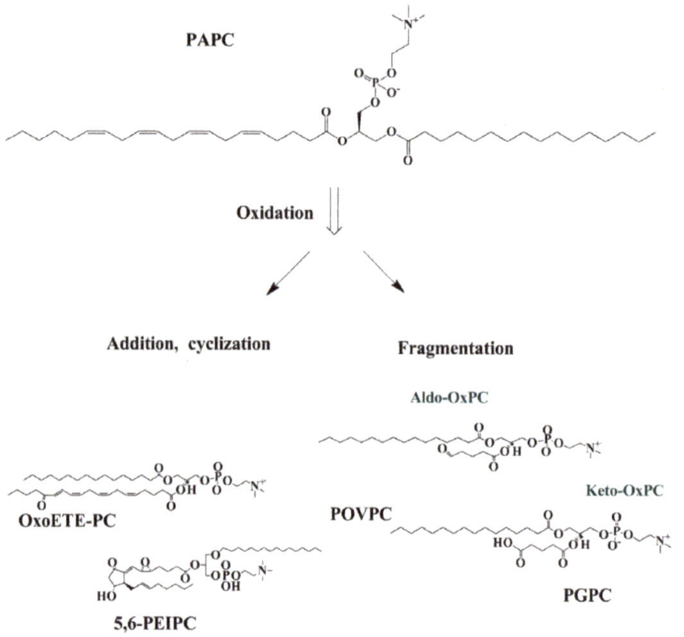

Figure 2. Fragmented and non-fragmented OxPC productions from PAPC. OxPLs can be classified as fragmented and non-fragmented species. Non- fragmented species are produced from the addition of peroxyl radicals where rearrangement/cyclization may happen. Fragmented species are comprised of aldehyde and carboxylic acid containing lipids. Abbreviations: Oxo-ETE-PC, oxoeicosatetraenoic acid phosphocholine; PEIPC, 1-palmitoyl-2-(5,6-epoxyisoprostane E2)-sn-glycero-3-phosphocholine; Aldo-OxPC, aldehyde containing oxidized phosphatidylcholine; Keto OxPC; carboxylic acid containing oxidized phosphatidylcholine; POVPC, 1-palmitoyl-2-(5′-oxo-valeroyl)-sn-glycero-3-phosphocholine; PGPC, 1-palmitoyl-2-glutaryl-sn-glycero-3-phosphocholine.

All PUFAs including omega-3 PUFAs are oxidized by the three main enzymes of cyclooxygenase (COX), lipoxygenase (LOX), and cytochrome P450 (CYP). The types of oxylipins produced from the PUFAs depend on the type/amounts of dietary PUFA, and the availability and affinity of the enzymes (COX, LOX, or CYP) for a specific substrate PUFA. The most well-known oxylipins are derived from AA and LA [1]. Half of the known oxylipins are derived from AA. However, other oxylipins can also be produced from PUFAs besides AA including both the omega-3 and omega-6 PUFA. It is important to mention that phospholipase-A2 (PL-A2), which has a key role in oxylipin production, has a preference for AA and eicosapentaenoic acid (EPA) [26]. These fatty acids may undergo enzymatic oxidation through cyclooxygenase (COX), lipoxygenase (LOX), and cytochrome P450 (CYP) pathways. Oxylipins are not stored in the cells and exert their biological roles through paracrine or autocrine mechanisms [27] before they are chemically inactivated or re-esterified into a glycerolipid pool [28] (Figure 3).

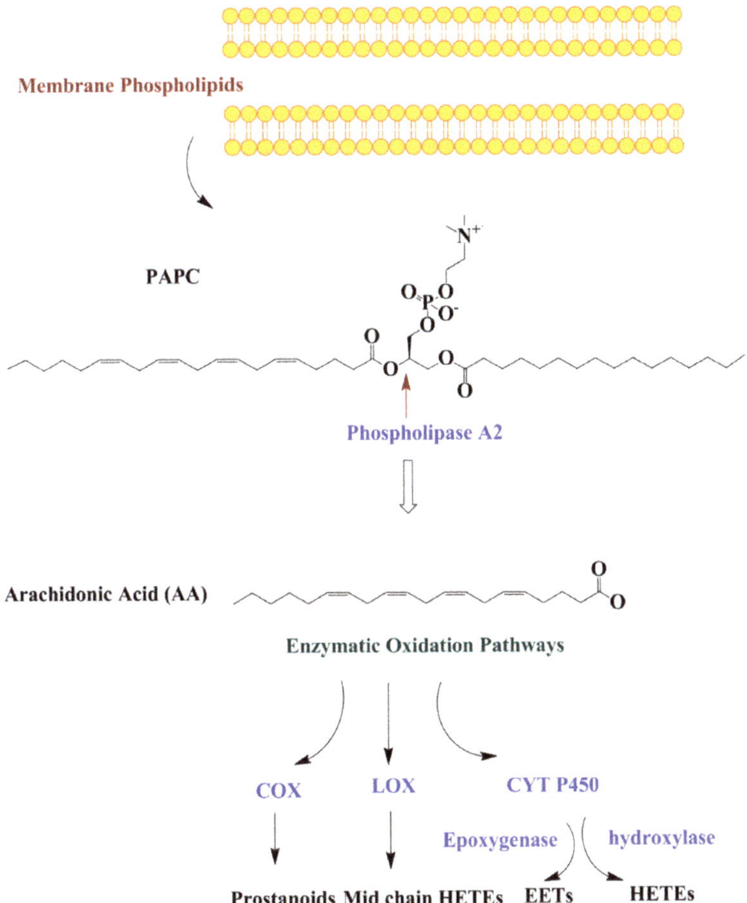

Figure 3. Enzymatic oxidation of membrane phospholipids. Fatty acids are released from the membrane PL by the phospholipase A2 enzyme and may undergo oxidation through three oxidation pathways including COX, LOX, and CYT P450. Abbreviations: COX, cyclooxygenase; LOX, lipoxygenase; CTY P450, cytochrome P450; HETE; hydroxyeicosatetraenoic acids; EET, epoxyeicosatrienoic acids.

Prostanoids (prostaglandines (PG) and thromboxanes (TX)) and some forms of hydroxy-metabolites such as 11-HETE are generated through the COX pathway from AA. LOX enzymes catalyze the generation of hydroxy fatty acids such as leukotrienes, lipoxins, resolvins, protectins, maresins, hepoxilins, and eoxins [3,29]. Mid chain (5-, 8-, 9-, 11-, 12-, and 15-) HETEs are also formed from AA through the LOX pathway [18,30]. CYP 450 enzymes have epoxygenase or ω-hydroxylase activity [29]. ω-terminal (16-, 17-, 18-, 19-, and 20-) HETEs are produced from AA and by ω-hydroxylase enzymes (CYP4A and CYP4F) and epoxyeicosatrienoic acid (EETs) are generated by CYPs with epoxygenase activity [28].

3. Measurement of Bioactive Lipids

3.1. 4-HNE and MDA

Free aldehydes can be identified and quantified by several analytical methods. Thiobarbituric acid reactive substance (TBARS)/spectrophotometry has been widely employed to measure MDA levels. Under acidic conditions and high temperatures, the aldehyde group of MDA reacts with the nucleophilic center of TBA and makes a red-colored derivative, which can be detected by spectrophotometric and spectrofluorometric approaches. The aldehyde group of HNE can also make derivatives with 2,4-dinitrophenylhydrazine (DNPH) that are detectable by spectrophotometry [12,14,31,32].

Kamiński et al. (2008) measured the HNE and MDA in the plasma of 15 STEMI patients and 10 patients with stable IHD as the control group by using derivatization/high performance lipid chromatography (HPLC)-fluorescence detection [33]. Solid phase extraction was applied to extract HNE and MDA [34]. MDA was also detected using the TBARS derivatization, and then separated and quantified by HPLC-spectrofluorometric assay [35].

Gas chromatography (GC) is the other main analytical method to measure MDA and HNE. MS is more specific and sensitive compared with other analytical methods as it can identify these aldehydes based on the mass to charge ratio and fragmentation pattern [36]. GC can also be coupled to MS. Tsikas et al. (2017) developed a method to measure the plasma concentrations of both MDA and HNE simultaneously by using GC/MS. They used pentafluorobenzyl hydroxylamine as a derivatization reagent, [1,3-2H2]-MDA (d2-MDA), and [9,9,9-2H3]-HNE (d3-HNE) as the internal standards. The ionization technique used here was hard ionization with high energy such as electron impact [37] and is different from soft ionization, which will be discussed later.

Syslová et al. (2009) developed a method using reverse phase HPLC/ESI-MS to assess the MDA and HNE in plasma, urine, and exhaled breath condensate [38]. HNE-d3 and Me-MDA was used as the internal standards, butylated hydroxytoluene (BHT) as the antioxidant, and acetonitrile was added to the plasma. Then, the plasma was sonicated and centrifuged to remove the precipitated proteins. The supernatant was dried under nitrogen gas, and re-suspended in acetonitrile to be injected into HPLC. HPLC with a Hypercarb Thermo 100 mm × 2.1 mm × 5 mm column and Hypercarb-precolumn was used. Water and ammonium hydroxide were used as solvent A, and methanol:acetonitrile with ammonium hydroxide was used as the co-solvent. Derivatization with 4-2-trimethylammonio ethoxy benzenaminiumhalide (4-APC) or cyclohexanedione (CHD) can be also done prior to extraction to increase the ionization of these aldehydes [39,40].

3.2. OxPL

Using monoclonal antibodies is one of the well-established methods to assess the OxPL levels on apoB100-containing lipoproteins, namely LDL, very low-density (VLDL), and lipoprotein (a) (LP (a)). To perform this assay, the murine monoclonal antibody MB47 must be added initially to capture all apoB-100 lipoproteins from the plasma. Then, by adding the E06 antibody, it can bind to apoB-100. This method has been applied extensively to measure the OxPL levels in CVD [41]. The limitation of this method is that only the OxPL species that are present on the apoB100 lipoproteins can be assessed, and not the total amount of OxPL in the plasma. In addition, this method cannot identify specific OxPL species among all types of OxPL (fragmented, non-fragmented OxPL) that are produced during lipid oxidation [19]. To overcome these limitations, LC/MS has been introduced as the best option for a detailed analysis of OxPLs.

Hassanaly et al. (2017) measured the levels of oxidized phosphatidylinositol (OxPI) in Ox-LDL and human atherosclerotic plaque by using reversed-phase HPLC/ESI-MS. Using this approach, they were able to identify and quantify 23 OxPI species in human Ox-LDL and atherosclerotic plaque. They found that levels of OxPI species increased significantly in Ox-LDL at 48 h when compared with the baseline. Moreover, non-fragmented hydroperoxides were the dominant species in Ox-LDL at 48 h, comprising 52.07% of the total OxPI species. Fragmented aldehyde and carboxylic acid containing OxPI comprised 17.32% and 0.89% of total the OxPI at the same time point. Likewise, in human atherosclerotic plaques, which were retrieved from patients who underwent saphenous vein graft (SVG) interventions, non-fragmented hydroperoxides were the most abundant OxPI compounds. Fragmented aldehyde and carboxylic acid containing OxPI comprised 18.6% and 1.5 % of the total OxPI compounds, respectively [42].

OxPC species have been identified in patients that have undergone percutaneous coronary and peripheral procedures by using normal phase HPLC/ESI-MS. In this study, the five most abundant OxPCs were in embolized material captured by distal protection filter devices during uncomplicated saphenous vein graft, carotid, renal, and superficial femoral artery interventions. 1-palmitoyl-2-(9-oxo-nonanoyl) PC (PONPC) was the most abundant fragmented OxPC, which comprised 50% of the total quantified fragmented OxPC compounds. POVPC, 1-palmitoyl-2-glutaroyl-sn-glycero-3-phosphocholine (PGPC), and 1-palmitoyl-2-(5-keto-6-octene-dioyl) PC (KOdiAPC) were the other fragmented OxPC species measured in this study [43].

Recently, we were able to identify and quantify 56 OxPC species including both fragmented and non-fragmented OxPCs in rat kidneys following ischemia/reperfusion (I/R) injury. 1-stearoyl-2-linoleoyl-phosphatidylcholine (SLPC-OH) and 1-palmitoyl-2-azelaoyl-sn-glycero-3-phosphocholine (PAzPC) were the most abundant non-fragmented and fragmented OxPC after I/R, respectively. The total levels of OxPC species (including fragmented and non-fragmented OxPC compounds) increased significantly after both 6 h and 24 h reperfusion when compared with the sham group. Concentrations of fragmented OxPCs were elevated significantly by increasing the time of reperfusion as their levels were significantly higher following 24 h reperfusion when compared to 6 h I/R and sham groups. However, no significant differences were observed between the sham and 6 h I/R groups. Changes in the levels of non-fragmented OxPCs were different to the fragmented compounds. Although the total levels of non-fragmented OxPC elevated significantly in the 6 h I/R group, no differences were observed in the 24 h I/R group. These data pointed to the importance of identifying the specific compounds, and not just the total concentrations of the oxidized species [44].

The first step in preparing samples for lipidomic analysis is extracting the lipids from the cell/tissue/plasma. Currently, conventional liquid–liquid extraction has been widely used for the extraction of OxPL [45]. Folch extraction, which uses chloroform/methanol, is one of the most common extraction approaches to extract OxPL. Adding antioxidants such as BHT is recommended to minimize further oxidation [46]. Recently, it has been suggested that enrichment strategies such as using gold nanoparticles (GNP) and anti-Ox-LDL antibodies on plasma samples [47] or lipid extracts [48] may increase the efficacy of OxPC identification. Hinterwirth et al. [47] used GNPS with four different Ox-LDL antibodies, namely the E06, anti-Cu Ox-LDL antibody, anti-MDA-LDL antibody, and anti-carboxymethyllysine-LDL antibody, to increase the detection of OxPC in plasma. Stübiger et al. [48] also reported that using 2-aminobenzoic acid (2-AA) as the reagent with GNP elevated the carbonyl-containing OxPC identification at subnanomolar concentrations, with up to 90% recoveries [49].

To separate the OxPLs species, reverse phase HPLC with C8 or C18 columns with either isocratic or gradient elutions has been widely used, although they can also be separated by normal phase HPLC [43]. By using HPLC, OxPL are separated based on polarity and molecular weight before interfacing with the MS, which increases the sensitivity of the assessment [50]. Reverse phase mobile phases are usually a mixture of water, methanol, or acetonitrile. Hexane or isopropanol can also be applied as co-solvents. Ammonium acetate, ammonium formate, or acetic or formic acid may also be added to the solvent to facilitate ionization in MS. There are no detruded internal standards for OxPLs analysis. Non-oxidized PLs and lyso PL (LPL) such as phosphatidylinositol (PI) (31:1) for OxPI analysis and PC (9:0)/LPC(17:0) for OxPC analysis have been used as internal standards [42,44]. These PLs have the same structures and fragmentation patterns and are not produced in the body. Therefore, they can be used as an internal standard to assess the extraction efficacy and instrument response [51].

ESI/MS and matrix-assisted laser desorption/ionization (MALDI) are two forms of soft ionization techniques. The soft ionization technique allows for the analysis of non-volatile compounds such as OxPL. ESI can readily interface with HPLC. This is very important when analyzing OxPLs as the levels of these compounds are considerably lower when compared with non-oxidized compounds. Therefore, separation techniques prevent ion suppression, which may occur with high abundant molecular ions. On the other hand, sample preparation is simpler with ESI when compared with MALDI as MALDI needs the co-crystallization of a matrix with the sample, which consequently may affect the quantification of the analytes. MALDI can examine solid state samples and is useful for MS imaging of tissue, while ESI needs tissue extraction as it requires a liquid sample [52]. Some studies have used MALDI to quantify chlorinated PL [53,54]. However, no study has measured the levels of oxidized lipids in tissue using MALDI. In a review by Ana Reis (2017), it was emphasized that MS imaging to assess the distribution of OxPL in tissue is challenging due to the low concentrations of OxPL/PL and the lack of fluorescent probes designed to bind to free OxPL in tissue samples [45].

3.3. Oxylipins

Like other oxidized lipids, traditional analytical methods have been widely used for assessing oxylipins [55]. Miler et al. (1985) developed enzyme-linked immunosorbent assays (ELISA) to assess LTC4, LTB4, 6-keto PGF1 alpha, and TXB2 [56]. The main drawback of this approach is that only one analyte can be targeted with one set of analysis. GC/MS has also been utilized to measure oxylipins, for instance, Tsukamoto et al. (2002) developed a method to measure oxylipins including PG, isoprostane and TXs with GC/MS [57]. Due to the complex sample preparation and thermal decomposition during derivatization, HPLC based methods have been used recently for oxylipin analysis [55].

HPLC/ESI-MS has been utilized to quantify plasma oxylipins in patients [18,58]. For instance, Caligiuri et al. (2017) quantified 39 plasma oxylipins in patients with PAD using HPLC/ESI-MS [18]. Among all the identified/quantified oxylipins, 4 oxylipins were significantly correlated with the presence of cardiovascular/cerebrovascular events. For instance, plasma levels of 8,9 DiHETrE were significantly elevated in patients with ACS when compared with ones without ACS (0.3 ± 0.1 versus 0.2 ± 0.0 nM, respectively). Plasma concentrations of PGE2 were significantly higher in patients with angina when compared with subjects without angina (0.4 ± 0.0 versus 0.3 ± 0.0 nM, respectively). Moreover, they found that 16-HETE, TRX B2, and 11,12- DiHETrE increased the odds of having cardiovascular/cerebrovascular events in this population.

To prepare samples for lipidomic analysis, oxylipins can be extracted using liquid–liquid extraction and/or solid-phase extraction procedures. Use of chloroform/methanol mixtures, according to Bligh and Dyer, is the most common liquid–liquid extraction protocol for oxylipins. In this method, oxylipins are dissolved in organic solvents, but hydrophilic materials such as proteins are eliminated following phase separation. Solid-phase extraction can be conducted using commercial columns pre-packed with various sorbents. Reverse-phase HPLC with a C18 column has been used widely to separate oxylipins [18]; however, Zu 2016 et al. used ultra-performance liquid chromatography (UPLC) (C18 column) to separate oxylipins before analysis with MS [15]. The UPLC column has better resolution, lower detection limits, and a shorter chromatographic run when compared with HPLC [51]. Deuterated oxylipins are commercially available, which are used as internal standards. These standards are matched with groups of endogenous oxylipin species in terms of chemistry, retention time, and ionization efficiencies [51]; as mentioned previously, internal standards are needed to assess the extraction efficacy and instrument response [51].

Tandem mass spectrometry is the most sensitive system for analyzing oxidized lipids, particularly when predetermined species are desired, and is known as the targeted approach. Multiple-reaction monitoring (MRM) is an acquisition mode that monitors the transition of a selected precursor ion, based on the mass/charge value, to a specific product ion using the fragmentation pattern. It has been reported that by using separation techniques such as UPLC and MRM transitions, more than hundreds of oxylipins can be identified/quantified in a single acquisition at picogram/fentomole levels [59].

Quantifications of oxidized lipids can be carried out by generating calibration curves for internal standards. As mentioned previously, deuterated oxylipins are commercially available, which can be used as oxylipin internal standards. However, non-oxidized PLs and lyso PL (LPL) such as PI (31:1), PC (9:0), and LPC (17:0) have been used as internal standards for OxPI and OxPC quantifications as there is no deuterated standard for OxPL analysis.

4. Role of Bioactive Lipids in ACS

4.1. 4-HNE and MDA

Previous studies have shown that 4-HNE may contribute to many cardiovascular diseases (CVD) [60–62]. It can be generated during Ox-LDL oxidation and makes apo B-adducts, which are identified by scavenger receptors, leading to elevated uptakes of Ox-LDL by macrophages and the formation of foam cells. Previous studies were able to identify HNE-adducts in human atherosclerosis lesions by using anti-HNE antibodies [63,64]. The role of HNE in ACS has not been studied well; however, a study by Gargiulo et al. (2015) showed that HNE may induce plaque instability by increasing the expression and synthesis of inflammatory cytokines such as interlukine-8 (IL-8), interlukine1-β (IL-1β), tumor necrosis factor-α (TNF-α), and matrix metalloproteinase-9 (MMP-9) via Toll-like Receptors 4/Nuclear Factor-κB (TLR4/NF-κB) signaling pathways [65]. In addition, a recent study showed that levels of HNE in coronary sinus were significantly higher in STEMI patients before and after percutaneous coronary intervention (PCI) when compared with patients with stable ischemic heart disease (IHD) who underwent elective PCI [33].

In the last 30 years, numerous studies have extensively shown that elevated levels of MDA are associated with CVD. Having traditional CVD risk factors such as cigarette smoking [21,66], hypertension [66], hyperlipidemia [67,68], and diabetes [69,70] were reported to be significantly correlated with higher MDA levels. Increased levels of MDA have been reported in plasma of patients with atherosclerotic diseases [71]. A nested case-control cohort showed that LDL-MDA was a strong predictor of carotid wall thickness in hypercholesterolemic men [72]. In a perspective study with 634 patients having CVD, serum levels of MDA were strong predictors of cardiovascular events (including MI, stroke, hospitalizations for non-fatal cardiovascular events mainly UA), and major vascular procedures (percutaneous transluminal coronary angioplasty (PTCA)/coronary artery bypass grafting (CABG)), independent of traditional risk factors such as blood pressure (BP), total cholesterol, high-density lipoprotein-cholesterol (HDL-cholesterol), LDL-cholesterol, triglycerides (TG), age, gender, body mass index (BMI), and inflammatory markers in patients with coronary heart disease (CHD) [73]. In a study by Bagatini et al. (2011), increased levels of MDA were observed in MI patients and subjects with CVD risk factors (including cigarette smoking, hypertension, and family history of CHD) when compared to healthy controls [74].

4.2. OxPL

Atherogenicity of OxPLs was shown by Hörkkö et al. (1999) as they found that OxPLs contribute to Ox-LDL recognition by macrophages. They also found that the monoclonal antibody E06, which binds to the phosphocholine head group of PLs on Ox-LDL, inhibits Ox-LDL uptakes by macrophages [75]. Since then, several studies have shown that OxPLs may have roles in various steps of atherosclerosis such as facilitating Ox-LDL uptake by macrophages [76], mediating cellular inflammatory responses [77], and stimulating angiogenesis [25].

1-palmitoyl-2-oxovaleroyl-sn-glycero-3-phosphorylcholine (POVPC), and 1-palmitoyl-2-glutaroyl-sn-glycero-3-phosphorylcholine (PGPC), which are derived from arachidonyl phosphatidylecholines, are produced during Ox-LDL modification and have been identified in atherosclerotic plaques [78,79]. These fragmented OxPLs are toxic and create tissue injury through inflammatory responses [77] and apoptosis [80]. LP (a) is the main carrier of OxPLs in plasma, although they can also be transferred by LDL and HDL [81]. Previous studies have demonstrated that levels of OxPLs are strongly correlated with LP (a) levels and the extent of coronary stenosis [82,83]. Therefore, it has been suggested that the atherogenicity of LP (a) can be attributed to OxPLs as its a carrier of proinflammatory oxidized molecules.

Tsimikas et al. (2003) developed a method to measure OxPL by using murine monoclonal antibody E06 [13]. This antibody binds to the phosphorylcholine (PC) head group of OxPL, particularly POVPC. Therefore, the amount of PC-OxPL per apoB-100 (OxPL/ApoB) containing lipoproteins can be calculated. By using this approach, they showed that OxPL levels increased significantly in MI patients after PCI, suggesting that these compounds are released and/or generated as a result of plaque rupture [13,84]. Prospective studies have shown that OxPLs levels can be considered as biomarkers of atherosclerosis progression, cardiovascular death, MI, and stroke. In a prospective Bruneck study, the 5-year follow-up of 700 participants aged 40 to 79 years old showed that OxPLs levels were strongly and significantly associated with the presence, extent, and development of carotid and femoral atherosclerosis, and predicted the presence of symptomatic CVD [85]. The ten-year follow-up of this population showed that risk of cardiovascular events, which was defined as cardiovascular death, MI, stroke, and transient ischemic attack (TIA), were significantly elevated in participants in the highest tertile of OxPLs/apoB than those in the lowest tertile independent of traditional risk factors, suggesting that OxPLs/apoB levels may predict the risk of 10-year CVD events [86].

4.3. Oxylipins

Pioneering studies have shown the association between oxylipins derived from AA with UA and atherosclerosis. Elevated levels of TXB2 have been reported in the coronary circulation of patients with unstable angina [87,88]. Moreover, Mallat Z et al. (1999) showed that HETEs levels were significantly higher in plaques obtained from symptomatic patients (with unstable plaque) versus patients with stable plaques [89]. Similarly, Waddington et al. (2001) found higher levels of 15-HETE and 11-HETE in atherosclerotic plaque retrieved from individuals undergoing carotid endarterectomy [90]. Recently, a targeted metabolomics study showed that among all identified metabolites, 20-HETE was the only compound that was significantly higher in patients with atheroma plaque when compared with healthy subjects [91].

New studies have also investigated the role of oxylipins in the diagnosis and prognosis of ACS and MI. A retrospective nested case-control study, comprised of 470 ACS patients and 39 subjects without CHD as a control group, was conducted in a Chinese population [15]. Among the ACS patients, subjects who had had a major adverse cardiovascular event (MACE) during the 1037 days of follow up period were identified as the MACE group, and ACS patients without MACE during this period were named as the non-MACE group. In this study, LTB4, 8-HETE, 11-HETE, 12-HETE, and 15-HETE were significantly elevated in the ACS patients (both the MACE and non- MACE groups) when compared with the controls. In addition, the levels of 5-HETE and 9-HETE were significantly higher in the MACE group when compared with the controls, suggesting the potential diagnostic value of these oxylipins in ACS. In addition, the levels of 20-HETE were significantly elevated in the STEMI group when compared with the non-STEMI group, indicating that the pathogenesis of STEMI and non-STEMI may be different. Moreover, the 19-HETE levels, a vasodilator oxylipin, were significantly lower in the MACE group than the non-MACE and control groups. ACS patients who had higher levels of 19-HETE (higher than 0.13 ng/mL) tended to have better prognosis (up to 72%) than those with lower levels [15]. In a prospective study by Sun et al. (2016) [92], the association between oxylipins and the incidence of MI was investigated in 744 AMI cases and 744 matched controls, aged 47–83 years within the Singapore Chinese Health Study. They found inverse correlations between pro-thrombotic TXB2 and AMI risk, and suggested that this unexpected association was more related to sample collection, processing, and storage conditions than biological differences. Moreover, in this study, only 19 oxylipins, which had potential roles in inflammation, blood pressure, and platelet degranulation were measured, and not the full spectrum. In a study by Caligiuri et al. (2017), the associations between oxylipins and the occurrence of cardiovascular/cerebrovascular events, defined as STEMI, non-STEMI, and UA, in 24 patients with peripheral artery disease (PAD) were assessed. They found that levels of 16-HETE, TXB2, and 11,12-DiHETrE were significantly associated with increased odds of cardiovascular/cerebrovascular events in PAD patients and showed that with every 1 nmol/L increase in 8,9-DiHETrE concentrations, the odds of ACS increased by 454-fold. In this particular study, 8,9-DiHETrE elevated the odds of ACS by 92-fold [18].

All of the clinical studies that have assessed these bioactive lipids in ACS patients are presented in Table 1.

Table 1. Clinical studies that have assessed bioactive lipids in ACS patients

Oxidized Lipids	Author	Year	Method of Detection	Population	Results
HNE/MDA	Aznar [4]	1983	TBARS-spectrophotometry	MI patients, patients with angina pectoris (AP), and normal control group	MDA values were normal in AP patients. MDA levels increased significantly following MI and reached maximum levels in 6–8 days
	De Scheerder [32]	1991	TBARS-spectrophotometry	CABG surgery candidates	Levels of MDA increased after repetitive coronary occlusions during coronary angioplasty. After 5-min of reperfusion, MAD levels further increased. Following 15-min of reperfusion, it returned to baseline levels
	Walter [73]	2004	HPLC-spectrophotometry	Patients with documented CAD	Baseline levels of MDA were higher in patients who had major/nonfatal MI, and major vascular procedures after three-year study
	Kaminski [33]	2009	HPLC-spectrophotometry	STEMI patient and stable angina patients (as controls)	Higher HNE and MDA levels in STEMI patients compared to controls
	More [14]	2017	TBARS-spectrophotometry	MI patients and normal healthy control	Higher MDA levels in MI patients compared to control
	Ismail [12]	2018	TBARS-spectrophotometry	MI patients and healthy controls	Higher MDA levels in MI patients compared to control
OxPL	Tsimikas [13]	2003	OxPL/ApoB	Patients with ACS (MI and unstable angina), stable angina and healthy subjects	Baseline levels of OxPL/ApoB were significantly higher in ACS patients compared with stable angina and healthy controls. In MI patients, OxPL/ApoB increased by 54% and 36% at hospital discharge and 30 days, respectively
	Tsimikas [84]	2004	OxPL/ApoB	Patients with stable angina pectoris undergoing PCI	OxPL/ApoB levels increased following PCI
	Tsimikas [83]	2005	OxPL/ApoB	CAD patients underwent coronary angiography	Percentage of stenosis was correlated with OxPL/apoB levels. OxPL/apoB levels predicted CAD independent of all other clinical markers except for LP (a)
	Tsimikas [86]	2006	OxPL/ApoB	Subjects aged 40 to 79 year-old followed for 5 years	OxPL/ApoB levels predicted the presence of symptomatic CVD
	Kaminski [33]	2007	OxPL/ApoB	Subjects aged 40 to 79 year-old followed for 10 years	OxPL/ApoB levels predicted future cardiovascular events independent of FRS
	Byun [16]	2015	OxPL/ApoB	Patients treated with intensive versus moderate atorvastatin therapy: the TNT trial	OxPL/apoB levels predicted secondary MACE
	Leibundgut [93]	2016	OxPL/plasminogen (PLG) and plasminogen	Patients with stable angina	OxPL/PLG and plasminogen decreased significantly immediately after PCI, rebounded to baseline at 6 h post-PCI, peaked at 3 days and slowly returned to baseline by 6 months
	Byun [17]	2017	OxPL/ApoB	Patients with prior stroke or TIA	Elevated baseline levels of OxPL/apoB predicted recurrent stroke and first major coronary events after five-year follow up
Oxylipins	Strassburg [58]	2012	HPLC-MS	Patient underwent cardiac surgery	Increased levels of 12-HETE and 12-HEPE at 24 h-post cardiac surgery
	Zu [15]	2016	UPLC-MS	ACS patients with or without MACE during follow up	20-HETE level was significantly higher in STEMI group comparing with NSTEMI. ACS patients with 19-HETE levels tended to have better prognosis for MACE.
	Auguet [91]	2018	HPLC-MS	Patients who underwent carotid endarterectomy	20-HETE levels were significantly higher in patients with atheroma plaque than healthy subjects
	Caligiuri [18]	2017	HPLC-MS	Patients with PAD	8,9-DiHETrE increased the odds of ACS. A positive relationship was observed between plasma concentrations of 18-HEPE and ACS.

5. Conclusions

There is accumulating evidence that bioactive lipids play roles in ischemic cardiovascular disease. We have made great strides in elucidating their activity by utilizing antibody based approaches. Given the advances in mass spectrometry, we were able to identify and quantitate individual oxidized lipids in plasma. It is important that we standardize the current mass spectrometric methods of quantitation and analysis, so that large cohorts of patients can be analyzed. This would lead to a better understanding of the specific contribution of each lipid molecule to the overall pathophysiology.

Author Contributions: Writing—Original Draft Preparation, Z.S.; Writing—Review & Editing, A.R.

Conflicts of Interest: The authors declare no conflicts of interest.

References

1. Vedanthan, R.; Seligman, B.; Fuster, V. Global perspective on acute coronary syndrome: A burden on the young and poor. *Circ. Res.* **2014**, *114*, 1959–1975. [CrossRef] [PubMed]
2. Ambrose, J.A.; Singh, M. Pathophysiology of coronary artery disease leading to acute coronary syndromes. *F1000Prime Rep.* **2015**, *7*, 8. [CrossRef] [PubMed]
3. Makki, N.; Brennan, T.M.; Girotra, S. Acute coronary syndrome. *J. Intensive Care Med.* **2015**, *30*, 186–200. [CrossRef] [PubMed]
4. Agewall, S. Acute and stable coronary heart disease: Different risk factors. *Eur. Heart J.* **2008**, *29*, 1927–1929. [CrossRef] [PubMed]
5. Shah, P.K. Mechanisms of plaque vulnerability and rupture. *J. Am. Coll. Cardiol.* **2003**, *41*, S15–S22. [CrossRef]
6. Kattoor, A.J.; Pothineni, N.V.K.; Palagiri, D.; Mehta, J.L. Oxidative Stress in Atherosclerosis. *Curr. Atheroscler. Rep.* **2017**, *19*, 42. [CrossRef] [PubMed]
7. Li, D.; Liu, L.; Chen, H.; Sawamura, T.; Mehta, J.L. LOX-1, an oxidized LDL endothelial receptor, induces CD40/CD40L signaling in human coronary artery endothelial cells. *Arterioscler. Thromb. Vasc. Biol.* **2003**, *23*, 816–821. [CrossRef] [PubMed]
8. Holvoet, P.; Vanhaecke, J.; Janssens, S.; Van de Werf, F.; Collen, D. Oxidized LDL and malondialdehyde-modified LDL in patients with acute coronary syndromes and stable coronary artery disease. *Circulation* **1998**, *98*, 1487–1494. [CrossRef] [PubMed]
9. Toshima, S.-I.; Hasegawa, A.; Kurabayashi, M.; Itabe, H.; Takano, T.; Sugano, J.; Shimamura, K.; Kimura, J.; Michishita, I.; Suzuki, T. Circulating oxidized low density lipoprotein levels: A biochemical risk marker for coronary heart disease. *Arterioscler. Thromb. Vasc. Biol.* **2000**, *20*, 2243–2247. [CrossRef] [PubMed]
10. Ehara, S.; Ueda, M.; Naruko, T.; Haze, K.; Itoh, A.; Otsuka, M.; Komatsu, R.; Matsuo, T.; Itabe, H.; Takano, T. Elevated levels of oxidized low density lipoprotein show a positive relationship with the severity of acute coronary syndromes. *Circulation* **2001**, *103*, 1955–1960. [CrossRef] [PubMed]
11. Förstermann, U.; Xia, N.; Li, H. Roles of vascular oxidative stress and nitric oxide in the pathogenesis of atherosclerosis. *Circ. Res.* **2017**, *120*, 713–735. [CrossRef] [PubMed]
12. Ismail, M.K.; Samera, M.; Abid, S. Oxidative stress markers and antioxidant activity in patients admitted to Intensive Care Unit with acute myocardial infarction. *Int. J. Health Sci.* **2018**, *12*, 14.
13. Tsimikas, S.; Bergmark, C.; Beyer, R.W.; Patel, R.; Pattison, J.; Miller, E.; Juliano, J.; Witztum, J.L. Temporal increases in plasma markers of oxidized low-density lipoprotein strongly reflect the presence of acute coronary syndromes. *J. Am. Coll. Cardiol.* **2003**, *41*, 360–370. [CrossRef]
14. More, H.; Pujari, K.; Jadkar, S.; Patil, C. Biochemical parameters in acute myocardial infarction with or without co-morbidities. *J. Med. Sci. Clin. Res.* **2017**, *5*, 17299–17304. [CrossRef]
15. Zu, L.; Guo, G.; Zhou, B.; Gao, W. Relationship between metabolites of arachidonic acid and prognosis in patients with acute coronary syndrome. *Thromb. Res.* **2016**, *144*, 192–201. [CrossRef] [PubMed]
16. Byun, Y.S.; Lee, J.-H.; Arsenault, B.J.; Yang, X.; Bao, W.; DeMicco, D.; Laskey, R.; Witztum, J.L.; Tsimikas, S.; Investigators, T.T. Relationship of oxidized phospholipids on apolipoprotein B-100 to cardiovascular outcomes in patients treated with intensive versus moderate atorvastatin therapy: The TNT trial. *J. Am. Coll. Cardiol.* **2015**, *65*, 1286–1295. [CrossRef] [PubMed]
17. Byun, Y.S.; Yang, X.; Bao, W.; DeMicco, D.; Laskey, R.; Witztum, J.L.; Tsimikas, S.; Investigators, S.T. Oxidized phospholipids on apolipoprotein B-100 and recurrent ischemic events following stroke or transient ischemic attack. *J. Am. Coll. Cardiol.* **2017**, *69*, 147–158. [CrossRef] [PubMed]
18. Caligiuri, S.P.; Aukema, H.M.; Ravandi, A.; Lavallée, R.; Guzman, R.; Pierce, G.N. Specific plasma oxylipins increase the odds of cardiovascular and cerebrovascular events in patients with peripheral artery disease. *Can. J. Physiol. Pharmacol.* **2017**, *95*, 961–968. [CrossRef] [PubMed]
19. Spickett, C.M.; Pitt, A.R. Oxidative lipidomics coming of age: Advances in analysis of oxidized phospholipids in physiology and pathology. *Antioxid. Redox Signal.* **2015**, *22*, 1646–1666. [CrossRef] [PubMed]
20. Spickett, C.M. The lipid peroxidation product 4-hydroxy-2-nonenal: Advances in chemistry and analysis. *Redox Biol.* **2013**, *1*, 145–152. [CrossRef] [PubMed]

21. Hevonoja, T.; Pentikäinen, M.O.; Hyvönen, M.T.; Kovanen, P.T.; Ala-Korpela, M. Structure of low density lipoprotein (LDL) particles: Basis for understanding molecular changes in modified LDL. *Biochim. Et Biophys. Acta (Bba)-Mol. Cell Biol. Lipids* **2000**, *1488*, 189–210. [CrossRef]
22. Yin, H.; Xu, L.; Porter, N.A. Free radical lipid peroxidation: Mechanisms and analysis. *Chem. Rev.* **2011**, *111*, 5944–5972. [CrossRef] [PubMed]
23. Papac-Milicevic, N.; Busch, C.-L.; Binder, C.J. Malondialdehyde epitopes as targets of immunity and the implications for atherosclerosis. *Adv. Immunol.* **2016**, *131*, 1–59. [PubMed]
24. Reis, A.; Spickett, C.M. Chemistry of phospholipid oxidation. *Biochim. Et Biophys. Acta (Bba)-Biomembr.* **2012**, *1818*, 2374–2387. [CrossRef] [PubMed]
25. Bochkov, V.; Gesslbauer, B.; Mauerhofer, C.; Philippova, M.; Erne, P.; Oskolkova, O.V. Pleiotropic effects of oxidized phospholipids. *Free Radic. Biol. Med.* **2017**, *111*, 6–24. [CrossRef] [PubMed]
26. Gabbs, M.; Leng, S.; Devassy, J.G.; Monirujjaman, M.; Aukema, H.M. Advances in Our Understanding of Oxylipins Derived from Dietary PUFAs. *Adv. Nutr. (BethesdaMd.)* **2015**, *6*, 513–540. [CrossRef] [PubMed]
27. Tourdot, B.E.; Ahmed, I.; Holinstat, M. The emerging role of oxylipins in thrombosis and diabetes. *Front. Pharmacol.* **2014**, *4*, 176. [CrossRef] [PubMed]
28. Shearer, G.C.; Walker, R. An overview of the biologic effects of omega-6 oxylipins in humans. *Prostaglandins Leukot. Essent. Fat. Acids* **2018**, *133*, 8–12. [CrossRef] [PubMed]
29. Virtanen, J.K. Randomized trials of replacing saturated fatty acids with n-6 polyunsaturated fatty acids in coronary heart disease prevention: Not the gold standard? *Prostaglandins Leukot. Essent. Fat. Acids* **2018**, *133*, 8–12. [CrossRef] [PubMed]
30. Oni-Orisan, A.; Edin, M.L.; Lee, J.A.; Wells, M.A.; Christensen, E.S.; Vendrov, K.C.; Lih, F.B.; Tomer, K.B.; Bai, X.; Taylor, J.M. Cytochrome P450-derived epoxyeicosatrienoic acids and coronary artery disease in humans: A targeted metabolomics study. *J. Lipid Res.* **2016**, *57*, 109–119. [CrossRef] [PubMed]
31. Aznar, J.; Santos, M.; Valles, J.; Sala, J. Serum malondialdehyde-like material (MDA-LM) in acute myocardial infarction. *J. Clin. Pathol.* **1983**, *36*, 712–715. [CrossRef] [PubMed]
32. De Scheerder, I.; Van de Kraay, A.; Lamers, J.; Koster, J.; de Jong, J.W.; Serruys, P. Myocardial malondialdehyde and uric acid release after short-lasting coronary occlusions during coronary angioplasty: Potential mechanisms for free radical generation. *Am. J. Cardiol.* **1991**, *68*, 392–395. [CrossRef]
33. Kamiński, K.; Bonda, T.; Wojtkowska, I.; Dobrzycki, S.; Kralisz, P.; Nowak, K.; Prokopczuk, P.; Skrzydlewska, E.; Kozuch, M.; Musial, W. Oxidative stress and antioxidative defense parameters early after reperfusion therapy for acute myocardial infarction. *Acute Card. Care* **2008**, *10*, 121–126. [CrossRef] [PubMed]
34. Yoshino, K.; Matsuura, T.; Sano, M.; Saito, S.-I.; Tomita, I. Fluorometric liquid chromatographic determination of aliphatic aldehydes arising from lipid peroxides. *Chem. Pharm. Bull.* **1986**, *34*, 1694–1700. [CrossRef] [PubMed]
35. Londero, D.; Greco, P.L. Automated high-performance liquid chromatographic separation with spectrofluorometric detection of a malondialdehyde-thiobarbituric acid adduct in plasma. *J. Chromatogr. A* **1996**, *729*, 207–210. [CrossRef]
36. Tsikas, D. Assessment of lipid peroxidation by measuring malondialdehyde (MDA) and relatives in biological samples: Analytical and biological challenges. *Anal. Biochem.* **2017**, *524*, 13–30. [CrossRef] [PubMed]
37. Tsikas, D.; Rothmann, S.; Schneider, J.Y.; Gutzki, F.-M.; Beckmann, B.; Frölich, J.C. Simultaneous GC-MS/MS measurement of malondialdehyde and 4-hydroxy-2-nonenal in human plasma: Effects of long-term L-arginine administration. *Anal. Biochem.* **2017**, *524*, 31–44. [CrossRef] [PubMed]
38. Syslová, K.; Kačer, P.; Kuzma, M.; Najmanová, V.; Fenclová, Z.; Vlčková, Š.; Lebedová, J.; Pelclová, D. Rapid and easy method for monitoring oxidative stress markers in body fluids of patients with asbestos or silica-induced lung diseases. *J. Chromatogr. B* **2009**, *877*, 2477–2486. [CrossRef] [PubMed]
39. Eggink, M.; Wijtmans, M.; Ekkebus, R.; Lingeman, H.; Esch, I.J.d.; Kool, J.; Niessen, W.M.; Irth, H. Development of a selective ESI-MS derivatization reagent: Synthesis and optimization for the analysis of aldehydes in biological mixtures. *Anal. Chem.* **2008**, *80*, 9042–9051. [CrossRef] [PubMed]
40. O'Brien-Coker, I.C.; Perkins, G.; Mallet, A.I. Aldehyde analysis by high performance liquid chromatography/tandem mass spectrometry. *Rapid Commun. Mass Spectrom.* **2001**, *15*, 920–928. [CrossRef] [PubMed]

41. Taleb, A.; Witztum, J.L.; Tsimikas, S. Oxidized phospholipids on apoB-100-containing lipoproteins: A biomarker predicting cardiovascular disease and cardiovascular events. *Biomark. Med.* **2011**, *5*, 673–694. [CrossRef] [PubMed]

42. Hasanally, D.; Edel, A.; Chaudhary, R.; Ravandi, A. Identification of oxidized phosphatidylinositols present in OxLDL and human atherosclerotic plaque. *Lipids* **2017**, *52*, 11–26. [CrossRef] [PubMed]

43. Ravandi, A.; Leibundgut, G.; Hung, M.-Y.; Patel, M.; Hutchins, P.M.; Murphy, R.C.; Prasad, A.; Mahmud, E.; Miller, Y.I.; Dennis, E.A. Release and capture of bioactive oxidized phospholipids and oxidized cholesteryl esters during percutaneous coronary and peripheral arterial interventions in humans. *J. Am. Coll. Cardiol.* **2014**, *63*, 1961–1971. [CrossRef] [PubMed]

44. Solati, Z.; Edel, A.L.; Shang, Y.; Karmin, O.; Ravandi, A. Oxidized phosphatidylcholines are produced in renal ischemia reperfusion injury. *PLoS ONE* **2018**, *13*, e0195172. [CrossRef] [PubMed]

45. Reis, A. Oxidative Phospholipidomics in health and disease: Achievements, challenges and hopes. *Free Radic. Biol. Med.* **2017**, *111*, 25–37. [CrossRef] [PubMed]

46. Li, M.; Fan, P.; Wang, Y. Lipidomics in health and diseases-beyond the analysis of lipids. *J. Glycom. Lipidom.* **2015**, *5*, 1.

47. Hinterwirth, H.; Stübiger, G.; Lindner, W.; Lämmerhofer, M. Gold nanoparticle-conjugated anti-oxidized low-density lipoprotein antibodies for targeted lipidomics of oxidative stress biomarkers. *Anal. Chem.* **2013**, *85*, 8376–8384. [CrossRef] [PubMed]

48. Stübiger, G.; Wuczkowski, M.; Bicker, W.; Belgacem, O. Nanoparticle-Based Detection of Oxidized Phospholipids by MALDI Mass Spectrometry: Nano-MALDI Approach. *Anal. Chem.* **2014**, *86*, 6401–6409. [CrossRef] [PubMed]

49. Haller, E.; Stübiger, G.; Lafitte, D.; Lindner, W.; Lämmerhofer, M. Chemical recognition of oxidation-specific epitopes in low-density lipoproteins by a nanoparticle based concept for trapping, enrichment, and liquid chromatography–tandem mass spectrometry analysis of oxidative stress biomarkers. *Anal. Chem.* **2014**, *86*, 9954–9961. [CrossRef] [PubMed]

50. Spickett, C.M.; Reis, A.; Pitt, A.R. Identification of oxidized phospholipids by electrospray ionization mass spectrometry and LC–MS using a QQLIT instrument. *Free Radic. Biol. Med.* **2011**, *51*, 2133–2149. [CrossRef] [PubMed]

51. Astarita, G.; Kendall, A.C.; Dennis, E.A.; Nicolaou, A. Targeted lipidomic strategies for oxygenated metabolites of polyunsaturated fatty acids. *Biochim. Et Biophys. Acta (Bba)-Mol. Cell Biol. Lipids* **2015**, *1851*, 456–468. [CrossRef] [PubMed]

52. Benabdellah, F.; Seyer, A.; Quinton, L.; Touboul, D.; Brunelle, A.; Laprévote, O. Mass spectrometry imaging of rat brain sections: Nanomolar sensitivity with MALDI versus nanometer resolution by TOF–SIMS. *Anal. Bioanal. Chem.* **2010**, *396*, 151–162. [CrossRef] [PubMed]

53. Flemmig, J.; Spalteholz, H.; Schubert, K.; Meier, S.; Arnhold, J. Modification of phosphatidylserine by hypochlorous acid. *Chem. Phys. Lipids* **2009**, *161*, 44–50. [CrossRef] [PubMed]

54. Flemmig, J.; Arnhold, J. Interaction of hypochlorous acid and myeloperoxidase with phosphatidylserine in the presence of ammonium ions. *J. Inorg. Biochem.* **2010**, *104*, 759–764. [CrossRef] [PubMed]

55. Massey, K.A.; Nicolaou, A. Lipidomics of oxidized polyunsaturated fatty acids. *Free Radic. Biol. Med.* **2013**, *59*, 45–55. [CrossRef] [PubMed]

56. Miller, D.K.; Sadowski, S.; DeSousa, D.; Maycock, A.L.; Lombardo, D.L.; Young, R.N.; Hayes, E.C. Development of enzyme-linked immunosorbent assays for measurement of leukotrienes and prostaglandins. *J. Immunol. Methods* **1985**, *81*, 169–185. [CrossRef]

57. Tsukamoto, H.; Hishinuma, T.; Mikkaichi, T.; Nakamura, H.; Yamazaki, T.; Tomioka, Y.; Mizugaki, M. Simultaneous quantification of prostaglandins, isoprostane and thromboxane in cell-cultured medium using gas chromatography–mass spectrometry. *J. Chromatogr. B* **2002**, *774*, 205–214. [CrossRef]

58. Strassburg, K.; Huijbrechts, A.M.; Kortekaas, K.A.; Lindeman, J.H.; Pedersen, T.L.; Dane, A.; Berger, R.; Brenkman, A.; Hankemeier, T.; Van Duynhoven, J. Quantitative profiling of oxylipins through comprehensive LC-MS/MS analysis: Application in cardiac surgery. *Anal. Bioanal. Chem.* **2012**, *404*, 1413–1426. [CrossRef] [PubMed]

59. Wang, Y.; Armando, A.M.; Quehenberger, O.; Yan, C.; Dennis, E.A. Comprehensive ultra-performance liquid chromatographic separation and mass spectrometric analysis of eicosanoid metabolites in human samples. *J. Chromatogr. A* **2014**, *1359*, 60–69. [CrossRef] [PubMed]

60. Csala, M.; Kardon, T.; Legeza, B.; Lizák, B.; Mandl, J.; Margittai, É.; Puskás, F.; Száraz, P.; Szelényi, P.; Bánhegyi, G. On the role of 4-hydroxynonenal in health and disease. *Biochim. Et Biophys. Acta (Bba)-Mol. Basis Dis.* **2015**, *1852*, 826–838. [CrossRef] [PubMed]

61. Poli, G.; Schaur, R.J.; Siems, W.a.; Leonarduzzi, G. 4-Hydroxynonenal: A membrane lipid oxidation product of medicinal interest. *Med. Res. Rev.* **2008**, *28*, 569–631. [CrossRef] [PubMed]

62. Chapple, S.J.; Cheng, X.; Mann, G.E. Effects of 4-hydroxynonenal on vascular endothelial and smooth muscle cell redox signaling and function in health and disease. *Redox Biol.* **2013**, *1*, 319–331. [CrossRef] [PubMed]

63. Poli, G.; Schaur, J.R. 4-Hydroxynonenal in the pathomechanisms of oxidative stress. *IUBMB Life* **2000**, *50*, 315–321. [CrossRef] [PubMed]

64. Milkovic, L.; Cipak Gasparovic, A.; Zarkovic, N. Overview on major lipid peroxidation bioactive factor 4-hydroxynonenal as pluripotent growth-regulating factor. *Free Radic. Res.* **2015**, *49*, 850–860. [CrossRef] [PubMed]

65. Gargiulo, S.; Gamba, P.; Testa, G.; Rossin, D.; Biasi, F.; Poli, G.; Leonarduzzi, G. Relation between TLR4/NF-κB signaling pathway activation by 27-hydroxycholesterol and 4-hydroxynonenal, and atherosclerotic plaque instability. *Aging Cell* **2015**, *14*, 569–581. [CrossRef] [PubMed]

66. Bridges, A.; Scott, N.; Parry, G.; Belch, J. Age, sex, cigarette smoking and indices of free radical activity in healthy humans. *Eur. J. Med.* **1993**, *2*, 205–208. [PubMed]

67. Zahavi, J.; Betteridge, J.D.; Jones, N.A.; Galton, D.J.; Kakkar, V.V. Enhanced in vivo platelet release reaction and malondialdehyde formation in patients with hyperlipidemia. *Am. J. Med.* **1981**, *70*, 59–64. [CrossRef]

68. Yalçın, A.S.; Sabuncu, N.; Kilinç, A.; Gülcan, G.; Emerk, K. Increased plasma and erythrocyte lipid-peroxidation in hyperlipidemic individuals. *Atherosclerosis* **1989**, *80*, 169–170. [PubMed]

69. Nacítarhan, S.; Özben, T. Serum and urine malondialdehyde levels in NIDDM patients with and without hyperlipidemia. *Free Radic. Biol. Med.* **1995**, *19*, 893–896. [CrossRef]

70. Noberasco, G.; Odetti, P.; Boeri, D.; Maiello, M.; Adezati, L. Malondialdehyde (MDA) level in diabetic subjects. Relationship with blood glucose and glycosylated hemoglobin. *Biomed. Pharmacother.* **1991**, *45*, 193–196. [CrossRef]

71. Sakuma, N.; Hibino, T.; Sato, T.; Ohte, N.; Akita, S.; Tamai, N.; Sasai, K.; Yoshimata, T.; Fujinami, T. Levels of Thiobarbituric Acid-Reactive Substance in Plasma from Coronary Artery Disease Patients-An update. *Clin. Biochem.* **1997**, *6*, 505–507. [CrossRef]

72. Salonen, J.T.; Nyyssönen, K.; Salonen, R.; Porkkala-Sarataho, E.; Tuomainen, T.-P.; Diczfalusy, U.; Björkhem, I. Lipoprotein oxidation and progression of carotid atherosclerosis. *Circulation* **1997**, *95*, 840–845. [CrossRef] [PubMed]

73. Walter, M.F.; Jacob, R.F.; Jeffers, B.; Ghadanfar, M.M.; Preston, G.M.; Buch, J.; Mason, R.P. Serum levels of thiobarbituric acid reactive substances predict cardiovascular events in patients with stable coronary artery disease: A longitudinal analysis of the PREVENT study. *J. Am. Coll. Cardiol.* **2004**, *44*, 1996–2002. [CrossRef] [PubMed]

74. Bagatini, M.D.; Martins, C.C.; Battisti, V.; Gasparetto, D.; Da Rosa, C.S.; Spanevello, R.M.; Ahmed, M.; Schmatz, R.; Schetinger, M.R.C.; Morsch, V.M. Oxidative stress versus antioxidant defenses in patients with acute myocardial infarction. *Heart Vessel.* **2011**, *26*, 55–63. [CrossRef] [PubMed]

75. Hörkkö, S.; Bird, D.A.; Miller, E.; Itabe, H.; Leitinger, N.; Subbanagounder, G.; Berliner, J.A.; Friedman, P.; Dennis, E.A.; Curtiss, L.K. Monoclonal autoantibodies specific for oxidized phospholipids or oxidized phospholipid–protein adducts inhibit macrophage uptake of oxidized low-density lipoproteins. *J. Clin. Investig.* **1999**, *103*, 117–128. [CrossRef] [PubMed]

76. Chang, M.-K.; Bergmark, C.; Laurila, A.; Hörkkö, S.; Han, K.-H.; Friedman, P.; Dennis, E.A.; Witztum, J.L. Monoclonal antibodies against oxidized low-density lipoprotein bind to apoptotic cells and inhibit their phagocytosis by elicited macrophages: Evidence that oxidation-specific epitopes mediate macrophage recognition. *Proc. Natl. Acad. Sci. USA* **1999**, *96*, 6353–6358. [CrossRef] [PubMed]

77. Que, X.; Hung, M.-Y.; Yeang, C.; Gonen, A.; Prohaska, T.A.; Sun, X.; Diehl, C.; Määttä, A.; Gaddis, D.E.; Bowden, K.; et al. Oxidized phospholipids are proinflammatory and proatherogenic in hypercholesterolaemic mice. *Nature* **2018**, *558*, 301–306. [CrossRef] [PubMed]

78. Watson, A.D.; Subbanagounder, G.; Welsbie, D.S.; Faull, K.F.; Navab, M.; Jung, M.E.; Fogelman, A.M.; Berliner, J.A. Structural identification of a novel pro-inflammatory epoxyisoprostane phospholipid in mildly oxidized low density lipoprotein. *J. Biol. Chem.* **1999**, *274*, 24787–24798. [CrossRef] [PubMed]

79. Watson, A.D.; Leitinger, N.; Navab, M.; Faull, K.F.; Hörkkö, S.; Witztum, J.L.; Palinski, W.; Schwenke, D.; Salomon, R.G.; Sha, W. Structural identification by mass spectrometry of oxidized phospholipids in minimally oxidized low density lipoprotein that induce monocyte/endothelial interactions and evidence for their presence in vivo. *J. Biol. Chem.* **1997**, *272*, 13597–13607. [CrossRef] [PubMed]

80. Stemmer, U.; Dunai, Z.A.; Koller, D.; Pürstinger, G.; Zenzmaier, E.; Deigner, H.P.; Aflaki, E.; Kratky, D.; Hermetter, A. Toxicity of oxidized phospholipids in cultured macrophages. *Lipids Health Dis.* **2012**, *11*, 110. [CrossRef] [PubMed]

81. Tselepis, A.D. Oxidized phospholipids and lipoprotein-associated phospholipase A2 as important determinants of Lp (a) functionality and pathophysiological role. *J. Biomed. Res.* **2018**, *32*, 13.

82. Wang, J.-J.; Zhang, C.-N.; Han, A.-Z.; Gong, J.-B.; Li, K. Percutaneous coronary intervention results in acute increases in native and oxidized lipoprotein (a) in patients with acute coronary syndrome and stable coronary artery disease. *Clin. Biochem.* **2010**, *43*, 1107–1111. [CrossRef] [PubMed]

83. Tsimikas, S.; Brilakis, E.S.; Miller, E.R.; McConnell, J.P.; Lennon, R.J.; Kornman, K.S.; Witztum, J.L.; Berger, P.B. Oxidized phospholipids, Lp (a) lipoprotein, and coronary artery disease. *N. Engl. J. Med.* **2005**, *353*, 46–57. [CrossRef] [PubMed]

84. Tsimikas, S.; Lau, H.K.; Han, K.-R.; Shortal, B.; Miller, E.R.; Segev, A.; Curtiss, L.K.; Witztum, J.L.; Strauss, B.H. Percutaneous coronary intervention results in acute increases in oxidized phospholipids and lipoprotein (a): Short-term and long-term immunologic responses to oxidized low-density lipoprotein. *Circulation* **2004**, *109*, 3164–3170. [CrossRef] [PubMed]

85. Tsimikas, S.; Witztum, J.L. The role of oxidized phospholipids in mediating lipoprotein (a) atherogenicity. *Curr. Opin. Lipidol.* **2008**, *19*, 369–377. [CrossRef] [PubMed]

86. Tsimikas, S.; Kiechl, S.; Willeit, J.; Mayr, M.; Miller, E.R.; Kronenberg, F.; Xu, Q.; Bergmark, C.; Weger, S.; Oberhollenzer, F. Oxidized phospholipids predict the presence and progression of carotid and femoral atherosclerosis and symptomatic cardiovascular disease: Five-year prospective results from the Bruneck study. *J. Am. Coll. Cardiol.* **2006**, *47*, 2219–2228. [CrossRef] [PubMed]

87. Hirsh, P.D.; Hillis, L.D.; Campbell, W.B.; Firth, B.G.; Willerson, J.T. Release of prostaglandins and thromboxane into the coronary circulation in patients with ischemic heart disease. *N. Engl. J. Med.* **1981**, *304*, 685–691. [CrossRef] [PubMed]

88. Montalescot, G.; Drobinski, G.; Maclouf, J.; Lellouche, F.; Ankri, A.; Moussallem, N.; Eugene, L.; Thomas, D.; Grosgogeat, Y. Early thromboxane release during pacing-induced myocardial ischemia with angiographically normal coronary arteries. *Am. Heart J.* **1990**, *120*, 1445–1447. [CrossRef]

89. Mallat, Z.; Nakamura, T.; Ohan, J.; Lesèche, G.; Tedgui, A.; Maclouf, J.; Murphy, R.C. The relationship of hydroxyeicosatetraenoic acids and F 2-isoprostanes to plaque instability in human carotid atherosclerosis. *J. Clin. Investig.* **1999**, *103*, 421–427. [CrossRef] [PubMed]

90. Waddington, E.; Sienuarine, K.; Puddey, I.; Croft, K. Identification and quantitation of unique fatty acid oxidation products in human atherosclerotic plaque using high-performance liquid chromatography. *Anal. Biochem.* **2001**, *292*, 234–244. [CrossRef] [PubMed]

91. Auguet, T.; Aragonès, G.; Colom, M.; Aguilar, C.; Martín-Paredero, V.; Canela, N.; Ruyra, X.; Richart, C. Targeted metabolomic approach in men with carotid plaque. *PLoS ONE* **2018**, *13*, e0200547. [CrossRef] [PubMed]

92. Sun, Y.; Koh, H.W.; Choi, H.; Koh, W.-P.; Yuan, J.-M.; Newman, J.W.; Su, J.; Fang, J.; Ong, C.N.; van Dam, R.M. Plasma fatty acids, oxylipins, and risk of myocardial infarction: The Singapore Chinese Health Study. *J. Lipid Res.* **2016**, *57*, 1300–1307. [CrossRef] [PubMed]

93. Leibundgut, G.; Lee, J.-H.; Strauss, B.H.; Segev, A.; Tsimikas, S. Acute and long-term effect of percutaneous coronary intervention on serially-measured oxidative, inflammatory, and coagulation biomarkers in patients with stable angina. *J. Thromb. Thrombolysis* **2016**, *41*, 569–580. [CrossRef] [PubMed]

International Journal of
Molecular Sciences

Review

Specialized Pro-Resolving Lipid Mediators in Cystic Fibrosis

Réginald Philippe and Valerie Urbach *

INSERM, U1151, Institut Necker Enfants Malades, 75993 Paris, France; reginald.philippe@inserm.fr
* Correspondence: valerie.urbach@inserm.fr; Tel.: +33-6-30-37-59-04

Received: 31 August 2018; Accepted: 15 September 2018; Published: 21 September 2018

Abstract: In cystic fibrosis (CF), impaired airway surface hydration (ASL) and mucociliary clearance that promote chronic bacterial colonization, persistent inflammation, and progressive structural damage to the airway wall architecture are typically explained by ion transport abnormalities related to the mutation of the gene coding for the Cystic Fibrosis Transmembrane Conductance Regulator (CFTR) channel. However, the progressive and unrelenting inflammation of the CF airway begins early in life, becomes persistent, and is excessive relative to the bacterial burden. Intrinsic abnormalities of the inflammatory response in cystic fibrosis have been suggested but the mechanisms involved remain poorly understood. This review aims to give an overview of the recent advances in the understanding of the defective resolution of inflammation in CF including the abnormal production of specialized pro-resolving lipid mediators (lipoxin and resolvin) and their impact on the pathogenesis of the CF airway disease.

Keywords: resolvin; lipoxin; cystic fibrosis

---

1. Cystic Fibrosis

Cystic fibrosis (CF) is due to the mutation of the gene coding for the CFTR Chloride channel. Among more than 2000 mutations of the *cftr* gene, the most common (F508del) results in a protein folding defect, its retention in the endoplasmic reticulum (ER), and degradation by the proteasome [1]. CF affects many organs where the CFTR protein is normally expressed, however, the progressive lung destruction is the main cause of morbidity and mortality. In the healthy lung, mucociliary clearance, which requires optimal hydration of the airway surface liquid (ASL), propels inhaled particles, micro-organisms, and allergens towards the pharynx so that they can be expectorated and removed from the lungs, thus protecting from infection and inflammation. Mutations of *cftr* result in defective Cl^- secretion and Na^+ hyperabsorption in airway epithelium. This ion transport dysfunction contributes to a reduction of the periciliary fluid volume and impairs mucociliary clearance, as seen in Figure 1, thus favoring bacterial colonization and sustained inflammation.

The focus on the function of CFTR in regulating epithelial ion transport and airway surface hydration provided a convincing explanation of the pathogenesis of airway disease in CF. The discovery in the CF epithelium of reduced ASL height and poor mucociliary clearance [2] was a significant step towards understanding this process in greater detail, as shown in Figure 1. However, the studies of altered ion and water transport alone failed to fully elucidate the mechanism by which the *cftr* gene mutation leads to pathogenesis in the CF lung. Furthermore, evidence began to emerge in the mid-1990s that airway inflammation begins much earlier than had previously been appreciated, even in children who are asymptomatic at the time of testing [3,4]. Overall, despite decades of research, the precise pathogenesis of the lung disease in CF remains not well understood.

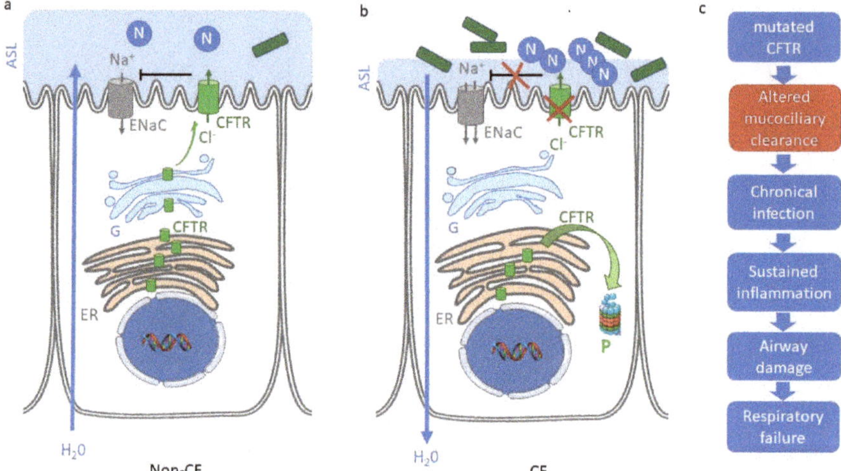

Figure 1. Cystic fibrosis (CF) and non-CF airway epithelial transport. (**a**) In non-CF airway epithelium, the airway surface liquid (ASL light blue) layer height is finely regulated by transepithelial chloride secretion and sodium absorption that involve the activities of the CFTR chloride channel (green arrow through CFTR) and the epithelial sodium channel (grey arrow through ENaC), respectively. In this tissue, the CFTR protein is normally expressed and trafficked (ER: endoplasmic reticulum, G: Golgi) toward the apical membrane of epithelial cells to play a role as a chloride channel and to down regulate ENaC activity. (**b**) In the case of the most common CF mutation, the CFTR protein is retained in the ER and degraded by the proteasome (P). This results in altered chloride secretion trough CFTR and regulation of ENaC (red cross) and leads to an ASL height decrease that favors bacterial (green rectangle) and neutrophil (N) accumulation in the airway. (**c**) Classical hypothesis for CF airway pathogenesis showing the central role of dehydration/ altered mucociliary clearance (red) that results from ion transport abnormality.

2. Persistent Inflammation in Cystic Fibrosis

The acute inflammatory response is a normal host-protective response to contain foreign invaders. In CF, the inflammatory reaction is characterized by a predominant neutrophilic component and increased concentrations of pro-inflammatory mediators, including TNF-α, IL-1β, IL-6, IL-8, IL-17, IL-33, GM-CSF, G-CSF, and HMGB-1 [5]. An excessive airway inflammation is developed in early life of patients with CF and persists even in absence of infection [3,6,7].

Two main factors lead to sustained airway inflammation: an increased neutrophil recruitment and their defective elimination (for recent reviews Chmiel et al. [5,8]). Recurrent infections result in neutrophil over-recruitment in CF patients' airways. In addition, it has been proposed that the excess of pro-inflammatory mediators' production by airway epithelial cells, which triggers neutrophil recruitment, is related to a CFTR-dependent activation of the nuclear factor kappa-light-chain-enhancer of activated B cells (NF-κB) pathway, as well as a decreased interferon-γ production [9–11]. Increased neutrophil count is also related to the defective mechanisms of elimination related to the altered airway mucociliary clearance resulting from CFTR loss of function and Na⁺ hyper-absorption, as seen in Figure 1. Furthermore, neutrophils from CF patients are less sensitive to apoptosis and CF macrophages show a decreased phagocytic activity [12–15].

Neutrophils also present functional defects. The loss of bacteria-killing capacity of CF neutrophils contributes to the maintenance of infection and persistence of inflammation [16,17]. The altered efferocytosis and mucociliary clearance result in the necrosis of neutrophils in the airways. This results in the liberation of toxic and pro-inflammatory compounds and to the recruitment of new neutrophils, leading to the persistence of inflammation [18,19].

Overall, it does not appear that the excessive inflammatory response in CF results from altered triggering mechanisms but is mainly due to its inefficiency to contain infection and to abnormality in its resolution phase [5,8].

3. Resolution of Inflammation

The acute inflammatory response is normally self-limited with an active resolution phase designed to restore tissue homeostasis. The resolution of inflammation was initially thought to be a passive process and that inflammatory mediators from the initiation of the acute response (e.g., chemoattractants, complement components, prostaglandins, chemokines and cytokines) would simply dilute and dissipate to stop the infiltration of leukocytes into the tissues. However, studies performed by Serhan and colleagues on mouse inflammatory exudates revealed that acute inflammatory responses involve an active resolution phase carried out by specialized pro-resolving eicosanoid mediators (SPM) such as Lipoxins (LX) [20], resolvins (Rv) [21], protectins (PD) [22], and maresins (Mar) [23] that are locally produced from polyunsaturated fatty acids.

The temporal evolution of acute inflammation and its active resolution involves the sequential biosynthesis and activity of characteristic classes of an eicosanoid mediator in a process termed "class switching" [24]. Leukotriene B_4 (LTB_4) plays its role in amplification and propagation of inflammation [24] acting in concert with the peptide Interleukin 8 (IL8) as a potent neutrophil chemo-attractant [25,26]. Both LTB_4 and IL8 are negatively correlated with pulmonary function in CF. Lipoxin A_4 (LXA_4) is the first eicosanoid mediator to be expressed in the active resolution phase and down-regulates neutrophil effector functions: limiting neutrophil chemotaxis, adherence and transmigration, and inhibiting superoxide anion and peroxynitrite generation [27]. Leukotrienes and lipoxins are closely related metabolites of arachidonic acid and can be synthesised from a common unstable intermediate [25]. They exhibit counter-regulatory effector functions and LXA_4 inhibits LTB_4 induced neutrophil transmigration in a dose dependent manner [28]. Along with lipoxins, other endogenous lipid mediators, including omega-3-derived resolvins, protectins, and maresins, stimulate and promote resolution of inflammation, clearance of microbes, tissue regeneration, and pain reduction, but do not evoke unwanted immunosuppression. SPMs are also involved in the regulation of adaptive immunity of B and T lymphocytes and can lower the amount of antibiotics needed to clear infections and reduce the potential for antibiotic resistance [29]. The abnormal process of resolution of inflammation is now considered a pathophysiologic basis associated with widely occurring diseases such as cardiovascular disease, neurodegenerative diseases, asthma, diabetes, and obesity, as well as arthritis, periodontal diseases, and cystic fibrosis.

4. Abnormalities of Lipid Levels in Cystic Fibrosis

Before the gene defect responsible for CF was identified, it was suggested that fatty acid metabolism abnormalities were responsible for the clinical symptoms of the CF disease. In 1962 Kuo et al. proposed that the disease was characterized by low linoleic acid (LA) concentrations in plasma and tissues [30]. Several groups also reported an important unbalance of unsaturated fatty acid with increased release of arachidonic acid (AA) and decreased levels of docosahexaenoic acid (DHA) in CF [31–33]. These lipid abnormalities were mainly explained by the fat malabsorption caused by the pancreatic insufficiency present in 85% of the patients [34,35]. In addition, CFTR dysfunction has been related to increased expression and activity of phospholipase A2 (PLA2) [36–41]. Since the release of AA by phospholipases is rate-limiting for the eicosanoid synthesis, an abnormal activity of PLA2 is a possible mechanism involved in increased AA levels in CF [42]. Increased oxidative stress, which is well documented in CF, has been also suggested as a cause of the fatty acid abnormalities seen in CF [43,44]. However, several studies were performed with supplementation of DHA (that should correct the unbalance between AA and DHA) to patients with CF, and failed to show significant clinical improvement [45–47].

A few groups have studied the levels of lipoxin, a metabolite of AA that could play a role in the pathophysiology of CF airway disease [48,49]. The mean content in LXA4, either absolute or adjusted to neutrophil count, did not significantly change in bronchoalveolar lavage (BAL) samples from patients with CF when compared to BAL samples from individuals without CF [48,50]. However, a significant decrease in LXA4 levels adjusted to neutrophil count was found in BAL from patients with CF when compared to BAL from patients with an airway inflammation but without CF [49]. Furthermore, a significant decrease in LXA4 levels adjusted to the pro-inflammatory cytokine IL8 (LXA4/IL8 ratio) was found in BAL samples of CF patients compared to control subjects [48,49]. In BAL and sputum samples from patients with CF, inter-individual changes in LXA4 content did not correlate with IL8 content [48]. The abnormality of eicosanoid class switching has been further investigated by analyzing LXA4/LTB4 ratios in BAL samples. In individuals without CF, LXA4/LTB4 ratio was related to infection, and was significantly higher in sterile samples than that measured in BAL samples from which pathogens were cultured. In contrast, in BAL samples from patients with CF, LXA4/LTB4 ratio did not vary with infection status and was uniformly depressed when compared with BAL from control children [48]. Consistent with reduced levels of DHA, RvD1/Il8 was also found to be decrease in sputa from CF patients, even in the absence of an acute infection or exacerbation [51]. Taken together, these data demonstrated that abnormalities in lipid metabolism, including SPM biosynthesis, characterize the CF airway disease.

5. Cellular Mechanisms Involved in the Defective Production of SPM in Cystic Fibrosis

To date, approximately 30 kinds of SPMs have been found in inflammatory exudates: lipoxins, D-series and E-series resolvins, protectins, and maresins. They are produced at inflammatory sites from essential fatty acids (AA, DHA and eicosapentaenoic acid (EPA)) by the interaction of lipoxygenase (LO) activities within several cell types including leukocytes, platelets and epithelium, as seen in Figure 2 [25,52]. The level of expression of 15LO in airway epithelial cells and macrophages and the levels of 12LO in platelets play a central role in the eicosanoid class-switching from leukotrienes B4 (LTB4), a pro-inflammatory lipid mediator, toward SPMs. Indeed, the activity of 15LO favors LXA4 (pro-resolving) synthesis at the expense of LTB4 (pro-inflammatory) synthesis that involves leukotriene A4 hydrolase (LTA4H) activity (Figure 2) [25]. Cellular location of 5LO also plays a crucial role in LTB4/SPM equilibrium. Inhibition of the CaM kinases by RvD1 favors extra-nuclear location of 5LO, and LXA4 biosynthesis at the expense of LTB4 [53]. A reduced expression of 12/15LO in the airway of a CF mouse model as well as a reduced 15LO2 level in the BAL of patients with F508del mutation was described [48,49]. A defective 12LO activity has been reported in platelets during CFTR inhibition [54]. The 15LO expression was shown to be decreased by 50% in nasal epithelium of patients with CF compared to non-CF [55]. The 5LO was initially thought to be restricted to leucocytes, but is also expressed in human bronchial epithelial cells [56]. The cellular mechanism by which CFTR could affect 5LO, 12LO, 15LO, 15LO2 and/or LTA4H level of expression, cellular localization and activities in CF airways remains unclear. However, several data suggested that the abnormal SPM production in CF is related to abnormal LO expression and/or activity in cells that fail to normally express CFTR, including CF airway epithelium and platelets [48,49,54,55].

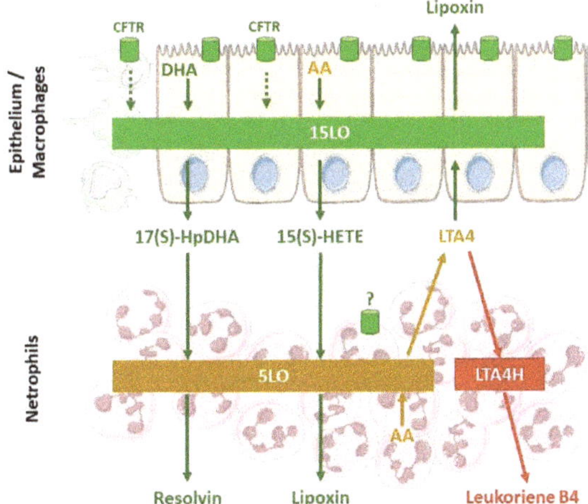

Figure 2. Schematic view of specialized pro-resolving eicosanoid mediators (SPM) biosynthesis pathways. Lipoxin and resolvin result from lipoxygenases interactions. Transcellular biosynthesis can occur via the interactions of two or more cell types expressing different enzymes. Epithelial cells and macrophages express the 15-lipoxygenase (15LO), that metabolizes docosahexaenoic acid (DHA) and arachidonic acid (AA) to produce 17(S)-HpDHA and 15-Hydroxyeicosatetraenoic acid (15(S)-HETE), respectively. These products of 15LO activity are the substrates for the 5LO expressed in neutrophils, to synthetize resolvin and lipoxin (pro-resolving mediators). The 5LO can also metabolize AA into leukotriene A4 (LTA4), which can be substrate of both, the 15LO to produce lipoxin or the LTA4H to produce leukotriene B4 (pro-inflammatory mediator). The pro-resolving pathways are illustrated in green, the pro-inflammatory pathway in red, and enzyme/product/substrate involved in both pathways in brown. The precise mechanism by which CFTR could affect LO activity is not known (green dotted arrow).

6. SPM Impact on CF Airway Epithelial Transport and ASL Height

In CF, the decreased levels in ASL height has been reported to be mainly due to a defect in trafficking and function of CFTR as a Cl^- channel as well as its impact on ENaC regulation leading to an increase in Na^+ absorption (Figure 2).

The calcium-activated chloride channel TMEM16A (Transmembrane member 16A) and the SLC26A9 (Solute Carrier Family 26 Member 9) are two other anion channels that provide the critical regulation of mucus hydration of airway epithelium [57]. Both TMEM16A and SLC26A9 activities which attenuate airway inflammation and prevent mucus obstruction during airway inflammation have been considered as alternative therapeutic targets to bypass CFTR dysfunction in the airway epithelia of CF patients However, strategies for increasing intracellular calcium concentration to stimulate calcium-activated chloride secretion have been plagued by the amplification of the calcium dependent pro-inflammatory response [58,59]. Therefore, research for pharmacological long-lasting stimulation of TMEM16A independent of intracellular Ca^{2+} has been intensified [60]. The SLC26A9, a member of the solute carrier 26 family of anion transporters was shown to act as a modifier of CF lung disease severity and the response to CFTR modulator therapy [61,62]. An alternative strategy to deliver artificial non-toxic Cl^- transporters to CF epithelia have emerged [60].

On another side, consistent with their effect in resolution of inflammation and stimulation of innate immunity, LXA4 and RvD1 treatments have been shown to target epithelium and to regulate airway epithelial ion transport [63]. LXA4 stimulates CFTR-independent chloride secretion and inhibit amiloride-sensitive Na^+ absorption resulting in an ASL height increase in human CF and non-CF airway epithelium [64–67]. This effect involves the apical ATP (adenosine triphosphate) secretion through pannexin channels induced by LXA4, leading to purino-receptor activation and further triggering of an intracellular calcium signal [63,65]. RvD1 also increases the CF ASL height of human bronchial epithelium in a calcium-dependent manner by decreasing the amiloride-sensitive Na$^+$ absorption and stimulating CFTR-independent Cl$^-$ secretion. In vivo studies reflect these in vitro data showing that LXA4 and RvD1 restored the nasal transepithelial potential difference in CF mice [67]. These observations, shown in Figure 3, support the concept that the abnormal production of SPM in CF contributes to the dehydration of the airway surface and the defective airway clearance.

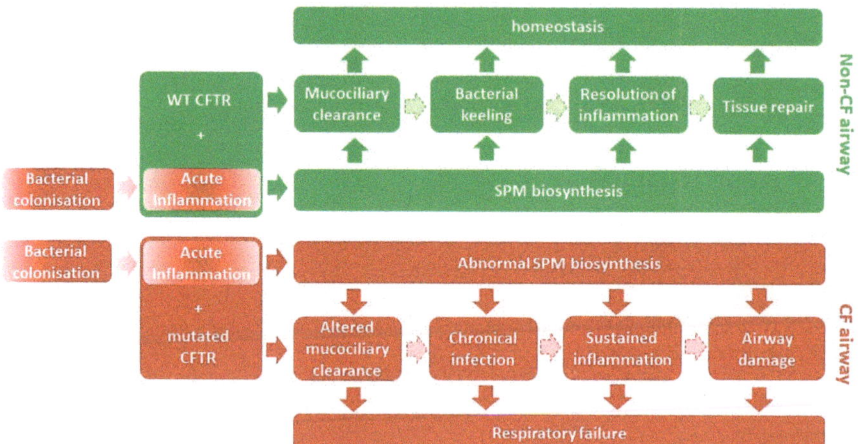

Figure 3. Novel hypothetical model for cystic fibrosis (CF) airway pathogenesis. In CF and non-CF airway bacterial infection triggers inflammation. In non-CF cells (top), self-limited inflammation involves the biosynthesis of specialized pro-resolving eicosanoid mediators (SPM) that enhance mucociliary clearance, bacterial killing, resolution of inflammation, and tissue repair, resulting in tissue returning to homeostasis (green). In CF cells (bottom), after initiation of inflammation, the eicosanoid class switching is altered and SPM are abnormally produced. This results into a defective mucociliary clearance, sustained inflammation, inefficient bacterial killing and tissue repair that lead to respiratory failure (red). This model enhances the role of SPM in the resolution phase of inflammation compared to the impact of dehydration and altered mucociliary clearance.

7. SPMs' Role in Bacterial Clearance and Limiting Tissue Damage

Resolution of acute inflammation has been shown to be crucial for ensuring bacterial clearance and limiting tissue damage. A few groups investigated the protective actions of LXA4 and RvD1 in lung infection induced by *Pseudomonas aeruginosa*. Patients with CF have a predisposition to chronic colonization and infection with *P. aeruginosa*, an organism whose presence in the CF lung is associated with progressive respiratory compromise. Antibiotics are insufficiently successful, because of resistance mechanisms developed by *P. aeruginosa* [68]. *P. aeruginosa* can escape killing by the immune system and establish chronic infections that do not resolve [69,70].

Karp et al. have shown that administration of a metabolically stable LXA4 analog in a mouse model of the chronic airway inflammation and infection associated with cystic fibrosis decreased pulmonary bacterial burden and attenuated disease severity [49]. In vitro, LXA4 has been reported to delay colonization of airway epithelium by *P. aeruginosa* by enhancing the airway transepithelial electrical resistance and the expression of the tight junction protein ZO-1 at the plasma membrane in human bronchial epithelial cells [64,71]. LXA4 also stimulates airway epithelial repair (migration and proliferation) via stimulation of ATP-sensitive potassium channel and mitogen-activated protein kinase pathway [72,73].

RvD1 decreased TNFα induced IL-8 secretion and enhanced the phagocytic and bacterial killing capacity of human CF alveolar macrophages [67]. RvD1 significantly diminished bacterial growth and neutrophil infiltration during acute pneumonia caused by a clinical strain of *P. aeruginosa*. Inoculum of *P. aeruginosa*, immobilized in agar beads, resulted in persistent lung infection and non-resolving inflammation. RvD1 significantly reduced bacterial load, leukocyte infiltration, and lung tissue damage. In murine lung macrophages sorted during *P. aeruginosa* chronic infection, RvD1 regulated the expression of Toll-like receptors, downstream genes, and microRNA (miR)-21 and 155, resulting in reduced inflammatory signaling [74]. Finally, Eickmeier et al. have reported that RvD1 levels in plasma and sputum samples from patients with CF showed a positive correlation with sputum inflammatory markers. The plasma concentrations of RvD1 were ten times higher than sputum concentrations and sputum RvD1/IL-8 levels showed a positive correlation with FEV1 [51].

These data, seen in Figure 3, provide evidence for a role of SPM in acute and chronic *P. aeruginosa* pneumonia, and for its potent pro-resolution and tissue protective properties.

8. Conclusions

Cystic fibrosis, the most common lethal genetic disease in Western populations, is characterized by a chronic inflammation and sustained neutrophilic inflammation of the airways. The most recent reports suggest that the persistent inflammation in CF airways appeared to be due to an altered resolution phase and efficiency to contain infection rather than abnormalities in its triggering. Early studies suggested that the abnormal fatty acid metabolism in CF played a central role in the CF disease, and more recent studies demonstrated the sustained airway inflammation to be related to an abnormal eicosanoid class switching between SPM (lipoxin and resolvin) and leukotrienes. Although the precise role of CFTR mutation on SPM biosynthesis remains unclear, the role of SPM in resolving inflammation, in increasing the airway surface liquid layer, in enhancing bacterial clearance and tissue repair strongly suggest that the altered production of pro-resolving lipid mediators play a central in CF pathogenesis, as shown in Figure 3.

Author Contributions: Writing—Review & Editing, R.P. and V.U. Funding Acquisition, V.U.

Funding: This research was funded by Vaincre la Mucoviscidose grant number [RF20170502010].

References

1. Pranke, I.M.; Sermet-Gaudelus, I. Biosynthesis of cystic fibrosis transmembrane conductance regulator. *Int. J. Biochem. Cell Biol.* **2014**, *52*, 26–38. [CrossRef] [PubMed]
2. Boucher, R.C. Airway surface dehydration in cystic fibrosis: Pathogenesis and therapy. *Annu. Rev. Med.* **2007**, *58*, 157–170. [CrossRef] [PubMed]
3. Khan, T.Z.; Wagener, J.S.; Bost, T.; Martinez, J.; Accurso, F.J.; Riches, D.W. Early pulmonary inflammation in infants with cystic fibrosis. *Am. J. Respir. Crit. Care Med.* **1995**, *151*, 1075–1082. [CrossRef] [PubMed]
4. Muhlebach, M.S.; Noah, T.L. Endotoxin activity and inflammatory markers in the airways of young patients with cystic fibrosis. *Am. J. Respir. Crit. Care Med.* **2002**, *165*, 911–915. [CrossRef] [PubMed]
5. Nichols, D.P.; Chmiel, J.F. Inflammation and its genesis in cystic fibrosis. *Pediatr. Pulmonol.* **2015**, *50*, S39–S56. [CrossRef] [PubMed]
6. Chmiel, J.F.; Berger, M.; Konstan, M.W. The role of inflammation in the pathophysiology of CF lung disease. *Clin. Rev. Allergy Immunol.* **2002**, *23*, 5–27. [CrossRef]
7. Armstrong, D.S.; Grimwood, K.; Carlin, J.B.; Carzino, R.; Gutièrrez, J.P.; Hull, J.; Olinsky, A.; Phelan, E.M.; Robertson, C.F.; Phelan, P.D. Lower airway inflammation in infants and young children with cystic fibrosis. *Am. J. Respir. Crit. Care Med.* **1997**, *156*, 1197–1204. [CrossRef] [PubMed]
8. Roesch, E.A.; Nichols, D.P.; Chmiel, J.F. Inflammation in cystic fibrosis: An update. *Pediatr. Pulmonol.* **2018**. [CrossRef] [PubMed]

9.   Weber, A.J.; Soong, G.; Bryan, R.; Saba, S.; Prince, A. Activation of NF-kappaB in airway epithelial cells is dependent on CFTR trafficking and Cl⁻ channel function. *Am. J. Physiol. Lung Cell Mol. Physiol.* **2001**, *281*, L71–L78. [CrossRef] [PubMed]

10.   Saadane, A.; Soltys, J.; Berger, M. Role of IL-10 deficiency in excessive nuclear factor-kappaB activation and lung inflammation in cystic fibrosis transmembrane conductance regulator knockout mice. *J. Allergy Clin. Immunol.* **2005**, *115*, 405–411. [CrossRef] [PubMed]

11.   Parker, D.; Cohen, T.S.; Alhede, M.; Harfenist, B.S.; Martin, F.J.; Prince, A. Induction of type I interferon signaling by Pseudomonas aeruginosa is diminished in cystic fibrosis epithelial cells. *Am. J. Respir. Cell Mol. Biol.* **2012**, *46*, 6–13. [CrossRef] [PubMed]

12.   Vandivier, R.W.; Fadok, V.A.; Hoffmann, P.R.; Bratton, D.L.; Penvari, C.; Brown, K.K.; Brain, J.D.; Accurso, F.J.; Henson, P.M. Elastase-mediated phosphatidylserine receptor cleavage impairs apoptotic cell clearance in cystic fibrosis and bronchiectasis. *J. Clin. Investig.* **2002**, *109*, 661–670. [CrossRef] [PubMed]

13.   Vandivier, R.W.; Richens, T.R.; Horstmann, S.A.; de Cathelineau, A.M.; Ghosh, M.; Reynolds, S.D.; Xiao, Y.Q.; Riches, D.W.; Plumb, J.; Vachon, E.; et al. Dysfunctional cystic fibrosis transmembrane conductance regulator inhibits phagocytosis of apoptotic cells with proinflammatory consequences. *Am. J. Physiol. Lung Cell Mol. Physiol.* **2009**, *297*, L677–L686. [CrossRef] [PubMed]

14.   Rowe, S.M.; Jackson, P.L.; Liu, G.; Hardison, M.; Livraghi, A.; Solomon, G.M.; McQuaid, D.B.; Noerager, B.D.; Gaggar, A.; Clancy, J.P.; et al. Potential role of high-mobility group box 1 in cystic fibrosis airway disease. *Am. J. Respir. Crit. Care Med.* **2008**, *178*, 822–831. [CrossRef] [PubMed]

15.   McKeon, D.J.; Condliffe, A.M.; Cowburn, A.S.; Cadwallader, K.C.; Farahi, N.; Bilton, D.; Chilvers, E.R. Prolonged survival of neutrophils from patients with Delta F508 CFTR mutations. *Thorax* **2008**, *63*, 660–661. [CrossRef] [PubMed]

16.   Painter, R.G.; Valentine, V.G.; Lanson, N.A.; Leidal, K.; Zhang, Q.; Lombard, G.; Thompson, C.; Viswanathan, A.; Nauseef, W.M.; Wang, G. CFTR Expression in human neutrophils and the phagolysosomal chlorination defect in cystic fibrosis. *Biochemistry* **2006**, *45*, 10260–10269. [CrossRef] [PubMed]

17.   Painter, R.G.; Bonvillain, R.W.; Valentine, V.G.; Lombard, G.A.; LaPlace, S.G.; Nauseef, W.M.; Wang, G. The role of chloride anion and CFTR in killing of Pseudomonas aeruginosa by normal and CF neutrophils. *J. Leukoc. Biol.* **2008**, *83*, 1345–1353. [CrossRef] [PubMed]

18.   Cheng, O.Z.; Palaniyar, N. NET balancing: A problem in inflammatory lung diseases. *Front. Immunol.* **2013**, *4*, 1. [CrossRef] [PubMed]

19.   Law, S.M.; Gray, R.D. Neutrophil extracellular traps and the dysfunctional innate immune response of cystic fibrosis lung disease: A review. *J. Inflamm.-Lond.* **2017**, *14*, 29. [CrossRef] [PubMed]

20.   Serhan, C.N.; Hamberg, M.; Samuelsson, B. Lipoxins: Novel series of biologically active compounds formed from arachidonic acid in human leukocytes. *Proc. Natl. Acad. Sci. USA* **1984**, *81*, 5335–5339. [CrossRef] [PubMed]

21.   Serhan, C.N.; Gotlinger, K.; Hong, S.; Arita, M. Resolvins, docosatrienes, and neuroprotectins, novel omega-3-derived mediators, and their aspirin-triggered endogenous epimers: An overview of their protective roles in catabasis. *Prostaglandins Other Lipid Mediat.* **2004**, *73*, 155–172. [CrossRef] [PubMed]

22.   Levy, B.D.; Kohli, P.; Gotlinger, K.; Haworth, O.; Hong, S.; Kazani, S.; Israel, E.; Haley, K.J.; Serhan, C.N. Protectin D1 is generated in asthma and dampens airway inflammation and hyperresponsiveness. *J. Immunol.* **2007**, *178*, 496–502. [CrossRef] [PubMed]

23.   Serhan, C.N.; Yang, R.; Martinod, K.; Kasuga, K.; Pillai, P.S.; Porter, T.F.; Oh, S.F.; Spite, M. Maresins: Novel macrophage mediators with potent antiinflammatory and proresolving actions. *J. Exp. Med.* **2009**, *206*, 15–23. [CrossRef] [PubMed]

24.   Levy, B.D.; Clish, C.B.; Schmidt, B.; Gronert, K.; Serhan, C.N. Lipid mediator class switching during acute inflammation: Signals in resolution. *Nat. Immunol.* **2001**, *2*, 612–619. [CrossRef] [PubMed]

25.   Haeggström, J.Z.; Funk, C.D. Lipoxygenase and leukotriene pathways: Biochemistry, biology, and roles in disease. *Chem. Rev.* **2011**, *111*, 5866–5898. [CrossRef] [PubMed]

26.   Mukaida, N.; Okamoto, S.; Ishikawa, Y.; Matsushima, K. Molecular mechanism of interleukin-8 gene expression. *J. Leukoc. Biol.* **1994**, *56*, 554–558. [CrossRef] [PubMed]

27.   Serhan, C.N.; Yacoubian, S.; Yang, R. Anti-inflammatory and proresolving lipid mediators. *Annu. Rev. Pathol.* **2008**, *3*, 279–312. [CrossRef] [PubMed]

28. Bonnans, C.; Fukunaga, K.; Levy, M.A.; Levy, B.D. Lipoxin A(4) regulates bronchial epithelial cell responses to acid injury. *Am. J. Pathol.* **2006**, *168*, 1064–1072. [CrossRef] [PubMed]

29. Chiang, N.; Serhan, C.N. Structural elucidation and physiologic functions of specialized pro-resolving mediators and their receptors. *Mol. Aspects Med.* **2017**, *58*, 114–129. [CrossRef] [PubMed]

30. Kuo, P.T.; Huang, N.N.; Bassett, D.R. The fatty acid composition of the serum chylomicrons and adipose tissue of children with cystic fibrosis of the pancreas. *J. Pediatr.* **1962**, *60*, 394–403. [CrossRef]

31. Underwood, B.A.; Denning, C.R.; Navab, M. Polyunsaturated fatty acids and tocopherol levels in patients with cystic fibrosis. *Ann. N. Y. Acad. Sci.* **1972**, *203*, 237–247. [CrossRef] [PubMed]

32. Freedman, S.D.; Blanco, P.G.; Zaman, M.M.; Shea, J.C.; Ollero, M.; Hopper, I.K.; Weed, D.A.; Gelrud, A.; Regan, M.M.; Laposata, M.; et al. Association of cystic fibrosis with abnormalities in fatty acid metabolism. *N. Engl. J. Med.* **2004**, *350*, 560–569. [CrossRef] [PubMed]

33. Gilljam, H.; Strandvik, B.; Ellin, A.; Wiman, L.G. Increased mole fraction of arachidonic acid in bronchial phospholipids in patients with cystic fibrosis. *Scand. J. Clin. Lab. Investig.* **1986**, *46*, 511–518. [CrossRef]

34. Elliott, R.B.; Robinson, P.G. Unusual clinical course in a child with cystic fibrosis treated with fat emulsion. *Arch. Dis. Child.* **1975**, *50*, 76–78. [CrossRef] [PubMed]

35. Elliott, R.B. A therapeutic trial of fatty acid supplementation in cystic fibrosis. *Pediatrics* **1976**, *57*, 474–479. [PubMed]

36. Medjane, S.; Raymond, B.; Wu, Y.; Touqui, L. Impact of CFTR DeltaF508 mutation on prostaglandin E2 production and type IIA phospholipase A2 expression by pulmonary epithelial cells. *Am. J. Physiol. Lung Cell Mol. Physiol.* **2005**, *289*, L816–L824. [CrossRef] [PubMed]

37. László, A.; Németh, M.; Gyukovits, K. Activity of phospholipase A in serum from patients with cystic fibrosis. *Clin. Chim. Acta* **1993**, *214*, 105–107. [CrossRef]

38. Berguerand, M.; Klapisz, E.; Thomas, G.; Humbert, L.; Jouniaux, A.M.; Olivier, J.L.; Béréziat, G.; Masliah, J. Differential stimulation of cytosolic phospholipase A2 by bradykinin in human cystic fibrosis cell lines. *Am. J. Respir. Cell Mol. Biol.* **1997**, *17*, 481–490. [CrossRef] [PubMed]

39. Borot, F.; Vieu, D.L.; Faure, G.; Fritsch, J.; Colas, J.; Moriceau, S.; Baudouin-Legros, M.; Brouillard, F.; Ayala-Sanmartin, J.; Touqui, L.; et al. Eicosanoid release is increased by membrane destabilization and CFTR inhibition in Calu-3 cells. *PLoS ONE* **2009**, *4*, e7116. [CrossRef] [PubMed]

40. Strandvik, B. Fatty acid metabolism in cystic fibrosis. *Prostaglandins Leukot. Essent. Fatty Acids* **2010**, *83*, 121–129. [CrossRef] [PubMed]

41. Strandvik, B.; O'Neal, W.K.; Ali, M.A.; Hammar, U. Low linoleic and high docosahexaenoic acids in a severe phenotype of transgenic cystic fibrosis mice. *Exp. Biol. Med.* **2018**, *243*, 496–503. [CrossRef] [PubMed]

42. Strandvik, B.; Brönnegård, M.; Gilljam, H.; Carlstedt-Duke, J. Relation between defective regulation of arachidonic acid release and symptoms in cystic fibrosis. *Scand. J. Gastroenterol. Suppl.* **1988**, *143*, 1–4. [CrossRef] [PubMed]

43. Brown, R.K.; Kelly, F.J. Role of free radicals in the pathogenesis of cystic fibrosis. *Thorax* **1994**, *49*, 738–742. [CrossRef] [PubMed]

44. Chen, J.; Kinter, M.; Shank, S.; Cotton, C.; Kelley, T.J.; Ziady, A.G. Dysfunction of Nrf-2 in CF epithelia leads to excess intracellular H2O2 and inflammatory cytokine production. *PLoS ONE* **2008**, *3*, e3367. [CrossRef] [PubMed]

45. Lloyd-Still, J.D.; Powers, C.A.; Hoffman, D.R.; Boyd-Trull, K.; Lester, L.A.; Benisek, D.C.; Arterburn, L.M. Bioavailability and safety of a high dose of docosahexaenoic acid triacylglycerol of algal origin in cystic fibrosis patients: A randomized, controlled study. *Nutrition* **2006**, *22*, 36–46. [CrossRef] [PubMed]

46. McKarney, C.; Everard, M.; N'Diaye, T. Omega-3 fatty acids (from fish oils) for cystic fibrosis. *Cochrane Database Syst. Rev.* **2007**, CD002201. [CrossRef] [PubMed]

47. Coste, T.C.; Armand, M.; Lebacq, J.; Lebecque, P.; Wallemacq, P.; Leal, T. An overview of monitoring and supplementation of omega 3 fatty acids in cystic fibrosis. *Clin. Biochem.* **2007**, *40*, 511–520. [CrossRef] [PubMed]

48. Ringholz, F.C.; Buchanan, P.J.; Clarke, D.T.; Millar, R.G.; McDermott, M.; Linnane, B.; Harvey, B.J.; McNally, P.; Urbach, V. Reduced 15-lipoxygenase 2 and lipoxin A4/leukotriene B4 ratio in children with cystic fibrosis. *Eur. Respir. J.* **2014**, *44*, 394–404. [CrossRef] [PubMed]

49. Karp, C.L.; Flick, L.M.; Park, K.W.; Softic, S.; Greer, T.M.; Keledjian, R.; Yang, R.; Uddin, J.; Guggino, W.B.; Atabani, S.F.; et al. Defective lipoxin-mediated anti-inflammatory activity in the cystic fibrosis airway. *Nat. Immunol.* **2004**, *5*, 388–392. [CrossRef] [PubMed]

50. Starosta, V.; Ratjen, F.; Rietschel, E.; Paul, K.; Griese, M. Anti-inflammatory cytokines in cystic fibrosis lung disease. *Eur. Respir. J.* **2006**, *28*, 581–587. [CrossRef] [PubMed]

51. Eickmeier, O.; Fussbroich, D.; Mueller, K.; Serve, F.; Smaczny, C.; Zielen, S.; Schubert, R. Pro-resolving lipid mediator Resolvin D1 serves as a marker of lung disease in cystic fibrosis. *PLoS ONE* **2017**, *12*, e0171249. [CrossRef] [PubMed]

52. Romano, M.; Cianci, E.; Simiele, F.; Recchiuti, A. Lipoxins and aspirin-triggered lipoxins in resolution of inflammation. *Eur. J. Pharmacol.* **2015**, *760*, 49–63. [CrossRef] [PubMed]

53. Fredman, G.; Ozcan, L.; Spolitu, S.; Hellmann, J.; Spite, M.; Backs, J.; Tabas, I. Resolvin D1 limits 5-lipoxygenase nuclear localization and leukotriene B4 synthesis by inhibiting a calcium-activated kinase pathway. *Proc. Natl. Acad Sci. USA* **2014**, *111*, 14530–14535. [CrossRef] [PubMed]

54. Mattoscio, D.; Evangelista, V.; De Cristofaro, R.; Recchiuti, A.; Pandolfi, A.; Di Silvestre, S.; Manarini, S.; Martelli, N.; Rocca, B.; Petrucci, G.; et al. Cystic fibrosis transmembrane conductance regulator (CFTR) expression in human platelets: Impact on mediators and mechanisms of the inflammatory response. *FASEB J.* **2010**, *24*, 3970–3980. [CrossRef] [PubMed]

55. Jeanson, L.; Guerrera, I.C.; Papon, J.F.; Chhuon, C.; Zadigue, P.; Prulière-Escabasse, V.; Amselem, S.; Escudier, E.; Coste, A.; Edelman, A. Proteomic analysis of nasal epithelial cells from cystic fibrosis patients. *PLoS ONE* **2014**, *9*, e108671. [CrossRef] [PubMed]

56. Jame, A.J.; Lackie, P.M.; Cazaly, A.M.; Sayers, I.; Penrose, J.F.; Holgate, S.T.; Sampson, A.P. Human bronchial epithelial cells express an active and inducible biosynthetic pathway for leukotrienes B4 and C4. *Clin. Exp. Allergy* **2007**, *37*, 880–892. [CrossRef] [PubMed]

57. Hahn, A.; Salomon, J.J.; Leitz, D.; Feigenbutz, D.; Korsch, L.; Lisewski, I.; Schrimpf, K.; Millar-Büchner, P.; Mall, M.A.; Frings, S.; et al. Expression and function of Anoctamin 1/TMEM16A calcium-activated chloride channels in airways of in vivo mouse models for cystic fibrosis research. *Pflugers Arch.* **2018**, *470*, 1335–1348. [CrossRef] [PubMed]

58. Kunzelmann, K.; Mall, M. Pharmacotherapy of the ion transport defect in cystic fibrosis: Role of purinergic receptor agonists and other potential therapeutics. *Am. J. Respir. Med.* **2003**, *2*, 299–309. [CrossRef] [PubMed]

59. Ribeiro, C.M. The role of intracellular calcium signals in inflammatory responses of polarised cystic fibrosis human airway epithelia. *Drugs R. D.* **2006**, *7*, 17–31. [CrossRef] [PubMed]

60. Li, H.; Salomon, J.J.; Sheppard, D.N.; Mall, M.A.; Galietta, L.J. Bypassing CFTR dysfunction in cystic fibrosis with alternative pathways for anion transport. *Curr. Opin. Pharmacol.* **2017**, *34*, 91–97. [CrossRef] [PubMed]

61. Strug, L.J.; Gonska, T.; He, G.; Keenan, K.; Ip, W.; Boëlle, P.Y.; Lin, F.; Panjwani, N.; Gong, J.; Li, W.; et al. Cystic fibrosis gene modifier SLC26A9 modulates airway response to CFTR-directed therapeutics. *Hum. Mol. Genet.* **2016**, *25*, 4590–4600. [CrossRef] [PubMed]

62. Bertrand, C.A.; Mitra, S.; Mishra, S.K.; Wang, X.; Zhao, Y.; Pilewski, J.M.; Madden, D.R.; Frizzell, R.A. The CFTR trafficking mutation F508del inhibits the constitutive activity of SLC26A9. *Am. J. Physiol. Lung Cell Mol. Physiol.* **2017**, *312*, L912–L925. [CrossRef] [PubMed]

63. Bonnans, C.; Mainprice, B.; Chanez, P.; Bousquet, J.; Urbach, V. Lipoxin A(4) stimulates a cytosolic Ca^{2+} increase in human bronchial epithelium. *J. Biol. Chem.* **2003**, *278*, 10879–10884. [CrossRef] [PubMed]

64. Higgins, G.; Fustero Torre, C.; Tyrrell, J.; McNally, P.; Harvey, B.J.; Urbach, V. Lipoxin A4 prevents tight junction disruption and delays the colonization of cystic fibrosis bronchial epithelial cells by Pseudomonas aeruginosa. *Am. J. Physiol. Lung Cell Mol. Physiol.* **2016**, *310*, L1053–L1061. [CrossRef] [PubMed]

65. Verrière, V.; Higgins, G.; Al-Alawi, M.; Costello, R.W.; McNally, P.; Chiron, R.; Harvey, B.J.; Urbach, V. Lipoxin A4 stimulates calcium-activated chloride currents and increases airway surface liquid height in normal and cystic fibrosis airway epithelia. *PLoS ONE* **2012**, *7*, e37746. [CrossRef] [PubMed]

66. Al-Alawi, M.; Buchanan, P.; Verriere, V.; Higgins, G.; McCabe, O.; Costello, R.W.; McNally, P.; Urbach, V.; Harvey, B.J. Physiological levels of lipoxin A4 inhibit ENaC and restore airway surface liquid height in cystic fibrosis bronchial epithelium. *Physiol. Rep.* **2014**, *2*. [CrossRef] [PubMed]

67. Ringholz, F.C.; Higgins, G.; Hatton, A.; Sassi, A.; Moukachar, A.; Fustero-Torre, C.; Hollenhorst, M.; Sermet-Gaudelus, I.; Harvey, B.J.; McNally, P.; et al. Resolvin D1 regulates epithelial ion transport and inflammation in cystic fibrosis airways. *J. Cyst. Fibros.* **2017**. [CrossRef] [PubMed]

68. Pier, G.B. The challenges and promises of new therapies for cystic fibrosis. *J. Exp. Med.* **2012**, *209*, 1235–1239. [CrossRef] [PubMed]

69. Bragonzi, A.; Worlitzsch, D.; Pier, G.B.; Timpert, P.; Ulrich, M.; Hentzer, M.; Andersen, J.B.; Givskov, M.; Conese, M.; Doring, G. Nonmucoid Pseudomonas aeruginosa expresses alginate in the lungs of patients with cystic fibrosis and in a mouse model. *J. Infect. Dis.* **2005**, *192*, 410–419. [CrossRef] [PubMed]

70. Winstanley, C.; O'Brien, S.; Brockhurst, M.A. Pseudomonas aeruginosa Evolutionary Adaptation and Diversification in Cystic Fibrosis Chronic Lung Infections. *Trends Microbiol.* **2016**, *24*, 327–337. [CrossRef] [PubMed]

71. Grumbach, Y.; Quynh, N.V.; Chiron, R.; Urbach, V. LXA4 stimulates ZO-1 expression and transepithelial electrical resistance in human airway epithelial (16HBE14o-) cells. *Am. J. Physiol. Lung Cell Mol. Physiol.* **2009**, *296*, L101–L108. [CrossRef] [PubMed]

72. Buchanan, P.J.; McNally, P.; Harvey, B.J.; Urbach, V. Lipoxin A_4-mediated KATP potassium channel activation results in cystic fibrosis airway epithelial repair. *Am. J. Physiol. Lung Cell Mol. Physiol.* **2013**, *305*, L193–L201. [CrossRef] [PubMed]

73. Higgins, G.; Buchanan, P.; Perriere, M.; Al-Alawi, M.; Costello, R.W.; Verriere, V.; McNally, P.; Harvey, B.J.; Urbach, V. Activation of P2RY11 and ATP release by lipoxin A4 restores the airway surface liquid layer and epithelial repair in cystic fibrosis. *Am. J. Respir. Cell Mol. Biol.* **2014**, *51*, 178–190. [CrossRef] [PubMed]

74. Codagnone, M.; Cianci, E.; Lamolinara, A.; Mari, V.C.; Nespoli, A.; Isopi, E.; Mattoscio, D.; Arita, M.; Bragonzi, A.; Iezzi, M.; et al. Resolvin D1 enhances the resolution of lung inflammation caused by long-term Pseudomonas aeruginosa infection. *Mucosal Immunol.* **2018**, *11*, 35–49. [CrossRef] [PubMed]

International Journal of
Molecular Sciences

Article

Biochemical Characterization of the *GBA2* c.1780G>C Missense Mutation in Lymphoblastoid Cells from Patients with Spastic Ataxia

Anna Malekkou [1,2,†], Maura Samarani [3,†], Anthi Drousiotou [1,2,*], Christina Votsi [2,4], Sandro Sonnino [3], Marios Pantzaris [2,5], Elena Chiricozzi [3], Eleni Zamba-Papanicolaou [2,6], Massimo Aureli [3,*], Nicoletta Loberto [3,‡] and Kyproula Christodoulou [2,4,‡]

[1] Biochemical Genetics Department, The Cyprus Institute of Neurology and Genetics, Nicosia 1683, Cyprus; annama@cing.ac.cy

[2] Cyprus School of Molecular Medicine, Nicosia 1683, Cyprus; votsi@cing.ac.cy (C.V.); pantzari@cing.ac.cy (M.P.); ezamba@cing.ac.cy (E.Z.-P.); roula@cing.ac.cy (K.C.)

[3] Department of Medical Biotechnology and Translational Medicine, University of Milano, 20122 Milano, Italy; maura.samarani@unimi.it (M.S.); sandro.sonnino@unimi.it (S.S.); elena.chiricozzi@unimi.it (E.C.); nicoletta.loberto@unimi.it (N.L.)

[4] Neurogenetics Department, The Cyprus Institute of Neurology and Genetics, Nicosia 1683, Cyprus

[5] Neurology Clinic C, The Cyprus Institute of Neurology and Genetics, Nicosia 1683, Cyprus

[6] Neurology Clinic D, The Cyprus Institute of Neurology and Genetics, Nicosia 1683, Cyprus

*   Correspondence: anthidr@cing.ac.cy (A.D.); massimo.aureli@unimi.it (M.A.);
    Tel.: +357-22392643 or +357-22358600 (A.D.); +39-0250330364 (M.A.);
    Fax: +357-22392768 (A.D.); +39-0250330365 (M.A.)

† These authors contributed equally to this work.

‡ These authors share the last authorship.

Received: 14 August 2018; Accepted: 8 October 2018; Published: 10 October 2018

Abstract: The *GBA2* gene encodes the non-lysosomal glucosylceramidase (NLGase), an enzyme that catalyzes the conversion of glucosylceramide (GlcCer) to ceramide and glucose. Mutations in *GBA2* have been associated with the development of neurological disorders such as autosomal recessive cerebellar ataxia, hereditary spastic paraplegia, and Marinesco-Sjogren-Like Syndrome. Our group has previously identified the *GBA2* c.1780G>C [p.Asp594His] missense mutation, in a Cypriot consanguineous family with spastic ataxia. In this study, we carried out a biochemical characterization of lymphoblastoid cell lines (LCLs) derived from three patients of this family. We found that the mutation strongly reduce NLGase activity both intracellularly and at the plasma membrane level. Additionally, we observed a two-fold increase of GlcCer content in LCLs derived from patients compared to controls, with the C_{16} lipid being the most abundant GlcCer species. Moreover, we showed that there is an apparent compensatory effect between NLGase and the lysosomal glucosylceramidase (GCase), since we found that the activity of GCase was three-fold higher in LCLs derived from patients compared to controls. We conclude that the c.1780G>C mutation results in NLGase loss of function with abolishment of the enzymatic activity and accumulation of GlcCer accompanied by a compensatory increase in GCase.

Keywords: GBA2; non-lysosomal β-glucosylceramidase; β-glucocerebrosidase; spastic ataxia; glucosylceramide; plasma membrane; lymphoblastoid cell lines

1. Introduction

Sphingolipids (SLs) are a class of lipids mainly associated with the external leaflet of the plasma membrane (PM) of all eukaryotic cells, playing an important role in the structural integrity of the PM and cellular signaling [1]. Glycosphingolipids (GSLs) are SLs with a head-group formed by a mono- or oligosaccharide moiety. Glucosylceramide (GlcCer) is the simplest member of GSLs and it is formed in the Golgi complex by the glycosylation of ceramide [2]. SLs play a fundamental role in cell physiology and this can be demonstrated by the numerous genetic diseases that arise from mutations in enzymes involved in SL metabolism and transport [3,4]. Cells can alter their lipid composition by the action of different hydrolases that are active at the lysosomes or at the PM, such as sphingomyelinase (SMase), β-hexosaminidase (β-Hex), β-galactosidase (β-gal), β-glucocerebrosidase (GCase), and the non-lysosomal β-glucosylceramidase (NLGase) [5,6].

GCase and NLGase are both involved in the catabolism of GlcCer to ceramide and glucose. GCase (EC 3.2.1.45) is encoded by the *GBA* gene (MIM_606463), located on chromosome 1q21. GCase is a membrane glycoprotein of 497 amino acids with a β-barrel structure, ubiquitously expressed in all tissues and mainly localized in lysosomes. The catalytic site of GCase contains two highly conserved residues of glutamic acid, which are necessary for the two-step mechanism of action, the nucleophilic attack, and the subsequent protonation. Loss of function mutations in the *GBA* gene cause Gaucher disease, the most common lysosomal storage disorder [7,8].

NLGase (EC 3.2.1.45) is encoded by the *GBA2* gene localized on chromosome 9p13.3. The catalytic site of the protein is characterized by both nucleophile and acid/base residues, Glu-528 and Asp-678 respectively. These residues define NLGase as a retaining β-glucosidase belonging to the CAZy glycosyl hydrolase family 116 (GH116) [9]. As previously described, the retaining β-glucosidases utilize a double-displacement mechanism [10]. In the first step of the reaction, called glycosylation, the nucleophile residue attacks the glucose anomeric center to create the glycosyl-enzyme intermediate, whereas the acid/base residue protonates the glycosydic oxygen, leading to the release of a glycone [11]. In the second step, the group working as an acid in the first step acts as a base catalyst that together with incoming water determines the de-glycosylation of the nucleophile. This mechanism allows the retention of the configuration at the anomeric carbon of the released glucose molecule [12].

The intracellular localization of NLGase has been controversial. It was reported to be a single pass transmembrane protein [13] and later was identified as a cytoplasmic membrane-associated protein of the ER and Golgi complex [14]. The activity of NLGase at the PM is directly modulated by the efflux of protons through the proton pumps associated with the cell surface [15,16].

The pathological involvement of NLGase was initially studied by generating *GBA2*-knockout mice [17]. Knockout mice present impairment in liver regeneration [18], and male infertility due to GlcCer accumulation. This accumulation causes dysregulation of lipid homeostasis due to a more ordered lipid composition of the PM [19,20]. As a result, the cytoskeletal dynamics are altered and the formation of sperm-head shaping and acrosome is affected [19].

Despite the abnormal GlcCer accumulation in brain, *GBA2*-knockout mice do not display any neurological symptoms or defects [19]. On the contrary, *GBA2*-knockdown zebrafish show abnormal motor neuron development [21], and mutations in the human *GBA2* gene have been found to lead to neurological disorders like spastic ataxia (SA) [22,23], hereditary spastic paraplegia (HSP) [21,24,25], and more recently Marinesco-Sjogren-Like Syndrome [26]. The molecular mechanism(s) leading to the development of disease are not currently known. Only one mutation of *GBA2* has been so far functionally characterized in vivo (i.e., zebrafish model) [21]. A more detailed biochemical analysis of the different mutants of *GBA2* in patient derived cells is still missing.

We have previously identified a *GBA2* missense mutation [c.1780G>C (p.Asp594His)] in a Cypriot family with progressive spastic ataxia [22]. In vitro characterization of this mutation in COS7 and HeLa cells showed that it causes a reduction at both the protein and enzyme activity levels [25]. In the present study we have undertaken the biochemical characterization of the *GBA2* c.1780G>C missense mutation in lymphoblastoid cell lines (LCLs) derived from spastic ataxia patients homozygous for the mutation. Our results contribute to the understanding of the biochemical consequences of mutations in the *GBA2* gene.

2. Results

2.1. The c.1780G>C Mutation Results in NLGase Loss of Activity and GlcCer Accumulation

GBA2 mRNA expression levels were measured in LCLs obtained from four healthy individuals and from three patients homozygous for the c.1780G>C mutation. As shown in Figure 1-panel A, the same level of *GBA2* transcript was found in control and patient LCLs, suggesting that the mutation does not affect the *GBA2* mRNA expression and stability.

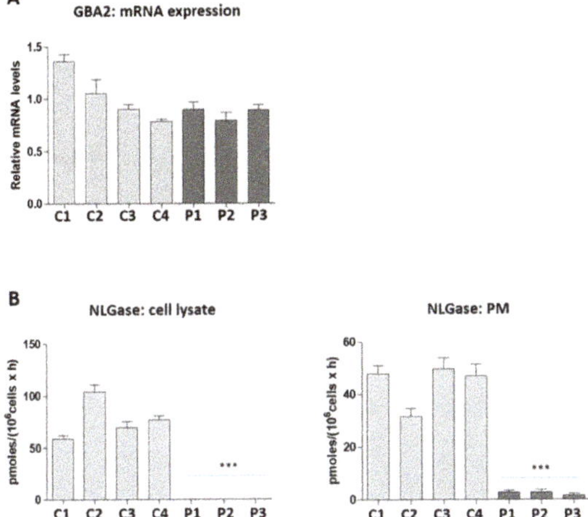

Figure 1. *GBA2* mRNA expression and non-lysosomal glucosylceramidase (NLGase) activity. (**A**) The graph represents the *GBA2* mRNA levels of controls (light grey) and patients (dark grey) relative to the average value of controls (n = 4) after normalization with the endogenous β-actin gene (*ACTB*). Values represent the mean ± SEM of two independent triplicate experiments. (**B**) NLGase activity associated with the total cell lysates and plasma membrane (PM) of controls (light grey) and patients (dark grey) derived lymphoblastoid cell lines (LCLs). Enzymatic activity was expressed as pmoles/10^6 cells/h. Data are expressed as mean ± SD of three independent triplicate experiments (*** $p < 0.0001$ vs. controls).

Subsequently, we measured the NLGase enzymatic activity on the total cell lysate of the same LCLs. We observed that NLGase activity was almost undetectable in patients' cells with respect to controls (Figure 1B), which are characterized by an average specific activity of 78 ± 19 pmoles/10^6 cells/h.

In several cell lines of different origin, NLGase was found to be associated with the external leaflet of the PM where it catalyzes the in situ hydrolysis of glucosylceramide (GlcCer) to ceramide [27–29]. For this reason, we measured the enzymatic activity of NLGase directly at the cell surface of control and patient derived lymphoblastoid living cells. In patients' cells NLGase activity was also strongly reduced at the cell surface, showing 2–4% of residual activity with respect to that found in controls (Figure 1B; controls 44 ± 8 pmoles/10^6 cells/h, patients 2.3 ± 0.5 pmoles/10^6 cells/h). We can exclude that the enzymatic activity measured at the PM was due to other β-glucocerebrosidases because, by adding AMP-DNM, a specific inhibitor of NLGase to the assay solution abolished the enzymatic activity at the cell surface. Our data demonstrate that the presence of the homozygous *GBA2* c.1780G>C mutation results in an important loss of NLGase activity.

Furthermore, we assessed the effect of the NLGase loss of function on the GlcCer content. To this purpose, total lipid extracts obtained from the same LCLs used for the evaluation of the enzymatic activity were subjected to SFC-MS/MS analysis. This method is capable of quantifying and also distinguishing GlcCer from galactosylceramide (GalCer).

As shown in Table 1, patient derived LCLs are characterized by a two-fold increase in the GlcCer content with respect to controls. No differences were found among controls and cells expressing the mutated protein in the level of GalCer. Moreover, no difference was found in cellular cholesterol content between patients and controls, suggesting that GBA2 loss-of-function does not affect cholesterol homeostasis. In Gaucher disease, where we have GCase loss-of-function, the accumulated GlcCer is converted to glucosphingosine (GlcShp) by the action of the acid ceramidase [30]. We therefore also analyzed the levels of GlcShp by SFC-MS/MS and these were found to be hardly detectable, without any significant difference between control and patient cells.

Table 1. Hexosylceramides of lymphoblastoid cell lines (LCLs) from controls and patients. Glucosylceramide (GlcCer) and galactosylceramide (GalCer) contents were evaluated by SFC-MS/MS in controls (WT, $n = 4$) and patients (c.1780G>C, $n = 3$) LCLs. Data are expressed as pmoles/mg of cell protein $\pm$ Error ($n = 3$).

Hexosylceramides of LCLs from Controls and Patients (Pmoles/mg Cell Proteins)		
	WT	**c.1780 G>C**
Glucosylceramide	1080 ± 107	2009 ± 114
Galactosylceramide	9 ± 3	6 ± 2

Quantitative analysis of the different molecular species of GlcCer showed that C16 is the most abundant species in both control and patient derived LCLs. In addition, it emerged that all the molecular species are doubled in pathological cells with respect to controls (Table 2).

Table 2. SFC-MS/MS analysis of glucosylceramide (GlcCer) molecular species in control (WT, $n = 4$) and patient (c.1780G>C, $n = 3$) derived LCLs. Data are the mean of three independent experiments and are expressed as pmoles/mg of cell proteins $\pm$ SEM.

		Glucosylceramide Molecular Species of LCLs from Controls and Patients (Pmoles/mg Cell Proteins)										
		C14	C16	C18	C18:1	C20	C22	C22:1	C24	C24:1	C26	C26:1
GlcCer	*WT*	13 ± 2	611 ± 100	24 ± 5	1 ± 0.1	28 ± 7	76 ± 14	6 ± 1	143 ± 22	150 ± 21	9 ± 1	15 ± 2
	c.1780 G>C	21 ± 1	1139 ± 103	51 ± 2	1.3 ± 0.2	63 ± 7	145 ± 7	11 ± 1	261 ± 20	278 ± 44	15 ± 2	21 ± 1

In order to show that the increase of GlcCer was the result of an impairment of its catabolism, we labeled the cell sphingolipids at steady state using radioactive sphingosine [1-^{3}H]Sph. This experimental procedure is based on the recycling of the radioactive precursor [1-^{3}H]Sph that is used by the anabolic pathways as the endogenous counterpart. When there is an impairment of a catabolic enzyme (i.e., NLGase), the [1-^{3}H]Sph is not further recycled and no degraded radioactive sphingolipid accumulates. As shown in Figure 2, despite the inter-individual variability in the content of lactosylceramide (LacCer) and globotriaosylceramide (Gb3), no significant differences were found between controls and patients for the other sphingolipids characteristic of LCLs. The only exception was the GlcCer and the ganglioside GM3 content. All patient derived LCLs showed an increase of more than two-fold of radioactive GlcCer with respect to control LCLs as well as a decrease of the radioactive GM3. The reduction in patient LCLs of the ganglioside GM3, which is a lipid typically associated with the external leaflet of the cell PM, with the concomitant increase in GlcCer, supports the hypothesis that the impairment of the catabolic pathway occurs at the cell PM level.

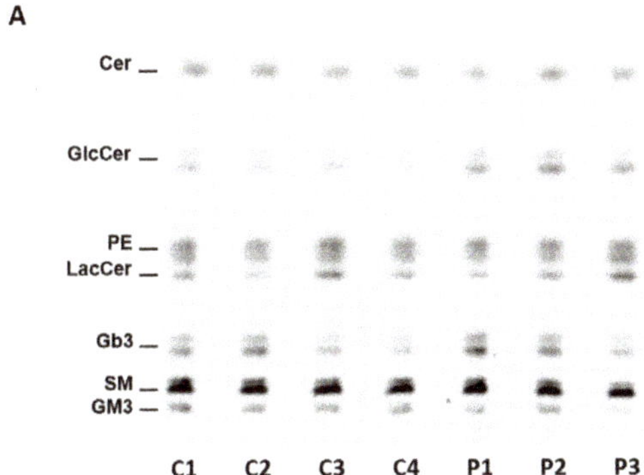
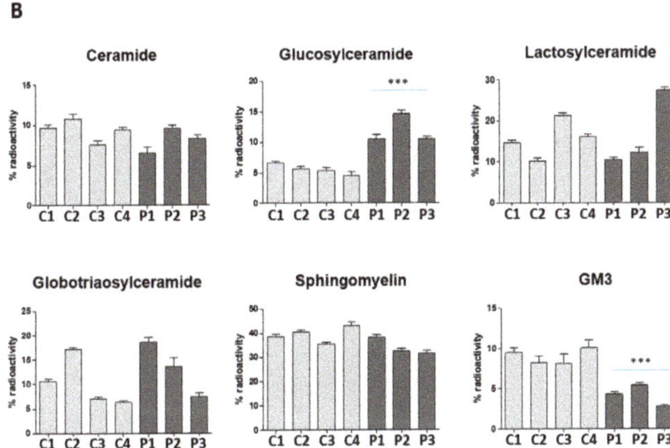

Figure 2. Radioactive sphingolipid pattern of control and patient LCLs. Total lipid extracts were separated by thin layer chromatography using the solvent system Chloroform/Methanol/Water 110:40:6 (*v:v:v*). (**A**) Representative digital autoradiogram obtained by the Beta-Imager ^{T}Racer equipment (BioSpace Lab). Same quantities of radioactivity were applied per lane. Ceramide (Cer), glucosylceramide (GlcCer), phosphatidylethanolamine (PE), lactosylceramide (LacCer), globotriaosylceramide (Gb3), sphingomyelin (SM), and ganglioside GM3. (**B**) Distribution of the radioactive sphingolipids associated with the total lipid extract expressed as % of the total radioactivity. Data are expressed as mean ± SD of three independent triplicate experiments (*** $p < 0.0003$ vs. controls).

2.2. GCase Activity is Up-Regulated in GBA2-Deficient LCLs Particularly at the PM

Several lines of evidence suggest the existence of a cross-talk among the enzymes involved in sphingolipid (SL) catabolism, since modification in the activity/expression of one enzyme could affect that of others [31]. Besides NLGase, another important enzyme involved in GlcCer catabolism is β-glucocerebrosidase (GCase) encoded by the *GBA* gene. Unlike NLGase, GCase is mainly a lysosomal enzyme and only partially associated with the external leaflet of the PM [15]. To investigate the possibility of a cross-talk between GCase and NLGase, we measured the activity of GCase both intracellularly and at the cell surface in control and patient derived LCLs. As shown in Figure 3–panel A, GCase activity was increased in total cell lysates derived from patients' cells with respect to controls. The average activity of GCase was 102 ± 14 pmoles/10^6 cells/h and 71 ± 17 pmoles/10^6 cells/h in patient and control derived LCLs, respectively. A marked increase, about three-fold, of GCase activity was also observed at the PM level in patients compared to control LCLs (patients: 33 ± 9 pmoles/10^6 cells/h; control: 11 ± 4 pmoles/10^6 cells/h). Interestingly, the augmented enzymatic activity is associated with an increase in the GCase protein levels but not in *GBA* mRNA expression (Figure 3B,C).

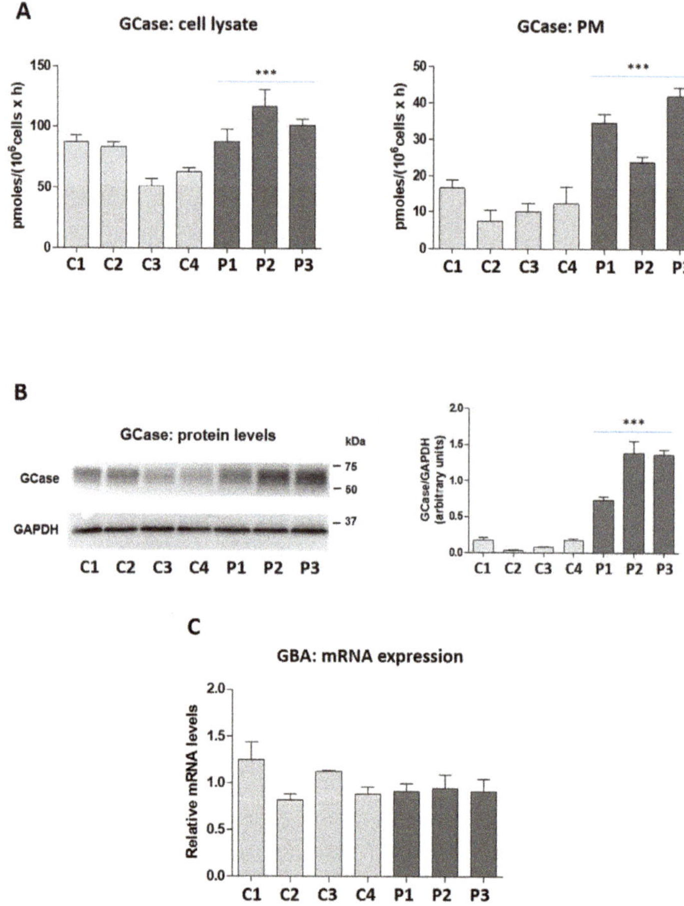

Figure 3. Activity and protein levels of β-glucocerebrosidase (GCase) and *GBA* mRNA expression. (**A**) GCase activity associated with the total cell lysate and plasma membrane of control (light grey) and patient (dark grey) derived LCLs. Activities were expressed as pmoles/10^6 cells/h. Data are expressed as mean $\pm$ SD ($n = 4$, *** $p < 0.0001$ vs. controls). (**B**) Immunoblot of GCase and control GAPDH accompanied by the semi-quantitative graph of normalized GCase/GAPDH. Data are expressed as mean $\pm$ SD ($n = 4$ *** $p < 0.0001$ vs. controls). (**C**) The graph represents the *GBA* mRNA levels of controls (light grey) and patients (dark grey) relative to the average value of controls ($n = 4$), after normalization with the endogenous β-actin gene (*ACTB*). Values represent the mean $\pm$ SEM of two independent triplicate experiments.

A substantial body of evidence shows the presence and action of mature and active lysosomal enzymes at the PM in addition to the lysosomes [15,32]. We decided to investigate the activity of two main hydrolases involved in SL catabolism, β-galactosidase and β-hexosaminidase, both intracellularly and at the PM level. We observed an increase only of the β-galactosidase activity associated with the cell surface of LCLs obtained from patients with respect to controls (Table 3).

Table 3. Enzymatic activities associated with total cell lysate and the plasma membrane (PM) of control (WT) and patient (c.1780 G>C) derived LCLs. Enzymatic activities were expressed as pmoles/10^6 cells/h ± error.

Enzymes	Cell Lysate		PM	
	WT	c.1780 G>C	WT	c.1780 G>C
β-Galactosidase	1284 ± 246	1399 ± 149	14 ± 4	30 ± 4
β-Hexosaminidase	1868 ± 233	1961 ± 206	24 ± 6	35 ± 15

3. Discussion

Several lines of evidence indicate that the regulation of GlcCer levels is important for cell homeostasis. GlcCer is a minor component of almost all membranes of eukaryotic cells suggesting an evolutionary strategy aimed to limit its presence. Indeed, de novo biosynthesized GlcCer is mainly used as a building block for the biosynthesis of complex GSLs [33].

In mammalian cells, GSL catabolism occurs by the sequential hydrolysis of the saccharidic chain by removing the reducing sugar. Lysosomes are involved in the catabolism of the endocytic portion of the cell PM and could be considered the principal site, together with the endoplasmic reticulum and Golgi complex, responsible for GSLs turnover [34]. On the other hand, the fine tuning of the GSLs composition is triggered directly at the cell PM by the action of specific glycohydrolases. In particular, the same enzymes that are associated with the lysosomes, such as sialidase Neu1, beta-hexosaminidase, beta-galactosiadase, and GCase, are present at the cell surface, even if in very small amounts, along with the sialidase Neu3 and NLGase, which are enzymes primarily residing at the PM [5]. Loss of function mutations in the lysosomal glycohydrolases determine the onset of lysosomal storage disorders, characterized by the accumulation of non-catabolized substrates.

NLGase is currently the most studied enzyme among the PM glycohydrolases. NLGase deficient mice, obtained both, by gene knockout and by pharmacological inhibition of the enzyme, showed an abnormal GlcCer accumulation in multiple tissues, including brain, liver, and testis. These data were quite surprising considering that the large amount of the GSL catabolism occurs in lysosomes. In addition, these mice were characterized by impaired liver regeneration and male infertility but no neurological involvement was observed [17–19]. However, in humans *GBA2* gene mutations are found in spastic ataxia and spastic paraplegia patients (SPastic Gait locus #46, SPG46) [23–25,35]. Among them, a c.1780G>C (p.Asp594His) missense mutation located in exon 11 of the *GBA2* gene was identified in a Cypriot consanguineous spastic ataxia family [22].

We used LCLs obtained from three patients of this family, who are homozygous for the *GBA2* c.1780G>C missense mutation, to evaluate the effect of the mutation on NLGase activity. By an in vitro enzymatic assay based on the use of CBE to block GCase activity, we found that the mutation strongly affects NLGase activity. Indeed, in pathological cells we were not able to detect any NLGase activity in the total cell lysate, and only a very low residual activity at the cell surface when compared with LCLs derived from four healthy controls. Similar to the results observed in mice, we found that the NLGase loss of function in patient derived LCLs is responsible for an increase of GlcCer content that reaches two-fold to that found in control cells. In addition, patient LCLs show an increased activity of GCase with respect to that measured in controls. In particular, GCase activity associated with the PM of patient derived LCLs is three-fold higher than that of control LCLs. This presumably compensatory effect has already been described in fibroblasts derived from patients affected by Gaucher disease, where the GCase loss of function induced an increase in NLGase activity [31]. Indeed, by the evaluation of the total cell β-glucocerebrosidase activity we did not find statistically significant differences between control and patient derived LCLs (Figure S1).

Taken together, the data, herein reported, further support an important role of NLGase in GlcCer metabolism and the existence of a cross-talk among the enzymes involved in GSL catabolism. Despite this new evidence, the challenges for future studies remain: (i) to explain why NLGase loss of function with concomitant increase of GCase could result in GlcCer accumulation and (ii) which are the molecular mechanisms linking the NLGase-dependent GlcCer accumulation with the onset of spastic ataxia.

4. Materials and Methods

4.1. Cell Culture

Seven LCLs (3 patients and 4 controls) were available, which were expanded and sub-cultured for the purposes of this study. Cells were grown in culture medium RPMI (Roswell Park Memorial Institute medium) supplemented with 10% FBS (fetal bovine serum), 1% penicillin/streptomycin and 1% glutamine, and expanded for a period of 6 weeks at 37 °C in a 5% CO_2 incubator. Genotypes of all patient derived LCLs sub-cultures were reconfirmed with Sanger sequencing and healthy controls were confirmed as homozygous wild-type.

All subjects gave their informed consent for inclusion before they participated in the study. The study was conducted in accordance with the Declaration of Helsinki, and approved by the National Bioethics Committee of Cyprus (EEBK/ΕΠ/2013/28, date of approval 14 May 2015).

4.2. Evaluation of Enzymatic Activities in Cell Lysates

The enzymatic activities associated with total cell lysates were determined by an assay based on the use of fluorogenic substrates as previously described [36]. To evaluate NLGase activity, cell lysates were pre-incubated for 30 min at room temperature in McIlvaine buffer (pH 6) with 1 mM CBE (Conduritol-B-epoxide, Merck, Darmstadt, Germany), a specific inhibitor of GCase [37]. For the measurement of GCase activity, cell lysates were pre-incubated for 30 min at room temperature in McIlvaine buffer (pH 5.2) containing 0.1% Triton X100(Merck, Darmstadt, Germany) with 5 nM AMP-dNM (adamantane-pentyl-dNM; *N*-(5-adamantane-1-yl-methoxy-pentyl) deoxynojirimycin, (A generous gift from Prof. Aerts JM form Leiden University) a specific inhibitor of NLGase. The total β-glucocerebrosidase assay was performed using the same procedure without inhibitors and detergents. At the end of the pre-incubation, the reactions were started by the addition of 25 μL of 4-Methylumbelliferyl β-D-glucopyranoside (MUB-β-Glc, Glycosynth, Warrington, UK) at a final concentration of 6 mM. To measure β-galactosidase and β-hexosaminidase activities, the fluorogenic substrates used were 4-Methylumbelliferyl β-D-galactopyranoside (MUB-β-Gal) and 4-Methylumbelliferyl *N*-acetyl-β-D-glucuronide (MUG) (all from Glycosynth, Warrington, UK), respectively. Aliquots of cell lysates were incubated with 25 μL of McIlvaine buffer 4× (0.4 M citric acid /0.8 M Na_2HPO_4) pH 5.2 and the specific fluorogenic substrates at a final concentration of 500 μM. Water was added to reach the final volumes of 100 μL. At different time points the reaction was stopped by adding 9 volumes of 0.25 M glycine pH 10.7 (Sigma-Aldrich, St. Louis, MO, USA). The fluorescence was detected by a Victor microplate reader (Perkin Elmer, Waltham, MA, USA). Standards of free 4-methylumbelliferone (MUB) were used to construct calibration curves. The enzymatic activities were expressed as pmoles of product/10^6 cells /h.

4.3. Evaluation of Enzymatic Activities at the Cell Surface of Living Cells

PM-associated activities of total β-glucocerebrosidase, GCase, NLGase, β-galactosidase and β-hexosaminidase were assessed in living cells, plated in a 96-well microplate at a density of 200,000 cells/well, by a high throughput live cell-based assay as previously described [15,27,31,36]. To distinguish between GCase and NLGase activities, cells were pre-incubated for 30 min at room temperature in DMEM-F12 without phenol red (Thermo Fisher Scientific, Waltham, MA, USA) containing 5 nM AMP-DNM or 1 mM CBE, respectively [38]. Total β-glucocerebrosidase assay was performed using the same procedure without any inhibitors. Activities were assayed using the artificial substrate MUB-β-Gal for β-galactosidase, MUG for β-hexosaminidase, and MUB-β-Glc for β-glucocerebrosidases GCase and NLGase. The fluorogenic substrates were solubilized in DMEM-F12 without phenol red at pH 6, with final concentrations of 250 μM, 1 mM, and 6 mM, respectively.

Aliquots of medium (10 µL) were analyzed at different time points by a Victor microplate reader (Perkin Elmer, Waltham, MA, USA), after adding 190 µL of 0.25 M glycine, with a pH of 10.7. Standards of free MUB were used to construct calibration curves. The enzymatic activities were expressed as pmoles of product/10^6 cells/h. The experimental design included internal controls. In particular, this method is based on the observation that the fluorogenic substrates commonly used for the in vitro assay of glycohydrolytic activities are not taken up by living cells. To assess that the substrate hydrolysis occurs only upon the activity of PM enzymes, a series of controls was performed. In the used experimental conditions, we did not observe any intracellular fluorescence, evaluated by both fluorescent microscopy and fluorimetric analysis of the cells lysed in 0.25 M glycine (pH 10.7), indicating that the substrates were not able to cross the cell membrane. Moreover, we verified that the artificial substrates did not undergo either spontaneous or secreted enzyme-driven hydrolysis by the establishment of an appropriate control without cells or by the solubilization of MUB-substrates directly in cell culture medium in the presence or not of cells.

4.4. Real-Time PCR

Total RNA was isolated from the LCLs using the RNeasy® Midi kit (Qiagen, Hilden, Germany) as described in the manufacturer's instructions. cDNA synthesis was performed using 1µg of total RNA according to the instructions of the Protoscript® M-MuLV II First Strand cDNA Synthesis Kit (New England Biolabs, Ipswich, Massachusetts, USA). Real-time PCR was carried out using the CFX96 Real-Time system (Bio-Rad, Hercules, CA, USA) and the amplification was done using the SsoFast EvaGreen Supermix according to the manufacturer's instructions (Bio-Rad, Hercules, CA, USA). The sequences of the primers are available upon request. *GBA1* and *GBA2* mRNA expression levels were normalized to the actin house-keeping gene (*ACTB*) and relative mRNA expression was calculated according to the ΔΔCT method.

4.5. Immunoblotting

Equivalent amounts of proteins associated with total cell lysates, determined by DC Protein Assay (Bio-Rad, Hercules, CA, USA), were separated on polyacrylamide gels and then transferred to PVDF (Polyvinylidene fluoride) membranes by electroblotting [29]. Blots were incubated with monoclonal rabbit anti-GCase (ab128879, Abcam, Cambridge, UK) or polyclonal rabbit anti-GAPDH (G9545, Sigma-Aldrich) primary antibodies at 4 °C overnight, followed by incubation with goat anti-rabbit HRP-conjugated (7074, Cell Signaling) secondary antibody and detection with a chemiluminescent kit (WESTAR ηC, Cyanagen, Bologna, Italy). Digital images were obtained by the chemiluminescence system Alliance Mini HD9 (UVItec, Cambridge, UK).

4.6. Cell Sphingolipid Labelling with [1-³H]-Sphingosine

[1-³H]-sphingosine was administered as tracer in non-bioactive concentration for 2 h (pulse) followed to 96 h (chase), to allow steady state metabolic labelling of all cell SLs [39]. Briefly, [1-³H]-sphingosine dissolved in methanol was transferred into a sterile glass tube, dried under a nitrogen stream and then solubilized in an appropriate volume of pre-warmed (37 °C) cell culture medium to obtain a final concentration of 30 nM. The correct solubilization was verified by measuring the radioactivity associated with an aliquot of the medium using a β-counter (PerkinElmer, Waltham, MA, USA). After 2 h of incubation (pulse) the medium was removed, and the cells were incubated for 96 h (chase) in fresh culture medium without radioactive sphingosine. After chase, cells were collected, lyophilized and subjected to lipid extraction and SLs analysis. Total lipids from lyophilized cells were extracted with chloroform:methanol:water 20:10:1 by volume, followed by a second extraction with chloroform:methanol: 2:1 by volume. The radioactivity associated with total lipid extract, was evaluated by liquid scintillation, using a β-counter system (PerkinElmer).

[³H]SLs of total extracts were separated by high performance thin layer chromatography (HPTLC), using the solvent system chloroform:methanol:water 110:40:6 by volume. [³H]SLs were identified by digital autoradiography using ^{T}Racer system (Biospace Lab) and quantified with M3vision software. The lipid identification was performed using purified radioactive standards.

4.7. Lipid Analysis by SFC-MS/MS

Quantitative analysis of glucosylceramide, galactosylceramide, glucosylsphingosine and galactosylsphingosine was performed by Lipidomics Shared Resources Analytical Unit (Medical University of South Carolina, Charleston, SC, USA) [40]. Briefly, quantitative analysis of sphingolipids is based on eight-point calibration curves generated for each target analyte. The synthetic standards, along with a set of internal standards, were spiked into an artificial matrix and subjected to an identical extraction procedure as the biological samples. These extracted standards were then analyzed by the SFC-MS/MS system operating in positive MRM mode employing a gradient elution. Peaks for the target analytes and internal standards were recorded and processed using the instrument's software system. The calibration curve for a particular analyte was generated by plotting the analyte/internal standard peak area ratio against analyte concentrations. Any sphingolipid for which no standards were available was quantitated using the calibration curve of its closest counterpart. Separation of galactosylceramide and glucosylceramide was performed by SFC-MS/MS. The equipment consisted of a Waters UPC^2 system coupled to a Thermo Scientific Quantum Access Max triple quadrupole mass spectrometer, equipped with an ESI (electrospray ionization) probe operating in the multiple reaction monitoring positive ion mode tuned and optimized for the Waters UPC^2 system. Chromatographic separations were obtained utilizing carbon dioxide gas and 1 mM ammonium formate in 0.2% formic acid in the methanol mobile phase.

Analytical results were expressed as pmoles of lipid/mg of total cellular proteins. Data were the mean of two independent triplicate experiments.

Supplementary Materials: Supplementary materials can be found at http://www.mdpi.com/1422-0067/19/10/3099/s1.

Author Contributions: A.M. performed biochemical and real-time PCR experiments, analyzed data and contributed to the writing, reviewing and editing of the manuscript; M.S. performed biochemical experiments and reviewed the manuscript; A.D. had a critical input in the interpretation of results, writing, reviewing and editing the manuscript; C.V. performed the Sanger sequencing confirmation of the LCLs and real-time PCR experiments, and reviewed the manuscript; S.S. had an input in the supervision and interpretation of the analyses related to the evaluation of the SL pattern; E.C. contributed to the preparation of the figures and writing of the discussion; M.P. and E.Z.-P. contributed to the patient samples as well as the clinical evaluation and reviewed the manuscript; M.A. and N.L. had a critical input in the conceptualization of the study, writing, reviewing and editing the manuscript; K.C. contributed in conceiving the study and reviewing the manuscript.

Funding: This study was funded by the Cyprus Institute of Neurology & Genetics and from the University of Milano.

Acknowledgments: We would like to thank the patients and healthy individuals who participated in this study.

Conflicts of Interest: The authors declare no conflict of interest.

References

1. Sonnino, G. Transport processes in magnetically confined plasmas in the nonlinear regime. *Chaos* **2006**. [CrossRef] [PubMed]
2. Jeckel, D.; Karrenbauer, A.; Burger, K.N.; van Meer, G.; Wieland, F. Glucosylceramide is synthesized at the cytosolic surface of various Golgi subfractions. *J. Cell Biol.* **1992**, *117*, 259–267. [CrossRef] [PubMed]
3. Lamari, F.; Mochel, F.; Sedel, F.; Saudubray, J.M. Disorders of phospholipids, sphingolipids and fatty acids biosynthesis: Toward a new category of inherited metabolic diseases. *J. Inherit. Metab. Dis.* **2013**, *36*, 411–425. [CrossRef] [PubMed]
4. Kuivenhoven, J.A.; Hegele, R.A. Mining the genome for lipid genes. *Biochim. Biophys. Acta* **2014**, *1842*, 1993–2009. [CrossRef] [PubMed]
5. Aureli, M.; Loberto, N.; Chigorno, V.; Prinetti, A.; Sonnino, S. Remodeling of sphingolipids by plasma membrane associated enzymes. *Neurochem. Res.* **2011**, *36*, 1636–1644. [CrossRef] [PubMed]

6. Magini, A.; Polchi, A.; Urbanelli, L.; Cesselli, D.; Beltrami, A.; Tancini, B.; Emiliani, C. TFEB activation promotes the recruitment of lysosomal glycohydrolases beta-hexosaminidase and beta-galactosidase to the plasma membrane. *Biochem. Biophys. Res. Commun.* **2013**, *440*, 251–257. [CrossRef] [PubMed]

7. Brady, R.O.; Kanfer, J.N.; Shapiro, D. Metabolism of Glucocerebrosides. Ii. Evidence of an Enzymatic Deficiency in Gaucher's Disease. *Biochem. Biophys. Res. Commun.* **1965**, *18*, 221–225. [CrossRef]

8. Grace, M.E.; Newman, K.M.; Scheinker, V.; Berg-Fussman, A.; Grabowski, G.A. Analysis of human acid beta-glucosidase by site-directed mutagenesis and heterologous expression. *J. Biol. Chem.* **1994**, *269*, 2283–2291. [PubMed]

9. Cobucci-Ponzano, B.; Aurilia, V.; Riccio, G.; Henrissat, B.; Coutinho, P.M.; Strazzulli, A.; Padula, A.; Corsaro, M.M.; Pieretti, G.; Pocsfalvi, G.; et al. A new archaeal beta-glycosidase from Sulfolobus solfataricus: seeding a novel retaining beta-glycan-specific glycoside hydrolase family along with the human non-lysosomal glucosylceramidase GBA2. *J. Biol. Chem.* **2010**, *285*, 20691–20703. [CrossRef] [PubMed]

10. Zecca, L.; Mesonero, J.E.; Stutz, A.; Poiree, J.C.; Giudicelli, J.; Cursio, R.; Gloor, S.M.; Semenza, G. Intestinal lactase-phlorizin hydrolase (LPH): the two catalytic sites; the role of the pancreas in pro-LPH maturation. *FEBS Lett.* **1998**, *435*, 225–228. [CrossRef]

11. Rempel, B.P.; Withers, S.G. Covalent inhibitors of glycosidases and their applications in biochemistry and biology. *Glycobiology* **2008**, *18*, 570–586. [CrossRef] [PubMed]

12. Koshland, D.E., Jr.; Clarke, E. Mechanism of hydrolysis of adenosinetriphosphate catalyzed by lobster muscle. *J. Biol. Chem.* **1953**, *205*, 917–924. [PubMed]

13. Matern, H.; Boermans, H.; Lottspeich, F.; Matern, S. Molecular cloning and expression of human bile acid beta-glucosidase. *J. Biol. Chem.* **2001**, *276*, 37929–37933. [CrossRef] [PubMed]

14. Korschen, H.G.; Yildiz, Y.; Raju, D.N.; Schonauer, S.; Bonigk, W.; Jansen, V.; Kremmer, E.; Kaupp, U.B.; Wachten, D. The non-lysosomal beta-glucosidase GBA2 is a non-integral membrane-associated protein at the endoplasmic reticulum (ER) and Golgi. *J. Biol. Chem.* **2013**, *288*, 3381–3393. [CrossRef] [PubMed]

15. Aureli, M.; Masilamani, A.P.; Illuzzi, G.; Loberto, N.; Scandroglio, F.; Prinetti, A.; Chigorno, V.; Sonnino, S. Activity of plasma membrane beta-galactosidase and beta-glucosidase. *FEBS Lett.* **2009**, *583*, 2469–2473. [CrossRef] [PubMed]

16. Aureli, M.; Loberto, N.; Bassi, R.; Ferraretto, A.; Perego, S.; Lanteri, P.; Chigorno, V.; Sonnino, S.; Prinetti, A. Plasma membrane-associated glycohydrolases activation by extracellular acidification due to proton exchangers. *Neurochem. Res.* **2012**, *37*, 1296–1307. [CrossRef] [PubMed]

17. Yildiz, Y.; Matern, H.; Thompson, B.; Allegood, J.C.; Warren, R.L.; Ramirez, D.M.; Hammer, R.E.; Hamra, F.K.; Matern, S.; Russell, D.W. Mutation of beta-glucosidase 2 causes glycolipid storage disease and impaired male fertility. *J. Clin. Invest.* **2006**, *116*, 2985–2994. [CrossRef] [PubMed]

18. Gonzalez-Carmona, M.A.; Sandhoff, R.; Tacke, F.; Vogt, A.; Weber, S.; Canbay, A.E.; Rogler, G.; Sauerbruch, T.; Lammert, F.; Yildiz, Y. Beta-glucosidase 2 knockout mice with increased glucosylceramide show impaired liver regeneration. *Liver Int.* **2012**, *32*, 1354–1362. [CrossRef] [PubMed]

19. Raju, D.; Schonauer, S.; Hamzeh, H.; Flynn, K.C.; Bradke, F.; Vom Dorp, K.; Dormann, P.; Yildiz, Y.; Trotschel, C.; Poetsch, A.; et al. Accumulation of glucosylceramide in the absence of the beta-glucosidase GBA2 alters cytoskeletal dynamics. *PLoS Genet.* **2015**, *11*, e1005063. [CrossRef] [PubMed]

20. Schonauer, S.; Korschen, H.G.; Penno, A.; Rennhack, A.; Breiden, B.; Sandhoff, K.; Gutbrod, K.; Dormann, P.; Raju, D.N.; Haberkant, P.; et al. Identification of a feedback loop involving beta-glucosidase 2 and its product sphingosine sheds light on the molecular mechanisms in Gaucher disease. *J. Biol. Chem.* **2017**, *292*, 6177–6189. [CrossRef] [PubMed]

21. Martin, E.; Schule, R.; Smets, K.; Rastetter, A.; Boukhris, A.; Loureiro, J.L.; Gonzalez, M.A.; Mundwiller, E.; Deconinck, T.; Wessner, M.; et al. Loss of function of glucocerebrosidase GBA2 is responsible for motor neuron defects in hereditary spastic paraplegia. *Am. J. Hum. Genet.* **2013**, *92*, 238–244. [CrossRef] [PubMed]

22. Votsi, C.; Zamba-Papanicolaou, E.; Middleton, L.T.; Pantzaris, M.; Christodoulou, K. A novel GBA2 gene missense mutation in spastic ataxia. *Ann. Hum. Genet.* **2014**, *78*, 13–22. [CrossRef] [PubMed]

23. Hammer, M.B.; Eleuch-Fayache, G.; Schottlaender, L.V.; Nehdi, H.; Gibbs, J.R.; Arepalli, S.K.; Chong, S.B.; Hernandez, D.G.; Sailer, A.; Liu, G.; et al. Mutations in GBA2 cause autosomal-recessive cerebellar ataxia with spasticity. *Am. J. Hum. Genet.* **2013**, *92*, 245–251. [CrossRef] [PubMed]

24. Citterio, A.; Arnoldi, A.; Panzeri, E.; D'Angelo, M.G.; Filosto, M.; Dilena, R.; Arrigoni, F.; Castelli, M.; Maghini, C.; Germiniasi, C.; et al. Mutations in CYP2U1, DDHD2 and GBA2 genes are rare causes of complicated forms of hereditary spastic paraparesis. *J. Neurol.* **2014**, *261*, 373–381. [CrossRef] [PubMed]

25. Sultana, S.; Reichbauer, J.; Schule, R.; Mochel, F.; Synofzik, M.; van der Spoel, A.C. Lack of enzyme activity in GBA2 mutants associated with hereditary spastic paraplegia/cerebellar ataxia (SPG46). *Biochem. Biophys. Res. Commun.* **2015**, *465*, 35–40. [CrossRef] [PubMed]

26. Haugarvoll, K.; Johansson, S.; Rodriguez, C.E.; Boman, H.; Haukanes, B.I.; Bruland, O.; Roque, F.; Jonassen, I.; Blomqvist, M.; Telstad, W.; et al. GBA2 Mutations Cause a Marinesco-Sjogren-Like Syndrome: Genetic and Biochemical Studies. *PLoS ONE* **2017**, *12*, e0169309. [CrossRef] [PubMed]

27. Aureli, M.; Bassi, R.; Prinetti, A.; Chiricozzi, E.; Pappalardi, B.; Chigorno, V.; Di Muzio, N.; Loberto, N.; Sonnino, S. Ionizing radiations increase the activity of the cell surface glycohydrolases and the plasma membrane ceramide content. *Glycoconj. J.* **2012**. [CrossRef] [PubMed]

28. Loberto, N.; Tebon, M.; Lampronti, I.; Marchetti, N.; Aureli, M.; Bassi, R.; Giri, M.G.; Bezzerri, V.; Lovato, V.; Cantu, C.; et al. GBA2-encoded beta-glucosidase activity is involved in the inflammatory response to Pseudomonas aeruginosa. *PLoS ONE* **2014**, *9*, e104763. [CrossRef] [PubMed]

29. Samarani, M.; Loberto, N.; Solda, G.; Straniero, L.; Asselta, R.; Duga, S.; Lunghi, G.; Zucca, F.A.; Mauri, L.; Ciampa, M.G.; et al. A lysosome-plasma membrane-sphingolipid axis linking lysosomal storage to cell growth arrest. *FASEB. J.* **2018**. [CrossRef] [PubMed]

30. Ferraz, M.J.; Marques, A.R.; Appelman, M.D.; Verhoek, M.; Strijland, A.; Mirzaian, M.; Scheij, S.; Ouairy, C.M.; Lahav, D.; Wisse, P.; et al. Lysosomal glycosphingolipid catabolism by acid ceramidase: formation of glycosphingoid bases during deficiency of glycosidases. *FEBS Lett.* **2016**, *590*, 716–725. [CrossRef] [PubMed]

31. Aureli, M.; Bassi, R.; Loberto, N.; Regis, S.; Prinetti, A.; Chigorno, V.; Aerts, J.M.; Boot, R.G.; Filocamo, M.; Sonnino, S. Cell surface associated glycohydrolases in normal and Gaucher disease fibroblasts. *J. Inherit. Metab. Dis.* **2012**, *35*, 1081–1091. [CrossRef] [PubMed]

32. Mencarelli, S.; Cavalieri, C.; Magini, A.; Tancini, B.; Basso, L.; Lemansky, P.; Hasilik, A.; Li, Y.T.; Chigorno, V.; Orlacchio, A.; et al. Identification of plasma membrane associated mature beta-hexosaminidase A, active towards GM2 ganglioside, in human fibroblasts. *FEBS Lett.* **2005**, *579*, 5501–5506. [CrossRef] [PubMed]

33. Merrill, A.H., Jr. Sphingolipid and glycosphingolipid metabolic pathways in the era of sphingolipidomics. *Chem. Rev.* **2011**, *111*, 6387–6422. [CrossRef] [PubMed]

34. Kolter, T.; Sandhoff, K. Principles of lysosomal membrane digestion: stimulation of sphingolipid degradation by sphingolipid activator proteins and anionic lysosomal lipids. *Annu. Rev. Cell Dev. Biol.* **2005**, *21*, 81–103. [CrossRef] [PubMed]

35. Walden, C.M.; Sandhoff, R.; Chuang, C.C.; Yildiz, Y.; Butters, T.D.; Dwek, R.A.; Platt, F.M.; van der Spoel, A.C. Accumulation of glucosylceramide in murine testis, caused by inhibition of beta-glucosidase 2: Implications for spermatogenesis. *J. Biol. Chem.* **2007**, *282*, 32655–32664. [CrossRef] [PubMed]

36. Aureli, M.; Loberto, N.; Lanteri, P.; Chigorno, V.; Prinetti, A.; Sonnino, S. Cell surface sphingolipid glycohydrolases in neuronal differentiation and aging in culture. *J. Neurochem.* **2011**, *116*, 891–899. [CrossRef] [PubMed]

37. Boot, R.G.; Verhoek, M.; Donker-Koopman, W.; Strijland, A.; van Marle, J.; Overkleeft, H.S.; Wennekes, T.; Aerts, J.M. Identification of the non-lysosomal glucosylceramidase as beta-glucosidase 2. *J. Biol. Chem.* **2007**, *282*, 1305–1312. [CrossRef] [PubMed]

38. Overkleeft, H.S.; Renkema, G.H.; Neele, J.; Vianello, P.; Hung, I.O.; Strijland, A.; van der Burg, A.M.; Koomen, G.J.; Pandit, U.K.; Aerts, J.M. Generation of specific deoxynojirimycin-type inhibitors of the non-lysosomal glucosylceramidase. *J. Biol. Chem.* **1998**, *273*, 26522–26527. [CrossRef] [PubMed]

39. Schiumarini, D.; Loberto, N.; Mancini, G.; Bassi, R.; Giussani, P.; Chiricozzi, E.; Samarani, M.; Munari, S.; Tamanini, A.; Cabrini, G.; et al. Evidence for the Involvement of Lipid Rafts and Plasma Membrane Sphingolipid Hydrolases in Pseudomonas aeruginosa Infection of Cystic Fibrosis Bronchial Epithelial Cells. *Mediators Inflamm.* **2017**, *2017*, 1730245. [CrossRef] [PubMed]

40. Bielawski, J.; Pierce, J.S.; Snider, J.; Rembiesa, B.; Szulc, Z.M.; Bielawska, A. Comprehensive quantitative analysis of bioactive sphingolipids by high-performance liquid chromatography-tandem mass spectrometry. *Methods Mol. Biol.* **2009**, *579*, 443–467. [CrossRef] [PubMed]

International Journal of
Molecular Sciences

Review

Kidney Lipidomics by Mass Spectrometry Imaging: A Focus on the Glomerulus

Imane Abbas [1,2], Manale Noun [1,2], David Touboul [2], Dil Sahali [3,4,5], Alain Brunelle [6] and Mario Ollero [3,4,*]

1   Lebanese Atomic Energy Commission, NCSR, Beirut 11-8281, Lebanon; imane.abbas@cnrs.edu.lb (I.A.); Manale.Noun@cnrs.edu.lb (M.N.)
2   Institut de Chimie des Substances Naturelles, CNRS UPR 2301, Univ. Paris-Sud, Université Paris-Saclay, Avenue de la Terrasse, 91198 Gif-sur-Yvette, France; David.TOUBOUL@cnrs.fr
3   Institut Mondor de Recherche Biomédicale, INSERM, U955 EQ21, 8, rue du Général Sarrail, 94010 Créteil, France; dil.sahali@inserm.fr
4   Université Paris Est Créteil, 61, avenue du Général de Gaulle, 94010 Créteil, France
5   Hôpital Henri Mondor, 51 avenue du Maréchal de Lattre de Tassigny, 94010 Créteil, France
6   Laboratoire d'Archéologie Moléculaire et Structurale, LAMS UMR8220, CNRS, Sorbonne Université, 4 place Jussieu, 75005 Paris, France; Alain.BRUNELLE@cnrs.fr
*   Correspondence: mario.ollero@inserm.fr; Tel.: +33-149813667

Received: 27 February 2019; Accepted: 28 March 2019; Published: 1 April 2019

Abstract: Lipid disorders have been associated with glomerulopathies, a distinct type of renal pathologies, such as nephrotic syndrome. Global analyses targeting kidney lipids in this pathophysiologic context have been extensively performed, but most often regardless of the architectural and functional complexity of the kidney. The new developments in mass spectrometry imaging technologies have opened a promising field in localized lipidomic studies focused on this organ. In this article, we revisit the main works having employed the Matrix Assisted Laser Desorption Ionization Time of Flight (MALDI-TOF) technology, and the few reports on the use of TOF-Secondary Ion Mass Spectrometry (TOF-SIMS). We also present a first analysis of mouse kidney cortex sections by cluster TOF-SIMS. The latter represents a good option for high resolution lipid imaging when frozen unfixed histological samples are available. The advantages and drawbacks of this developing field are discussed.

Keywords: MALDI-TOF; TOF-SIMS; glomerulopathies; nephrotic syndrome; MSI

---

1. Introduction: The Kidney Glomerulus as a Witness and Target of Lipid Disorders

The renal glomerulus is a highly specialized anatomic structure containing the filtration barrier that separates the blood and urine compartments. This barrier comprises a fenestrated endothelium, a basement membrane, and a layer of podocytes, highly differentiated postmitotic epithelial cells. Alterations in this barrier lead to proteinuria, which is one of the hallmarks of nephrotic syndrome (NS), along with hypoalbuminemia, hyperlipidemia, lipiduria, and edema. NS characterizes a group of diseases of either genetic or non-genetic—most likely immune—origin. The latter are known as idiopathic nephrotic syndrome (INS), which comprehends two main histologically-defined forms, namely minimal change nephrotic syndrome (MCNS) and focal and segmental glomerulosclerosis (FSGS). Other glomerulopathies include the immune-mediated membranous nephropathy (MN), IgA nephropathy, C3 glomerulopathies, lupus nephritis, anti-neutrophil circulating antibody-associated vasculitis, and different forms of glomerulonephritis. In addition, a number of diseases, such as diabetes, obesity and cancer, the exposure to infectious agents and toxic molecules, are associated with glomerular dysfunction and secondary NS.

NS has been associated with a defect in lipoprotein metabolism and major changes in lipoprotein profiling and content (reviewed in [1]). These alterations are attributed to impaired chylomicron and very low-density lipoprotein (VLDL) clearance [2–4], which is associated, at least in part, with decreased abundance of lipoprotein lipase, hepatic lipase, and glycosylphosphatidylinositol-anchored binding protein 1 (GPI-BP1) in several tissues of animal models of NS. Other alterations contributing to NS dyslipidemia are the high serum levels of angiopoietin-like protein 4 in NS patients. The latter has been suggested to play a direct role in the onset of proteinuria [5]. Other alterations include higher levels of cholesterol and LDL-cholesterol, due to increased synthesis and decreased catabolism of cholesterol in the liver [6], and lower levels of HDL-cholesterol to cholesterol ratio in NS patients [7]. Moreover, mutations in the *APOL1* gene, encoding apolipoprotein L1, a key component of high-density lipoproteins (HDL), have been linked to FSGS susceptibility [8,9]. All these alterations lead to higher risk of cardiovascular disease, and to local functional defects in the kidney, such as nephron loss and lipotoxicity in proximal tubular cells. These observations parallel the report of accumulated cholesterol in glomeruli from FSGS patients [10] and in the renal cortex of diabetic nephropathy models [11]. Globally this suggests the presence of profound alterations in the renal and the glomerular lipidome in these pathologic conditions.

Nevertheless, the glomerulus and the podocyte have been scarcely the target of global lipid analyses [12,13]. In an elegant study, Jin and collaborators used a targeted lipidomic approach to unveil the lipid interactors of soluble vascular endothelial growth factor (VEGF) receptor Flt1 at the podocyte surface [13]. They identified the glycosphingolipid ganglioside M3 species as key interactors, and this binding necessary to the functional integrity of podocyte and glomerular filtration barrier. More recently, in a very different setup, lipidomic analysis of urine from pediatric FSGS patients led to the conclusion that these individual display higher free fatty acids and lysophosphatidylcholine (LPC) levels and lower phosphatidylcholine (PC) with respect to controls, suggesting an increased phospholipase A2 activity associated with the FSGS glomerulus [12].

Several pathologies characterized by lipid disorders associated with glomerular alterations have been described in the literature, such as Thay-Sachs, Gaucher, Niemann-Pick, Sandhoff, Fabry, Myopathy 2, HIV-associated nephropathy, and diabetic nephropathy, among others. These disorders are very often characterized by accumulation of different sphingolipid (SP) species (reviewed in [14]). For example, in Fabry disease, the glycosphingolipids globotriaosylceramide (Gb3) and digalactosylceramide (Ga2) significantly accumulate in several tissues, including the kidney, which has been associated with podocyte dysfunction and proteinuria [15]. Sphingomyelin levels, along with those of cholesterol, are increased in the kidney of Niemann-Pick disease, a genetic disorder due to mutations in *NPC1/NPC2* (encoding intracellular cholesterol transporters) and *SMPD1* (encoding an acid sphingomyelinase) genes, the latter associated with proteinuria [16]. Likewise, in genetic forms of nephrotic syndrome due to mutations in the gene *NPHS1* (encoding nephrin, a protein located at the podocyte slit diaphragm, a raft-like membranous structure), there is distal tubular and parietal glomerular accumulation of disyaloganglioside GD3 [17]. Moreover, post-transplant recurrence of FSGS has been found linked to decreased expression of the acid sphingomyelinase-like phosphodiesterase 3b (SMPDL3b) in podocytes, leading to increased sphingomyelin and decreased ceramide [18]. Strikingly, in many of these pathologies the glycosphingolipid metabolic pathway is altered.

2. Lipid Imaging of the Kidney: MALDI-MSI

Analysis of kidney lipids by mass spectrometry imaging (MSI) is a relatively recent approach that has made a still discreet yet significant contribution by accompanying histological evaluation, global multiomic (including lipidomics) studies, and further proteomic analyses by MSI. MSI lipidomics has extensively been performed on rodent material by matrix-assisted laser desorption/ionization (MALDI), the mainstream method for MSI, which is characterized by the addition of an organic chemical matrix on biological specimens. This technology has been used on normal tissue [19–23] and in different pathophysiologic contexts, such as acute renal injury [24], diabetes/obesity [25], renal acidosis [26] polycystic kidney [27], drug-induced nephrotoxicity [28,29], bisphenol toxicity [30], polymyxin toxicity [31], Fabry disease [32,33], and in glomerulopathies, such as diabetic nephropathy [34] and IgA nephropathy [35].

Some of these studies have provided valuable information about the spatial localization of lipids in kidney anatomical structures. One example is the thorough study that was performed to determine the distribution of sulfated glycosphingolipids (sulfatides) by MALDI-MSI in combination with source-decay LC-MS/MS in a mouse model of renal acidosis. The latter consisted of invalidation of ceramide synthase, one of the key enzymes in the glycosphingolipid biosynthetic pathway. The technical approach was able to discern the sulfatides characteristic of cortex, medulla, and papillae, according to their sphingoid bases, C18-phytosphingosine, C18-sphingosine, and C20-sphingosine, respectively. In particular, ceramide synthase deficiency was correlated with a depletion in C18-phytosphingosine in the cortex region [26].

In other cases, although not strictly in the context of lipidomics, MSI has been employed to trace an administered drug and its metabolites [28]. The reported study was performed on frozen and paraffin sections of rabbit kidney, and the targeted small molecule, an inhibitor of c-Met tyrosine kinase. This study was interesting in that samples were formaldehyde-fixed, and two technologies, namely MALDI-MSI and DESI-MSI, were used. The authors examined cortex sections and identified unambiguously crystal deposits containing the molecule and some of its metabolites in tubular areas. This underlines the potential of MSI in this type of studies and also the feasibility of analyzing previously-fixed paraffin material, even considering the ion suppression effects of paraffin and of formaldehyde cross-linking.

2.1. MSI Analysis of Human Tissue: Lessons from Renal Cell Carcinoma

Much effort has been made in applying the spatial information provided by MSI techniques to the study of human specimens in the context of kidney cancer, the rationale being the search for differences in lipid composition between normal and tumor tissue, which could be used eventually as diagnostic or prognostic markers. Recently, a MALDI variant consisting of gold nanoparticle-enhanced target-based surface-assisted desorption/ionization (SALDI) has been performed to analyze biopsies of renal cell carcinoma [36]. The study resulted in the identification of two markers of tumor tissue, a sodium adduct of diglyceride DG 38:1 and protonated octadecylamide. Moreover, MALDI-MSI was used in parallel to validate the results obtained by HILIC/ESI-MS when comparing patient tumor material with adjacent normal tissue. By targeting polar lipids, such as gangliosides and sulfoglycosphingolipids, and acidic phospholipids [37], MSI results confirmed the findings by HILIC/ESI-MS, consisting of increased phosphatidylinositols, PI 40:5 and PI 40:4, and gangliosides, GM3 34:1, and GM3 42:2, in carcinoma. In another study comparing normal and tumor tissue, MALDI was coupled to an Orbitrap analyzer. The MALDI–Orbitrap method was initially focused on sulfoglycosphingolipids and allowed the identification of more than 120 molecules. The use of multivariate analyses led to the description of a discriminant signature and this correlated with the result of the parallel MSI analysis. The latter showed two distinct markers, namely phosphatidylethanolamine (PE) 36:4 and sulfodihexosyl-ceramide 42:1, decreased and increased in carcinoma tissue, respectively [38]. Another study compared both types of tissues using two parallel approaches, namely touch-spray ionization and DESI-MSI (desorption electrospray ionization-MSI). Multivariate analysis enlightened a signature of discriminant lipid molecules encompassing ion forms of phosphatidylserine (PS 36:1, PS 38:4) and phosphatidylinositol (PI 38:4) from MSI results, while touch-spray led to discriminant phosphatidylcholine (PC)-derived ions [39].

In these examples, MSI often confirmed the results obtained by a LC-MS setup, but in some cases unveiled new differences between both types of tissues, demonstrating the complementarity of both strategies.

2.2. Glomerular Disease: The Spatial Limits

A few of those MSI studies targeting the kidney have attempted to address the lipid composition of glomeruli, but the low resolution (i.e., most often in the 30–70 µm range), which roughly corresponds to the mouse glomerular diameter, limits the observations to the renal cortex, as it is unable to discriminate between glomerular and tubular structures. Nevertheless, some remarkable findings have been reported to date.

One of these reports corresponds to a study on IgA nephropathy, the most prevalent glomerular disease, characterized by deposition of immunoglobulin A on the glomerular mesangium. In a spontaneous mouse model of the disease, the authors used a MALDI source coupled to two tandem analyzers, a quadrupole ion trap, and a time-of-flight [35]. The analysis was able to distinguish the renal cortex and the hilum, marked respectively by phosphatidylcholines (PC 38:6 and PC 40:6) and triglycerides (TAG 52:3 and TAG 54:4). A signature characteristic of the diseased mice was composed of PC O-38:6 and PC O-40:7 (ether forms of PC), an analog of platelet activating factor, and a plasmalogen, respectively. In all cases, the presence of 22:6 (docosahexaenoate, DHA) moieties was surprising, since this fatty acid is not known as particularly abundant in kidney tissue. In addition, some ions were detected as characteristic of the IgA model, *m/z* 854.6, 856.6, 880.6, and 882.6, but could not be identified. This was a pioneering study in several ways. On the one side, it is the first reported MSI study focused on a glomerular disease. On the other side, the finding of a relevant molecule containing a 22:6 moiety in the kidney, pointing at a particular function for this fatty acid in this organ that is still remaining to be characterized. This is not an irrelevant issue, as in a chronic kidney disease cohort study, low plasma levels of polyunsaturated fatty acids were found associated with increased renal insufficiency [40]. Moreover, oral administration of n-3 fatty acids (DHA and eicosapentaenoate, EPA) was shown to reduce proteinuria in FSGS patients [41], while the DHA metabolite Resolvin D1 protected podocytes in an adriamycin-induced mouse model of glomerular disease [42].

The latter (adriamycin/doxorubicin injection in rodents) represents one of the few available in vivo models of nephrotic syndrome, as one single injection leads to the development of delayed proteinuria. Hara and collaborators [43] combined doxorubicin injection with hypercholesterolemia, in a mouse invalidated for LDL-receptor. Mice exhibited glomerular injuries resembling FSGS, with podocyte alterations and foam cell infiltration in glomeruli. In this model, the authors found by MALDI-MSI at an 80 µm resolution, two ions at *m/z* 518.3 and 543.3, identified as Na⁺ adducts of lysophosphatidylcholines LPC 16:0 and LPC 18:0, some of the main components of oxidized LDL, that were abnormally increased in glomerular regions and colocalizing with oxidized phospholipids. These molecules, when administered in vitro to cultured cells induced the expression of adhesion molecules and cytokines, as well as adhesion and migration of macrophages. MSI in this study contributed to unveiling an interesting lipid peroxidation mechanism associated with this mouse model.

Lysophospholipids were also found increased in a mouse model of diabetic nephropathy, along with PE, gangliosides, and sulfoglycosphingolipids (sulfatides) [34]. Glomerular and tubular lesions accompany proteinuria and happen in a high proportion (about one third) of diabetic individuals. The lipid changes reported in this study were, like in the FSGS model cited above, attributed, at least in part, to increased oxidation. This was proven by the use of pyridoxamine, an inhibitor of oxidative processes, as a control condition. The changes were presented as happening in glomeruli and/or tubuli, in spite of a relatively high resolution (between 10 and 40 µm). Interestingly, ganglioside GM3 forms, *m/z* 1151.7 (NeuAc-GM3) and *m/z* 1167.7 (NeuGc-GM3), showed a pointed pattern in the kidney, specifically colocalized with glomeruli (identified by PAS staining). NeuGc-GM3, the oxidized form, was found increased in diabetic kidneys. The same pattern was found for sulfoglycolipid SB1a t18:0/22:0, *m/z* 1407.8, as compared with its non-oxidized form, as well as glucose-modified PE (Amadori-PE, a plasma marker of diabetes). In all cases the increase was corrected with the pyridoxamine treatment. Modified PE; however, was not specific of glomerular regions. These outstanding results indicate that GM3 and sulfated sphingolipids are particularly abundant in glomeruli, most likely in podocytes, and point at this particular lipid class as a target of glomerular alterations in pathologic conditions.

Modifications in the lipid content of glomeruli in diabetes has been the result of another study, devoted to a MALDI-MSI metabolomic profiling of diabetic mice exposed to high-fat diet. In this case the initial search for nucleotide distribution in tissues ended up in a lipid finding. Thus, along with the finding of an altered ATP/AMP ratio, untargeted analysis revealed an accumulation of sphingomyelin SM d34:1 in the glomeruli of those mice as compared to controls [25]. It was argued that the sphingomyelin accumulation could be responsible for the increased ATP production, due to activation of the glycolytic pathway. In this study, the results in glomeruli were obtained with a relatively high-resolution setting (30 μm). Glomeruli were identified using a phosphatidic acid species (PA 36:1) as a glomerular signature. This represents an interesting strategy to be used in the selection of regions of interest in those MSI studies addressing glomerular lipid composition.

Another condition that leads to glomerular dysfunction is renal toxicity due to bisphenol exposure. Experimentally, bisphenol targets glomeruli inducing necrosis, in addition to other lesions, such as cloudy swelling of medulla and interstitial collapsing of renal pelvis, in mice exposed to different concentrations of the compound [30]. The modifications in the cortex were associated with differential distribution of seven lipids, including accumulation of SM d22:0/20:4, TAG 16:0/14:0/16:0, PS 18:0/22:6, and PG 16:0/16:0 and a decrease in DAG (18:0/22:6), PE (20:1/20:4), and PI (16:1/18:1) when compared with control mice. The results suggest that the renal cortex is the most sensitive area to the toxic, and a 2D and 3D model construction coupled with a multivariate analysis pointed to a signature specific of bisphenol exposure, based on increased SM d22:0/20:4 and Cer d18:2/24:1, which could eventually be considered as a marker of toxicity.

Finally, a matrix-free variant of MALDI, such as NALDI-MSI has been used to target renal lipids in normal mouse kidney [44], showing increased presence of a K^+ adduct of PC 32:0 in the renal cortex as compared to the medulla [45]. NALDI technology, in this case consisting of a Fourier transform ion cyclotron resonance mass spectrometer, was applied in parallel to study the brain and kidney, and compared with MALDI. The authors claimed NALDI to improve mass accuracy and detection efficiency. It could represent an alternative to conventional MALDI-MSI in kidney-based studies.

3. The TOF-SIMS Alternative

Imaging by cluster time-of-flight secondary ion mass spectrometry (TOF-SIMS) represents a powerful tool for localized lipidomics. SIMS was first developed in the 1960s and 1970s for surface analysis [46,47]. This technology consists of the bombardment of the sample by a beam of polyatomic ions, which induces desorption/ionization of secondary ions from the sample surface. Polyatomic ion focused beams have been used successfully in the analysis of organic surfaces [48–53]. The current development of the technology allows a high lateral resolution of some hundreds of nanometers, which makes the technology particularly well fitted for the analysis of tissue sections [48,52–62]. In addition, no matrix is required for ion generation, which avoids all the drawbacks associated with matrix deposition on tissue.

The analysis of kidney tissue by TOF-SIMS was first performed in rats by gold particle bombardment on silver-coated sections [63–65]. In their first report, the authors found a two-fold improvement in sensitivity by silver-coating and attained a 200 nm lateral resolution. They reported a clear patchy distribution of silver-cationized cholesterol ions (m/z 493.4 and m/z 495.4). These cholesterol-rich areas were identified as nuclear regions of distal tubular epithelial cells [64]. In another report they attributed high cholesterol levels to glomerular areas [63]. In their third work, Na^+ ions were found increased in glomerular areas, while phosphocholine (m/z 184) was uniformly localized, along with K^+ ions.

The only report so far of TOF-SIMS analysis of human kidney biopsies was performed by our group in the context of Fabry disease [66], an X-linked disorder characterized by accumulation of Gb3 and Ga2 in the renal cortex due to a deficiency in α-galactosidase A. Our approach was able to detect both sphingolipid molecules characteristic of this pathology in human material [67]. More recently, a MALDI-MSI approach has detected the same defect in the kidney of α-galactosidase

A knockout mice [33]. Apart from these pioneering studies, no systematic analysis of glomerular lipids has been addressed by TOF-SIMS imaging.

4. Cluster-TOF-SIMS Analysis of Mouse Kidney Cortex

Here we report the use of cluster-TOF-SIMS to analyze the kidney cortex of normal mice. Three-month old C57B6 mice were sacrificed, kidneys were harvested, embedded in optimal cutting temperature (OCT), and immersed in frozen in liquid nitrogen at −195 °C. Samples were stored at −80 °C. Before analysis, 14 µm-thick tissue sections were obtained at −20 °C using a cryostat (CM3050–S Cryostat LEICA Microsystems, SAS, Nanterre, France), then deposited onto a conductive indium thin oxide (ITO) slide and dried for 15 min under a pressure of a few hectopascals.

We used a TOF-SIMS IV mass spectrometer (ION-TOF GmbH, Münster, Germany) with bismuth ion source and a reflectron time of flight analyzer. Bi_3^+ cluster primary ions hit the surface of the tissue section with a kinetic energy of 25 keV and an incidence angle of 45°. The focusing mode which is used are the so-called "high current bunched" mode. Secondary ions are extracted with an energy of 2 keV and are post-accelerated to 10 keV just before hitting the detector. A low-energy electron flood gun is used to neutralize the surface during the experiments. For each sample, two areas of 500×500 µm with 256×256 pixels were analyzed with a primary ion dose of 2×10^{12} ions/cm^2. The stability of the ion emission of different components is controlled along the analysis, most of ions kept the same intensity during the analysis, so that the analysis can be considered as being done below the so-called static limit. An internal mass calibration has been made with H^+, H_2^+, H_3^+, CH_3^+, and $C_{29}H_{50}O_2^+$ ([Vitamin E]$^{+\bullet}$) ions in the positive ion mode and with H^-, C^-, CH^-, C_2^-, C_2H^-, C_3^-, C_3H^-, C_4^-, C_4H^-, and $C_{29}H_{49}O_2^-$ ([Vitamin E-H]$^-$) ions in the negative ion mode. On each ion image, "MC" represents the maximal number of counts in a pixel and "TC" the total number of counts. The color scales correspond to the interval [0, MC].

The global spectra of glomerular and non-glomerular areas were compared. A total of 136 distinct ions (67 in the positive ion mode and 69 in the negative ion mode) were detected and attributed with one or more potential identities by comparison with the *m/z* values of bona fide standards and/or compared with standards in the ION-TOF database (https://www.iontof.com). Some of these standards have been analyzed in our laboratory and the spectra published [51,57,60,68]. Examples of standard spectra are shown in Figure S1. The *m/z* and attributed identities are presented in Supplementary Tables S1 and S2. Among these ions, 128 were identified as lipids, lipid adducts or lipid fragments, belonging to six different lipid categories (according to Lipid MAPS classification). All of them were detected both in glomerular and non-glomerular areas.

Figure 1 shows the relative intensity maps of five ions or ion combinations corresponding to the sum of cholesterol peaks, of SM 34:1 peaks, α-tocopherol (vitamin E) peaks, sulfatides ST 40:1, ST 40:2, ST 42:2, and PC 32:0, PC 34:2, PC 36:0. Relative intensity is represented by a color scale, in which yellow corresponds to higher and blue to lower abundances. Glomerular areas were surrounded by a white dashed line. We observed a consistent increased intensity of both cholesterol ions and SM 34:1 in glomerular areas. In the lower panels, one artificial color (red or green) was attributed to each compound, which allowed visualizing the co-localization of either cholesterol and α-tocopherol or cholesterol and SM 34:1. As it can be observed, cholesterol co-localized with SM 34:1 and counter-localized with α-tocopherol (yellowish signal) in all cortical tissue, but intensity was higher in glomerular areas. Sulfatides presented a distinct patchy distribution in non-glomerular regions, while PC ions were distributed more or less homogeneously throughout the renal cortex, but showing a slightly higher abundance (red color) in non-glomerular areas. Figure 2 shows the distribution of cholesterol (in positive and negative mode) in glomerular and non-glomerular regions.

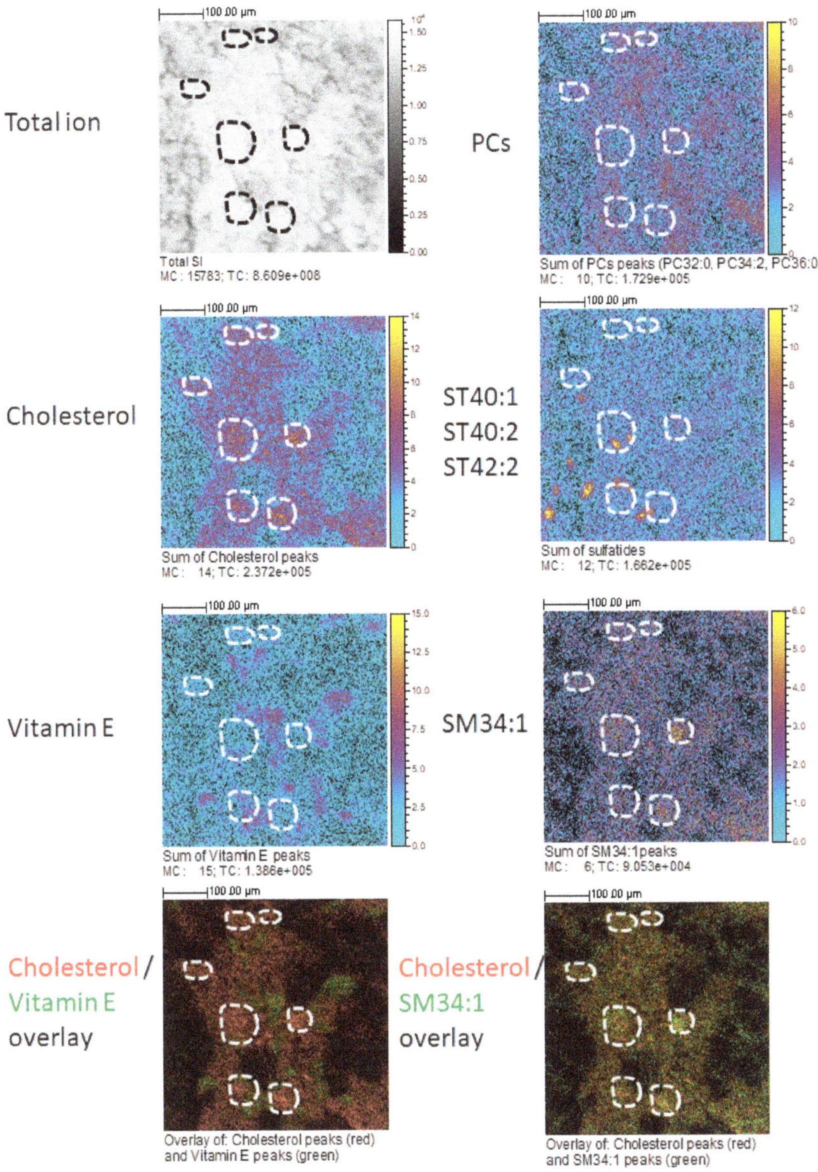

Figure 1. Representative section of a mouse kidney cortex, showing the spatial distribution of several ions (sum of cholesterol peaks, sum of α-tocopherol peaks, sum of phosphatidylcholine (PC) peaks, sum of sulfatides (ST 40:1, ST 40:2, ST 40:3), sum of sphingomyelin (SM) 34:1 peaks, and the total ion current) at a spatial resolution of 2 μm per pixel. The scale on the right (gray levels or colors) indicates the relative abundance. Yellow or white denote higher abundances, while blue or black denote lower abundances. The two lower panels have been constructed by attributing one color (red or green) to each compound. Glomerular regions of interest are selected and depicted by the dashed white line. Bar length: 100 μm. The name of the compound, the maximal number of counts in a pixel (MC), and the total number of counts (TC) are written below each image.

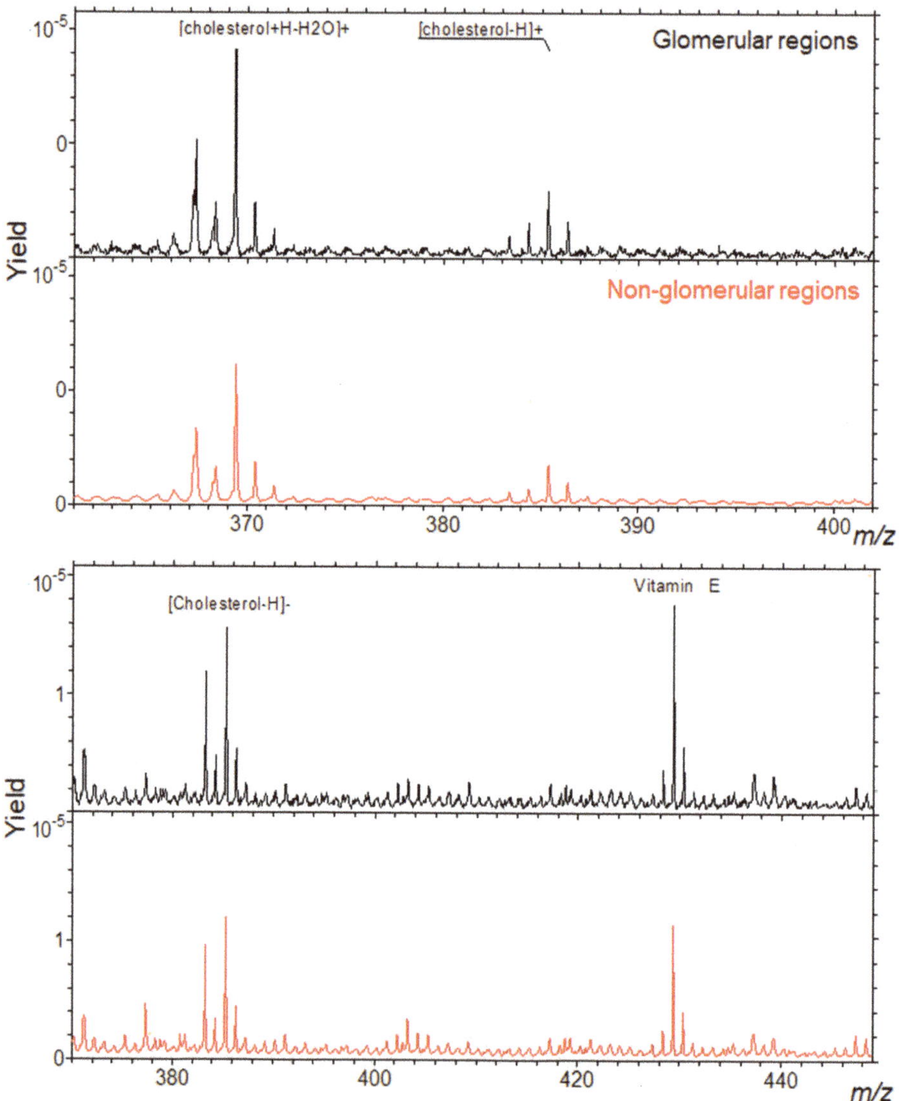

Figure 2. Representative spectra of a mouse kidney cortex, showing the relative intensities of peaks corresponding to cholesterol ions (in positive-top- and negative-bottom-mode) in glomerular and non-glomerular regions. The spectra are normalized to the total primary ion dose.

In agreement with our result, a MALDI-MSI-based analysis of rat kidney combined with histological data has reported cholesterol and squalene as specific glomerular metabolites [19]. Cholesterol is; therefore, enriched in glomeruli, but not exclusively. Additionally, we found an enrichment in SM d34:1 in glomerular regions, which could correspond to the same molecule that has been found increased in the glomeruli of a mouse model of diabetic nephropathy in comparison with control animals [25]. Ideally, an exclusive ion could be used for ROI selection in only-glomerular studies. The studies by other groups have proposed interesting glomerular markers, such as PA (36:1). Unfortunately, our TOF-SIMS setup was unable to detect this ion. Glomeruli are, nevertheless, an anatomical structure relatively easy to identify in adjacent sections by conventional staining, and even on ion images. Lipid ions found in glomerular areas can reflect the lipid composition of podocytes, parietal epithelial cells, or glomerular endothelial cells. Our setup can be used to establish modifications in the lipid composition of glomeruli in mouse models of glomerular diseases or in biopsies from patients. Potential changes found in these areas would be complementary to those found in global lipidomic analyses on isolated glomeruli, or on primary podocytes. The setup presented above provides the combination of a good spatial resolution, the unbiased identification of a large number of molecules, and the potential for glomerular ROI selection and subsequent relative quantitation. This setup will be used to explore variations in the localization and intensity of these ions in models of glomerular dysfunction.

5. Concluding Remarks: Possible Evolution of Kidney MSI. Application to Glomerulopathies

The advantages of MSI are, mainly, the possibility of a multiplexed analysis of several ions in a single acquisition, and the information about their spatial localization. The drawbacks, the insufficient resolution as compared to immunofluorescence and electron microscopy, and the limited availability of technical platforms, especially TOF-SIMS. Nonetheless kidney MSI has been proven to support and complement other untargeted lipidomic studies.

INS, and in particular MCNS and FSGS, unlike other glomerulopathies, require renal biopsy to establish the right diagnosis. Therefore, these forms are defined histologically, meaning that little information is provided by regular optical microscopy evaluation of tissue sections. MSI could be helpful both in diagnosis, and as an unbiased way of finding molecular mechanisms that could clarify their uncertain pathogenesis. To date, two main technical approaches for MSI of lipids have been used in the context of renal disease. MALDI-TOF and TOF-SIMS cannot be considered as competing approaches, since they can provide complementary information at different levels (Table 1). Firstly, the type of compounds covered by TOF-SIMS are under m/z 1500, including small molecules, lipids, metabolites, and elements, while the MALSDI-TOF range is over m/z 200, which includes many lipids along with peptides. TOF-SIMS presents some advantages, like a higher spatial resolution, higher sensitivity for low molecular weight compounds, no need for homogeneous matrix coating, and the potential for 3D mapping. However, it also presents some drawbacks, mainly associated with large fragmentation, often making identification of original molecules problematic. MALDI-TOF is; therefore, more suitable for unambiguous compound identification. In both cases the analysis is relatively long and complex, and always semiquantitative.

Table 1. Comparison of the features provided by Time of Flight Secondary Ion Mass Spectrometry (TOF-SIMS) and Matrix Assisted Laser Desorption Ionization - TOF (MALDI-TOF) in the context of mass spectrometry imaging (MSI).

	TOF-SIMS	MALDI-TOF
Analysis	Elemental and molecular analysis	High molecular weight covering a large range of molecules
Compounds	Lipids, glycosphingolipids, cyclopeptides, drugs, metabolites, minerals	Proteins, peptides, lipids, drugs, metabolites
Mass range	$m/z \leq 1500$	$m/z > 200$
Sample	Dehydrated, no fixation, no matrix	Dehydrated, homogeneous matrix coating
Imaging	Elemental and chemical imaging and mapping in 2D and 3D	Possibility to characterize and visualize in 2D
Spatial resolution	Down to 100–400 nm	5–50 µm
Sensitivity	High sensitivity for trace elements or compounds, in order of ppm to ppb for most species	Low sensitivity for low molecular weight molecules
Overall	Long and complex semi-quantitative analysis	Long and complex semi-quantitative analysis

New developments in MALDI-MSI aim at increasing spatial resolution. An example is a reported atmospheric-pressure (AP-MALDI) setup applied to brain and kidney imaging, leading to a lateral resolution of 1.4 μm and coupled to an Orbitrap analyzer [69]. This setup has efficiently shown the tissue localization of PC 38:6, PC 40:6, PC 38:1, PC 32:0, PS 40:0, and SM d36:0. On the other side, recent developments in TOF-SIMS aim to give access to a much higher mass resolution, with the recent launch of the Orbi-SIMS instrument [70] to in situ structural analysis of secondary ions through a tandem TOF analyzer (TOF/TOF) [71].

Interestingly, most of the studies published so far on kidney MSI point at alterations of the glycosphingolipid profile in different pathologies involving glomerular dysfunction (Table 2). This lipid class seems to play a significant role in the correct function of the organ, and of the glomerular barrier in particular. In fact, gangliosides are especially abundant at the surface of podocytes [13]. Integrity of the slit diaphragm, a lipid raft-like membranous structure establishing the interaction between foot processes of adjacent podocytes, is a key aspect in filtration barrier function. Sphingolipids (SP) are main components of this structure, and their metabolism seems to play a major role. In our setting, we have found an increased presence of a sphingomyelin species (SM d34:1) in mouse glomeruli. Nevertheless, other relevant lipid classes have been identified throughout MSI lipidomic analyses, such as glycerophospholipids (GP) and glycero-lysophospholipids, and, most importantly, their differential esterifying acyl chains have been unambiguously identified. Glomerular lipidomics is still underexplored, and MSI constitutes a promising strategy to address this point.

Table 2. Synthesis of reported findings by lipid MSI in renal cortex and glomerular areas. MALDI: matrix-assisted laser desorption ionization. TOF-SIMS: time of flight–secondary ion mass spectrometry. SP: sphingolipids. GP: glycerophospholipids. SL: sterol lipids.

Reference	Species	Technology	Pathology/Condition	Main Finding (ion/molecule)	Lipid MAPS Category	Anatomical Region
[30]	mouse	MALDI	Bisphenol toxicity	SM d22/20:4 Cer d18:2/24:1	SP	Cortex
[35]	mouse	MALDI	IgA nephropathy	PC O-16:0/22:6 PC O-18:1/22:6	GP	Cortex
[34]	mouse	MALDI	Diabetic nephropathy	NeuGc-GM3	SP	Glomeruli
[25]	mouse	MALDI	Diabetic nephropathy	SM d18:0/16:0	SP	Glomeruli
[28]	mouse	MALDI	FSGS (doxorubicin injection model)	LPC 16:0 LPC 18:0	GP	Glomeruli
[33]	mouse	MALDI	Fabry disease model	Gb3; Ga2	SP	Glomeruli
[66]	Human biopsies	TOF-SIMS	Fabry disease	Gb3; Ga2	SP	Glomeruli
[44,45]	mouse	MALDI	normal	PC 32:0	GP	Cortex
[25]	mouse	MALDI	normal	PA 36:1	GP	Glomeruli
[63–65]	mouse	TOF-SIMS	normal	Cholesterol	SL	Glomeruli
this report	mouse	TOF-SIMS	normal	Cholesterol SM d34:1	SL-SP	Glomeruli

Supplementary Materials: Supplementary materials can be found at http://www.mdpi.com/1422-0067/20/7/1623/s1. Table S1: Negative ions detected by TOF-SIMS analysis of mouse renal cortex. Table S2: Positive ions detected by TOF-SIMS analysis of mouse renal cortex. Figure S1: Upper panel, partial positive spectrum of phosphatidylcholine PC32: 0 standard from Sigma. Lower panel, partial negative spectrum of sulfatide 42:2(18:1/24:1) standard from Sigma. The analyses were done at ICSN-CNRS laboratory. Experimental spectra were compared with standard spectra for identity attributions.

Author Contributions: I.A. and M.N. performed the technical part and wrote the manuscript. D.T., D.S., A.B. and M.O. wrote the manuscript.

Funding: The research was funded by Fondation du Rein, grant "Prix jeune chercheur 2013" (M.O.).

Acknowledgments: The authors would like to acknowledge the International Atomic Energy Agency (IAEA) for providing the necessary expenses for I.A. and M.N. fellowship (Code No.: C6/LEB/14010), and the Lebanese Atomic Energy Commission of the National Council of Scientific Research in Lebanon.

Conflicts of Interest: The authors declare no conflicts of interest.

Abbreviations

Cer	Ceramide
DHA	Docosahexaenoate
DESI	Desorption electrospray ionization
GL	Glycerolipid category
GP	Glycerophospholipid category
GPI-BP1	Glycosylphosphatidylinositol-anchored binding protein 1
HILIC	Hydrophilic interaction chromatography
INS	Idiopathic nephrotic syndrome
LC-MS	Liquid chromatography-mass spectrometry
LPC	Lysophosphatidylcholine
MALDI	Matrix-assisted laser desorption ionization
MCNS	Minimal change nephrotic syndrome
MN	Membranous nephropathy
MSI	Mass spectrometry imaging
NALDI	Nanostructure-assisted laser desorption ionization
NS	Nephrotic syndrome
PC	Phosphatidylcholine
PR	Prenol lipid category
SALDI	Surface-assisted laser desorption ionization
SIMS	Secondary ion mass spectrometry
SL	Sterol lipid category
SM	Sphingomyelin
ST	Sulfatide
SP	Sphingolipid category
TOF	Time of flight
VLDL	Very low-density lipoprotein

References

1. Vaziri, N.D. Disorders of lipid metabolism in nephrotic syndrome: Mechanisms and consequences. *Kidney Int.* **2016**, *90*, 41–52. [CrossRef] [PubMed]
2. Davies, R.W.; Staprans, I.; Hutchison, F.N.; Kaysen, G.A. Proteinuria, not altered albumin metabolism, affects hyperlipidemia in the nephrotic rat. *J. Clin. Investig.* **1990**, *86*, 600–605. [CrossRef] [PubMed]
3. Garber, D.W.; Gottlieb, B.A.; Marsh, J.B.; Sparks, C.E. Catabolism of very low density lipoproteins in experimental nephrosis. *J. Clin. Investig.* **1984**, *74*, 1375–1383. [CrossRef] [PubMed]
4. Shearer, G.C.; Stevenson, F.T.; Atkinson, D.N.; Jones, H.; Staprans, I.; Kaysen, G.A. Hypoalbuminemia and proteinuria contribute separately to reduced lipoprotein catabolism in the nephrotic syndrome. *Kidney Int.* **2001**, *59*, 179–189. [CrossRef]
5. Clement, L.C.; Mace, C.; Avila-Casado, C.; Joles, J.A.; Kersten, S.; Chugh, S.S. Circulating angiopoietin-like 4 links proteinuria with hypertriglyceridemia in nephrotic syndrome. *Nat. Med.* **2013**, *20*, 37–46. [CrossRef]
6. Vaziri, N.D.; Sato, T.; Liang, K. Molecular mechanisms of altered cholesterol metabolism in rats with spontaneous focal glomerulosclerosis. *Kidney Int.* **2003**, *63*, 1756–1763. [CrossRef] [PubMed]
7. Muls, E.; Rosseneu, M.; Daneels, R.; Schurgers, M.; Boelaert, J. Lipoprotein distribution and composition in the human nephrotic syndrome. *Atherosclerosis* **1985**, *54*, 225–237. [CrossRef]
8. Kao, W.H.; Klag, M.J.; Meoni, L.A.; Reich, D.; Berthier-Schaad, Y.; Li, M.; Coresh, J.; Patterson, N.; Tandon, A.; Powe, N.R.; et al. MYH9 is associated with nondiabetic end-stage renal disease in African Americans. *Nat. Genet.* **2008**, *40*, 1185–1192. [CrossRef]
9. Kopp, J.B.; Smith, M.W.; Nelson, G.W.; Johnson, R.C.; Freedman, B.I.; Bowden, D.W.; Oleksyk, T.; McKenzie, L.M.; Kajiyama, H.; Ahuja, T.S.; et al. MYH9 is a major-effect risk gene for focal segmental glomerulosclerosis. *Nat. Genet.* **2008**, *40*, 1175–1184. [CrossRef] [PubMed]
10. Lee, H.S.; Kruth, H.S. Accumulation of cholesterol in the lesions of focal segmental glomerulosclerosis. *Nephrology* **2003**, *8*, 224–223. [CrossRef]

11. Merscher-Gomez, S.; Guzman, J.; Pedigo, C.E.; Lehto, M.; Aguillon-Prada, R.; Mendez, A.; Lassenius, M.I.; Forsblom, C.; Yoo, T.; Villarreal, R.; et al. Cyclodextrin protects podocytes in diabetic kidney disease. *Diabetes* **2013**, *62*, 3817–3827. [CrossRef] [PubMed]

12. Erkan, E.; Zhao, X.; Setchell, K.; Devarajan, P. Distinct urinary lipid profile in children with focal segmental glomerulosclerosis. *Pediatr. Nephrol.* **2016**, *31*, 581–588. [CrossRef] [PubMed]

13. Jin, J.; Sison, K.; Li, C.; Tian, R.; Wnuk, M.; Sung, H.K.; Jeansson, M.; Zhang, C.; Tucholska, M.; Jones, N.; et al. Soluble FLT1 binds lipid microdomains in podocytes to control cell morphology and glomerular barrier function. *Cell* **2012**, *151*, 384–399. [CrossRef]

14. Merscher, S.; Fornoni, A. Podocyte pathology and nephropathy - sphingolipids in glomerular diseases. *Front. Endocrinol.* **2014**, *5*, 127. [CrossRef]

15. Najafian, B.; Svarstad, E.; Bostad, L.; Gubler, M.C.; Tondel, C.; Whitley, C.; Mauer, M. Progressive podocyte injury and globotriaosylceramide (GL-3) accumulation in young patients with Fabry disease. *Kidney Int.* **2011**, *79*, 663–670. [CrossRef]

16. Zhang, H.; Wang, Y.; Gong, Z.; Li, X.; Qiu, W.; Han, L.; Ye, J.; Gu, X. Identification of a distinct mutation spectrum in the SMPD1 gene of Chinese patients with acid sphingomyelinase-deficient Niemann-Pick disease. *Orphanet J. Rare Dis.* **2013**, *8*, 15. [CrossRef]

17. Haltia, A.; Solin, M.L.; Jalanko, H.; Holmberg, C.; Miettinen, A.; Holthofer, H. Sphingolipid activator proteins in a human hereditary renal disease with deposition of disialogangliosides. *Histochem. J.* **1996**, *28*, 681–687. [CrossRef] [PubMed]

18. Fornoni, A.; Sageshima, J.; Wei, C.; Merscher-Gomez, S.; Aguillon-Prada, R.; Jauregui, A.N.; Li, J.; Mattiazzi, A.; Ciancio, G.; Chen, L.; et al. Rituximab targets podocytes in recurrent focal segmental glomerulosclerosis. *Sci. Transl. Med.* **2011**, *3*, 85ra46. [CrossRef]

19. Frohlich, S.; Putz, B.; Schachner, H.; Kerjaschki, D.; Allmaier, G.; Marchetti-Deschmann, M. Renopathological Microstructure Visualization from Formalin Fixed Kidney Tissue by Matrix-Assisted Laser/Desorption Ionization-Time-of-Flight Mass Spectrometry Imaging. *Balkan J. Med. Genet.* **2012**, *15*, 13–16. [CrossRef] [PubMed]

20. Muller, L.; Kailas, A.; Jackson, S.N.; Roux, A.; Barbacci, D.C.; Schultz, J.A.; Balaban, C.D.; Woods, A.S. Lipid imaging within the normal rat kidney using silver nanoparticles by matrix-assisted laser desorption/ionization mass spectrometry. *Kidney Int.* **2015**, *88*, 186–192. [CrossRef] [PubMed]

21. Marsching, C.; Eckhardt, M.; Grone, H.J.; Sandhoff, R.; Hopf, C. Imaging of complex sulfatides SM3 and SB1a in mouse kidney using MALDI-TOF/TOF mass spectrometry. *Anal. Bioanal. Chem.* **2011**, *401*, 53–64. [CrossRef] [PubMed]

22. Marsching, C.; Jennemann, R.; Heilig, R.; Grone, H.J.; Hopf, C.; Sandhoff, R. Quantitative imaging mass spectrometry of renal sulfatides: Validation by classical mass spectrometric methods. *J. Lipid Res.* **2014**, *55*, 2343–2353. [CrossRef] [PubMed]

23. Patterson, N.H.; Thomas, A.; Chaurand, P. Monitoring time-dependent degradation of phospholipids in sectioned tissues by MALDI imaging mass spectrometry. *J. Mass Spectrom.* **2014**, *49*, 622–627. [CrossRef] [PubMed]

24. Rao, S.; Walters, K.B.; Wilson, L.; Chen, B.; Bolisetty, S.; Graves, D.; Barnes, S.; Agarwal, A.; Kabarowski, J.H. Early lipid changes in acute kidney injury using SWATH lipidomics coupled with MALDI tissue imaging. *Am. J. Physiol. R. Physiol.* **2016**, *310*, F1136–F1147. [CrossRef]

25. Miyamoto, S.; Hsu, C.C.; Hamm, G.; Darshi, M.; Diamond-Stanic, M.; Decleves, A.E.; Slater, L.; Pennathur, S.; Stauber, J.; Dorrestein, P.C.; et al. Mass Spectrometry Imaging Reveals Elevated Glomerular ATP/AMP in Diabetes/obesity and Identifies Sphingomyelin as a Possible Mediator. *EBioMedicine* **2016**, *7*, 121–134. [CrossRef]

26. Marsching, C.; Rabionet, M.; Mathow, D.; Jennemann, R.; Kremser, C.; Porubsky, S.; Bolenz, C.; Willecke, K.; Grone, H.J.; Hopf, C.; et al. Renal sulfatides: Sphingoid base-dependent localization and region-specific compensation of CerS2-dysfunction. *J. Lipid Res.* **2014**, *55*, 2354–2369. [CrossRef] [PubMed]

27. Ruh, H.; Salonikios, T.; Fuchser, J.; Schwartz, M.; Sticht, C.; Hochheim, C.; Wirnitzer, B.; Gretz, N.; Hopf, C. MALDI imaging MS reveals candidate lipid markers of polycystic kidney disease. *J. Lipid Res.* **2013**, *54*, 2785–2794. [CrossRef] [PubMed]

28. Bruinen, A.L.; van Oevelen, C.; Eijkel, G.B.; Van Heerden, M.; Cuyckens, F.; Heeren, R.M. Mass Spectrometry Imaging of Drug Related Crystal-Like Structures in Formalin-Fixed Frozen and Paraffin-Embedded Rabbit Kidney Tissue Sections. *J. Am. Soc. Mass Spectrom.* **2016**, *27*, 117–123. [CrossRef]

29. Moreno-Gordaliza, E.; Esteban-Fernandez, D.; Lazaro, A.; Aboulmagd, S.; Humanes, B.; Tejedor, A.; Linscheid, M.W.; Gomez-Gomez, M.M. Lipid imaging for visualizing cilastatin amelioration of cisplatin-induced nephrotoxicity. *J. Lipid Res.* **2018**, *59*, 1561–1574. [CrossRef] [PubMed]

30. Zhao, C.; Xie, P.; Yong, T.; Wang, H.; Chung, A.C.K.; Cai, Z. MALDI-MS Imaging Reveals Asymmetric Spatial Distribution of Lipid Metabolites from Bisphenol S-Induced Nephrotoxicity. *Anal. Chem.* **2018**, *90*, 3196–3204. [CrossRef] [PubMed]

31. Nilsson, A.; Goodwin, R.J.; Swales, J.G.; Gallagher, R.; Shankaran, H.; Sathe, A.; Pradeepan, S.; Xue, A.; Keirstead, N.; Sasaki, J.C.; et al. Investigating nephrotoxicity of polymyxin derivatives by mapping renal distribution using mass spectrometry imaging. *Chem. Res. Toxicol.* **2015**, *28*, 1823–1830. [CrossRef]

32. Kuchar, L.; Faltyskova, H.; Krasny, L.; Dobrovolny, R.; Hulkova, H.; Ledvinova, J.; Volny, M.; Strohalm, M.; Lemr, K.; Kryspinova, L.; et al. Fabry disease: Renal sphingolipid distribution in the alpha-gal a knockout mouse model by mass spectrometric and immunohistochemical imaging. *Anal. Bioanal. Chem.* **2015**, *407*, 2283–2291. [CrossRef] [PubMed]

33. Vens-Cappell, S.; Kouzel, I.U.; Kettling, H.; Soltwisch, J.; Bauwens, A.; Porubsky, S.; Muthing, J.; Dreisewerd, K. On-Tissue Phospholipase C Digestion for Enhanced MALDI-MS Imaging of Neutral Glycosphingolipids. *Anal. Chem.* **2016**, *88*, 5595–5599. [CrossRef] [PubMed]

34. Grove, K.J.; Voziyan, P.A.; Spraggins, J.M.; Wang, S.; Paueksakon, P.; Harris, R.C.; Hudson, B.G.; Caprioli, R.M. Diabetic nephropathy induces alterations in the glomerular and tubule lipid profiles. *J. Lipid Res.* **2014**, *55*, 1375–1385. [CrossRef] [PubMed]

35. Kaneko, Y.; Obata, Y.; Nishino, T.; Kakeya, H.; Miyazaki, Y.; Hayasaka, T.; Setou, M.; Furusu, A.; Kohno, S. Imaging mass spectrometry analysis reveals an altered lipid distribution pattern in the tubular areas of hyper-IgA murine kidneys. *Exp. Mol. Pathol.* **2011**, *91*, 614–621. [CrossRef] [PubMed]

36. Niziol, J.; Ossolinski, K.; Ossolinski, T.; Ossolinska, A.; Bonifay, V.; Sekula, J.; Dobrowolski, Z.; Sunner, J.; Beech, I.; Ruman, T. Surface-Transfer Mass Spectrometry Imaging of Renal Tissue on Gold Nanoparticle Enhanced Target. *Anal. Chem.* **2016**, *88*, 7365–7371. [CrossRef] [PubMed]

37. Hajek, R.; Lisa, M.; Khalikova, M.; Jirasko, R.; Cifkova, E.; Student, V., Jr.; Vrana, D.; Opalka, L.; Vavrova, K.; Matzenauer, M.; et al. HILIC/ESI-MS determination of gangliosides and other polar lipid classes in renal cell carcinoma and surrounding normal tissues. *Anal. Bioanal. Chem.* **2017**, *410*, 6585–6594. [CrossRef]

38. Jirasko, R.; Holcapek, M.; Khalikova, M.; Vrana, D.; Student, V.; Prouzova, Z.; Melichar, B. MALDI Orbitrap Mass Spectrometry Profiling of Dysregulated Sulfoglycosphingolipids in Renal Cell Carcinoma Tissues. *J. Am. Soc. Mass Spectrom.* **2017**, *28*, 1562–1574. [CrossRef] [PubMed]

39. Alfaro, C.M.; Jarmusch, A.K.; Pirro, V.; Kerian, K.S.; Masterson, T.A.; Cheng, L.; Cooks, R.G. Ambient ionization mass spectrometric analysis of human surgical specimens to distinguish renal cell carcinoma from healthy renal tissue. *Anal. Bioanal. Chem.* **2016**, *408*, 5407–5414. [CrossRef] [PubMed]

40. Lauretani, F.; Semba, R.D.; Bandinelli, S.; Miller, E.R., 3rd; Ruggiero, C.; Cherubini, A.; Guralnik, J.M.; Ferrucci, L. Plasma polyunsaturated fatty acids and the decline of renal function. *Clin. Chem.* **2008**, *54*, 475–481. [CrossRef]

41. De Caterina, R.; Caprioli, R.; Giannessi, D.; Sicari, R.; Galli, C.; Lazzerini, G.; Bernini, W.; Carr, L.; Rindi, P. n-3 fatty acids reduce proteinuria in patients with chronic glomerular disease. *Kidney Int.* **1993**, *44*, 843–850. [CrossRef]

42. Zhang, X.; Qu, X.; Sun, Y.B.; Caruana, G.; Bertram, J.F.; Nikolic-Paterson, D.J.; Li, J. Resolvin D1 protects podocytes in adriamycin-induced nephropathy through modulation of 14-3-3beta acetylation. *PLoS ONE* **2013**, *8*, e67471.

43. Hara, S.; Kobayashi, N.; Sakamoto, K.; Ueno, T.; Manabe, S.; Takashima, Y.; Hamada, J.; Pastan, I.; Fukamizu, A.; Matsusaka, T.; et al. Podocyte injury-driven lipid peroxidation accelerates the infiltration of glomerular foam cells in focal segmental glomerulosclerosis. *Am. J. Pathol.* **2015**, *185*, 2118–2131. [CrossRef]

44. Vidova, V.; Novak, P.; Strohalm, M.; Pol, J.; Havlicek, V.; Volny, M. Laser desorption-ionization of lipid transfers: Tissue mass spectrometry imaging without MALDI matrix. *Anal. Chem.* **2010**, *82*, 4994–4997. [CrossRef] [PubMed]

45. Goodwin, R.J.; Mackay, C.L.; Nilsson, A.; Harrison, D.J.; Farde, L.; Andren, P.E.; Iverson, S.L. Qualitative and quantitative MALDI imaging of the positron emission tomography ligands raclopride (a D2 dopamine antagonist) and SCH 23390 (a D1 dopamine antagonist) in rat brain tissue sections using a solvent-free dry matrix application method. *Anal. Chem.* **2011**, *83*, 9694–9701. [CrossRef]

46. Benninghoven, A.; Loebach, E. Tandem Mass spectrometer for secondary ion studies. *Rev. Sci. Instrum.* **1971**, *42*, 49–52. [CrossRef]

47. Castaing, R.; Slodzian, G.J. Microanalyse par émission ionique secondaire. *J. Microsc.* **1962**, *1*, 395–410.

48. Benabdellah, F.; Seyer, A.; Quinton, L.; Touboul, D.; Brunelle, A.; Laprevote, O. Mass spectrometry imaging of rat brain sections: Nanomolar sensitivity with MALDI versus nanometer resolution by TOF-SIMS. *Anal. Bioanal. Chem.* **2010**, *396*, 151–162. [CrossRef]

49. Biddulph, G.X.; Piwowar, A.M.; Fletcher, J.S.; Lockyer, N.P.; Vickerman, J.C. Properties of C84 and C24H12 molecular ion sources for routine TOF-SIMS analysis. *Anal. Chem.* **2007**, *79*, 7259–7266. [CrossRef] [PubMed]

50. Sjovall, P.; Lausmaa, J.; Johansson, B. Mass spectrometric imaging of lipids in brain tissue. *Anal. Chem.* **2004**, *76*, 4271–4278. [CrossRef] [PubMed]

51. Touboul, D.; Brunelle, A.; Halgand, F.; De La Porte, S.; Laprevote, O. Lipid imaging by gold cluster time-of-flight secondary ion mass spectrometry: Application to Duchenne muscular dystrophy. *J. Lipid Res.* **2005**, *46*, 1388–1395. [CrossRef]

52. Touboul, D.; Halgand, F.; Brunelle, A.; Kersting, R.; Tallarek, E.; Hagenhoff, B.; Laprevote, O. Tissue molecular ion imaging by gold cluster ion bombardment. *Anal. Chem.* **2004**, *76*, 1550–1559. [CrossRef]

53. Weibel, D.; Wong, S.; Lockyer, N.; Blenkinsopp, P.; Hill, R.; Vickerman, J.C. A C60 primary ion beam system for time of flight secondary ion mass spectrometry: Its development and secondary ion yield characteristics. *Anal. Chem.* **2003**, *75*, 1754–1764. [CrossRef] [PubMed]

54. Brunelle, A.; Laprevote, O. Recent advances in biological tissue imaging with Time-of-flight Secondary Ion Mass Spectrometry: Polyatomic ion sources, sample preparation, and applications. *Curr. Pharm. Des.* **2007**, *13*, 3335–3343. [CrossRef] [PubMed]

55. Brunelle, A.; Laprevote, O. Lipid imaging with cluster time-of-flight secondary ion mass spectrometry. *Anal. Bioanal. Chem.* **2009**, *393*, 31–35. [CrossRef]

56. Brunelle, A.; Touboul, D.; Laprevote, O. Biological tissue imaging with time-of-flight secondary ion mass spectrometry and cluster ion sources. *J. Mass Spectrom.* **2005**, *40*, 985–999. [CrossRef] [PubMed]

57. Debois, D.; Bralet, M.P.; Le Naour, F.; Brunelle, A.; Laprevote, O. In situ lipidomic analysis of nonalcoholic fatty liver by cluster TOF-SIMS imaging. *Anal. Chem.* **2009**, *81*, 2823–2831. [CrossRef] [PubMed]

58. Le Naour, F.; Bralet, M.P.; Debois, D.; Sandt, C.; Guettier, C.; Dumas, P.; Brunelle, A.; Laprevote, O. Chemical imaging on liver steatosis using synchrotron infrared and ToF-SIMS microspectroscopies. *PLoS ONE* **2009**, *4*, e7408. [CrossRef]

59. Mas, S.; Touboul, D.; Brunelle, A.; Aragoncillo, P.; Egido, J.; Laprevote, O.; Vivanco, F. Lipid cartography of atherosclerotic plaque by cluster-TOF-SIMS imaging. *Analyst* **2007**, *132*, 24–26. [CrossRef]

60. Tahallah, N.; Brunelle, A.; De La Porte, S.; Laprevote, O. Lipid mapping in human dystrophic muscle by cluster-time-of-flight secondary ion mass spectrometry imaging. *J. Lipid Res.* **2008**, *49*, 438–454. [CrossRef]

61. Touboul, D.; Kollmer, F.; Niehuis, E.; Brunelle, A.; Laprevote, O. Improvement of biological time-of-flight-secondary ion mass spectrometry imaging with a bismuth cluster ion source. *J. Am. Soc. Mass Spectrom.* **2005**, *16*, 1608–1618. [CrossRef]

62. Touboul, D.; Laprevote, O.; Brunelle, A. Medical and biological applications of cluster ToF-SIMS. In *ToF-SIMS: Materials Analysis by Mass Spectrometry*, 2nd ed.; IM Publications and SurfaceSpectra: Manchester, UK, 2013; p. 583.

63. Nygren, H.; Johansson, B.R.; Malmberg, P. Bioimaging TOF-SIMS of tissues by gold ion bombardment of a silver-coated thin section. *Microsc. Res. Tech.* **2004**, *65*, 282–286. [CrossRef] [PubMed]

64. Nygren, H.; Malmberg, P.; Kriegeskotte, C.; Arlinghaus, H.F. Bioimaging TOF-SIMS: Localization of cholesterol in rat kidney sections. *FEBS Lett.* **2004**, *566*, 291–293. [CrossRef]

65. Nygren, H.; Borner, K.; Malmberg, P.; Tallarek, E.; Hagenhoff, B. Imaging TOF-SIMS of rat kidney prepared by high-pressure freezing. *Microsc. Res. Tech.* **2005**, *68*, 329–334. [CrossRef]

66. Touboul, D.; Roy, S.; Germain, D.P.; Chaminade, P.; Brunelle, A.; Laprevote, O. MALDI-TOF and cluster-TOF-SIMS imaging of Fabry disease biomarkers. *Int. J. Mass Spectrom.* **2007**, *260*, 158–165. [CrossRef]

67. Touboul, D.; Brunelle, A.; Germain, D.P.; Laprevote, O. A new imaging technique as a diagnostic tool: Mass spectrometry. *Presse Med.* **2007**, *36*, 1S82-87.

68. Heim, C.; Sjovall, P.; Lausmaa, J.; Leefmann, T.; Thiel, V. Spectral characterisation of eight glycerolipids and their detection in natural samples using time-of-flight secondary ion mass spectrometry. *Rapid Commun. Mass Spectrom.* **2009**, *23*, 2741–2753. [CrossRef] [PubMed]

69. Kompauer, M.; Heiles, S.; Spengler, B. Atmospheric pressure MALDI mass spectrometry imaging of tissues and cells at 1.4-mum lateral resolution. *Nat. Methods* **2016**, *14*, 90–96. [CrossRef] [PubMed]

70. Passarelli, M.K.; Pirkl, A.; Moellers, R.; Grinfeld, D.; Kollmer, F.; Havelund, R.; Newman, C.F.; Marshall, P.S.; Arlinghaus, H.; Alexander, M.R.; et al. The 3D OrbiSIMS-label-free metabolic imaging with subcellular lateral resolution and high mass-resolving power. *Nat. Methods* **2017**, *14*, 1175–1183. [CrossRef] [PubMed]

71. Fu, T.; Touboul, D.; Della-Negra, S.; Houel, E.; Amusant, N.; Duplais, C.; Fisher, G.L.; Brunelle, A. Tandem Mass Spectrometry Imaging and in Situ Characterization of Bioactive Wood Metabolites in Amazonian Tree Species Sextonia rubra. *Anal. Chem.* **2018**, *90*, 7535–7543. [CrossRef] [PubMed]

International Journal of
Molecular Sciences

Article

Oncogenic H-Ras Expression Induces Fatty Acid Profile Changes in Human Fibroblasts and Extracellular Vesicles

Krizia Sagini [1,†], Lorena Urbanelli [1,†], Eva Costanzi [1], Nico Mitro [2], Donatella Caruso [2], Carla Emiliani [1,3] and Sandra Buratta [1,*]

1   Department of Chemistry, Biology and Biotechnology, University of Perugia, 06123 Perugia, Italy; krizia.sagini@studenti.unipg.it (K.S.); lorena.urbanelli@unipg.it (L.U.); eva.costanzi@studenti.unipg.it (E.C.); carla.emiliani@unipg.it (C.E.)
2   Department of Pharmacological and Biomolecular Sciences, University of Milan, 20133 Milan, Italy; nico.mitro@unimi.it (N.M.); donatella.caruso@unimi.it (D.C.)
3   CEMIN-Center of Excellence for Innovative Nanostructured Material, 06123 Perugia, Italy
*   Correspondence: sandra.buratta@unipg.it; Tel.: +39-075-585-7440
†   These authors contributed equally to this work.

Received: 15 September 2018; Accepted: 5 November 2018; Published: 8 November 2018

Abstract: Extracellular vesicles (EVs) are lipid bilayer surrounded particles that are considered an additional way to transmit signals outside the cell. Lipids have not only a structural role in the organization of EVs membrane bilayer, but they also represent a source of lipid mediators that may act on target cells. Senescent cells are characterized by a permanent arrest of cell proliferation, but they are still metabolically active and influence nearby tissue secreting specific signaling mediators, including those carried by EVs. Notably, cellular senescence is associated with increased EVs release. Here, we used gas chromatography coupled to mass spectrometry to investigate the total fatty acid content of EVs released by fibroblasts undergoing H-RasV12-induced senescence and their parental cells. We find that H-RasV12 fibroblasts show increased level of monounsaturated and decreased level of saturated fatty acids, as compared to control cells. These changes are associated with transcriptional up-regulation of specific fatty acid-metabolizing enzymes. The EVs released by both controls and senescent fibroblasts show a higher level of saturated and polyunsaturated species, as compared to parental cells. Considering that fibroblasts undergoing H-RasV12-induced senescence release a higher number of EVs, these findings indicate that senescent cells release via EVs a higher amount of fatty acids, and in particular of polyunsaturated and saturated fatty acids, as compared to control cells.

Keywords: extracellular vesicles; oncogene-induced senescence; H-Ras; fatty acids; desaturases; elongases; acyl-coenzyme A synthetases

---

1. Introduction

Extracellular vesicles (EVs) have been implicated in many physiological processes [1]. The first observation considered EVs as a cellular mean to discard unneeded material during cell differentiation [2]. Later, it emerged that EVs transmit signals and they are now considered an alternative manner of cell signal transmission [3,4]. Three main types of EVs have been described, i.e., microvesicles budding from the plasma membrane (100–1000 nm), exosomes originating from the inward budding of late endosomes (30–150 nm), and apoptotic bodies released by cells undergoing apoptosis [5]. Despite their apparent simple classification, their similar and overlapping biochemical properties make it difficult to obtain preparation containing exclusively microvesicles or exosomes, so EVs are preferentially indicated as small EVs and large EVs, enriched either in exosomes or microvesicles, respectively [6]. EVs have been found in every fluid of the body, including easily accessible ones such as saliva or discard products such as urine. For this reason, and because EVs often maintain important properties of the cells secreting them, EVs have gained considerable attention for diagnostic and therapeutic purposes [7]. EVs contain lipids, proteins and nucleic acids, namely ncRNA such as miRNAs and lncRNA and a few databases have been developed, such as EVpedia [8], listing the biochemical components retrieved in EVs.

Despite their relevance, the number of studies on EV lipid composition is limited as compared to studies on protein or nucleic acid composition [9–12]. There is a certain consensus that EVs are enriched in cholesterol, sphingomyelin, ether-linked phospholipids and lysophospholipids [12,13]. EVs contain not only lipids forming the bulk of their membranes, but also lipid mediators acting as carriers between cells [14,15]. Studies have shown the presence within EVs of lipid mediators and enzymes involved in the release of their precursors [16], demonstrating that EVs may be important players in the so-called "transcellular biosynthesis" of eicosanoids [17,18]. Fatty acid analyses provide evidence that EVs mainly contain saturated fatty acids (SFA), even if monounsaturated (MUFA) and polyunsaturated (PUFA) ones are also present. In fact, EVs can be also considered an additional mechanism to transport fatty acids across the plasma membrane to target cells [19]. Thus, understanding the fatty acid composition of EVs may allow us to identify what kind of fatty acids are preferentially discarded by cells, influencing the metabolism of the surrounding tissue [20].

Senescence is a condition characterized by a permanent arrest of cell proliferation [21,22]. Senescent cells are metabolically active and release in the extracellular environment specific biochemical mediators, collectively known as Senescence Associated Secretory Phenotype (SASP) [23]. SASP can deeply influence surrounding cells, namely through the action of cytokines and chemokines stimulating the clearance of senescent cells by innate immune system and maintaining autocrine senescence signals. Although the absence of proliferation by senescent cells is considered a barrier towards oncogenic transformation [24–26], many SASP factors also have pro-tumorigenic properties [27] and EVs are currently considered additional SASP components [28].

Oncogene-induced senescence (OIS) is due to the activation of oncogenes in normal cells in the absence of additional oncogenic activation and/or inactivation of tumor suppressor genes that are necessary to fully transform cells. OIS is phenotypically indistinguishable from cellular senescence and cells undergoing are characterized by flattened morphology and positivity to well established senescence markers, namely as senescence-associated β-galactosidase [29]. The Ras family of oncogenes encodes small monomeric GTP-binding proteins that transduce mitogenic stimuli. It is well established that expression of constitutively active H-RasV12 induces senescence [30]. We have previously demonstrated that fibroblasts undergoing H-RasV12-induced senescence release a higher number of EVs [12], in agreement with studies on other senescence models [31,32]. Further, EVs released by fibroblasts undergoing H-RasV12-induced senescence show a peculiar lipid signature, as they are enriched in lysosphospholipids, ether-linked lipids and sulfatides [12]. Nevertheless, the total fatty acid composition of EVs released by OIS cells is not known, although it is a key assessment to understand what kind of fatty acids are preferentially discarded by the cells via EVs and how they influence the metabolism of neighbouring cells.

Here, we investigate the total fatty acid composition of EVs released by H-RasV12 and control fibroblasts, and their parental cells. Our findings provide evidence that EVs released by senescent and control fibroblasts share a similar fatty acid profile, which is enriched in SFA and PUFA as compared to releasing cells. Taking into account that senescent fibroblasts release a higher number of EVs, these findings indicate that senescent cells release a higher amount of fatty acids, in particular SFA and PUFA, with respect to controls. Besides this, we observed a significant increase of MUFA in cells undergoing senescence, in association with changes in the expression profile of enzymes involved in fatty acid desaturation and release from membrane phospholipids.

2. Results

2.1. Analysis of H-RasV12 and Control Fibroblasts Released EVs by Immunoblotting and Immuno-TEM

H-RasV12 was expressed in HuDe fibroblasts by transfection and cells were pharmacologically selected with blasticidin-S to get rid of untransfected cells. pcDNA6 empty vector was transfected as control. H-RasV12 expression induced an arrest of cell proliferation, changes in cell morphology accompanied by senescence-associated β-galactosidase staining and DNA damage demonstrated by γH2AX immunoreactivity (Figure 1). H-RasV12 expression was checked by immunoblotting with anti-H-Ras antibody (Figure 1). EVs were isolated from fibroblasts expressing H-RasV12 and control cells using the polymer co-precipitation method (Exoquick-TC was used as reagent) and their main features characterized as previously described [12]. Immunoblotting and immuno-TEM analysis showed the presence in our preparations of proteins specifically enriched in EVs (Figure 1). Positive markers CD9, CD63 and Tsg101 were clearly detectable in EVs, in agreement with guidelines [5]. Calnexin, an endoplasmic reticulum protein usually not present in EVs, and actin were used as negative markers. As expected, they were not detectable in EVs, although they were clearly present in equal amount in cell samples (Figure 1). In addition, H-RasV12 was clearly present in EVs released by fibroblasts over-expressing it.

The structural characterization of EVs was carried out by immuno-TEM (Figure 1). Image analysis detected small EVs of less than 100 nm size in H-RasV12 and control samples, compatible with an enrichment in small EVs. The presence of CD63 on their membrane bilayer was confirmed using immunogold labelling with an anti-CD63. These results confirmed an enrichment of small membranous vesicles in our preparation, consisting of exosomes and small microvesicles [12].

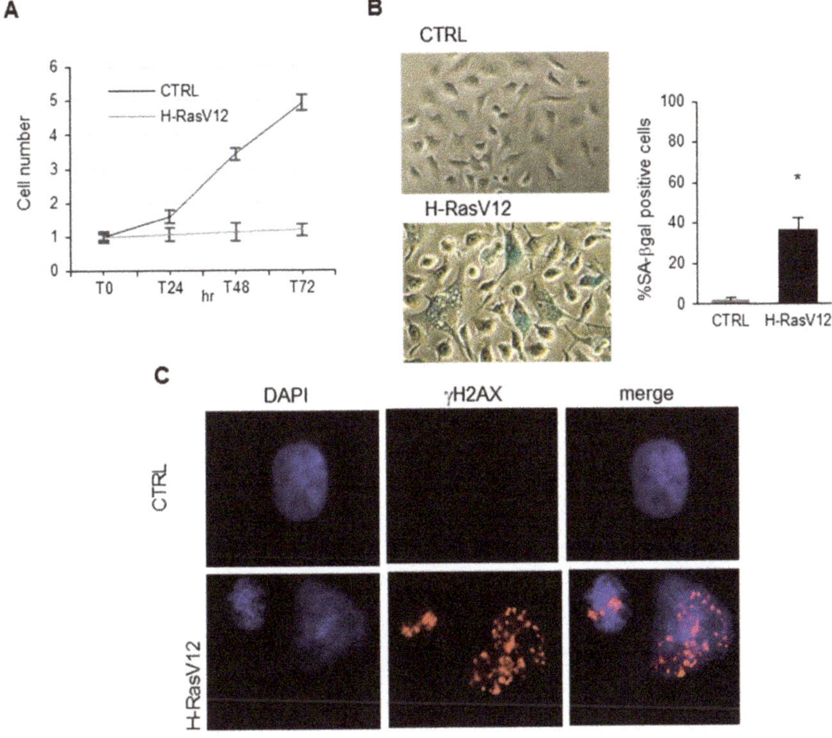

Figure 1. *Cont.*

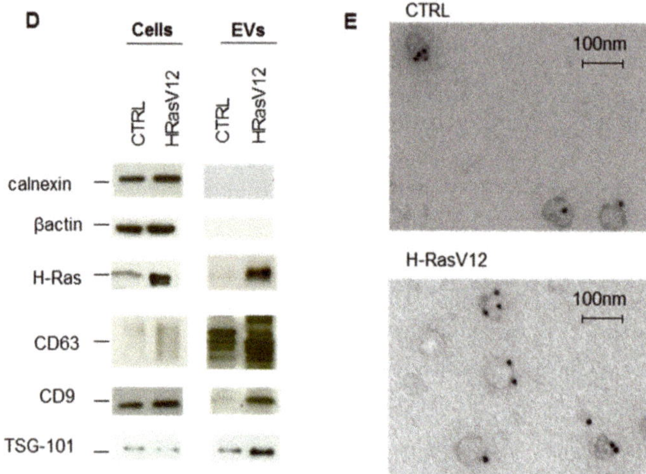

Figure 1. Analysis of H-RasV12-induced senescence in HuDe fibroblasts and characterization of EVs released by control and H-RasV12 expressing cells. (**A**) Growth curve of HuDe fibroblasts expressing H-RasV12 or transfected with the vector alone as control (CTRL). Mean values were calculated on 3 replicates and mean ± S.E.M. is indicated as fold increase with respect to the number of seeded cells (set 1). (**B**) Senescence-associated β-galactosidase staining and quantification of senescence-associated β-galactosidase (SA-β-gal) positive cells. Microscopy images (40×) of H-RasV12 and CTRL fibroblasts transfected with the vector alone. SA-β-gal positive cells were counted on three different fields in three separate experiments experiments (* $p < 0.05$, CTRL vs. H-RasV12). (**C**) Immunostaining for γH2AX. Cells were fixed in 4% paraformaldehyde, permeabilized in PBS/0.1% Triton X-100, incubated with an anti-γH2AX and labelled with an anti-rabbit Alexa-Fluor 594 antibody. Nuclei were stained with 1 μg/mL DAPI. Fluorescence microscopy analysis was carried out with a Nikon TE2000 microscope through a 60× oil immersion objective. (**D**) Immunoblotting. Cell extracts and EVs samples were separated by SDS-PAGE, electrotransferred, and probed with positive and negative markers indicated. (**E**) Immuno-transmission electron micrographs of EVs. Samples were fixed, dropped directly onto formvar/carbon coated grids, blocked and incubated with mouse anti-CD63 primary antibody, rabbit anti-mouse secondary antibody and gold-labelled Protein A.

2.2. Analysis of Fatty Acids Content

The GC-MS analysis of fatty acids in both cells and EVs highlighted significant differences between cells and EVs (Figure 2). First, EVs had a higher fatty acids/protein ratio with respect to cells (Figure 2A,B) and the content of total fatty acids normalized for proteins was lower for EVs prepared from H-RasV12 cells as compared to controls (Figure 2B). The high lipid/protein ratio in EVs with respect to cells agrees with previous studies [10,22,33]. In addition, when we grouped fatty acids in saturated (SFA), monounsaturated (MUFA) and polyunsaturated (PUFA), we clearly observed that H-RasV12 expressing fibroblasts were enriched in MUFA (~33% of the total detected fatty acids as compared to 17% of control samples) (Figure 2A). This increase was associated with the decrease of SFA (~65% of the total detected fatty acids as compared to 80% of control), whereas the content of PUFA was similar. EVs were characterised by a similar and elevated SFA level in both samples (Figure 2B), which is consistent with previous studies [9,10].

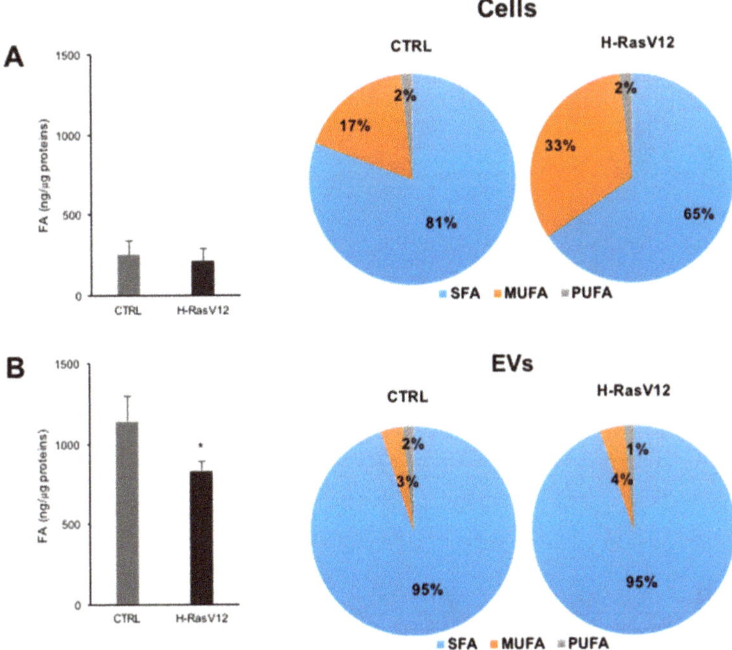

Figure 2. Fatty acid content and distribution of SFA, MUFA and PUFA in control and H-RasV12 cells (**A**) and their released EVs (**B**). Lipids were extracted and total fatty acids analysis was carried out by GC-MS. In the graphs are reported the amounts of total fatty acids relative to protein content. Data are expressed as ng of FA/µg of proteins and are presented as mean $\pm$ SD ($n = 6$) (* $p < 0.05$, control vs. H-RasV12). In the pie charts are reported the proportion of fatty acids grouped on the basis of their unsaturation level; SFA: saturated fatty acids; MUFA: Mono-unsaturated fatty acids; PUFA: Poly-unsaturated fatty acids.

When the fatty acids profile was analyzed in detail (Figure 3A), significant modifications were observed in cells undergoing H-RasV12-inducing senescence. The most relevant ones were the significant decrease of all SFA species and the significant increase of palmitoleic (C16:1) and oleic (C18:1) acids, leading to a general increase of MUFA in senescent cells. Regarding PUFA, in H-RasV12 fibroblasts we observed decreased levels of γ-linolenic acid (C18:3 n6) and an increased level of γ-linoleic (C18:2 n6), docosahexaenoic (C22:6 n3) and eicosapentaenoic (C20:5 n3) acids (Figure 3A).

The comparison between cells and EVs revealed that the higher fatty acids/protein ratio in EVs with respect to cells was due to a high SFA and PUFA content (Table 1). As for fibroblasts, palmitic (C16:0) and stearic (C18:0) acids were the most abundant species in EVs (Figure 3B). Besides, comparison of MUFA species in cells and in EVs revealed that two monounsaturated species, C22:1 and C24:1, were not detectable in EVs (Figure 3). As mentioned above, a greater amount of PUFA in EVs with respect to parental cells was observed. In fact, even if eicosapentaenoic acid was not detected in EVs, they were enriched in each PUFA with respect to parental cells (Figure 3B).

The comparison of the composition of EVs released from H-RasV12 and control fibroblasts revealed a significant decrease of the two most abundant SFA species (C16:0 and C18:0) that was in line with the decrease in ng of total fatty acids/µg of proteins observed for H-RasV12 EVs (Figure 2B, left). No additional changes in the levels of MUFA and PUFA species were observed (Figure 3B), thus indicating the H-RasV12 expressing cells release EVs with a similar fatty acid profile as compared to control cells.

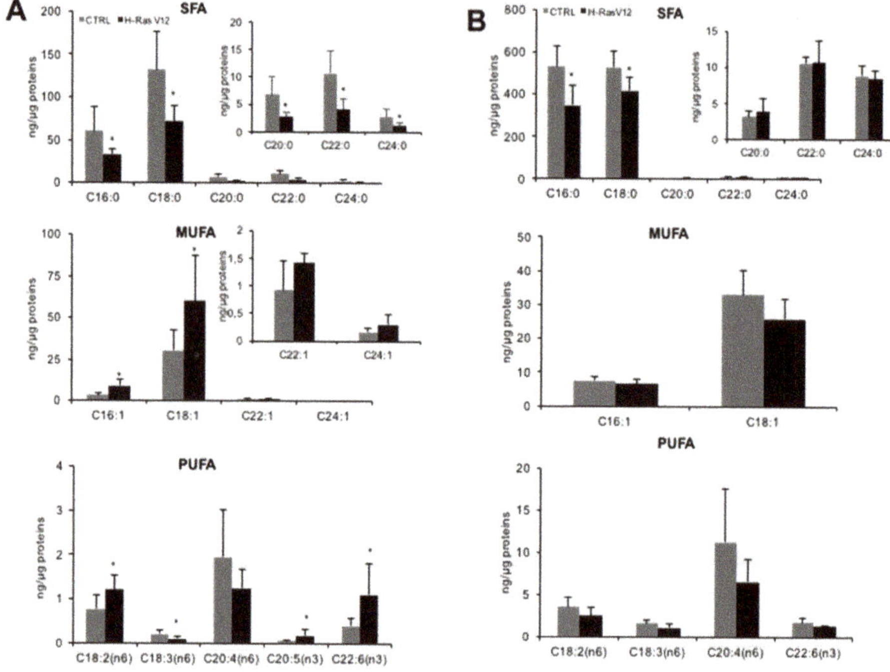

Figure 3. Fatty acid profiles of control and H-RasV12 expressing fibroslasts (**A**) and their released EVs (**B**). Lipids were extracted, and total fatty acids composition were analysed by GC-MS. Data, are expressed as ng fatty acids/μg proteins and are presented as mean ± S.D (n = 6). * p < 0.05 (control vs. H-RasV12). SFA: saturated fatty acids; MUFA: mono-unsaturated fatty acids; PUFA: poly-unsaturated fatty acids.

Table 1. SFA, MUFA and PUFA content in control and H-RasV12 expressing fibroblasts and their released EVs. Data are expressed as amount of fatty acids relative to protein content (ng fatty acids/μg proteins). Mean values ± SD are shown (n = 6) (* p < 0.05, control vs. H-RasV12). SFA: saturated fatty acids; MUFA: mono-unsaturated fatty acids; PUFA: poly-unsaturated fatty acids.

ng Lipid/μg Protein	Fibroblasts		EVs	
	CTRL	**RasV12**	**CTRL**	**RasV12**
SFA	212.39 ± 66.30	126.19 ± 47.05 *	1084.78 ± 147.63	787.4 ± 65.65 *
MUFA	35.14 ± 12.29	70.60 ± 31.65 *	40.38 ± 8.19	32.28 ± 7.22
PUFA	3.35 ± 1.54	3.38 ± 1.48	18.35 ± 8.46	11.67 ± 4.19

2.3. Expression Analysis of Genes Involved in Fatty Acids Remodelling and Phospholipases A2

To gain insight into transcriptional changes underlying the different fatty acid profile of H-RasV12 vs. control fibroblasts, we analyzed the expression of several genes involved in fatty acid metabolism, i.e., desaturases (stearoyl-CoA desaturase, *SCD* and fatty acids desaturase, *FADS*), elongases (*ELOVL*) and acyl-coenzyme A synthetases (*ACSL*) by qRT-PCR. Among genes expressed in fibroblasts (Supplementary Table S1), we observed an up-regulation of *SCD* and notably, of 2 out of 3 *ACSL* genes, i.e., *ACSL3* and *ACSL4* (Figure 4A). The increased expression of *SCD* gene (stearoyl-CoA desaturase 1 or delta-9 desaturase) in H-RasV12 fibroblasts (Figure 4A) well correlated with the observation that our model of oncogene-induced senescence is characterized by higher content of C16:1 and C18:1 (Figure 3A). Furthermore, we investigated the expression of phospholipase A2 transcripts, which are involved in the release of fatty acid precursor of lipid mediators from membrane phospholipid and found that *PL2G3* and *PLA2G6B* were significantly up-regulated (Figure 4B).

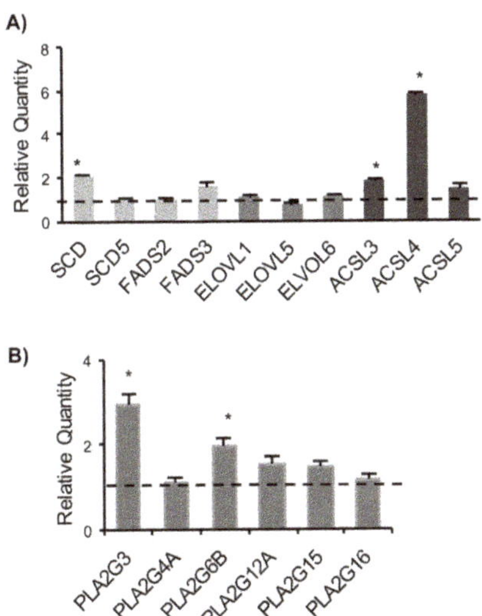

Figure 4. Gene expression analysis of fatty acid-metabolizing enzymes in control and H-RasV12 expressing fibroblasts by qRT-PCR. (**A**) Gene expression analysis of desaturases (*SCD* and *FADS*), elongases (*ELOVL*) and acyl-coenzyme A synthetases (*ACSL*). (**B**) Gene expression analysis of phospholipases A2 (*PLA2*). Ten ng of each cDNA were used as template. Reactions were performed in triplicate, using SYBR green technology StepOne RT-PCR machine to detect amplification. The GAPDH gene was used as endogenous control. The fold expression in H-RasV12 fibroblasts with respect to control is displayed, expressed as Relative Quantity (RQ). The analysis was repeated three times in triplicate and the mean ± S.D. is reported (* $p < 0.05$).

3. Discussion

Extracellular vesicles (EVs) have been recognized as an additional way to transmit cell-to-cell signals. Much evidence has been also provided that their lipid content not only has a structural function, but that they can also represent conveyors of membrane-derived bioactive lipids [14,15]. Analysis of fatty acids released extracellularly via EVs is an important factor in the determination of the amount and type of lipid mediator precursors available to the surrounding tissue and to extracellular enzymes, in order to spread lipid-mediated signals. To gain insight on how H-RasV12 expression in human fibroblasts influences fatty composition of cells and EVs, we have investigated the fatty acid profile of H-RasV12 and control fibroblasts and their released EVs.

Fatty acid composition confirmed that EVs have a higher fatty acid content per μg of protein respect to parental cells. This is consistent with other studies that have shown a higher lipid to protein ratio for EVs with respect to the releasing cells [10,12]. Lipid/protein ratio has been also proposed as a marker for EVs [33]. Grouping fatty acids into SFA, MUFA and PUFA category showed that H-RasV12 fibroblasts are characterized by a significantly increased level of MUFA and a decreased level of SFA as compared to control cells, whereas the content of PUFA was similar.

The level of SFA, MUFA and PUFA in EVs revealed that they were mostly made up of saturated species (about 95%). This was consistent with previous studies [9,10]. Comparing fatty acid content between cells and EVs, we observed an enrichment of PUFA in EVs, as well as SFA, whereas MUFA were present at a similar level. PUFA are precursor of lipid mediators which can have either a pro-inflammatory or a pro-resolving function [34]. Their presence in EVs has been previously described, as well as the presence of enzymes responsible for lipid mediator biosynthesis, such as prostaglandins [16,18,35]. Considering that fibroblasts undergoing H-RasV12-induced senescence release a higher number of EVs [12], these findings indicate that senescent cells release via EVs a higher amount of fatty acids, and in particular of PUFA and SFA, as compared to control cells. As discussed above, PUFA availability could affect the biosynthesis of lipid mediators in recipient cells following EVs uptake. As for SFA, their increased availability could represent an alternative source of fatty acids for recipient cells in neighbor tissue. This might be relevant for tissue homeostasis, as it has been previously shown that SFA-enriched EVs may induce pro-inflammatory responses [36,37].

From a structural point of view, the high level of SFA, and saturated phospholipid species, has been previously correlated with an increased membrane rigidity relative to parent cell membranes [9,10,38,39] and with an increased stability of EVs in biological fluids [40].

In fibroblasts, the increased level of MUFA in cells undergoing H-RasV12-induced senescence was mainly due to an increase of palmitoleic acid (C16:1) and oleic acid (C18:1), whereas their corresponding saturated fatty acids palmitic (C16:0) and stearic (C18:0) were decreased. The conversion of C16:0 and C18:0 in C16:1 and C18:1 is due to stearoyl-CoA desaturase (*SCD*), that introduces a cis-Δ9 double bond into saturated fatty acids. Gene expression analysis showed a significant up-regulation of *SCD* gene, whereas the other desaturases were not affected. The conversion of palmitic acid (C16:0) into stearic acid (C18:0), which was the most abundant species, requires the activity of elongases, but none of the expressed elongases showed a significantly altered transcript level. We also tested the expression of fatty acid activating enzymes acyl CoA synthetases, which are key enzymes converting long-chain fatty acids into fatty acyl-CoA esters, which then serve as a substrate for both lipid synthesis and β-oxidation [41]. We detected a significantly increased level of two enzymes, *ACSL3* and *ACSL4*. Even if these two enzymes catalyze the same reaction, there is increasing evidence for a specialization in the substrates. In fact, *ACSL3* prefers palmitate and *ACSL4* arachidonic acid (C20:4) and eicosapentaenoic acid (C20:5), thus modulating prostaglandin E(2) release from human smooth muscle cells [42]. These evidences suggest that H-RasV12-induced senescence may be associated with pro-inflammatory fatty acid metabolism. *ACSL3* and *ACSL4* expression levels have been reported to be frequently altered in cancer [43] but here we provide evidence that they have also an altered expression profile in senescence induced by oncogenic H-Ras. This finding is consistent with previous studies, demonstrating the up-regulation of *ACSL3* in NeuT oncogene-induced senescence [44] and *ACSL4* in BJ fibroblasts undergoing replicative senescence [45].

The first step of fatty acid release from membrane (to make lipid mediator precursors available for the biosynthetic reactions) is the hydrolysis of fatty acids from the sn-2 position of phospholipids by phospholipase A2, to give free fatty acids and lysophospholipids [46]. We tested a panel of human phospholipase A2 and detected an up-regulation of *PL2G3* and *PLA2G6B*. *PLA2G3* encodes for a secreted phospholipase A2 [47]. *PLA2G3* has a relevant role in male reproduction and is an endogenous regulator of mast cells [48], but no role in H-Ras induced senescence has not been reported so far. *PLA2GB6B* encodes for an isoform of inducible phospholipase A2 also known as iPLA2γ. This enzyme promotes cellular membrane hydrolysis and prostaglandin production [49], reinforcing the evidence that H-RasV12 induced senescence may be associated with an increased release of pro-inflammatory lipid mediator precursors such as arachidonic acid.

Lipid metabolism changes induced by H-Ras activation have been often investigated with a specific attention on intracellular metabolic alterations. Here, we provide evidence that H-RasV12-induced senescence is associated with changes in fatty acids content and in the level of some fatty acid metabolizing enzymes transcripts. Whether these changes are a specific feature of oncogene induced senescence or are associated with senescence induced by other types of stimuli such as telomere shortening or DNA damage remains to be determined. However, the changes in the expression profile of the fatty acids-related enzymes have an impact not only on cellular lipid composition, but also on the levels of saturated and polyunsaturated fatty acids released extracellularly via EVs. These fatty acids represent a factor that contributes to the properties of the senescence-associated secretory phenotype, influencing metabolism and cell signaling to nearby tissue.

4. Materials and Methods

4.1. HuDe Fibroblasts Transfection and H-RasV12 Expression

HuDe (human dermal fibroblasts) were purchased from the Istituto Zooprofilattico Sperimentale, Brescia, Italy as previously reported [25]. Cells were cultured in DMEM (Dulbecco's modified Eagle's medium) with 10% (*v/v*) heat-inactivated FBS, 2 mM glutamine, 100 U/mL penicillin, 100 mg/mL streptomycin at 37 °C in a 5% CO_2. Cell viability was assessed by trypan blue staining exclusion (0.1% in 0.9% NaCl) and cell growth was determined by counting cell numbers in a hemocytometer. Cells were transiently transfected with the pcDNA6 plasmid encoding the constitutively active mutant H-RasV12 and with the empty vector as control using Lipofectamine LTX, then selected for 5 to 7 days with 4 µg/mL blasticidin-S as selective agent, as previously described [12].

4.2. Staining of Senescence Associated β-Galactosidase (SA-β-Gal)

Cells were washed with PBS, fixed with 0.5% glutaraldehyde for 10 min at RT, washed with PBS and incubated at 37 °C overnight with the staining solution (5 mM potassium ferrocyanide, 5 mM potassium ferricyanide, 2 mM $MgCl_2$ in 100 mM citric acid and 200 mM Na_2HPO_4 solutions pH 6.0, 1 mg/mL 5-bromo-4-chloro-3-indolyl β-D-galactopyranoside). The development of blue color was checked under microscope using 20× total magnification. Positive fibroblasts were counted and results expressed as mean ± S.E.M. of the percentage of SA-β-gal positive fibroblasts with respect to the total number of fibroblasts.

4.3. Immunostaining

Cells were washed with PBS, fixed in 4% paraformaldehyde for 20 min at RT, permeabilized with 0.1% Triton X-100 in PBS for 10 min at RT, then incubated with an anti-γH2AX (Santa Cruz Biotechnology, Santa Cruz, CA, USA) in 2% FBS/0.01% Triton X-100/PBS and labelled with an anti-rabbit Alexa-Fluor 594 antibody (Thermofisher Scientific, Waltham, MA, USA). Nuclei were stained with 1 µg/mL DAPI (Sigma-Aldrich, St. Louis, MO, USA). Fluorescence microscopy analysis was carried out using a Nikon TE2000 microscope (Minato, Tokyo, Japan) through a 60× oil immersion objective.

4.4. Extracellular Vesicles Preparation

After selection with 4 µg/mL blasticidin-S, HuDe fibroblasts were further incubated for 72 h in serum free medium containing 4 µg/mL blasticidin-S to avoid any contamination by FBS lipoproteins. Then, medium was collected and centrifuged to remove cells, cell debris and large EVs (300× *g*, 10 min; 2000× *g*, 10 min), adding a filtration step at 0.22 µm with cellulose filter (Millipore, Burlington, MA, USA) to enrich for small EVs. Vesicles were isolated by polymer co-precipitation using Exoquick-TC precipitation method (System Biosciences, Palo Alto, CA, USA) as previously described [12]. Pelleted EVs were stored at −80 °C in PBS. EV protein content was determined by the Bradford method, using bovine serum albumin as standard.

4.5. Immunoblotting

Cell extracts were prepared as previously described [12]. Cell extracts (30 µg) or EVs (3 µg) were mixed with sample buffer 5× (1M Tris-HCl pH 6.8, 5% (*w/v*) SDS, 6% (*v/v*) glycerol, 0.01% (*w/v*) Bromophenol blue) without DTT (non-reducing conditions, used for CD9 and CD63 antibodies) or with 125 mM DTT (other antibodies). Samples were electrophoresed on 12% acrylamide gel at 150 V for 1 h and transferred to PVDF membrane at 100 V for 1 h. As internal control, actin or calnexin were used. Rabbit polyclonal anti-H-Ras antibody, mouse monoclonal anti-Tsg101 antibody, goat polyclonal anti-calnexin antibody were from Santa Cruz Biotechnology, mouse monoclonal anti-CD9 and mouse monoclonal anti-CD63 were from Abcam (Cambridge, UK), mouse monoclonal anti β-actin was from Sigma-Aldrich. Sheep anti-goat (Sigma), donkey anti-rabbit and sheep anti-mouse HRP-linked secondary antibodies (Sigma Aldrich) were probed according to manufacturer's instructions. Immunoblots were detected by chemiluminescence using ECL system.

4.6. Immunogold

For immuno-transmission electron microscopy (immuno-TEM), EVs were fixed in 4% formaldehyde and 0.1% glutaraldehyde for 15 min then dropped directly onto formvar/carbon coated grids and left for 20 min at room temperature. Grids were washed with PBS, blocked for 10 min with 0.5% bovine serum albumin in PBS, incubated for 20 min with mouse anti-CD63 primary antibody, incubated 20 min with rabbit anti-mouse secondary antibody and, finally, incubated with gold-labelled protein A for 10 min. Between each antibody incubation, grids were washed twice with PBS. Two washes, one in PBS and one in water, were performed before grids stained/embedded with 0.4% uranyl acetate/1.8% methyl cellulose for 10 min at 4 °C. Grids were allowed to dry at least for 20 min and samples were observed on a JEOL-JEM 1230 at 80 kV and images were recorded with a Morada digital camera.

4.7. Cells and Vesicles Preparation for Fatty Acid Analysis

For fatty acids analysis, about 3.6×10^6 of H-RasV12 expressing cells or control cells were washed twice with PBS, pelleted and stored at -80 °C. Total cellular lipids were extracted from 6 cell pellets from 3 different preparations and protein concentration was determined in each sample to normalize lipid content. For EVs, they were obtained from cell culture medium of the same preparations used for cell lipid extraction, pooled, and 8 μg of proteins used for each analysis. Lipid extraction was carried out as previously reported [50,51]. Extracts were dried under nitrogen and resuspended in methanol prior to be submitted for analyses. An acid hydrolysis was carried out by dissolving the methanol fraction in chloroform/methanol 1:1 (v/v). Then, 1 M HCl:methanol (1:1, v/v) solution was added to the total lipid extract and the mixed solution shaken for 2 h. Chloroform:water (1:1 v/v) was added and the organic phase collected and dried under nitrogen flow. The residue was dissolved in methanol. For fatty acid quantification, the MS analysis was carried out with a selective ion monitoring-tandem mass spectrometry (SIM-MS/MS) method. Quantitative analysis was performed with calibration curves. An ESI source connected with an API 4000 triple quadrupole instrument (AB Sciex, Old Connecticut Path, Framingham, MA USA) was used. The mobile phases were: Water/10 mM isopropylethylamine/15 mM acetic acid (phase A) and methanol (phase B). MultiQuant™ software version 3.0.2 (AB Sciex, Old Connecticut Path, Framingham, MA, USA) was used for data analysis and peak review of chromatograms. Quantitative evaluation of phospholipid families was performed based on standard curves. Quantitative data were normalized on the protein content of cells or vesicles. An external standard for each phospholipid family was used for the semi-quantitative analysis. Data were normalized on protein content.

4.8. qRT-PCR

RNA was extracted and retro-transcribed as previously described [12]. cDNA was used to determine the expression of genes listed in Supplementary Table S1. cDNA was used to determine transcript levels by qRT-PCR in a StepOne RT-PCR machine (Applied Biosystems, Foster City, CA, USA) using SYBR® Select Master Mix (Life Technologies, Carlsbad, CA, USA). Reactions were performed in triplicate and GAPDH used as endogenous control. Data were analysed using the ΔΔCt method. ΔCt was calculated subtracting the average Ct value of *GAPDH* to the average Ct value of a specific gene for each sample, then ΔΔCt as the difference between the ΔCt for each sample and the ΔCt of empty vector transfected fibroblasts as control. The reported fold expression, expressed as Relative Quantity, was calculated by $2^{-\Delta\Delta Ct}$.

4.9. Statistical Analysis

Statistical comparison was performed using Student's *t*-test. Differences were considered statistically significant when $p < 0.05$. For lipid analysis, 6 independent experiments for cells and EVs were carried out.

Supplementary Materials: Supplementary materials can be found at http://www.mdpi.com/1422-0067/19/11/3515/s1.

Author Contributions: S.B., L.U. and K.S. designed and carried out the experiments. E.C. carried out the experiments. D.C. and N.M. carried out lipid analysis and evaluated data. C.E., S.B. and L.U. analysed results and wrote the paper. All authors reviewed and critically revised the paper.

Funding: This work funded by Fondazione Cassa di Risparmio di Perugia Grant No. 2016.0050.021 to Carla Emiliani and by University of Perugia Grant FONDO D'ATENEO PER LA RICERCA DI BASE 2015.

Conflicts of Interest: The authors declare no conflicts of interest.

References

1. Yáñez-Mó, M.; Siljander, P.R.; Andreu, Z.; Zavec, A.B.; Borràs, F.E.; Buzas, E.I.; Buzas, K.; Casal, E.; Cappello, F.; Carvalho, J.; et al. Biological properties of extracellular vesicles and their physiological functions. *J. Extracell. Vesicles* **2015**, *4*, 27066. [CrossRef] [PubMed]

2. Johnstone, R.M.; Adam, M.; Hammond, J.R.; Orr, L.; Turbide, C. Vesicle formation during reticulocyte maturation. Association of plasma membrane activities with released vesicles (exosomes). *J. Biol. Chem.* **1987**, *262*, 9412–9420. [PubMed]

3. Urbanelli, L.; Magini, A.; Buratta, S.; Brozzi, A.; Sagini, K.; Polchi, A.; Tancini, B.; Emiliani, C. Signaling pathways in exosomes biogenesis, secretion and fate. *Genes* **2013**, *4*, 152–170. [CrossRef] [PubMed]

4. Tkach, M.; Thery, C. Communication by Extracellular Vesicles: Where We Are and Where We Need to Go. *Cell* **2016**, *164*, 1226–1232. [CrossRef] [PubMed]

5. Lotvall, J.; Hill, A.F.; Hochberg, F.; Buzas, E.I.; Di Vizio, D.; Gardiner, C.; Gho, Y.S.; Kurochkin, I.V.; Mathivanan, S.; Quesenberry, P.; et al. Minimal experimental requirements for definition of extracellular vesicles and their functions: A position statement from the International Society for Extracellular Vesicles. *J. Extracell. Vesicles* **2014**, *3*, 26913. [CrossRef] [PubMed]

6. Kowal, J.; Arras, G.; Colombo, M.; Jouve, M.; Morath, J.P.; Primdal-Bengtson, B.; Dingli, F.; Loew, D.; Tkach, M.; Théry, C. Proteomic comparison defines novel markers to characterize heterogeneous populations of extracellular vesicle subtypes. *Proc. Natl. Acad. Sci. USA* **2016**, *113*, 968–977. [CrossRef] [PubMed]

7. Urbanelli, L.; Buratta, S.; Sagini, K.; Ferrara, G.; Lanni, M.; Emiliani, C. Exosome-based strategies for diagnosis and therapy. *Recent Pat. CNS Drug Discov.* **2015**, *10*, 10–27. [CrossRef] [PubMed]

8. Kim, D.K.; Lee, J.; Kim, S.R.; Choi, D.S.; Yoon, Y.J.; Kim, J.H.; Go, G.; Nhung, D.; Hong, K.; Jang, S.C.; et al. EVpedia: A community web portal for extracellular vesicles research. *Bioinformatics* **2015**, *31*, 933–939. [CrossRef] [PubMed]

9. Laulagnier, K.; Motta, C.; Hamdi, S.; Roy, S.; Fauvelle, F.; Pageaux, J.F.; Kobayashi, T.; Salles, J.P.; Perret, B.; Bonnerot, C.; et al. Mast cell- and dendritic cell-derived exosomes display a specific lipid composition and an unusual membrane organization. *Biochem. J.* **2004**, *380*, 161–171. [CrossRef] [PubMed]

10. Llorente, A.; Skotland, T.; Sylvänne, T.; Kauhanen, D.; Róg, T.; Orłowski, A.; Vattulainen, I.; Ekroos, K.; Sandvig, K. Molecular lipidomics of exosomes released by PC-3 prostate cancer cells. *Biochim. Biophys. Acta* **2013**, *1831*, 1302–1309. [CrossRef] [PubMed]

11. Lydic, T.A.; Townsend, S.; Adda, C.G.; Collins, C.; Mathivanan, S.; Reid, G.E. Rapid and comprehensive 'shotgun' lipidome profiling of colorectal cancer cell derived exosomes. *Methods* **2015**, *87*, 83–95. [CrossRef] [PubMed]

12. Buratta, S.; Urbanelli, L.; Sagini, K.; Giovagnoli, S.; Caponi, S.; Fioretto, D.; Mitro, N.; Caruso, D.; Emiliani, C. Extracellular vesicles released by fibroblasts undergoing H-Ras induced senescence show changes in lipid profile. *PLoS ONE* **2017**, *12*, e0188840. [CrossRef] [PubMed]

13. Record, M.; Silvente-Poirot, S.; Poirot, M.; Wakelam, M.J.O. Extracellular vesicles: Lipids as key components of their biogenesis and functions. *J. Lipid Res.* **2018**, *59*, 1316–1324. [CrossRef] [PubMed]

14. Sagini, K.; Costanzi, E.; Emiliani, C.; Buratta, S.; Urbanelli, L. Extracellular Vesicles as Conveyors of Membrane-Derived Bioactive Lipids in Immune System. *Int. J. Mol. Sci.* **2018**, *19*, 1227. [CrossRef] [PubMed]

15. Boilard, E. Extracellular vesicles and their content in bioactive lipid mediators: More than a sack of microRNA. *J. Lipid Res.* **2018**. [CrossRef] [PubMed]

16. Subra, C.; Grand, D.; Laulagnier, K.; Stella, A.; Lambeau, G.; Paillasse, M.; De Medina, P.; Monsarrat, B.; Perret, B.; Silvente-Poirot, S.; et al. Exosomes account for vesicle-mediated transcellular transport of activatable phospholipases and prostaglandins. *J. Lipid Res.* **2010**, *51*, 2105–2120. [CrossRef] [PubMed]

17. Esser, J.; Gehrmann, U.; D'Alexandri, F.L.; Hidalgo-Estévez, A.M.; Wheelock, C.E.; Scheynius, A.; Gabrielsson, S.; Rådmark, O. Exosomes from human macrophages and dendritic cells contain enzymes for leukotriene biosynthesis and promote granulocyte migration. *J. Allergy Clin. Immunol.* **2010**, *126*, 1032–1040. [CrossRef] [PubMed]

18. Deng, Z.B.; Zhuang, X.; Ju, S.; Xiang, X.; Mu, J.; Liu, Y.; Jiang, H.; Zhang, L.; Mobley, J.; McClain, C.; et al. Exosome-like nanoparticles from intestinal mucosal cells carry prostaglandin E2 and suppress activation of liver NKT cells. *J. Immunol.* **2013**, *190*, 3579–3589. [CrossRef] [PubMed]

19. Record, M.; Carayon, K.; Poirot, M.; Silvente-Poirot, S. Exosomes as new vesicular lipid transporters involved in cell-cell communication and various pathophysiologies. *Biochim. Biophys. Acta* **2014**, *1841*, 108–120. [CrossRef] [PubMed]

20. Lazar, I.; Clement, E.; Attane, C.; Muller, C.; Nieto, L. A new role for extracellular vesicles: How small vesicles can feed tumors' big appetite. *J. Lipid Res.* **2018**. [CrossRef] [PubMed]

21. López-Otín, C.; Blasco, M.A.; Partridge, L.; Serrano, M.; Kroemer, G. The hallmarks of aging. *Cell* **2013**, *153*, 1194–1217. [CrossRef] [PubMed]

22. He, S.; Sharpless, N.E. Senescence in Health and Disease. *Cell* **2017**, *169*, 1000–1011. [CrossRef] [PubMed]

23. Childs, B.G.; Baker, D.J.; Kirkland, J.L.; Campisi, J.; van Deursen, J.M. Senescence and apoptosis: Dueling or complementary cell fates? *EMBO Rep.* **2014**, *15*, 1139–1153. [CrossRef] [PubMed]

24. Lecot, P.; Alimirah, F.; Desprez, P.Y.; Campisi, J.; Wiley, C. Context-dependent effects of cellular senescence in cancer development. *Br. J. Cancer* **2016**, *114*, 1180–1184. [CrossRef] [PubMed]

25. Armeni, T.; Ercolani, L.; Urbanelli, L.; Magini, A.; Magherini, F.; Pugnalono, A.; Piva, F.; Modesti, A.; Emiliani, C.; Principato, G. Cellular redox imbalance and changes of protein S-glutathionylation patterns are associated with senescence induced by oncogenic H-Ras. *PLoS ONE* **2012**, *7*, e52151. [CrossRef] [PubMed]

26. Urbanelli, L.; Magini, A.; Ercolani, L.; Sagini, K.; Polchi, A.; Tancini, B.; Brozzi, A.; Armeni, T.; Principato, G.; Emiliani, C. Oncogenic H-Ras up-regulates acid β-hexosaminidase by a mechanism dependent on the autophagy regulator TFEB. *PLoS ONE* **2014**, *9*, e89485. [CrossRef] [PubMed]

27. Georgilis, A.; Klotz, S.; Hanley, C.J.; Herranz, N.; Weirich, B.; Morancho, B.; Leote, A.C.; D'Artista, L.; Gallage, S.; Seehawer, M.; et al. PTBP1-mediated alternative splicing regulates the inflammatory secretome and the pro-tumorigenic effects of senescent cells. *Cancer Cell* **2018**, *34*, 85–102. [CrossRef] [PubMed]

28. Urbanelli, L.; Buratta, S.; Sagini, K.; Tancini, B.; Emiliani, C. Extracellular Vesicles as New Players in Cellular Senescence. *Int. J. Mol. Sci.* **2016**, *17*, 1408. [CrossRef] [PubMed]

29. Di Micco, R.; Fumagalli, M.; d'Adda di Fagagna, F. Breaking news: High-speed race ends in arrest—How oncogenes induce senescence. *Trends Cell Biol.* **2007**, *17*, 529–536. [CrossRef] [PubMed]

30. Serrano, M.; Lin, A.W.; McCurrach, M.E.; Beach, D.; Lowe, S.W. Oncogenic ras provokes premature cell senescence associated with accumulation of p53 and p16INK4a. *Cell* **1997**, *88*, 593–602. [CrossRef]

31. Lehmann, B.D.; Paine, M.S.; Brooks, A.M.; McCubrey, J.A.; Renegar, R.H.; Wang, R.; Terrian, D.M. Senescence-associated exosome release from human prostate cancer cells. *Cancer Res.* **2008**, *68*, 7864–7871. [CrossRef] [PubMed]

32. Yu, X.; Harris, S.L.; Levine, A.J. The regulation of exosome secretion: A novel function of the p53 protein. *Cancer Res.* **2006**, *66*, 4795–4801. [CrossRef] [PubMed]

33. Osteikoetxea, X.; Balogh, A.; Szabó-Taylor, K.; Németh, A.; Szabó, T.G.; Pálóczi, K.; Sódar, B.; Kittel, Á.; György, B.; Pállinger, É.; et al. Improved Characterization of EV Preparations Based on Protein to Lipid Ratio and Lipid Properties. *PLoS ONE* **2015**, *10*, e0121184. [CrossRef] [PubMed]

34. Marion-Letellier, R.; Savoye, G.; Ghosh, S. Polyunsaturated fatty acids and inflammation. *IUBMB Life* **2015**, *67*, 659–667. [CrossRef] [PubMed]

35. Xiang, X.; Poliakov, A.; Liu, C.; Liu, Y.; Deng, Z.B.; Wang, J.; Cheng, Z.; Shah, S.V.; Wang, G.J.; Zhang, L.; et al. Induction of myeloid-derived suppressor cells by tumor exosomes. *Int. J. Cancer* **2009**, *124*, 2621–2633. [CrossRef] [PubMed]

36. Kakazu, E.; Mauer, A.S.; Yin, M.; Malhi, H. Hepatocytes release ceramide-enriched pro-inflammatory extracellular vesicles in an IRE1α-dependent manner. *J. Lipid Res.* **2016**, *57*, 233–245. [CrossRef] [PubMed]

37. Hirsova, P.; Ibrahim, S.H.; Krishnan, A.; Verma, V.K.; Bronk, S.F.; Werneburg, N.W.; Charlton, M.R.; Shah, V.H.; Malhi, H.; Gores, G.J. Lipid-induced signaling causes release of inflammatory extracellular vesicles from hepatocytes. *Gastroenterology* **2016**, *150*, 956–967. [CrossRef] [PubMed]

38. Subra, C.; Laulagnier, K.; Perret, B.; Record, M. Exosome lipidomics unravels lipid sorting at the level of multivesicular bodies. *Biochimie* **2007**, *89*, 205–212. [CrossRef] [PubMed]

39. Trajkovic, K.; Hsu, C.; Chiantia, S.; Rajendran, L.; Wenzel, D.; Wieland, F.; Schwille, P.; Brügger, B.; Simons, M. Ceramide triggers budding of exosome vesicles into multivesicular endosomes. *Science* **2008**, *319*, 1244–1247. [CrossRef] [PubMed]

40. Alvarez-Erviti, L.; Seow, Y.; Yin, H.; Betts, C.; Lakhal, S.; Wood, M.J. Delivery of siRNA to the mouse brain by systemic injection of targeted exosomes. *Nat. Biotechnol.* **2011**, *29*, 341–345. [CrossRef] [PubMed]

41. Yan, S.; Yang, X.F.; Liu, H.L.; Fu, N.; Ouyang, Y.; Qing, K. Long-chain acyl-CoA synthetase in fatty acid metabolism involved in liver and other diseases: An update. *World J. Gastroenterol.* **2015**, *21*, 3492–3498. [CrossRef] [PubMed]

42. Golej, D.L.; Askari, B.; Kramer, F.; Barnhart, S.; Vivekanandan-Giri, A.; Pennathur, S.; Bornfeldt, K.E. Long-chain acyl-CoA synthetase 4 modulates prostaglandin E$_2$ release from human arterial smooth muscle cells. *J. Lipid Res.* **2011**, *52*, 782–793. [CrossRef] [PubMed]

43. Radif, Y.; Ndiaye, H.; Kalantzi, V.; Jacobs, R.; Hall, A.; Minogue, S.; Waugh, M.G. The endogenous subcellular localisations of the long chain fatty acid-activating enzymes ACSL3 and ACSL4 in sarcoma and breast cancer cells. *Mol. Cell. Biochem.* **2018**. [CrossRef] [PubMed]

44. Cadenas, C.; Vosbeck, S.; Hein, E.M.; Hellwig, B.; Langer, A.; Hayen, H.; Franckenstein, D.; Büttner, B.; Hammad, S.; Marchan, R.; et al. Glycerophospholipid profile in oncogene-induced senescence. *Biochim. Biophys. Acta* **2012**, *1821*, 1256–1268. [CrossRef] [PubMed]

45. Lizardo, D.Y.; Lin, Y.L.; Gokcumen, O.; Atilla-Gokcumen, G.E. Regulation of lipids is central to replicative senescence. *Mol. Biosyst.* **2017**, *13*, 498–509. [CrossRef] [PubMed]

46. Dennis, E.A.; Cao, J.; Hsu, Y.H.; Magrioti, V.; Kokotos, G. Phospholipase A2 enzymes: Physical structure, biological function, disease implication, chemical inhibition, and therapeutic intervention. *Chem. Rev.* **2011**, *111*, 6130–6185. [CrossRef] [PubMed]

47. Murakami, M.; Sato, H.; Miki, Y.; Yamamoto, K.; Taketomi, Y. A new era of secreted phospholipase A$_2$. *J. Lipid Res.* **2015**, *56*, 1248–1261. [CrossRef] [PubMed]

48. Murakami, M.; Masuda, S.; Shimbara, S.; Ishikawa, Y.; Ishii, T.; Kudo, Y. Cellular distribution, post-translational modification, and tumorigenic potential of human group III secreted phospholipase A$_2$. *J. Biol. Chem.* **2005**, *280*, 24987–24998. [CrossRef] [PubMed]

49. Murakami, M.; Masuda, S.; Ueda-Semmyo, K.; Yoda, E.; Kuwata, H.; Takanezawa, Y.; Aoki, J.; Arai, H.; Sumimoto, H.; Ishikawa, Y.; et al. Group VIB Ca^{2+}-independent phospholipase A$_2$γ promotes cellular membrane hydrolysis and prostaglandin production in a manner distinct from other intracellular phospholipases A$_2$. *J. Biol. Chem.* **2005**, *280*, 14028–14041. [CrossRef] [PubMed]

50. Cermenati, G.; Giatti, S.; Audano, M.; Pesaresi, M.; Spezzano, R.; Caruso, D.; Mitro, N.; Melcangi, R.C. Diabetes alters myelin lipid profile in rat cerebral cortex: Protective effects of dihydroprogesterone. *J. Steroid Biochem. Mol. Biol.* **2017**, *168*, 60–70. [CrossRef] [PubMed]

51. Cermenati, G.; Audano, M.; Giatti, S.; Carozzi, V.; Porretta-Serapiglia, C.; Pettinato, E.; Ferri, C.; D'Antonio, M.; De Fabiani, E.; Crestani, M.; et al. Lack of sterol regulatory element binding factor-1c imposes glial Fatty Acid utilization leading to peripheral neuropathy. *Cell Metab.* **2015**, *21*, 571–583. [CrossRef] [PubMed]

International Journal of
Molecular Sciences

Article

Stable Isotope Labeling Highlights Enhanced Fatty Acid and Lipid Metabolism in Human Acute Myeloid Leukemia

Lucille Stuani [1,2], Fabien Riols [3], Pierre Millard [4], Marie Sabatier [1,2], Aurélie Batut [3], Estelle Saland [1,2], Fanny Viars [3], Laure Tonini [2,5], Sonia Zaghdoudi [1,2], Laetitia K. Linares [6], Jean-Charles Portais [2,7], Jean-Emmanuel Sarry [1,2] and Justine Bertrand-Michel [2,3,*]

[1]   Centre de Recherches en Cancérologie de Toulouse, UMR1037, Inserm, Equipe Labellisée LIGUE 2018, F-31037 Toulouse, France; lucille.stuani@inserm.fr (L.S.); marie.sabatier@inserm.fr (M.S.); estelle.saland@inserm.fr (E.S.); sonia.zaghdoudi@inserm.fr (S.Z.); jean-emmanuel.sarry@inserm.fr (J.-E.S.)
[2]   Université de Toulouse, 31000 Toulouse, France; laure.tonini@inserm.fr (L.T.); jean-charles.portais@insa-toulouse.fr (J.-C.P.)
[3]   MetaToul-Lipidomic Core Facility, MetaboHUB, I2 MC, Inserm, 31100 Toulouse, France; fabien.riols@gmail.com (F.R.); aurelie.batut@inserm.fr (A.B.); viars.fanny1@gmail.com (F.V.)
[4]   LISBP, Université de Toulouse, CNRS, INRA, INSA, F-31077 Toulouse, France; pierre.millard@insa-toulouse.fr
[5]   Centre de Recherches en Cancérologie de Toulouse, UMR1037, Inserm, Pôle Technologique, F-31037 Toulouse, France
[6]   Institut de Recherche en Cancérologie de Montpellier, U1194 Inserm, Université de Montpellier, Equipe Labellisée LIGUE 2017, F-34090 Montpellier, France; laetitia.linares@inserm.fr
[7]   MetaToul Core Facility, INSA, LISBP, F-31077 Toulouse, France
*   Correspondence: justine.bertrand-michel@inserm.fr

Received: 7 September 2018; Accepted: 22 October 2018; Published: 25 October 2018

Abstract: Background: In Acute Myeloid Leukemia (AML), a complete response to chemotherapy is usually obtained after conventional chemotherapy but overall patient survival is poor due to highly frequent relapses. As opposed to chronic myeloid leukemia, B lymphoma or multiple myeloma, AML is one of the rare malignant hemopathies the therapy of which has not significantly improved during the past 30 years despite intense research efforts. One promising approach is to determine metabolic dependencies in AML cells. Moreover, two key metabolic enzymes, isocitrate dehydrogenases (IDH1/2), are mutated in more than 15% of AML patient, reinforcing the interest in studying metabolic reprogramming, in particular in this subgroup of patients. **Methods**: Using a multi-omics approach combining proteomics, lipidomics, and isotopic profiling of [U-^{13}C] glucose and [U-^{13}C] glutamine cultures with more classical biochemical analyses, we studied the impact of the IDH1 R132H mutation in AML cells on lipid biosynthesis. **Results**: Global proteomic and lipidomic approaches showed a dysregulation of lipid metabolism, especially an increase of phosphatidylinositol, sphingolipids (especially few species of ceramide, sphingosine, and sphinganine), free cholesterol and monounsaturated fatty acids in IDH1 mutant cells. Isotopic profiling of fatty acids revealed that higher lipid anabolism in IDH1 mutant cells corroborated with an increase in lipogenesis fluxes. **Conclusions**: This integrative approach was efficient to gain insight into metabolism and dynamics of lipid species in leukemic cells. Therefore, we have determined that lipid anabolism is strongly reprogrammed in IDH1 mutant AML cells with a crucial dysregulation of fatty acid metabolism and fluxes, both being mediated by 2-HG (2-Hydroxyglutarate) production.

Keywords: lipidomics; isotopic profiling; metabolic reprogramming; IDH mutation; leukemia

---

1. Introduction

Cancer cells, including acute myeloid leukemia (AML) cells, grow and divide faster and more efficiently than normal cells, which increase their demand for energy, biosynthetic precursors, and macromolecular synthesis [1–4]. Most of them reprogram their metabolism from oxidative phosphorylation to aerobic glycolysis. The finding of this phenomenon, termed the "Warburg effect", stimulated much research on tumorigenesis [4–6]. Over the past twenty years, advanced developments in genetic, omics and high-throughput screening methods have revealed that many of oncogenic signaling pathways regulate cell metabolism in cancer. Therefore, changes in cell metabolism represent a key hallmark in cancer biology [3] and it has been largely demonstrated that metabolic reprogramming in cancer cells occurs far beyond the Warburg Effect [7,8]. Indeed, cancer cells activate various metabolic pathways—e.g., glutaminolysis, amino acid degradation, fatty acid β-oxidation (FAO)—to generate the numerous precursors that are required for macromolecule biosynthesis, such as ribose for nucleic acids and glycerol for lipid synthesis. Dysregulation of lipid-associated pathways is increasingly described in tumors [9–12], and different studies have demonstrated that lipogenesis is significantly up-regulated in human cancers, in particular to respond to higher demands for membrane biogenesis [13–15] or to serve as energy source when nutrients are limited [16,17].

Few years ago, mutations in two key metabolic enzymes, isocitrate dehydrogenases (IDH1 and IDH2) have been discovered in gliomas and myeloid malignancies [18–21]. In AML, 15–20% of patients carry mutations in IDH1 or IDH2 [19,22–25]. This finding has reinforced the interest in studying cell metabolism in this pathology. IDH mutations induce a neomorphic activity resulting from a rearrangement of the enzyme active site favoring the reduction of α-ketoglutarate (α-KG) to D-2-hydroxyglutarate (2-HG) oncometabolite [26,27]. 2-HG strongly structurally resembles α-KG and can function as a potent competitive inhibitor of α-ketoglutarate-dependent enzyme reactions, including dehydrogenases, transaminases, and dioxygenases [28–32]. On the other side, the wild type enzyme catalyzes the interconversion between isocitrate and α-KG and produces NADPH, an essential cofactor that is required for numerous anabolic pathways (nucleotide, fatty acid elongation, lipid synthesis, and cholesterol synthesis) to sustain cell growth and proliferation [2], especially in cancer cells exhibiting aerobic glycolysis (Warburg phenotype), in hypoxia or with defective mitochondria [33–35]. IDH mutations are heterozygous with the conservation of a wild-type allele, suggesting the importance of this wild-type protein to favor the mutant activity. However, while the role of wild type IDH is well documented in normal and cancer cells, the impact of IDH mutation on lipid metabolism, and especially on its respective metabolic fluxes in cancer, is still largely unknown.

In this study we investigated lipid metabolism in AML cells harboring IDH1 R132H mutation, the most common IDH1 mutation. Lipids represent a very large class of molecules that show strong structural diversity (e.g., various combinations of fatty acyls and functional headgroups in phospholipids or various positions for hydroxyl groups on sterol). This chemical heterogeneity, together with the occurrence of many isomeric and isobaric lipid species and the large concentration range over which lipids are found, preclude the measurement of complete lipidomic profiles with a single analytical method. Specific methods are then used for each class of lipids: neutral lipids and fatty acids were analyzed by GC-FID, phospholipids and sphingolipids by LC-MS. Furthermore, the lack of analytical standards for complex lipids hampers the absolute quantification of many molecular species [36]. In this project, lipids were analyzed through different complementary approaches to get a broad coverage of the lipidome [37–39]. The data indicated significant changes in the lipidomic profile of IDH1 R132H cells as compared to WT cells, especially with the increase of phosphatidylinositol, ceramide, and monounsaturated fatty acid. These results encouraged us to investigate the dynamics of lipid synthesis in these cells. This was achieved by using ^{13}C-labeling strategies in which the incorporation of ^{13}C-label from [U-^{13}C]-labeled glucose or [U-^{13}C]-labeled glutamine into fatty acids was measured by GC-MS. The results showed increased de novo synthesis of fatty acids in IDH mutants through the production of 2-HG. Altogether, our investigations show that IDH1 mutation results in significant reprogramming of lipid metabolism in AML cells and could represent an interesting therapeutic target for this subgroup of patients.

2. Results and Discussion

2.1. Lipid Metabolism is Dysregulated in IDH1 Mutant Cells

We first compared the proteome of IDH1 mutant HL60 cells to the one of IDH1 WT cells. A list of proteins that are significantly more abundant (fold change higher than 1.5 and FDR lower than 0.06) have been established (Supplementary Table S1B). Data mining of this specific protein set with Genomatix software revealed major changes in proteins that are associated to pathways of lipid biosynthesis and degradation (Figure 1), while proliferation rates remained unchanged and no significant differences in size, morphology, or doubling time for these cells have been observed. IDH1 R132H cells showed higher content in proteins that are involved in lipid synthesis, including cholesterol and sterol biosynthesis (IDI1, LSS, EBP; Supplementary Table S1). Interestingly, proteins involved in fatty acids (FA) oxidation were also significantly increased in IDH1 R132H cells (ACOX1/ACOX2, HSD17B4, Figure 1), suggesting higher FA catabolism to produce acetyl-CoA and feed the TCA cycle. FAO and lipogenesis are traditionally not described as being operating synchronously because they have opposite functions and are both regulated by ACC activity in opposite ways. However, some studies have demonstrated that FAO was essential to cell survival and metastasis in highly lipogenic solid cancers [40,41]. Of note, German and colleagues [42] have investigated this feature in AML cells. In fact, the authors demonstrated that under nutrient abundance, prolyl-hydroxylase 3 (PHD3) activates specifically ACC2 by hydroxylation, hence favoring malonyl-CoA formation and consequently inhibiting FAO. PHD3 does not act on ACC1, which could therefore maintain lipid synthesis while FAO is upregulated. Furthermore, 2-HG inhibition of α-KG-dependent dioxygenases, including PHDs, has been mainly described [28,43]. As a result, inhibition of PHD3 by 2-HG could prevent ACC2 hydroxylation and malonyl-CoA production to favor FAO in IDH1 mutant AML cells.

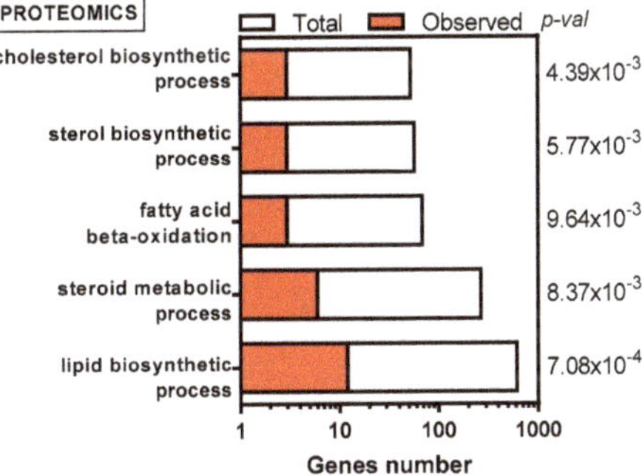

Figure 1. Enrichment in proteins associated with lipid metabolism in HL60 IDH1 R132H cells compared to WT cells (*n* = 4), based on GO biological processes. (Total) means all the genes encoding the proteins corresponding to the pathways described in the literature while (observed) refers to the genes encoding the proteins found more abundant in IDH1 R132H cells.

Moreover, we and others have shown that FAO and FA translocase/receptor CD36 played a crucial role in cell survival and drug resistance in AML in vitro and in vivo [44,45], thus reinforcing the key role of lipid metabolism in AML.

The comparison of the proteomes of IDH1 R132H and WT AML cells suggests a major reprogramming in the pathways of lipid degradation and biosynthesis, with potentially different impacts considering the diverse classes of lipids affected: mainly with sterol and fatty acids metabolism.

A better understanding of this reprogramming process could give a precious insight into the biology of AML and the consequences of IDH1 mutation.

2.2. Lipidomic Experiments Confirmed Changes in Fatty Acid Reorganization in IDH1 R132H AML Cells

The lipidome of IDH1 R132H cells was measured using a combination of LC-MS and GC methods and compared to the one of WT cells. Lipids were extracted using a universal liquid-liquid extraction method [46] in the presence of internal standards (one per family studied) to follow the sample preparation and to perform relative quantification of the molecular species. Relative quantification of most lipids has been achieved by expressing the intensities of the peaks of interest relative to the area of the internal standard (Arbitrary Unit AU/million of cells). Absolute quantification (µg/million of cells) has been performed for LPC and sphingoïd bases due to the availability of pure standards. Phospholipids were measured by LC-MS/MS [37]. The different classes (PE, PC, PI, PS) were separated by polar head on a HILIC column (except for LysoPC, which were analyzed on an apolar column) [47]. Molecular species (with their number of carbons and number of double bonds) were discriminated by MS/MS based on specific MRM transitions.

The relative amounts of PS, PE, and PC were similar in IDH1 WT and R132H cells (Figure 2A). This result is different from observations reported for gliomas harboring IDH mutation, suggesting a potential metabolic specificity of IDH mutation in AML cells. Indeed, Izquierdo-Garcia et al. [48] measured reduced PC levels while Reitman et al. [49] observed a decrease in PE levels in IDH1/2 mutant gliomas cells compared to IDH1 WT cells. More recently, Viswanath et al. [50] demonstrated that reduced PC and PE amounts in IDH mutant gliomas were due to a decrease in choline kinase and ethanolamine kinase, the enzymes that catalyze the production of PC and PE, respectively. However, we observed a significant increase for LysoPC (+15%) and mainly PI (+82%) families in AML mutant cells. The majority of the PI species were significantly increased, except 32:1; 34:2; 38:3; 38:4; 38:5; 40:5; and 40:6 (details of molecular species profiled for PI are listed in Supplementary Figure S1A). Modifications in intermediates of glycerophospholipid metabolism, such as LysoPC and PI, suggest that membrane trafficking and lipid signaling are stimulated in these cells [51].

Sphingolipids d18 (Cer and SM) were analyzed with the same method as phospholipids, while sphingoïds bases (sphinganine and sphingosine) were analyzed on a C8 column, and could be quantified due to appropriate standards [52]. Interestingly, the total amount of sphingolipids was increased by more than 40% in mutant cells and all of the four sphingolipids classes were enhanced (Figure 2B). For ceramides, if a global tendency to be increased in IDH1 mutant cells has been observed, significant changes have specifically been measured in *N*-(hexadecanoyl)-sphing-4-enine (Cer(d18:1/16:0)) and *N*-(docosanoyl)-sphing-4-enine (Cer(d18:1/22:0)) amounts (details of molecular species profiled for Cer are listed in Supplementary Figure S1B). Sphingomyelins are specific components of the cell membranes as they can form lipid rafts [53] that are essential for membrane protein dynamics and trafficking [51,54]. It is also well known that sphingolipids are key metabolites in oncogenic transformations [55].

Free and esterified cholesterol, as well as triacylglycerides (TG), were analyzed by GC-FID. The total amount of neutral lipids was unchanged between IDH1 WT and R132H cells, but the distribution of the molecular species was different. Indeed, we observed a decrease in esterified cholesterol in mutant cell counterbalanced by higher proportion of free cholesterol, while TG remained stable (Figure 2C).

Finally, the total FAs were profiled. Esterified FAs of the total extract (i.e., glycerolipids) were hydrolyzed in basic conditions and they were derivatized to be analyzed by GC-FID. A slight but significant increase in total FAs was observed in IDH1 mutant cells (+8%), which was mainly due to higher amounts of monounsaturated FAs (+17%). Polyunsaturated and saturated FAs remained stable (Figure 2D).

As expected regarding proteomics experiments, quantitative lipidomics data confirmed that IDH1 mutation leads to a re-organization of lipid metabolism, especially sphingolipids, lysoPC, the balance between cholesterol and cholesterol esters, and total FAs. In order to better understand how IDH1 mutation could be involved in the accumulation of lipids, we decided to apply stable isotope labeling experiments to identify pathways leading to total FA accumulation.

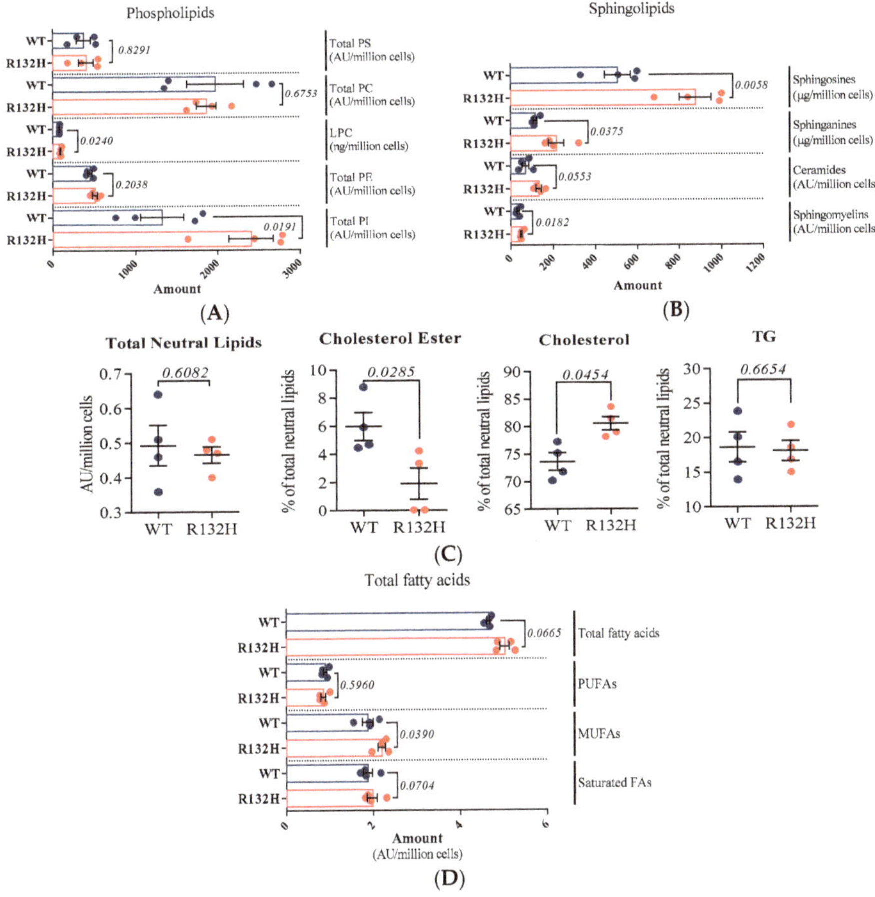

Figure 2. Lipidomic analysis in HL60 AML IDH1 WT (blue dots) and IDH1 R132H cells (red dots) (*n* = 4). (**A**) Phospholipids; (**B**) Sphingolipids; (**C**) Neutral lipids and percentages of each of its constituents; and, (**D**) Total Fatty Acids.

2.3. Isotopic Measurements of FAs Revealed Enhanced Lipid Anabolic Fluxes in IDH1 Mutant AML Cells

Due to the size and structural diversity of lipids, methods that are based on stable isotopes are not so common to investigate the metabolism of FAs, glycerophospholipids, or sphingolipids species [Ecker, Progress in lipid research, 2014]. Current and main applications used labeled FAs to track metabolism of longer chain FA [56,57]. Here, to investigate the relationship between IDH1 mutation and FA production (Figure 3A,B), we investigated whether the mutation modified the conversion of the main nutrients that support cell proliferation, glucose, and glutamine, into FAs (Figure 3C). The conversion was measured from ^{13}C-labeling experiments in which the incorporation of label into FAs from [U-^{13}C]-labelled glucose or [U-^{13}C]-labelled glutamine was measured by GC-MS. This approach provides quantitative information on the contribution of each carbon source to lipid biosynthesis [58] and further insight into specific pathways (e.g., reductive glutamine) by which the nutrient is converted into FA.

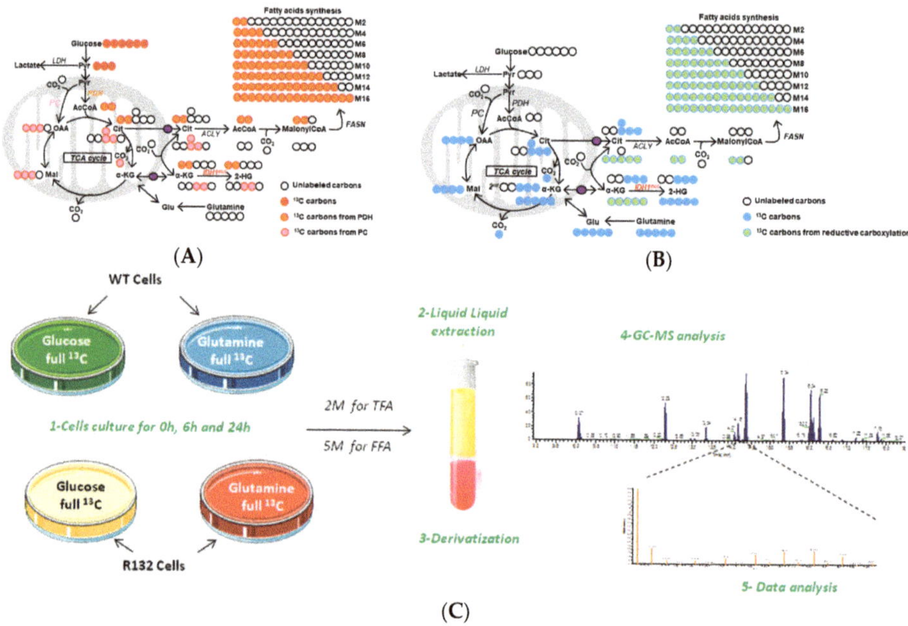

Figure 3. Simplified schematic of carbon atom (circles) transitions and tracers used to detect labeled fatty acids (FAs). (**A**) Isotopic label from [U-¹³C]glucose (red) to 2-HG and FAs synthesis through PC (Pyruvate Carboxylase; pink) or PDH (Pyruvate DeHydrogenase; orange); (**B**) Isotopic label from [U-¹³C]glutamine (blue) to 2-HG and FAs synthesis through classical TCA cycle (blue) or reductive glutamine metabolism (green). (**C**) Experimental design of the isotopic measurement of FAs on WT and IDH1 mutant cells.

Analysis of FAs is commonly performed by GC-MS after derivatization of the carboxyl group of FAs with methyl ester (FAMEs) [38]. For FA quantification, the GC-MS is operated with electronic ionization, a high-energy ionization method that results in extensive fragmentation of molecules, and the molecular ion is too low to be detected [59]. This is detrimental for the purpose of isotopic profiling, for which the extent of ¹³C-atoms incorporation in FAs is derived from the isotopic cluster of the molecular ions. Hence, EI-GC-MS is not adapted to isotopic profiling of FAs. Indeed, our attempts to profile labelled FAs by classical methylation and EI-GC-MS were not conclusive. As an alternative, chemical ionization (CI) leads to much lower fragmentation, and thereby, better detection of (labeled) molecular ions than EI, though being slightly less robust [60]. We used a method in which FAs were derivatized with pentaflurobenzyl (PTF) and analysed by negative CI-GC-MS. Interestingly, the PTF group (m/z = 181 g/mol) is lost in the ionization source so that the major peak is a clear fragment corresponding to the molecular ion of the considered FA, providing high sensitivity. This major peak was followed by single ion monitoring (SIM) for each FA (Supplementary Table S2).

IDH1 WT and R132H cells were grown in medium containing uniformly ¹³C-labeled glucose or ¹³C-labeled glutamine and dialyzed serum to avoid the dilution of label from traces of ¹²C-glucose or ¹²C-glutamine. To monitor incorporation of label into FAs, cells were sampled at different cultivation time points (0, 6, and 24 h). At each time-point, five millions of cells were collected and their lipids were extracted with a classical acidic extraction method adapted from the Bligh and Dyer protocol [46]. Extracted lipids were hydrolyzed with TFA (to hydrolyze esterified FFAs), derivatized with PTFBr after hydrolysis for TFA, and analysed by CI-GC-MS. The GC-MS profiles showed eight different FAs, with five of them (C14:0, C16:0, C16:1, C18:0, and C18:1) giving signals exploitable for isotopic profiling. For these FAs, the intensity of each isotopologue (M0, M + 1, M + 2, . . . , M + n) in the isotopic cluster of the molecular ion was measured (Table S2), and the distribution of carbon isotopologues (i.e., the fraction of molecules having incorporated 0, 1, 2, etc ^{13}C atoms) was derived from these intensities after correction for ^{13}C natural abundance using the sofware IsoCor [61]. Then, the molecular enrichment (average % of ^{13}C-atoms in the molecule) was calculated (Figure 4).

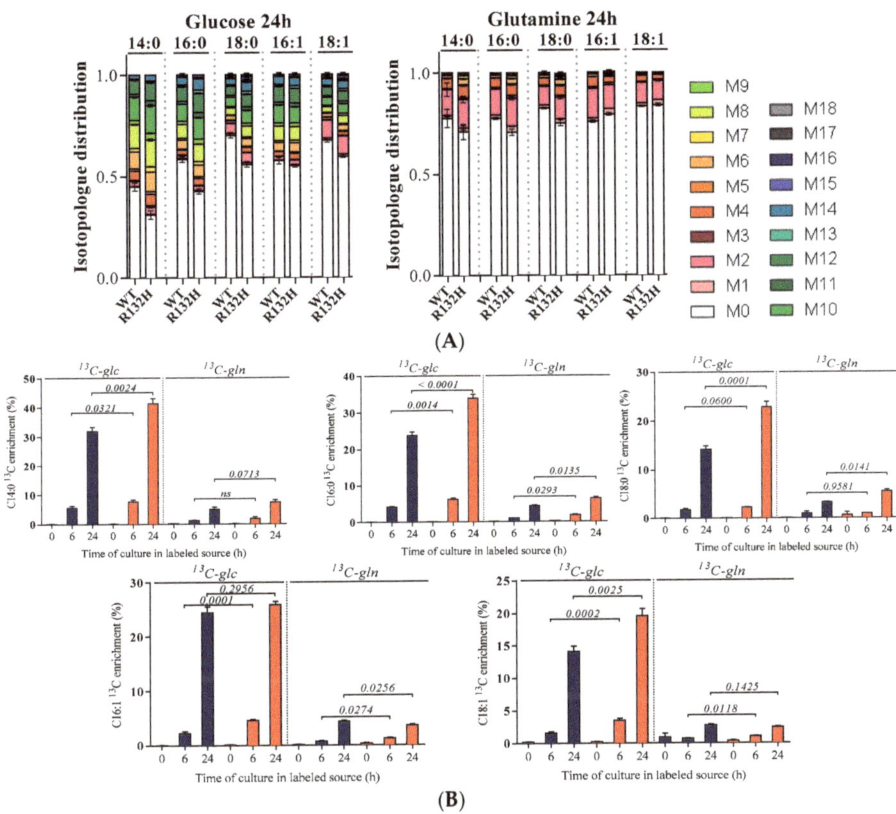

Figure 4. Isotopologues distribution (**A**) in C14:0; C16:0; C18:0; C16:1; and, C18:1 in HL60 IDH1 WT (in blue) and R132H (in red) at 24 h following and (**B**) ^{13}C enrichment at 0, 6, or 24 h cultures on [U-^{13}C]-glucose or [U-^{13}C]-glutamine ($n > 2$).

For all conditions, we observed that the most abundant FA isotopologues contain an even number of ^{13}C atoms, with very low fractions of isotopologues containing an odd number of ^{13}C atoms (<0.03). These isotopic profiles are thus consistent with the known elongation mechanism of FAs by successive incorporation of C2 blocks from the acetyl moiety of AcCoA. For all FAs, molecular ^{13}C-enrichments were higher on ^{13}C-glucose than on ^{13}C-glutamine, indicating that acetyl-CoA is mainly produced from glucose. From the enrichment data at 24h, it can be estimated that glucose contributed to FA biosynthesis 4–7 times more than glutamine. This ratio was similar in both cells, indicating no impact of IDH mutation on the contribution of the two carbon sources to FA biosynthesis. Importantly, the dynamics of ^{13}C-incorporation was significantly faster for all FAs in R132H mutant cells as compared to WT cells. This was surprising since total FAs pools were increased in IDH1 R132H AML cells (Figure 2D), which was expected to result in lower relative label incorporation. The faster labeling dynamics in IDH1 R132H AML cells therefore revealed a significant increase in the rate of FA biosynthesis as compared to the WT cells, which resulted in an increased turnover of intermediates despite higher pools. This enhanced lipid anabolism in mutant cells demonstrates that the upregulation of the protein machinery for FA biosynthesis observed in R132H cells actually translates in terms of metabolic fluxes. While inferring absolute flux values from these data would require mathematical models of FA biosynthesis, these results demonstrated the applicability of the proposed workflow to infer flux information of lipid metabolism in mammalian cells.

2.4. Lipogenesis is Regulated by 2-HG Production in IDH1 Mutant Cells

As all of the experiments described above showed that lipid biosynthesis is enhanced in IDH1 R132H cells, it was important to establish a more direct link between lipogenesis, 2-HG and IDH1 mutation. Therefore, we pharmacologically manipulated the amount of 2-HG using IDH1 mutation inhibitors AGI-5198 (the preclinical version of AG-120) and newly FDA approved AG-120 [62–64] during 24 h and one week for IDH1 R132H culture and we observed that the decrease in IDH1 R132H protein abundance correlated with the reduction in Fatty Acid Synthase (FAS) protein amount (Figure 5).

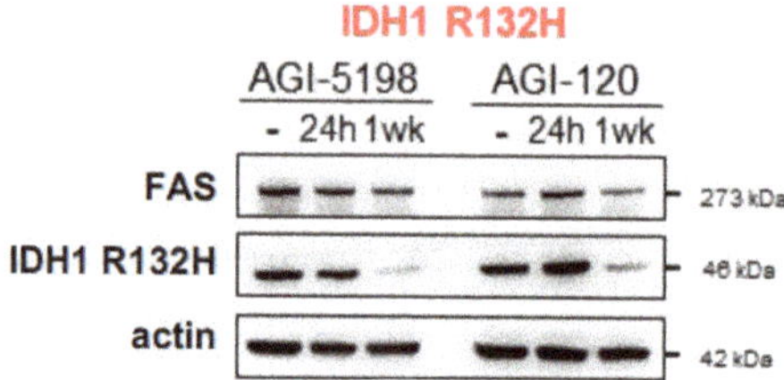

Figure 5. Fatty Acid Synthase (FAS) is linked to 2-HG production in IDH1 mutant cells. Lysates of IDH1 R132H AML cells in basal and following AGI treatments were immunoblotted with the indicated antibodies.

Mechanistically, several transcriptional factors, such as SREBP1/2, LXR, ChREBP, or CEBPα/β regulate de novo lipogenesis and lipid metabolism in various cell types. Interestingly, Ricoult et al. [65] have shown that, in two solid cancers, fibrosarcoma and colorectal carcinoma with IDH1 mutation, genetic invalidation of SREBP1/2 reduced 2-HG production. Moreover, knockdown of SREBP1/2 decreased FASN protein levels, mainly in fibrosarcoma. The differences observed into the two cell lines suggest that other transcription factors could regulate mutant IDH1 and lipogenesis depending on the oncogenic context. Notably, we have previously showed that IDH1 mutation and its oncometabolite (R)-2-HG induced an increase in CEBPα expression in epigenetic-dependent manner, and an activation to prime these cells to myeloid differentiation [66]. It would be of particular interest to study the regulation of lipid synthesis and IDH1 mutation by these different transcriptional factors in AML and to address, whereas IDH1 mutant inhibitors reverse this regulation to determine potential combinatory therapies.

Proteomic experiments on IDH1 mutant AML cells showed an upregulation of protein implicated in cholesterol and sterol biosynthesis and proteins that are involved in fatty acids oxidation. These modifications suggest a reprogramming in the pathways of lipid degradation and biosynthesis. It was then really interesting to characterize the lipidome of these cells versus the wild type one. The lipidomic approach used in this study showed an increase in phosphadidylinositol, ceramide, sphingosine and sphinganine, free cholesterol, and monounsaturated fatty acid (MUFA) species amounts and a decrease in cholesterol esters level in IDH1 mutant cells. In order to understand how IDH1 mutation could be involved in the increase of MUFA, we applied a stable isotope labeling experiments using ^{13}C labeling after growing the cancer cells on uniformly ^{13}C-labeled glutamine or labeled glucose. Dynamics of ^{13}C-incorporation were clearly faster in IDH1 mutant cells. This enhanced lipid anabolism demonstrates that the upregulation of the protein machinery for FA biosynthesis observed in IDH1 mutant AML cells actually translates in terms of metabolic fluxes. Further questions will need to be investigated in order to evaluate the therapeutic possibilities of these findings. Is lipids' dysregulation an Achilles' heel of IDH1 mutant AML cells? Can we exploit it with specific inhibitors such as FASN inhibitors like C75 or orlistat? Are the lipid fluxes reversed by IDH1 mutant inhibitors or could the combination between lipolysis inhibitors and IDH1 mutant inhibitors lead to anti-leukemic effects? Our study highlighted the importance of lipids reprogramming in IDH1 mutant AML cells and paved the way for further studies that could lead to new therapeutic alternatives for this subgroup of AML patients.

3. Materials and Methods

3.1. Chemicals and Reagents

Acetonitrile (ACN) was HPLC-grade and purchased from Acros Organics (Geel, Belgium). Methanol HPLC-grade (MeOH), Dichloromethane (CH_2Cl_2), Ammonium Formate (>99%) (AF), Boron trifluoride-methanol solution 14% (BF3-MeOH), Heptane, Ethyl acetate (EtOAc), potassium hydroxide (KOH), pentafluorobenzyl bromide (PFB-Br), and diisopropylethylamine (DIPEA), iodoacetamide, ammonium bicarbonate, trifluoroacetic acid, and trypsin was supplied by Sigma Aldrich Chemicals Co. (Saint Quentin Fallavier, France), acetic acid (AA) from Honeywell Fluka. Ultrapure water (18.2 MΩ) was obtained from a milliQ apparatus from Millipore (Guyancourt, France).

Internal synthetic standards of phospholipids (PL: PE 12:0/12:0, PC 13:0/13:0, PS 12:0/12:0), Ceramides (Cer: Cer d18:1/15:0), sphingomyelins (SM: SM d18:1/12:0), and sphingosine (So: So d17:1) and sphinganine (Sa: Sa d17:0) were purchased from Avanti Polar Lipids (Alabaster, AL, USA). Synthetic internal standard PI 16:0/17:0 was supplied by J. Clark (Cambridge). Synthetic internal standards for neutral lipid (LN: stigmasterol, cholesteryl heptadecanoate, glyceryl trinonadecanoate) and for free FAs (FFA: heptadecanoate) and total FAs (TFA: glyceryl triheptadecanoate, glyceryl trinonadecanoate) were purchased from Sigma Aldrich (St Quentin Fallavier, France).

3.2. Cell Culture

Clones from the HL60 cell line expressing either IDH1 WT (#2, #4) or IDH1 R132H (#5, #11) were generated by our team [66]. These cell lines have been routinely tested for Mycoplasma contamination in the laboratory. They were maintained in minimum essential medium-α (MEMα, 22561-021, Gibco (Illkirch, France) supplemented with 10% FBS (Invitrogen, Villebon sur Yvette, France) in the presence of 100 U/mL of penicillin and 100 μg/mL of streptomycin (1% P/S), and were incubated at 37 °C with 5% CO_2. The cultured cells were split every two to three days and maintained in an exponential growth phase. For starvations and ^{13}C culture experiments, a specific MEMα media was ordered with the same formula than the one usually used in this paper (MEMα, 22561-021, Gibco), except that glucose, glutamate, glutamine, and pyruvate were removed. The media was first supplemented with 1% P/S and 5% dialyzed FBS (F0392, Thermo, Illkirch, France). Then, the media was supplemented with 5.6 mM ^{12}C or ^{13}C glucose, 1 mM ^{12}C and 2 mM ^{12}C or ^{13}C glutamine.

3.3. Proteomics

3.3.1. Protein Preparation

Three million cells of two independent experiments of two IDH1 WT (#2, #4) and two IDH1 R132H (#5, #11) clones ($n = 4$) were lysed using Tris buffer 50 mM pH 7.4, NaCl 150 mM and Chaps 1% during 15 min on ice. The lysates were then centrifuged 12,000 rpm, 15 min, 4 °C and the supernatants were collected. Proteins were first reduced using Laemmli Buffer (40 mM DTT final) at 95 °C during 5 min, then alkylated with iodoacetamide 90 mM for 30 min at RT in the dark. Next, protein migration was performed on 7.5% SDS PAGE and the gels were stained by Coomassie Blue. A unique band was cut and washed several times in ACN 100%, ammonium bicarbonate 100 mM and dried in vacuo. Gel pieces were rehydrated with 20 ng μL^{-1} trypsin prepared in ammonium bicarbonate 100 mM, and submitted to in gel-digestion overnight at 37 °C. Peptides were extracted and purified from gel and then subjected to mass spectrometry analysis.

3.3.2. Analysis, Identification and Quantification of Proteins

Analysis of proteins was performed using a microLC system Ultimate 3000 (Dionex, Villebon sur Yvette, France) coupled to a Triple-TOF 5600+ (AB Sciex, Les Ullis, France) in the positive ion mode. Samples were first dissolved in 16 µL of buffer (5% ACN, 0.05% trifluoroacetic acid) and spiked with iRT calibration mix (Biognosys, Schlieren, France). The totality of the samples was then injected on a YMC-Pack Pro C18 column (3.0 mm × 150 mm; 3 µm particle size) at a flow rate of 5 µL.min^{-1}. The run length was over 90 min with a gradient from 7% to 45% buffer B (buffer A: 0.1% formic acid, buffer B: 90% ACN, and 0.1% formic acid) in 70 min.

The MS data were acquired with a SWATH mode. The source parameters were set as follows: IS at 5500V, Cur gas at 25, GS1 at 5. The acquisition parameters were as follows: one 50 msec accumulation time MS scan followed by 50 variable SWATH windows each at 40 msec accumulation time for m/z 400–1235.

Identification was determined using an in-house SWATH library created from AML IDH1 WT and mutant cells with MaxQuant software, Les Ulis, France) (FDR 1%). A mass accuracy of 20 ppm on precursor ions was used, and 0.5 Da on the fragments. Cysteine carbamidomethylation, methionine oxidation, proline hydroxylation and serine, threonine and tyrosine phosphorylations were taken into account. Data treatment was done with Spectronaut Software 9.0, Les Ulis, France). First, relative abundance was calculated for each peptide (background noise normalized to 1) and the mean of the three most intense peptides for each protein were measured. Wilcoxon t-test was performed to determine differences between the two groups.

3.3.3. Data Exploration and Mining

List of proteins (FC > 1.5 and FDR < 0.06) obtained throughout this study was uploaded in the Genome Analyzer bioinformatics tool (Genomatix, Les Ulis, France) for further functional analyses (GO term and small molecules) based on the Genomatix literature mining with a special interest in metabolic-linked pathways. The significance of the association between each list and functions or canonical pathways was measured by the Fisher's exact test. As a result, a p-value was obtained, determining the probability that the association between the genes in our dataset and a function or canonical pathway can be explained by chance alone.

3.4. Lipidomic Analysis

3.4.1. Preparation of Total Lipid Extracts

Cell pellet (of one million cells) was extracted adapted from Bligh and Dyer (B&D) [46] in CH_2Cl_2/MeOH with 2% AA/H_2O (2.5: 2.5: 2, $v/v/v$), in the presence of the suitable internal standards. For PL, Cer, and SM relative quantification: 16 ng of Cer (d18:1/15:0), 180 ng of PE (12:0/12:0), 16 ng of PC (13:0/13:0), 16 ng of SM (d18:1/12:0), 30 ng of PI (16:0/17:0), and 156.25 ng of PS (12:0/12:0). For sphingoid bases: 5 ng of So (d17:1) and 5 ng of Sa (d17:0). For neutral lipid: 4 µg of stigmasterol, 4 µg of cholesteryl heptadecanoate, 8 µg of glyceryl trinonadecanoate, and for total FA: 2 µg of glyceryl triheptadecanoate.

3.4.2. Phospholipid and Sphingolipid Relative Quantification

The lipid extract was dried, dissolved in 50 µL of MeOH then stored at −20 °C prior to analysis. Analysis were performed on an Agilent 1290 UPLC system coupled to a G6460 triple quadripole spectrometer (Agilent Technologies, Les Ulis, France)) using a Kinetex HILIC column (Phenomenex, Le Pecq, France), 50 × 4.6 mm, 2.6 µm). The column temperature was controlled at 40 °C. The mobile phases A and B were ACN and 10 mM AF in H_2O at pH 3.2, respectively. The gradient was as follows:

from 10% to 30% B in 10 min; 10–12 min, 100% B; and, then back to 10% B at 13 min for 1 min. The flow rate of mobile phase was 0.3 mL/min and the injection volume was 5 μL. Electrospray ionization (ESI) was employed at 325 °C in positive (for Cer, PE, PC, and SM analysis) and negative ion mode (for PI and PS analysis). The collision gas was nitrogen. Needle voltage was set at +4000 V. SRM transitions were used for relative quantification with a precursor ion scan of 184 m/z, 241 m/z, and 264 m/z to PC/SM, PI, and Cer, respectively; and, a neutral loss scan of 141 m/z and 87 m/z to PE and PS, respectively. In each family, individual molecular species were scanned with suitable SRM scan mode, the area of each peak were measured via Mass Hunter software. The relative quantitative calculations were based on the peak area ratios relative to the internal standards [37].

3.4.3. Sphingoid Bases

The lipid extract was dried and dissolved in 50 μL of MeOH then stored at −20 °C prior to analysis. Analysis were performed on an Agilent 1290 UPLC system coupled to a G6460 triple quadripole spectrometer (Agilent Technologies, Les Ulis, France) an Acquity UPLC BEH-C8 (Waters, Issy les Moulineaux, France), 100 × 2.1 mm, 1.7 μm) maintained at 35 °C. The mobile phases A and B were H$_2$O, FA (99.9:0.1; v/v), and ACN, FA (99.9:0.1, v/v), respectively. The gradient was as follows: 50% B at 0 min, 60% B at 2 min, 60% B at 3 min, 100% B at 4 min, 100% B at 8.5 min, and 50% B at 9 min. The flow rate of mobile phase was 0.3 mL/min and the injection volume was 5 μL. ESI was performed in positive ion mode at 300 °C. The collision gas was nitrogen. Needle voltage was set at + 4000 V. SRM transitions in neutral loss scan were used. For quantitative analysis, calibration samples (500 to 0.976 ng) were prepared with commercial sphingolipid standards. All of the quantitative calculations were based on the peak area ratios relative to the internal standards [adapted from Sikora [52].

3.4.4. Neutral Lipid Relative Quantification

The lipid extract was dried, dissolved in 30 μL of EtOAc, and then stored at −20 °C prior to analysis. 1 μL of the lipid extract was analyzed by gas chromatography on a FOCUS Thermo Electron system using a Zebron-1 fused silica capillary column (Phenomenex, 5 m × 0.32 mm, 0.50 μm film thickness). Oven temperature was programmed from 200 °C to 350 °C at a rate of 5 °C/min and the carrier gas was hydrogen (0.5 bar). The injector and the detector temperatures were at 315 °C and 345 °C, respectively. All of the quantitative calculations were based on the chromatographic peak area relative to the internal standards [39].

3.4.5. Total FA Profiling

The lipid extract was hydrolysed in KOH (0.5 M in MeOH) at 50 °C for 30 min, and transmethylated in 1 mL of BF$_3$-MeOH and 1 mL of heptane at 80 °C for 1h. After the addition of 1 mL H$_2$O to the crude, FAs methyl esters (FAMEs) extract was extracted with 3 mL of heptane, dried, and dissolved in 20 μL of EtOAc. 1 μL of FAMEs extract was analyzed by gas chromatography (GC) on a Clarus 600 Perkin Elmer system using a Famewax RESTEK fused silica capillary column (30 m × 0.32 mm, 0.25 μm film thickness). Oven temperature was programmed from 110 °C to 220 °C at a rate of 2 °C/min and the carrier gas was hydrogen (0.5 bar). The injector and the detector temperatures were at 225 °C and 245 °C, respectively. All of the quantitative calculations were based on the chromatographic peak area relative to the internal standards [38].

3.5. FA Isotopic Labeling Profiling

3.5.1. Sample Preparation

Cell pellet (5 M) was extracted like previously in the presence of the internal standards: glyceryl trinonadecanoate (0.2 μg) for TFA and heptadecanoate (0.2 μg) for FFA. The equivalent of 4 M of cell was collected for the direct derivatization of FFA. Concerning TFA, the equivalent of 1 M of cells was hydrolysed in KOH (0.5 M in MeOH) at 50 °C for 30 min. Free and total FA were derivatized in pentafluorobenzyl esters with 1% PFB-Br and 1% DIPEA in ACN (50 µL), at RT for 20 min. Samples were dried and dissolved in EtOAc (10 µL).

3.5.2. GC-MS Analysis

Labelled total FA analysis were performed on a Thermo Fisher Trace GC system that was connected to a ThermoFisher TSQ8000 (Les Ulis, France)) triple quadrupole detector using a HP-5 MS capillary column (30 m × 0.25 mm, 0.25 µm film thickness). Oven temperature was programmed, as follows: 150 °C for 1 min, 8 °C/min to 350 °C, then the temperature is kept constant for 2 min. The carrier gas was helium (0.8 mL/min). The injector, the transfer line, and the ion source temperature were at 250 °C, 330 °C, and 300 °C, respectively. The TSQ8000 was operated in negative ionization mode (Methane at 1 mL/min) in selected ion monitoring (SIM) mode and 1 µL of sample was injected in splitless mode.

3.5.3. Data Processing

GC-MS analysis produced a mass spectrum for each FA, which contains the abundance of each isotopologue. For each FA, the lightest (unlabeled) isotopologue is denoted M + 0; e.g., PFB-palmitate M + 0 has a mass of 255.3, whereas the isotopologue with 1 atom [^{13}C] PFB-palmitate (M + 1) has a mass of 256.3, etc., (Supplementary Table S1). Isotopic clusters were obtained by integrating gas chromatographic signals for each isotopologue. Isotopologue distributions were obtained from the corresponding isotopic clusters after correction for natural abundance of carbon and non-tracer elements (oxygen and hydrogen) using the software IsoCor, and purity of the tracer was corrected assuming 99% ^{13}C-purity. Finally, the ^{13}C-enrichment, which represents the mean content in tracer atoms (^{13}C) within the molecule, was calculated from the corresponding IDs, as detailed in Millard et al. [61].

3.6. Immunoblotting of Total Proteins

For immunoblot assay, cells were first subjected to lysis in NuPAGE LDS Sample Buffer. Total proteins content from every samples was measure using Pierce BCA Protein Assay Kit according to the manufacturer's recommendations. Samples lysates were then loaded onto NuPAGE 4–12% Bis-Tris Protein Gels (10, 12, or 15 wells). After electrophoresis, proteins were transferred to nitrocellulose membranes. The transblotted membranes were blocked for 1 h and then probed with appropriate primary antibodies (dilution as recommended by manufacturers) overnight at 4 °C. Next, the membranes were washed three times for a total of 30 min and then incubated with secondary antibodies at room temperature for 1 hr. After another three washes, proteins were detected using SuperSignal West Pico PLUS chemiluminescent Substrate, PXi imager (Syngene), and GeneSys software (Syngene, Paris, France), according to the manufacturer's manual. Proteins expression were quantified using GeneTool software (Syngene) and normalized to their corresponding loading control. Antibodies for immunoblotting were purchased from the following sources: FASN (#3180S) from Cell Signaling Technology; Actin (MAB1501) from Millipore; IDH1 R132H (DIA-H09) from Dianova. HRP conjugate anti-rabbit (W4011) and anti-mouse (W4021) secondary antibodies were purchased from Promega (Charbonnières les Bains, France).

3.7. ChIP Assays

To perform chip assays briefly, 10^7 cells were cross-linked in 1% formaldehyde/1% paraformaldehyde for 5 min, followed by addition of 125 mM Glycine to stop the reaction. Cells were then washed in PBS, resuspended in lysis buffer (10 mM Tris pH 8, 140 mM NaCl, 0.5 mM EGTA, 0.1% SDS, 0.5% Triton X-100, 0.05% NaDoc and protease inhibitors) and chromatin was shared by sonication. qChIPs were carried out by incubating cell lysates (Input) with 20 µL of protein G-Dynabeads and 5 ug of antibody. The same amount of rabbit IgGs (Santa Cruz, Boulogne Billancourt, France) was used for control ChIP experiments. After O/N incubation, washing, reverse cross-linking, and treatment with both RNase A and Proteinase K, proteins were removed with phenol/chloroform extraction and DNA was recovered using the NucleoSpin Extract II kit. Input and immunoprecipitated DNA were then analyzed by QPCR using the SYBR Green Master mix on a LightCycler 480 SW 1.5 apparatus (Roche, Boulogne Billancourt, France). Results are represented as the mean value of at least three independent experiments of immunoprecipitated chromatin (calculated as a percentage of the input) with the indicated antibodies after normalization by a control ChIP performed with rabbit IgGs.

3.8. Statistical Analysis

Statistical analyses were conducted using Prism software v6.0 (GraphPad Software, La Jolla, CA, USA). Statistical significance was determined by the two-tailed unpaired Student's *t*-test. A *p*-value < 0.05 was considered to be statistically significant. The statistical parameters (i.e., exact value of *n*, *p*-values) have been noted in the figures and figure legends. Unless otherwise indicated, all data represent the mean $\pm$ standard error of the mean (SEM) from at least three independent experiments.

4. Conclusions

The combination of proteomics, lipidomics, and isotopic profiling experiments allowed us to uncover a profound reprogramming of lipid metabolism in IDH1 mutant AML cells through a simultaneous increase of both FA oxidation and de novo lipogenesis. This reprogramming is—at least partly- dependent on 2-HG production, which controls FAS expression. Integration of all of these omics data in an AML metabolic network could allow fluxes calculations to gain even more insight into the metabolic regulation of IDH WT and R132H cells.

Supplementary Materials: The following are available online at http://www.mdpi.com/1422-0067/19/11/3325/s1.

Author Contributions: Conceptualization, J.-E.S., J.-C.P. and J.B.-M.; methodology, J.B.-M., F.R. and L.S.; software, P.M.; validation, F.R., A.B. and L.S.; formal analysis, M.S. and F.R.; investigation, F.R, A.B., F.V., L.T., S.Z., E.S., L.K.L. and L.S.; resources, L.S., P.M. and E.S.; data curation, P.M.; writing—original draft preparation, L.S. and J.B.-M.; writing—review and editing, L.S, P.M., J.-C.P., J.-E.S. and J.B.-M.; visualization, L.S., M.S. and J.B.-M.; supervision, J.-E.S. and J.B.-M. project administration, J.B.-M. and L.S.; funding acquisition, J.-E.S. and J.-C.P.

Funding: This work was supported by grants from the Région Midi-Pyrénées (CRLE; J.-E.S.), Plan Cancer 2014-BioSys (FLEXAML; J.-E.S.) and the Institut national de la santé et de la recherche médicale (Inserm). MetaToul is part of the national infrastructure MetaboHUB (The French National infrastructure for metabolomics and fluxomics, www.metabohub.fr) and is supported by MetaboHUB-ANR-11-INBS-0010, by the Région Occitanie, the European Regional Development Fund, the SICOVAL, the Infrastructures en Biologie Sante et Agronomie (IBiSa, France), the Centre National de la Recherche Scientifique (CNRS), the Institut National de la Recherche Agronomique (INRA) and Inserm.

Acknowledgments: MetaToul (Metabolomics & Fluxomics Facitilies, Toulouse, France, www.metatoul.fr) is gratefully acknowledged for carrying out metabolome et lipdiomic analysis.

Conflicts of Interest: The authors declare no conflict of interest.

Abbreviations

AA	Acetic acid
ACN	Acetonitrile HPLC-grade
AML	Accute Myeloide Leukemia
AraC	Cytarabine
AU	Arbitrary Unit
B&D	Bligh and Dyer
BF_3-MeOH	Boron trifluoride-methanol solution 14%
Cer	Ceramides
CH_2Cl_2	Dichloromethane HPLC-grade
ESI	Electrospray ionization
FA	Fatty acids

FAMEs	Fatty acid methyl esters
FAO	Fatty acids oxidation
FID	Flame Ionization Detector
GC	Gas chromatography
GO	Gene Ontology
H_2O	Ultrapure water
HPLC	High performance liquid chromatography
IDH	Isocitrate dehydrogenase
2-HG	2-hydroxyglatarate
KOH	Acetic acid potassium hydroxide
LC-MS/MS	Liquid chromatography coupled to detector MS/MS
LOD	Limit of detection
LOQ	Limit of quantification
m/z	Mass-to-charge ratio
MeOH	Methanol HPLC-grade
MS	Mass spectrometry
PC	Phosphatidylcholine
PE	Phosphatidylethanolamine
PI	Phosphatidylinositol
PL	Glycerophospholipids
PS	Phosphatidylserine
PTFB-Br	Pentafluorobenzyl-bromide
PUFAs	Polyunsaturated fatty acids
RLCs	Resistant leukemic cells
S/N	Signal to noise ratio
Sa	Sphinganine
SIM	Selected ion monitoring
SM	Sphingomyelins
So	Sphingosine
SRM	Selected-reaction monitoring
MRM	Multiple-reaction monitoring
MS	Mass SPectrometry
UPLC	Ultra performance liquid chromatography
WT	Wild Type

References

1. Boroughs, L.K.; DeBerardinis, R.J. Metabolic pathways promoting cancer cell survival and growth. *Nat. Cell Biol.* **2015**, *17*, 351–359. [CrossRef] [PubMed]
2. DeBerardinis, R.J.; Mancuso, A.; Daikhin, E.; Nissim, I.; Yudkoff, M.; Wehrli, S.; Thompson, C.B. Beyond aerobic glycolysis: Transformed cells can engage in glutamine metabolism that exceeds the requirement for protein and nucleotide synthesis. *Proc. Natl. Acad. Sci. USA* **2007**, *104*, 19345–19350. [CrossRef] [PubMed]
3. Hanahan, D.; Weinberg, R.A. Hallmarks of cancer: The next generation. *Cell* **2011**, *144*, 646–674. [CrossRef] [PubMed]
4. Vander Heiden, M.G.; DeBerardinis, R.J. Understanding the intersections between metabolism and cancer biology. *Cell* **2017**, *168*, 657–669. [CrossRef] [PubMed]
5. Warburg, O. On respiratory impairment in cancer cells. *Science* **1956**, *124*, 269–270. [PubMed]
6. Warburg, O. On the origin of cancer cells. *Science* **1956**, *123*, 309–314. [CrossRef] [PubMed]
7. Hsu, P.P.; Sabatini, D.M. Cancer cell metabolism: Warburg and beyond. *Cell* **2008**, *134*, 703–707. [CrossRef] [PubMed]
8. Liberti, M.V.; Locasale, J.W. The warburg effect: How does it benefit cancer cells? *Trends Biochem. Sci.* **2016**, *41*, 211–218. [CrossRef] [PubMed]

9.	Ackerman, D.; Simon, M.C. Hypoxia, lipids, and cancer: Surviving the harsh tumor microenvironment. *Trends Cell Biol.* **2014**, *24*, 472–478. [CrossRef] [PubMed]

10.	Baenke, F.; Peck, B.; Miess, H.; Schulze, A. Hooked on fat: The role of lipid synthesis in cancer metabolism and tumour development. *Dis. Model Mech.* **2013**, *6*, 1353–1363. [CrossRef] [PubMed]

11.	Cheng, C.; Geng, F.; Cheng, X.; Guo, D. Lipid metabolism reprogramming and its potential targets in cancer. *Cancer Commun. (Lond.)* **2018**, *38*, 27. [CrossRef] [PubMed]

12.	Rohrig, F.; Schulze, A. The multifaceted roles of fatty acid synthesis in cancer. *Nat. Rev. Cancer* **2016**, *16*, 732–749. [CrossRef] [PubMed]

13.	Holthuis, J.C.; Menon, A.K. Lipid landscapes and pipelines in membrane homeostasis. *Nature* **2014**, *510*, 48–57. [CrossRef] [PubMed]

14.	Van Meer, G.; Voelker, D.R.; Feigenson, G.W. Membrane lipids: Where they are and how they behave. *Nat. Rev. Mol. Cell Biol.* **2008**, *9*, 112–124. [CrossRef] [PubMed]

15.	Yoon, S.; Lee, M.Y.; Park, S.W.; Moon, J.S.; Koh, Y.K.; Ahn, Y.H.; Park, B.W.; Kim, K.S. Up-regulation of acetyl-CoA carboxylase alpha and fatty acid synthase by human epidermal growth factor receptor 2 at the translational level in breast cancer cells. *J. Biol. Chem.* **2007**, *282*, 26122–26131. [CrossRef] [PubMed]

16.	Efeyan, A.; Comb, W.C.; Sabatini, D.M. Nutrient-sensing mechanisms and pathways. *Nature* **2015**, *517*, 302–310. [CrossRef] [PubMed]

17.	Menendez, J.A.; Lupu, R. Fatty acid synthase and the lipogenic phenotype in cancer pathogenesis. *Nat. Rev. Cancer* **2007**, *7*, 763–777. [CrossRef] [PubMed]

18.	Dang, L.; White, D.W.; Gross, S.; Bennett, B.D.; Bittinger, M.A.; Driggers, E.M.; Fantin, V.R.; Jang, H.G.; Jin, S.; Keenan, M.C.; et al. Cancer-associated IDH1 mutations produce 2-hydroxyglutarate. *Nature* **2009**, *462*, 739–744. [CrossRef] [PubMed]

19.	Mardis, E.R.; Wilson, R.K. Cancer genome sequencing: A review. *Hum. Mol. Genet.* **2009**, *18*, R163–R168. [CrossRef] [PubMed]

20.	Parsons, D.W.; Jones, S.; Zhang, X.; Lin, J.C.; Leary, R.J.; Angenendt, P.; Mankoo, P.; Carter, H.; Siu, I.M.; Gallia, G.L.; et al. An integrated genomic analysis of human glioblastoma multiforme. *Science* **2008**, *321*, 1807–1812. [CrossRef] [PubMed]

21.	Yan, H.; Parsons, D.W.; Jin, G.; McLendon, R.; Rasheed, B.A.; Yuan, W.; Kos, I.; Batinic-Haberle, I.; Jones, S.; Riggins, G.J.; et al. IDH1 and IDH2 mutations in gliomas. *N. Engl. J. Med.* **2009**, *360*, 765–773. [CrossRef] [PubMed]

22.	Marcucci, G.; Maharry, K.; Wu, Y.Z.; Radmacher, M.D.; Mrozek, K.; Margeson, D.; Holland, K.B.; Whitman, S.P.; Becker, H.; Schwind, S.; et al. IDH1 and IDH2 gene mutations identify novel molecular subsets within de novo cytogenetically normal acute myeloid leukemia: A cancer and leukemia group b study. *J. Clin. Oncol.* **2010**, *28*, 2348–2355. [CrossRef] [PubMed]

23.	Paschka, P.; Schlenk, R.F.; Gaidzik, V.I.; Habdank, M.; Kronke, J.; Bullinger, L.; Spath, D.; Kayser, S.; Zucknick, M.; Gotze, K.; et al. IDH1 and IDH2 mutations are frequent genetic alterations in acute myeloid leukemia and confer adverse prognosis in cytogenetically normal acute myeloid leukemia with NPM1 mutation without FLT3 internal tandem duplication. *J. Clin. Oncol.* **2010**, *28*, 3636–3643. [CrossRef] [PubMed]

24.	Chou, W.C.; Chou, S.C.; Liu, C.Y.; Chen, C.Y.; Hou, H.A.; Kuo, Y.Y.; Lee, M.C.; Ko, B.S.; Tang, J.L.; Yao, M.; et al. TET2 mutation is an unfavorable prognostic factor in acute myeloid leukemia patients with intermediate-risk cytogenetics. *Blood* **2011**, *118*, 3803–3810. [CrossRef] [PubMed]

25.	Boissel, N.; Nibourel, O.; Renneville, A.; Gardin, C.; Reman, O.; Contentin, N.; Bordessoule, D.; Pautas, C.; de Revel, T.; Quesnel, B.; et al. Prognostic impact of isocitrate dehydrogenase enzyme isoforms 1 and 2 mutations in acute myeloid leukemia: A study by the acute leukemia french association group. *J. Clin. Oncol.* **2010**, *28*, 3717–3723. [CrossRef] [PubMed]

26.	Gross, S.; Cairns, R.A.; Minden, M.D.; Driggers, E.M.; Bittinger, M.A.; Jang, H.G.; Sasaki, M.; Jin, S.; Schenkein, D.P.; Su, S.M.; et al. Cancer-associated metabolite 2-hydroxyglutarate accumulates in acute myelogenous leukemia with isocitrate dehydrogenase 1 and 2 mutations. *J. Exp. Med.* **2010**, *207*, 339–344. [CrossRef] [PubMed]

27. Ward, P.S.; Patel, J.; Wise, D.R.; Abdel-Wahab, O.; Bennett, B.D.; Coller, H.A.; Cross, J.R.; Fantin, V.R.; Hedvat, C.V.; Perl, A.E.; et al. The common feature of leukemia-associated IDH1 and IDH2 mutations is a neomorphic enzyme activity converting alpha-ketoglutarate to 2-hydroxyglutarate. *Cancer Cell* **2010**, *17*, 225–234. [CrossRef] [PubMed]

28. Xu, W.; Yang, H.; Liu, Y.; Yang, Y.; Wang, P.; Kim, S.H.; Ito, S.; Yang, C.; Wang, P.; Xiao, M.T.; et al. Oncometabolite 2-hydroxyglutarate is a competitive inhibitor of alpha-ketoglutarate-dependent dioxygenases. *Cancer Cell* **2011**, *19*, 17–30. [CrossRef] [PubMed]

29. Elkashef, S.M.; Lin, A.P.; Myers, J.; Sill, H.; Jiang, D.; Dahia, P.L.M.; Aguiar, R.C.T. IDH mutation, competitive inhibition of FTO, and RNA methylation. *Cancer Cell* **2017**, *31*, 619–620. [CrossRef] [PubMed]

30. Figueroa, M.E.; Abdel-Wahab, O.; Lu, C.; Ward, P.S.; Patel, J.; Shih, A.; Li, Y.; Bhagwat, N.; Vasanthakumar, A.; Fernandez, H.F.; et al. Leukemic IDH1 and IDH2 mutations result in a hypermethylation phenotype, disrupt TET2 function, and impair hematopoietic differentiation. *Cancer Cell* **2010**, *18*, 553–567. [CrossRef] [PubMed]

31. Jiang, S.; Zou, T.; Eberhart, C.G.; Villalobos, M.A.V.; Heo, H.Y.; Zhang, Y.; Wang, Y.; Wang, X.; Yu, H.; Du, Y.; et al. Predicting IDH mutation status in grade II gliomas using amide proton transfer-weighted (APTW) MRI. *Magn. Reson. Med.* **2017**, *78*, 1100–1109. [CrossRef] [PubMed]

32. Chowdhury, R.; Yeoh, K.K.; Tian, Y.M.; Hillringhaus, L.; Bagg, E.A.; Rose, N.R.; Leung, I.K.; Li, X.S.; Woon, E.C.; Yang, M.; et al. The oncometabolite 2-hydroxyglutarate inhibits histone lysine demethylases. *EMBO Rep.* **2011**, *12*, 463–469. [CrossRef] [PubMed]

33. Metallo, C.M. Expanding the reach of cancer metabolomics. *Cancer Prev. Res. (Phila.)* **2012**, *5*, 1337–1340. [CrossRef] [PubMed]

34. Mullen, A.R.; DeBerardinis, R.J. Genetically-defined metabolic reprogramming in cancer. *Trends Endocrinol. Metab.* **2012**, *23*, 552–559. [CrossRef] [PubMed]

35. Wise, D.R.; Ward, P.S.; Shay, J.E.; Cross, J.R.; Gruber, J.J.; Sachdeva, U.M.; Platt, J.M.; DeMatteo, R.G.; Simon, M.C.; Thompson, C.B. Hypoxia promotes isocitrate dehydrogenase-dependent carboxylation of alpha-ketoglutarate to citrate to support cell growth and viability. *Proc. Natl. Acad. Sci. USA* **2011**, *108*, 19611–19616. [CrossRef] [PubMed]

36. Burla, B.; Arita, M.; Arita, M.; Bendt, A.K.; Cazenave-Gassiot, A.; Dennis, E.A.; Ekroos, K.; Han, X.; Ikeda, K.; Liebisch, G.; et al. MS-based lipidomics of human blood plasma: A community-initiated position paper to develop accepted guidelines. *J. Lipid Res.* **2018**, *59*, 2001–2017. [CrossRef] [PubMed]

37. Chiappini, F.; Coilly, A.; Kadar, H.; Gual, P.; Tran, A.; Desterke, C.; Samuel, D.; Duclos-Vallee, J.C.; Touboul, D.; Bertrand-Michel, J.; et al. Metabolism dysregulation induces a specific lipid signature of nonalcoholic steatohepatitis in patients. *Sci. Rep.* **2017**, *7*, 46658. [CrossRef] [PubMed]

38. Lillington, J.M.; Trafford, D.J.; Makin, H.L. A rapid and simple method for the esterification of fatty acids and steroid carboxylic acids prior to gas-liquid chromatography. *Clin. Chim. Acta* **1981**, *111*, 91–98. [CrossRef]

39. Barrans, A.; Jaspard, B.; Barbaras, R.; Chap, H.; Perret, B.; Collet, X. Pre-beta HDL: Structure and metabolism. *Biochim. Biophys. Acta* **1996**, *1300*, 73–85. [CrossRef]

40. Daniels, V.W.; Smans, K.; Royaux, I.; Chypre, M.; Swinnen, J.V.; Zaidi, N. Cancer cells differentially activate and thrive on de novo lipid synthesis pathways in a low-lipid environment. *PLoS ONE* **2014**, *9*, e106913. [CrossRef] [PubMed]

41. Kamphorst, J.J.; Cross, J.R.; Fan, J.; de Stanchina, E.; Mathew, R.; White, E.P.; Thompson, C.B.; Rabinowitz, J.D. Hypoxic and ras-transformed cells support growth by scavenging unsaturated fatty acids from lysophospholipids. *Proc. Natl. Acad. Sci. USA* **2013**, *110*, 8882–8887. [CrossRef] [PubMed]

42. German, N.J.; Yoon, H.; Yusuf, R.Z.; Murphy, J.P.; Finley, L.W.; Laurent, G.; Haas, W.; Satterstrom, F.K.; Guarnerio, J.; Zaganjor, E.; et al. PHD3 loss in cancer enables metabolic reliance on fatty acid oxidation via deactivation of ACC2. *Mol. Cell* **2016**, *63*, 1006–1020. [CrossRef] [PubMed]

43. Zhao, S.; Lin, Y.; Xu, W.; Jiang, W.; Zha, Z.; Wang, P.; Yu, W.; Li, Z.; Gong, L.; Peng, Y.; et al. Glioma-derived mutations in IDH1 dominantly inhibit IDH1 catalytic activity and induce HIF-1alpha. *Science* **2009**, *324*, 261–265. [CrossRef] [PubMed]

44. Farge, T.; Saland, E.; de Toni, F.; Aroua, N.; Hosseini, M.; Perry, R.; Bosc, C.; Sugita, M.; Stuani, L.; Fraisse, M.; et al. Chemotherapy-resistant human acute myeloid leukemia cells are not enriched for leukemic stem cells but require oxidative metabolism. *Cancer Discov.* **2017**, *7*, 716–735. [CrossRef] [PubMed]

45. Samudio, I.J.; Duvvuri, S.; Clise-Dwyer, K.; Watt, J.C.; Mak, D.; Kantarjian, H.; Yang, D.; Ruvolo, V.; Borthakur, G. Activation of p53 signaling by MI-63 induces apoptosis in acute myeloid leukemia cells. *Leuk. Lymphoma* **2010**, *51*, 911–919. [CrossRef] [PubMed]

46. Bligh, E.G.; Dyer, W.J. A rapid method of total lipid extraction and purification. *Can. J. Biochem. Physiol.* **1959**, *37*, 911–917. [CrossRef] [PubMed]

47. Saulnier-Blache, J.S.; Feigerlova, E.; Halimi, J.M.; Gourdy, P.; Roussel, R.; Guerci, B.; Dupuy, A.; Bertrand-Michel, J.; Bascands, J.L.; Hadjadj, S.; et al. Urinary lysophopholipids are increased in diabetic patients with nephropathy. *J. Diabetes Complicat.* **2017**, *31*, 1103–1108. [CrossRef] [PubMed]

48. Izquierdo-Garcia, J.L.; Viswanath, P.; Eriksson, P.; Chaumeil, M.M.; Pieper, R.O.; Phillips, J.J.; Ronen, S.M. Metabolic reprogramming in mutant IDH1 glioma cells. *PLoS ONE* **2015**, *10*, e0118781. [CrossRef] [PubMed]

49. Reitman, Z.J.; Jin, G.; Karoly, E.D.; Spasojevic, I.; Yang, J.; Kinzler, K.W.; He, Y.; Bigner, D.D.; Vogelstein, B.; Yan, H. Profiling the effects of isocitrate dehydrogenase 1 and 2 mutations on the cellular metabolome. *Proc. Natl. Acad. Sci. USA* **2011**, *108*, 3270–3275. [CrossRef] [PubMed]

50. Viswanath, P.; Radoul, M.; Izquierdo-Garcia, J.L.; Luchman, H.A.; Gregory Cairncross, J.; Pieper, R.O.; Phillips, J.J.; Ronen, S.M. Mutant IDH1 gliomas downregulate phosphocholine and phosphoethanolamine synthesis in a 2-hydroxyglutarate-dependent manner. *Cancer Metab.* **2018**, *6*, 3. [CrossRef] [PubMed]

51. Beloribi-Djefaflia, S.; Vasseur, S.; Guillaumond, F. Lipid metabolic reprogramming in cancer cells. *Oncogenesis* **2016**, *5*, e189. [CrossRef] [PubMed]

52. Sikora, J.; Dworski, S.; Jones, E.E.; Kamani, M.A.; Micsenyi, M.C.; Sawada, T.; Le Faouder, P.; Bertrand-Michel, J.; Dupuy, A.; Dunn, C.K.; et al. Acid ceramidase deficiency in mice results in a broad range of central nervous system abnormalities. *Am. J. Pathol.* **2017**, *187*, 864–883. [CrossRef] [PubMed]

53. Lingwood, D.; Simons, K. Lipid rafts as a membrane-organizing principle. *Science* **2010**, *327*, 46–50. [CrossRef] [PubMed]

54. Mollinedo, F.; Gajate, C. Lipid rafts as major platforms for signaling regulation in cancer. *Adv. Biol. Regul.* **2015**, *57*, 130–146. [CrossRef] [PubMed]

55. Don, A.S.; Lim, X.Y.; Couttas, T.A. Re-configuration of sphingolipid metabolism by oncogenic transformation. *Biomolecules* **2014**, *4*, 315–353. [CrossRef] [PubMed]

56. Fielding, B. Tracing the fate of dietary fatty acids: Metabolic studies of postprandial lipaemia in human subjects. *Proc. Nutr. Soc.* **2011**, *70*, 342–350. [CrossRef] [PubMed]

57. Persson, X.M.; Blachnio-Zabielska, A.U.; Jensen, M.D. Rapid measurement of plasma free fatty acid concentration and isotopic enrichment using lc/ms. *J. Lipid Res.* **2010**, *51*, 2761–2765. [CrossRef] [PubMed]

58. Buescher, J.M.; Antoniewicz, M.R.; Boros, L.G.; Burgess, S.C.; Brunengraber, H.; Clish, C.B.; DeBerardinis, R.J.; Feron, O.; Frezza, C.; Ghesquiere, B.; et al. A roadmap for interpreting (13)c metabolite labeling patterns from cells. *Curr. Opin. Biotechnol.* **2015**, *34*, 189–201. [CrossRef] [PubMed]

59. Argus, J.P.; Yu, A.K.; Wang, E.S.; Williams, K.J.; Bensinger, S.J. An optimized method for measuring fatty acids and cholesterol in stable isotope-labeled cells. *J. Lipid Res.* **2017**, *58*, 460–468. [CrossRef] [PubMed]

60. Oosterveer, M.H.; van Dijk, T.H.; Tietge, U.J.; Boer, T.; Havinga, R.; Stellaard, F.; Groen, A.K.; Kuipers, F.; Reijngoud, D.J. High fat feeding induces hepatic fatty acid elongation in mice. *PLoS ONE* **2009**, *4*, e6066. [CrossRef] [PubMed]

61. Millard, P.; Letisse, F.; Sokol, S.; Portais, J.C. Isocor: Correcting ms data in isotope labeling experiments. *Bioinformatics* **2012**, *28*, 1294–1296. [CrossRef] [PubMed]

62. DiNardo, C.D.; Stein, E.M.; de Botton, S.; Roboz, G.J.; Altman, J.K.; Mims, A.S.; Swords, R.; Collins, R.H.; Mannis, G.N.; Pollyea, D.A.; et al. Durable remissions with ivosidenib in IDH1-mutated relapsed or refractory AML. *N. Engl. J. Med.* **2018**, *378*, 2386–2398. [CrossRef] [PubMed]

63. Popovici-Muller, J.; Lemieux, R.M.; Artin, E.; Saunders, J.O.; Salituro, F.G.; Travins, J.; Cianchetta, G.; Cai, Z.; Zhou, D.; Cui, D.; et al. Discovery of ag-120 (ivosidenib): A first-in-class mutant IDH1 inhibitor for the treatment of IDH1 mutant cancers. *ACS Med. Chem. Lett.* **2018**, *9*, 300–305. [CrossRef] [PubMed]

64. Rohle, D.; Popovici-Muller, J.; Palaskas, N.; Turcan, S.; Grommes, C.; Campos, C.; Tsoi, J.; Clark, O.; Oldrini, B.; Komisopoulou, E.; et al. An inhibitor of mutant IDH1 delays growth and promotes differentiation of glioma cells. *Science* **2013**, *340*, 626–630. [CrossRef] [PubMed]

65. Ricoult, S.J.; Dibble, C.C.; Asara, J.M.; Manning, B.D. Sterol regulatory element binding protein regulates the expression and metabolic functions of wild-type and oncogenic IDH1. *Mol. Cell Biol.* **2016**, *36*, 2384–2395. [CrossRef] [PubMed]
66. Boutzen, H.; Saland, E.; Larrue, C.; de Toni, F.; Gales, L.; Castelli, F.A.; Cathebas, M.; Zaghdoudi, S.; Stuani, L.; Kaoma, T.; et al. Isocitrate dehydrogenase 1 mutations prime the all-trans retinoic acid myeloid differentiation pathway in acute myeloid leukemia. *J. Exp. Med.* **2016**, *213*, 483–497. [CrossRef] [PubMed]

International Journal of
Molecular Sciences

Review

A Role for Lipid Mediators in Acute Myeloid Leukemia

Andreas Loew [1,†], Thomas Köhnke [2,†], Emma Rehbeil [1], Anne Pietzner [1] and Karsten-H. Weylandt [1,3,*]

[1]   Department of Medicine B, Ruppin General Hospital, Brandenburg Medical School, 16816 Neuruppin, Germany; andreas.loew@mhb-fontane.de (A.L.); emma.rehbeil@mhb-fontane.de (E.R.); anne.pietzner@mhb-fontane.de (A.P.)

[2]   Department of Internal Medicine III, University of Munich, 81377 Munich, Germany; thomas.koehnke@med.uni-muenchen.de

[3]   Medical Department, Campus Virchow Klinikum, Charité-Universitätsmedizin Berlin, 13353 Berlin, Germany

*   Correspondence: karsten.weylandt@mhb-fontane.de or karsten.weylandt@charite.de; Tel.: +49-3391-39-3200

†   These authors contributed equally to this work.

Received: 12 March 2019; Accepted: 6 May 2019; Published: 16 May 2019

Abstract: In spite of therapeutic improvements in the treatment of different hematologic malignancies, the prognosis of acute myeloid leukemia (AML) treated solely with conventional induction and consolidation chemotherapy remains poor, especially in association with high risk chromosomal or molecular aberrations. Recent discoveries describe the complex interaction of immune effector cells, as well as the role of the bone marrow microenvironment in the development, maintenance and progression of AML. Lipids, and in particular omega-3 as well as omega-6 polyunsaturated fatty acids (PUFAs) have been shown to play a vital role as signaling molecules of immune processes in numerous benign and malignant conditions. While the majority of research in cancer has been focused on the role of lipid mediators in solid tumors, some data are showing their involvement also in hematologic malignancies. There is a considerable amount of evidence that AML cells are targetable by innate and adaptive immune mechanisms, paving the way for immune therapy approaches in AML. In this article we review the current data showing the lipid mediator and lipidome patterns in AML and their potential links to immune mechanisms.

Keywords: AML; immune therapy; PGE2; omega-3; omega-6; lipidomics

---

1. Acute Myeloid Leukemia

Acute myeloid leukemia (AML) is a complex and biological heterogenous disease. Different mutations lead to alterations in the differentiation of hematopoietic stem cells and are responsible for the accumulation of immature leukemic blast cells in the bone marrow and peripheral blood. AML accounts for approximately 20% of all deaths due to hematologic malignancies, while only comprising 12% of all new cases [1].

The relapse rate after conventional induction chemotherapy is high, particularly in association with adverse chromosomal or molecular aberrations. Therapeutic advances in AML in recent years are mainly attributed to progress in hematopoietic stem cell transplantation techniques and advances in supportive care.

Increasing evidence suggests that AML as well as other malignancies are sustained by a minor subpopulation with self-renewal potential, referred to as "leukemic stem cells" (LSC) [2], which have been shown to be more quiescent than the bulk of leukemic cells [3]. Current treatments utilizing cytotoxic agents aimed at proliferation might therefore not target LSCs adequately, which in turn can survive treatment and ultimately lead to relapse. Gene expression analyses have shown that LSCs have a similar gene expression profile compared to hematopoietic stem cells (HSC) [2] and that a stem cell rich expression signature in AML blasts correlates with worse prognosis [4]. The knowledge concerning biology and pathophysiology of LSCs has drastically improved over the past decades [5]. It has become especially clear that the microenvironment surrounding tumor cells plays a vital role in carcinogenesis, and growing evidence suggests that it also plays a central role in how tumor cells interact with the immune system [6].

The concept of the elimination of minimal residual disease by immunotherapy has shown to be successful—as a proof of principle—in allogeneic hematopoietic stem cell transplantation for postremission therapy, leading to long lasting remissions in a significant proportion of AML cases.

For patients ineligible for transplantation, alternative therapeutic strategies are mandatory. Immunotherapeutic approaches for clearing of evading AML cells from the stem cell niche involve different monoclonal antibodies including check point inhibitors, adoptive transfer of NK and T cells, T-cell engineering, systemic cytokine administration, and vaccinations with different approaches such as peptides, modified leukemic cells, and dendritic cells [7–10].

In this context, there has been increased interest in research aimed at lipid mediators such as prostaglandins, as well as other lipid species and their associated regulatory networks, as these can be critical components affecting tumor cell biology, tumor microenvironment, and thus immune mechanisms affecting AML biology as well as response to treatment approaches.

In the following sections we aim to highlight aspects in the field of lipid and lipid mediator biology. In this context immune mechanisms affected will be addressed in order to explore potential links to immunotherapy in the context of hematologic malignancies in general and in AML in particular.

2. Lipids and Fatty Acids in Hematologic Malignancies

As reviewed before, lipid species and the lipidome are highly abundant and essential components of human cells and tissues [11]. Many of these lipid species (e.g., eicosanoids, sphingolipids, glycerolipids) were shown to be changed in the context of tumor disease and might serve as markers as well as targets for new treatment approaches in malignant disorders. Particularly in the context of the tumor surrounding microenvironment lipid species could be important—and modifiable—targets in oncology [12].

Beside an increased de novo synthesis of fatty acids that is required for membrane synthesis and therefore for cell growth and proliferation, AML cells might have an increased lipid catabolism. Fatty acid oxidation (FAO) has been recognized as a relevant component of the metabolic switch in cancer cells where FAO is used for ATP production in conditions of metabolic stress [13]. Indeed, recent in vitro studies have shown that distinct genetic changes in AML are associated with enhanced dynamics and metabolism of lipid species in AML cells [14].

Data from the late 1970s found altered lipid compositions of AML cells with a decreased total cholesterol and cholesterol-to-phospholipid ratio, and an increased percentage of unsaturated fatty acids when compared to normal mature neutrophils, but these patterns might be shared by normal immature myeloid cells [15].

Recent studies also demonstrated wide-ranging changes in the plasma [16] as well as bone marrow [17] lipidome in patients with AML. Total plasma fatty acids were found to be depressed in plasma from AML patients, with the attenuation of plasma phosphocholines, triglycerides, and cholesterol esters [16]. However, free fatty acids such as arachidonic acid (AA) 20:4 n-6 and the corresponding precursors gamma-linolenic acid 18:3 n-6 and 8,11,14-eicosatrienoic acid 20:3 n-6 were increased, while many prostaglandins such as PGE2 and 15-keto-PGF2α were reduced in these plasma analyses. Interestingly, AA as well as gamma-linolenic acid 18:3 n-6 and 8,11,14-eicosatrienoic acid 20:3 n-6 tended to be increased slightly more in patients with higher blast counts [16]. While only observed in plasma, and in a very heterogeneous patient population, these observations might indicate a role for AA in the malignant phenotype of AML.

3. Omega-6 and Omega-3-Polyunsaturated Fatty Acids and Their Derived Lipid Mediators in Inflammation

Of particular importance with regard to immune processes are lipid mediators derived from long-chain polyunsaturated fatty acids (PUFA) and in particular arachidonic acid (AA). The PUFA are grouped according to the position of the first double bound, counting from the first methyl-, or "omega"-group. Two groups of PUFAs are important for human physiology: omega-6-PUFA and omega-3-PUFA. They are termed *essential*, since mammals cannot synthesize them and they have to be ingested with the diet in sufficient amounts.

Concerning omega-6-PUFAs, arachidonic acid (AA) and linoleic acid (LA) are the main components. Most mammals can synthesize AA from LA through enzymatic conversion by desaturases. The most important omega-3-PUFAs are α-linolenic acid (ALA) 18:3n-3, eicosapentaenoic acid (EPA) 20:5 n-3 and docosahexaenoic acid (DHA) 22:6 n-3.

From these PUFA numerous potent lipid mediators are formed (Figure 1). Especially those lipid mediators derived from the omega-6-PUFA arachidonic acid (AA) have been studied intensively. AA is cleaved from its site within phospholipids in the cellular membrane by phospholipase C and A2. Then, AA is further metabolized by two main groups of enzymes: the cyclooxygenases (COX-1 and COX-2) and the lipoxygenases (LOX-5, LOX-12, LOX-15) [18].

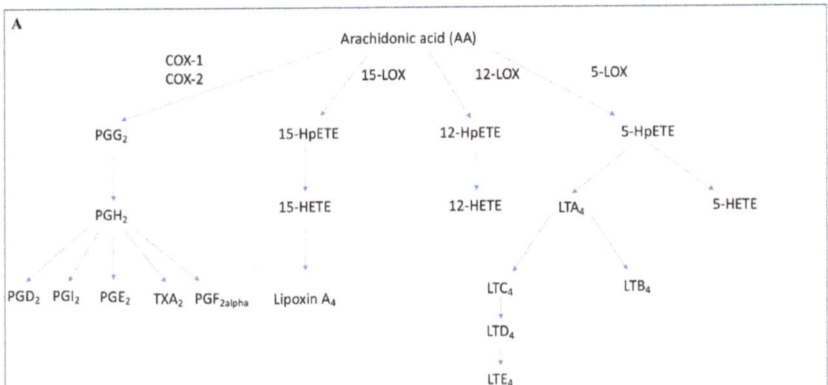

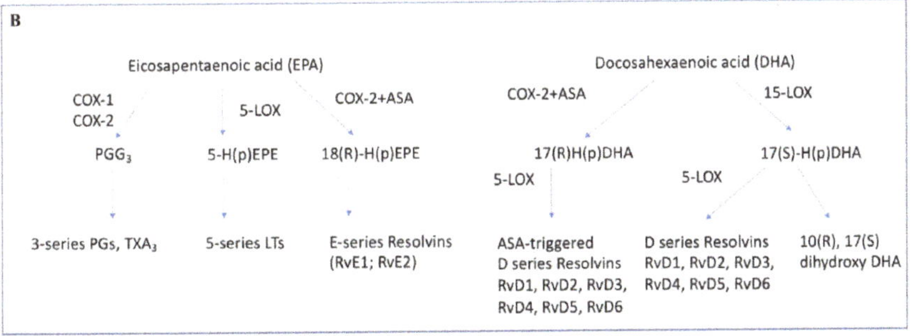

Figure 1. Lipid mediators formed from arachidonic acid (**A**) or eicosapentaenoic acid and docosahexaenoic acid (**B**). COX-1/2: Cyclooxygenase-1/2, 5-LOX: 5-Lipoxygenase, 15-LOX: 15-Lipoxygenase, 12-LOX: 12-Lipoxygenase, ASA: acetylsalicylic acid, PG: prostaglandin, LT: Leukotriene, Rv: Resolvin, HpETE: hydroperoxyeicosatetraenoic acid, HETE: hydroxyeicosatetraenoic acid, H(p)EPE: hydro(pero)xyeicosapentaenoic acid, H(p)DHA: hydro(pero)xydocosahexaenoic acid.

The most important enzymes in this pathway are the cyclooxygenases (COX) or prostaglandin endoperoxide H synthases. Two isoforms were identified in the late 1980s and early 1990s. The two cyclooxygenases, COX-1 and COX-2, though derived from different genes of different size, are highly homologous in sequence and three-dimensional structure [19]. They are capable of converting arachidonic acid into prostaglandin (H2), which then is transformed further into prostaglandin E2 (PGE2), amongst others.

PGE2 is the most abundant eicosanoid and has been shown to be a crucial mediator of inflammation, fever, cancer and numerous other physiological systems [20–22]. Elevated PGE2 concentrations can be detected in inflamed tissue and the injection of PGE2 causes inflammation [23]. Another lipid mediator derived from AA is thromboxane A2 (TXA2), which is important for platelet function. The effects of PGE2 are mediated through four membrane-bound G-protein coupled receptors—EP1, EP2, EP3, and EP4 [24]. EP1 induces intracellular calcium level variation [25]. EP2 as well as EP4 stimulate cAMP production, which leads to gene regulation. In contrast, EP3 is coupled to Gi and inhibits cAMP production [24]. These receptors differ slightly in their binding characteristics for PGE2 (and to some extent other prostaglandins), as well as their signaling mechanisms, further contributing to a differential biological response due to PGE2 [26,27]. Local amounts of PGE2 are controlled by 15-hydroxyprostaglandin dehydrogenase (15-PGDH) mediated degradation. Overexpression of 15-PGDH can protect from carcinogenesis [28,29] while downregulation of this enzyme can contribute to tumor progression [30,31].

Further groups of lipid mediators derived from AA are the leukotrienes and lipoxins. These are formed by the lipoxygenases. Leukotrienes have pro-inflammatory properties and contribute mainly to allergic reactions, but also play a role in infections and carcinogenesis [18]. Lipoxins are mediators with anti-inflammatory properties [32,33]. Interestingly, it could be shown that under the influence of acetylsalicylic acid (ASA) also the COX enzymes can synthesize potent lipoxins, the so called Aspirin-triggered-lipoxins (ATLs) [32,34].

Omega-3-PUFAs were first postulated to act as anti-inflammatory compounds through the competitive inhibition of PGE2 formation and to a certain degree, EPA and DHA do inhibit the formation of AA derived lipid mediators [35]. Studies have shown increased formation of omega-3-PUFA derived prostaglandins (i.e., PGE3) and decreased formation of AA derived mediators (i.e., PGE2) caused by increased intake of dietary omega-3 PUFA [35,36]. Mechanistically, eicosanoids derived from omega-3-PUFA seem to have a lower biological effect than their omega-6-PUFA derived counterparts [37,38]. However, there is also evidence for some distinct functionality, since PGE3 could be shown to have an inhibitory effect on tumor cell growth in vitro [39]. The same is true for the leukotrienes derived from omega-3-PUFA. For example, for leukotriene B5, which is formed through enzymatic conversion by 5-lipoxygenase. Asthmatic subjects receiving omega-3 supplements showed decreased formation of leukotriene B4 (omega-6-PUFA) and increased formation of leukotriene B5 while displaying improved pulmonary function compared to the control group [40].

In addition to these prostaglandin and leukotriene mediators further omega-3-PUFA derived lipid mediators also play important roles in the course of inflammation. Through enzymatic conversion by lipoxygenases, COX enzymes or cytochrome P450 enzymes, the omega-3-PUFAs DHA and EPA can be converted into potent anti-inflammatory oxylipin mediators [41–43]. Particularly the Specialized Proresolving Mediators (SPM) derived from omega-3 PUFA, the resolvins, maresins and protectins were characterized in detail since their initial discovery in 2000 [44] and were found to widely regulate immune cell function [33].

Beside the receptors on the cell membrane like EP1-3, transcription factors from cytoplasm and nucleus play an important role in the signaling of inflammatory process and their key mediators. The Peroxisome proliferator-activated receptors (PPARs) are ligand-activated transcription factors belonging to the nuclear receptor family. Their three subtypes, PPARα, PPARβ/δ, and PPARγ have different expression levels in various tissues, biological activity and ligand affinity [45,46]. PPARs are important players in the lipid signaling network between the cell surface and the nucleus. Fatty acids and eicosanoids which also signal through membrane receptors are natural PPAR ligands. For example, PPAR-α is activated by different compounds including arachidonic acid metabolites (LTB4), fibrates and eicosanoids or prostaglandin J2 (15d- PGJ2) is a ligand of PPAR-γ. The activation of PPAR was shown to inhibit the transcription of inflammatory response genes (such as IL-2, IL-6, IL-8, TNF-α) by negatively interfering with the NF-κB, STAT and AP-1 signaling pathways [47]. It is suggested that PPARγ as a transcription factor and its ligands contribute in regulation of a variety of factors related to tumorigenicity [48]. PPARγ could be a target for AML treatment, several ligands with potential anti-leukemic effects have been identified [49].

The nuclear factor NF-κB is part of this lipid signaling network. NF-κB influences, as a rapid-acting transcription factor, many processes including immune response and inflammation. Five different proteins (IkBs) inhibit NF-κB in unstimulated cells. NF-κB Proteins are activated through phosphorylation of IkB proteins by the ikB kinase complex, the result is the translocation of NF-κB to the nucleolus. Via TNF- and IL-receptors on the cell surface, proinflammatory cytokines like TNF-α and IL6 activate NF-κB and Stat-3 System. NF-κB itself induces the transcription of TNF-α and with the expression of COX-2 the release of PGE2 [50].

The involvement of TNF-α/NF-κB and IL6/Stat3 pathways in tumorigenesis have been confirmed in a series of mouse models of GI malignancy focusing on inflammatory network of the tumor microenvironment [51,52].

4. Inflammatory Mediators, Immune Function, and Tumor Progression

In the tumor microenvironment, a variety of inflammatory mediators, such as cytokines (IL-6, IL-10, VEGF, TNFα, and TGFβ), chemokines (CCL20 and CXCL8) as well as lipid mediators (such as PGE2) are continuously produced [53]. These mediators are postulated to form a critical interface between immune and neoplastic compartments. Not only do they continuously support tumor survival and expansion, but suppress the function of immune cells, notably, dendritic cells (DCs)—the powerful antigen presenting cells that are crucial for induction of tumor-specific immune responses [53]. In a study from Sombroek et al. examining the supernatants of primary tumor cells (colon, breast, renal cell carcinoma, and melanoma), negative impact on DC development by the factors contained in the supernatants could be demonstrated. Among the factors for which hampering of the differentiation of DCs is known (IL-10, TGF-β1, VEGF, IL-6, M-CSF, and PGE2), only PGE2 was present in such concentrations in the tumor supernatants to show inhibitory effects on the acquisition of DC morphology [54]. Paradoxically, PGE2 also enhances the maturation, migration, and antigen-presenting capacity of DCs. In an effort to explain these seemingly contradictory effects a recent study by Shimabukuro-Vornhagen et al. suggests that whether PGE2-treatment results in inhibition or stimulation of T-cells is dependent on the DC to T-cell ratio during their interaction, showing an inhibitory effect at high DC to T-cell ratios [55]. The authors go on to speculate that this mechanism could serve as a counter-regulatory response in the context of physiologic immune response: Further T-cell activation then would be limited once a large number of mature DCs have accumulated [55].

However, other cell types actively contribute to the immunosuppressive environment within tumors. Myeloid-derived suppressor cells (MDSC) have been found in various cancers. MDSC consist of immature myeloid cells and display a diversity of phenotypes, whereby factors contained in the tumor microenvironment seem to have a major effect on their phenotype and function [56]. They are capable of suppressing adaptive and innate anti-tumor immune responses [57]. PGE2 has emerged as a key molecule in MDSC biology [58]. It not only induces the formation of MDSC (through the EP2 receptor) [59], but also promotes MDSC recruitment to the tumor microenvironment and stabilizes the MDSC phenotype [58].

It has been shown recently that the COX2/PGE2 pathway is involved in the regulation of immune checkpoints by influencing the programmed cell death ligand 1 (PD-L1) expression in tumor-infiltrating bone marrow derived myeloid cells, primarily MDSC and macrophages, and that the inhibition of PGE2 formation is able to attenuate the tumor induced PD-L1 expression [60].

Aside from MDSC, regulatory T-cells or Tregs play a role in tumor immune escape. These cells infiltrate the tumor microenvironment and dampen anti-tumor immune responses by inhibiting effector

T-cell function [61]. Though the specific mechanisms are yet to be elucidated. Further, Tregs seem to suppress T-cell activity in a PGE2-dependent manner, which can be reversed by COX-2 inhibitors or EP-receptor antagonists [62]. Beside the mediation of suppressive functions COX-2 derived PGE2 from DCs enhances the generation of Tregs and their expansion [63,64]. In peripheral blood of AML patients the frequency of Tregs is significantly higher in comparison to healthy individuals [65].

Data from a murine AML model show that PD-1 signaling and regulatory T-cells collaborate to resist the function of cytotoxic T lymphocytes in advanced AML [66].

One report investigated the role of COX-2 inhibition on indoleamine 2,3-dioxygenase 1 (IDO1) mediated immune dysfunction in AML [67]. IDO1 has been shown to contribute to activation of Tregs, which in turn hamper anti-cancer immunity. In the report by Iachininoto et al., the authors were able to show in vitro that inhibition of the COX-2/PGE2 pathway reduced the expression of IDO1 and inhibits the formation of Tregs [67]. These data, together with the observation that those AML-patients presenting with a high frequency of Tregs at diagnosis were shown to have worse responses to induction chemotherapy, have potential implications to optimize immunotherapeutic approaches [68]. PGE2 thus has a central role in the modulation of immune function as is summarized in Figure 2.

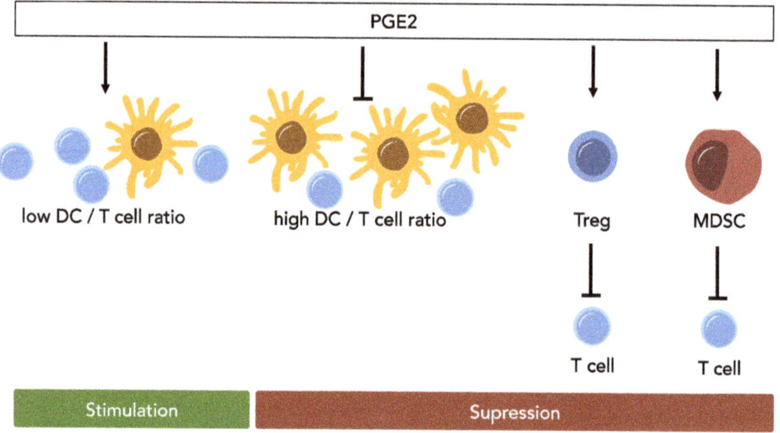

Figure 2. Effect of PGE2 on anti-cancer T cell activity. PGE2 has a differential impact on T cell activity, showing stimulatory effects at low DC/T cell ratios, but suppressive effects as DC numbers increase. PGE2 increases activation of Tregs and is involved in MDSC formation, which in turn hampers anti-cancer immunity. MDSC: myeloid derived suppressor cell; Treg: regulatory T cell; DC: dendritic cell.

Another approach to modify immunotherapeutic approaches could be based on omega-3 PUFA-derived SPM, which have recently been shown to decrease tumor debris-associated inflammation in an experimental model of tumor debris-stimulated tumor cell proliferation and macrophage-associated inflammation. Compounds such as resolvin D1 (RvD1), RvD2, and RvE1 were able to increase macrophage phagocytosis of tumor cell debris and to decrease the release of cytokines/chemokines from human macrophages stimulated with cell debris [69].

5. PUFA-Derived Lipid Mediators in Benign Hematopoiesis

Aside from the presence and effects in terminally differentiated blood cells, the expression and the function of COX isoenzymes and lipid mediators formed by these enzymes in hematopoietic progenitors and precursors remain subject of investigation [70]. Studies in the last decade have provided some insights into the role of the eicosanoid PGE2 in hematopoietic regulation [20]. In particular the stable PGE2-derivative 16,16-dimethyl-PGE2 (dmPGE2) was shown to increase the frequency of long-term repopulating hematopoietic stem cells (HSCs) in irradiated murine bone marrow [71]. This effect was further enhanced by combining dmPGE2 treatment with DPP-4 inhibition using sitagliptin in a mouse model [72].

Furthermore, HSCs pulsed with PGE2 were shown to display a higher (short term) competitiveness, as determined by a head-to-head comparison in a murine competitive transplantation model [20,21,73]. In the context of HSCT, trafficking of HSCs from the peripheral blood to bone marrow niches in the recipient patient, i.e., HSC homing, has been shown to increase under the influence of PGE2 [20].

Improving engraftment is especially relevant in the context of umbilical cord blood (UCB) transplantation. UCB transplantations offer some advantages over other sources of HSC, such as lower immune-matching requirements and to some degree a higher availability as UCB is cryopreserved [74]. However, the main pitfall of UCB transplantation is less efficient engraftment than in HSCT from other sources. Utilizing dmPGE2-treatment, Cutler et al. could show promising results in a phase I study by ex vivo-pulsing of UCB with dmPGE2 [75]. Furthermore, also inhibition of 15-PGDH, and thus increase of local PGE2 concentration can contribute to bone marrow transplant recovery [76].

Within the bone marrow, PGE2 is secreted by osteoblasts in large amounts, and given their close physical proximity to HSCs in the bone marrow niche, PGE2 is available to HSCs for the paracrine regulation of stem and progenitor function [20].

Historically, however, there has been conflicting data on whether PGE2 stimulates or inhibits the growth of hematopoietic progenitor cells. Older studies demonstrated an inhibitory effect of PGE2 on mouse and human myeloid progenitor cells in vitro [77,78]. Further studies revealed that dose, timing, and duration of PGE2-exposure are critical for positive or negative effects on proliferation. Since PGE2 is also produced by the hematopoietic cells themselves, it is therefore postulated that PGE2 might act as a feedback regulator of myelopoiesis [20]. Together, these data suggest that in benign hematopoiesis, PGE2 plays a central role in the HSC niche (Figure 3).

In addition to these data, recent studies in zebrafish and mice have identified the arachidonic-acid derived cytochrome P 450 metabolite 11,12-eipoxyeicosatrienoic acid (11,12-EET) as potent factor to increase embryonic hematopoiesis and adult marrow engraftment [79,80].

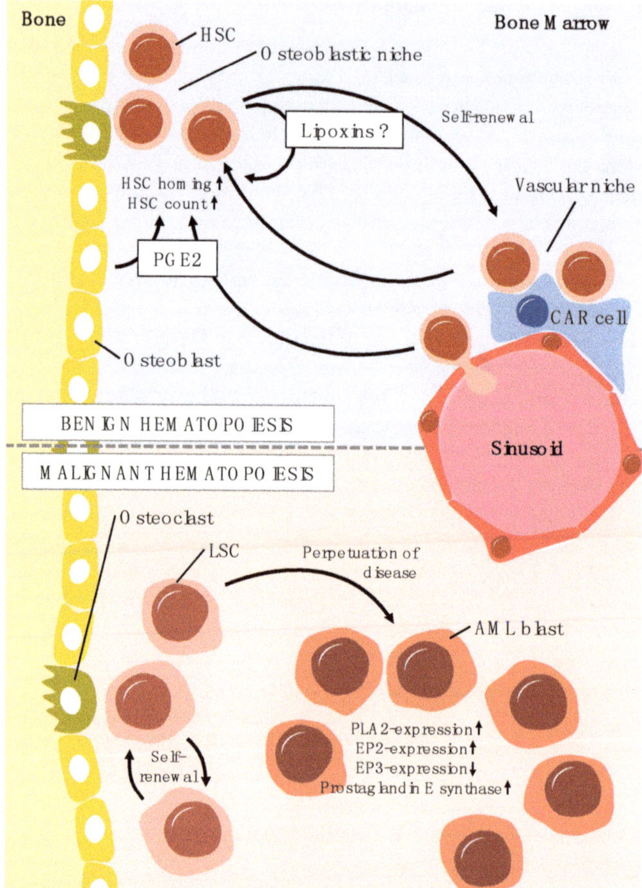

Figure 3. Lipid mediators in benign and malignant hematopoiesis. In benign hematopoiesis, PGE2 is secreted in large amounts in the osetoblastic niche and increases stem cell homing, and long-term LSC numbers. The role of lipoxins is still not fully understood, however lipoxins are required for stem cell quiescence and long-term renewal. In AML, LSC are considered to be chemoresistant and responsible for disease relapse. Self-renewal and maintenance of LSC in the bone marrow niche are increasingly better understood and growing data show alterations of lipid pathway enzymes suggesting eicosanoid pathways are active in leukemic blasts. HSC: hematopoietic stem cell, LSC: Leukemic stem cell; CAR cell: CXCL12-abundant reticular cell; PGE2: Prostaglandin E2; EP2: Prostaglandin E receptor 2; EP3: Prostaglandin E receptor 3; PLA2: Phospholipase A2.

6. PUFA-Derived Lipid Mediators in Malignant Hematopoiesis

While omega-3 PUFA have been widely implicated in anti-tumor effects in a variety of solid tumors, albeit with mixed results [81], data in hematological malignancies are sparse. In a review published by Betiati et al., a Scopus and PubMed database search between 1998 and 2012 returned 6 studies published on the subject of the effect of omega-3-PUFAs on hematological malignancies [82]. Since then, only few publications regarding effects of omega-3 PUFAs in hematological malignancies have been published. There is some evidence showing a lower incidence of non-Hodgkin lymphomas (NHLs) in patients on a diet high in omega-3-PUFAs [83]. Another study established higher omega-3-PUFAs in NHL patients in remission as compared to those with active disease [84] and recent data implicate low plasma omega-3-PUFAs as marker of inferior prognosis in diffuse large B-cell lymphoma [85]. However, no reduced risk with higher omega-3 PUFAs for lymphoid and myeloid leukemia could be detected so far [86]. In the context of ALL omega-3-PUFAs were shown to be able to lower treatment-related hypertriglyceridemia [87]. Noteworthy is a small recent study with a total of 22 leukemia or lymphoma patients in which a prolonged overall survival time of patients receiving fish oil (2g/d) was shown. These observations might indicate that EPA and DHA improve the response to treatment with conventional chemotherapy in hematological malignancies [88]. In vitro data show an inhibition of cell growth in AML cell lines by EPA and DHA [89,90]. In the erythrocytes of multiple myeloma (MM) patients, Jurczyszyn et al. demonstrated a decreased n-3/n-6 ratio and lower levels of EPA, despite higher levels of its precursor, α-linolenic acid, were measured. This might suggest an impaired functionality of desaturase and elongase enzymes in these patients [91]. A recent systematic review has assessed the current knowledge regarding the omega-3 PUFA EPA and DHA in the context of cells and models of malignant hematopoiesis [92]. There is a wealth of data gained in different cell models, but generally accepted mechanisms, and applicability in vivo and in humans, are still uncertain.

In contrast to this rather limited experimental evidence regarding omega-3 PUFA in the context of malignant hematopoiesis, the main focus of research in this field has been the omega-6 PUFA derived PGE2. Transcript levels of soluble phospholipase A2 (PLA2) subtypes IB and X have been shown to be upregulated in AML blasts compared to control blood mononuclear cells [93]. This finding might be significant since the enzymatic activity of PLA2 releases the eicosanoid precursor, arachidonic acid (AA) from membrane phospholipids for the generation of COX- and lipoxygenase-derived lipid mediators and is in keeping with the higher levels of free AA observed recently in plasma from AML patients [94].

PGE2 might normally act as a feedback regulator of myelopoiesis as described above [20]. However leukemic cells were shown to be resistant to this feedback mechanism seen in benign hematopoiesis [77,95,96]. Furthermore, insensitivity to PGE2-mediated growth inhibition in marrow cultures from patients with myelodysplastic syndrome preceded the patient's progression to acute leukemia [97]. In keeping with these results, the overall incidence of hematologic malignancies seems not to be reduced by long-term intake of the COX-inhibitor acetylsalicylic acid [98].

In contrast, treatment of mice with indomethacin (a COX inhibitor and thus inhibiting PGE2 synthesis) prior to injection of erythroleukemia cells significantly reduced the number of leukemic cells in both spleen as well as bone marrow [99]. Additionally, Wang et al. could demonstrate a marked reduction of leukemia-initiating cells in a murine limiting dilution transplant assay after treatment with indomethacin [100]. These data indicate a role for COX-metabolites in the proliferation of leukemic cells.

While COX-1 and COX-2 transcripts can be detected by polymerase chain reaction, the COX-2 protein is not present in primary AML and ALL blasts which is concordance with in vitro data from the human promyelocytic leukemia cell line HL-60 [70,101]. However, AML blasts were shown to express COX-1 [101]. Interestingly, constitutive expression of COX-1 can be upregulated by tumor necrosis factor-related apoptosis-inducing ligand (TRAIL) in HL-60 cells. This is accompanied by an increase of PGE2 synthesis and shows a protective effect towards TRAIL-induced apoptosis [70]. Similarly, AML cells treated with doxorubicin showed overexpression of multidrug transporter MDR1 triggered by increased PGE2-formation, and thereby decreased cytostatic efficacy of doxorubicin [102]. These studies suggest that protective mechanisms of the leukemic blasts to avoid eradication are—at least in part—mediated by PGE2.

In vitro data with AML-mesenchymal stroma cells (AML-MSC) co-cultures show a greatly increased COX-2 expression in MSC and induced PGE2 production in dependence of IL1β and ARC (apoptosis repressor with caspase recruitment domain). ARC is a protein that regulates leukemia microenvironment interactions through NFκB/IL1β and was shown to be an adverse prognostic marker in AML [103]. The COX-2 derived elevation of PGE2 from stromal cells seem to support AML chemoresistance through the expression of β-catenin which regulates ARC [104]. These data indicate that PGE2 production in the microenvironment takes part in a mechanism of an antiapoptotic action and microenvironment-mediated chemoresistance in certain subgroups of AML.

Downstream of the COX enzymes the last step in the synthesis of PGE2 is performed by the prostaglandin E synthase. This enzyme was shown to be present in normal tissues in minor amounts but is strongly upregulated in neoplastic cells [105,106]. In AML cell lines the prostaglandin E synthase was shown to be upregulated and specific inhibition of the enzyme resulted in an inhibition of proliferation [107] (Figure 2).

As shown for solid tumors, the inhibition of PGE2 receptors EP1, EP2, and EP4 allows for inhibition of cancer-associated inflammation and tumor growth. Mice deficient of these PGE2 receptors display decreased tumorigenesis as demonstrated in various experimental settings mainly for solid tumors [57,108–110]. Concordantly, expression levels of EP1 and EP2 have been demonstrated to be increased in cancerous tissues [111]. In AML, Ross et al. as well as Yagi et al. could demonstrate elevated transcript levels of EP2 in AML blasts in a pediatric cohort [112,113] and Denizot et al. could show that AML blasts express functional EP2 receptors [25,114] (Figure 2).

Interestingly, an omega-3 PUFA-derived lipid mediator has been implicated in anti-leukemia effects: The cyclopentenone prostaglandin Δ12-PGJ3, produced through cyclooxygenase action from the omega-3 PUFA EPA, was able to decrease leukemia burden in two murine models of leukemia [115] by selectively targeting leukemia stem cells (LSCs).

Concerning other eicosanoids and their role in hematopoiesis, only limited data has been published. For instance, lipoxins, which are produced by the lipoxygenases (as well as by the COX enzymes if acetylated by ASA), were shown to suppress tumor angiogenesis [116]. Actually, defective lipoxin synthesis was found in leukemia [117], indicating a stabilizing role for lipoxin in benign hematopoiesis (Figure 2). Additionally, a landmark study by Kode et al. demonstrated that an activating β-catenin mutation in osteoblasts can induce the development of leukemia by activating Notch signaling in hematopoietic precursors [118]. Here, osteoblasts exhibited increased expression of the Notch ligand Jagged 1. Conversely, there is some evidence suggesting that lipoxin A4 might decrease the expression of Jagged 1 [119]. Generally, however, the relevance of the lipoxygenase-pathway in benign as well as malignant hematopoiesis is yet to be explored further.

Concerning 5-lipoxygenase, Gal et al. could show elevated transcript levels of 5-lipoxygenase in the CD34+/CD38+ fraction of AML blasts compared to the less mature CD34+/CD38- fraction of the same patients [2]. In a model utilizing mice deficient in 12/15-lipoxygenase it was found that this enzyme is required for the maintenance of long-term HSC quiescence as well as self-renewal [120].

7. Conclusions

Recent treatment approaches in AML focus increasingly on immune therapy. One of the challenges in the field is to eliminate or reprogram the immune suppressive microenvironment often created by tumors [56]. Interestingly, the otherwise pro-inflammatory lipid mediator PGE2 seems to play a major role in mediating some of these suppressive effects by either direct inhibition of effector T-cell function or indirectly by increasing the frequency of immunosuppressive cell types. Particularly approaches to lower PGE2 might thus serve to enhance immune therapy approaches Current experimental data confirm the importance of this aspect also in the context of AML.

Author Contributions: Conceptualization, T.K. and K.H.W.; writing—original draft preparation, T.K. and A.L.; writing—review and editing, A.P., E.R. and K.H.W.; visualization, T.K.

Funding: This research received no external funding.

References

1. Siegel, R.; Ma, J.; Zou, Z.; Jemal, A. Cancer statistics, 2014. *CA: Cancer J. Clin.* **2014**, *64*, 9–29. [CrossRef]
2. Gal, H.; Amariglio, N.; Trakhtenbrot, L.; Jacob-Hirsh, J.; Margalit, O.; Avigdor, A.; Nagler, A.; Tavor, S.; Ein-Dor, L.; Lapidot, T.; et al. Gene expression profiles of AML derived stem cells; similarity to hematopoietic stem cells. *Leukemia* **2006**, *20*, 2147–2154. [CrossRef] [PubMed]
3. Guan, Y.; Gerhard, B.; Hogge, D.E. Detection, isolation, and stimulation of quiescent primitive leukemic progenitor cells from patients with acute myeloid leukemia (AML). *Blood* **2003**, *101*, 3142–3149. [CrossRef]
4. Eppert, K.; Takenaka, K.; Lechman, E.R.; Waldron, L.; Nilsson, B.; van Galen, P.; Metzeler, K.H.; Poeppl, A.; Ling, V.; Beyene, J.; et al. Stem cell gene expression programs influence clinical outcome in human leukemia. *Nat. Med.* **2011**, *17*, 1086–1093. [CrossRef]
5. Dick, J.E. Stem cell concepts renew cancer research. *Blood* **2008**, *112*, 4793–4807. [CrossRef]
6. Cavallo, F.; de Giovanni, C.; Nanni, P.; Forni, G.; Lollini, P.-L. 2011: The immune hallmarks of cancer. *Cancer Immunol. Immunother.* **2011**, *60*, 319–326. [CrossRef]
7. Lichtenegger, F.S.; Krupka, C.; Köhnke, T.; Subklewe, M. Immunotherapy for Acute Myeloid Leukemia. *Semin. Hematol.* **2015**, *52*, 207–214. [CrossRef] [PubMed]
8. Yang, D.; Zhang, X.; Zhang, X.; Xu, Y. The progress and current status of immunotherapy in acute myeloid leukemia. *Ann. Hematol.* **2017**, *96*, 1965–1982. [CrossRef] [PubMed]
9. Acheampong, D.O.; Adokoh, C.K.; Asante, D.-B.; Asiamah, E.A.; Barnie, P.A.; Bonsu, D.O.M.; Kyei, F. Immunotherapy for acute myeloid leukemia (AML): A potent alternative therapy. *Biomed. Pharmacother.* **2018**, *97*, 225–232. [CrossRef]
10. Lichtenegger, F.S.; Schnorfeil, F.M.; Hiddemann, W.; Subklewe, M. Current strategies in immunotherapy for acute myeloid leukemia. *Immunotherapy* **2013**, *5*, 63–78. [CrossRef]
11. Quehenberger, O.; Dennis, E.A. The human plasma lipidome. *N. Engl. J. Med.* **2011**, *365*, 1812–1823. [CrossRef]
12. Sulciner, M.L.; Gartung, A.; Gilligan, M.M.; Serhan, C.N.; Panigrahy, D. Targeting lipid mediators in cancer biology. *Cancer Metastasis Rev.* **2018**, *37*, 557–572. [CrossRef]
13. Carracedo, A.; Cantley, L.C.; Pandolfi, P.P. Cancer metabolism: Fatty acid oxidation in the limelight. *Nat. Rev. Cancer* **2013**, *13*, 227–232. [CrossRef]
14. Stuani, L.; Riols, F.; Millard, P.; Sabatier, M.; Batut, A.; Saland, E.; Viars, F.; Tonini, L.; Zaghdoudi, S.; Linares, L.K.; et al. Stable Isotope Labeling Highlights Enhanced Fatty Acid and Lipid Metabolism in Human Acute Myeloid Leukemia. *Int. J. Mol. Sci.* **2018**, *19*, 3325. [CrossRef]
15. Klock, J.C.; Pieprzyk, J.K. Cholesterol, phospholipids, and fatty acids of normal immature neutrophils: Comparison with acute myeloblastic leukemia cells and normal neutrophils. *J. Lipid Res.* **1979**, *20*, 908–911.
16. Pabst, T.; Kortz, L.; Fiedler, G.M.; Ceglarek, U.; Idle, J.R.; Beyoğlu, D. The plasma lipidome in acute myeloid leukemia at diagnosis in relation to clinical disease features. *BBA Clin.* **2017**, *7*, 105–114. [CrossRef]
17. Stefanko, A.; Thiede, C.; Ehninger, G.; Simons, K.; Grzybek, M. Lipidomic approach for stratification of acute myeloid leukemia patients. *PLoS ONE* **2017**, *12*, e0168781. [CrossRef]
18. Stables, M.J.; Gilroy, D.W. Old and new generation lipid mediators in acute inflammation and resolution. *Prog. Lipid Res.* **2011**, *50*, 35–51. [CrossRef]
19. Bakhle, Y.S. Structure of COX-1 and COX-2 enzymes and their interaction with inhibitors. *Drugs Today* **1999**, *35*, 237–250. [CrossRef]
20. Hoggatt, J.; Pelus, L.M. Eicosanoid regulation of hematopoiesis and hematopoietic stem and progenitor trafficking. *Leukemia* **2010**, *24*, 1993–2002. [CrossRef]
21. Hoggatt, J.; Mohammad, K.S.; Singh, P.; Pelus, L.M. Prostaglandin E2 enhances long-term repopulation but does not permanently alter inherent stem cell competitiveness. *Blood* **2013**, *122*, 2997–3000. [CrossRef]
22. Miller, S.B. Prostaglandins in health and disease: An overview. *Semin Arthritis Rheum.* **2006**, *36*, 37–49. [CrossRef]
23. Tilley, S.L.; Coffman, T.M.; Koller, B.H. Mixed messages: Modulation of inflammation and immune responses by prostaglandins and thromboxanes. *J. Clin. Investig.* **2001**, *108*, 15–23. [CrossRef]
24. Harris, S.G.; Padilla, J.; Koumas, L.; Ray, D.; Phipps, R.P. Prostaglandins as modulators of immunity. *Trends Immunol.* **2002**, *23*, 144–150. [CrossRef]

25. Malissein, E.; Reynaud, S.; Bordessoule, D.; Faucher, J.L.; Turlure, P.; Trimoreau, F.; Denizot, Y. PGE(2) receptor subtype functionality on immature forms of human leukemic blasts. *Leuk. Res.* **2006**, *30*, 1309–1313. [CrossRef]

26. Sugimoto, Y.; Narumiya, S. Prostaglandin E receptors. *J. Biol. Chem.* **2007**, *282*, 11613–11617. [CrossRef]

27. Kawahara, K.; Hohjoh, H.; Inazumi, T.; Tsuchiya, S.; Sugimoto, Y. Prostaglandin E2-induced inflammation: Relevance of prostaglandin E receptors. *Biochim. Biophys. Acta* **2015**, *1851*, 414–421. [CrossRef]

28. Hirsch, F.R.; Lippman, S.M. Advances in the biology of lung cancer chemoprevention. *J. Clin. Oncol.* **2005**, *23*, 3186–3197. [CrossRef] [PubMed]

29. Na, H.-K.; Park, J.-M.; Lee, H.G.; Lee, H.-N.; Myung, S.-J.; Surh, Y.-J. 15-Hydroxyprostaglandin dehydrogenase as a novel molecular target for cancer chemoprevention and therapy. *Biochem. Pharmacol.* **2011**, *82*, 1352–1360. [CrossRef] [PubMed]

30. Arima, K.; Komohara, Y.; Bu, L.; Tsukamoto, M.; Itoyama, R.; Miyake, K.; Uchihara, T.; Ogata, Y.; Nakagawa, S.; Okabe, H.; et al. Downregulation of 15-hydroxyprostaglandin dehydrogenase by interleukin-1β from activated macrophages leads to poor prognosis in pancreatic cancer. *Cancer Sci.* **2018**, *109*, 462–470. [CrossRef] [PubMed]

31. Arima, K.; Ohmuraya, M.; Miyake, K.; Koiwa, M.; Uchihara, T.; Izumi, D.; Gao, F.; Yonemura, A.; Bu, L.; Okabe, H.; et al. Inhibition of 15-PGDH causes Kras-driven tumor expansion through prostaglandin E2-ALDH1 signaling in the pancreas. *Oncogene* **2019**, *38*, 1211–1224. [CrossRef] [PubMed]

32. Serhan, C.N.; Hamberg, M.; Samuelsson, B. Lipoxins: Novel series of biologically active compounds formed from arachidonic acid in human leukocytes. *Proc. Natl. Acad. Sci. USA* **1984**, *81*, 5335–5339. [CrossRef]

33. Wei, J.; Gronert, K. Eicosanoid and Specialized Proresolving Mediator Regulation of Lymphoid Cells. *Trends Biochem. Sci.* **2019**, *44*, 214–225. [CrossRef] [PubMed]

34. McMahon, B.; Godson, C. Lipoxins: Endogenous regulators of inflammation. *Am. J. Physiol. Renal Physiol.* **2004**, *286*, F189–F201. [CrossRef] [PubMed]

35. Calder, P.C. Polyunsaturated fatty acids and inflammatory processes: New twists in an old tale. *Biochimie* **2009**, *91*, 791–795. [CrossRef]

36. Rees, D.; Miles, E.A.; Banerjee, T.; Wells, S.J.; Roynette, C.E.; Wahle, K.W.; Calder, P.C. Dose-related effects of eicosapentaenoic acid on innate immune function in healthy humans: A comparison of young and older men. *Am. J. Clin. Nutr.* **2006**, *83*, 331–342. [CrossRef]

37. Bagga, D.; Wang, L.; Farias-Eisner, R.; Glaspy, J.A.; Reddy, S.T. Differential effects of prostaglandin derived from omega-6 and omega-3 polyunsaturated fatty acids on COX-2 expression and IL-6 secretion. *Proc. Natl. Acad. Sci. USA* **2003**, *100*, 1751–1756. [CrossRef] [PubMed]

38. Seki, H.; Sasaki, T.; Ueda, T.; Arita, M. Resolvins as regulators of the immune system. *ScientificWorldJournal* **2010**, *10*, 818–831. [CrossRef]

39. Yang, P.; Chan, D.; Felix, E.; Cartwright, C.; Menter, D.G.; Madden, T.; Klein, R.D.; Fischer, S.M.; Newman, R.A. Formation and antiproliferative effect of prostaglandin E(3) from eicosapentaenoic acid in human lung cancer cells. *J. Lipid Res.* **2004**, *45*, 1030–1039. [CrossRef]

40. Mickleborough, T.D.; Lindley, M.R.; Ionescu, A.A.; Fly, A.D. Protective effect of fish oil supplementation on exercise-induced bronchoconstriction in asthma. *Chest* **2006**, *129*, 39–49. [CrossRef]

41. Fischer, R.; Konkel, A.; Mehling, H.; Blossey, K.; Gapelyuk, A.; Wessel, N.; von Schacky, C.; Dechend, R.; Muller, D.N.; Rothe, M.; et al. Dietary omega-3 fatty acids modulate the eicosanoid profile in man primarily via the CYP-epoxygenase pathway. *J. Lipid Res.* **2014**, *55*, 1150–1164. [CrossRef]

42. Schunck, W.-H.; Konkel, A.; Fischer, R.; Weylandt, K.-H. Therapeutic potential of omega-3 fatty acid-derived epoxyeicosanoids in cardiovascular and inflammatory diseases. *Pharmacol. Ther.* **2018**, *183*, 177–204. [CrossRef]

43. Weylandt, K.H.; Chiu, C.-Y.; Gomolka, B.; Waechter, S.F.; Wiedenmann, B. Omega-3 fatty acids and their lipid mediators: Towards an understanding of resolvin and protectin formation. *Prostaglandins Other Lipid Mediat.* **2012**, *97*, 73–82. [CrossRef]

44. Serhan, C.N.; Clish, C.B.; Brannon, J.; Colgan, S.P.; Chiang, N.; Gronert, K. Novel functional sets of lipid-derived mediators with antiinflammatory actions generated from omega-3 fatty acids via cyclooxygenase 2-nonsteroidal antiinflammatory drugs and transcellular processing. *J. Exp. Med.* **2000**, *192*, 1197–1204. [CrossRef]

45. Mangelsdorf, D.J.; Thummel, C.; Beato, M.; Herrlich, P.; Schütz, G.; Umesono, K.; Blumberg, B.; Kastner, P.; Mark, M.; Chambon, P.; et al. The nuclear receptor superfamily: The second decade. *Cell* **1995**, *83*, 835–839. [CrossRef]

46. Auboeuf, D.; Rieusset, J.; Fajas, L.; Vallier, P.; Frering, V.; Riou, J.P.; Staels, B.; Auwerx, J.; Laville, M.; Vidal, H. Tissue Distribution and Quantification of the Expression of mRNAs of Peroxisome Proliferator-Activated Receptors and Liver X Receptor- in Humans: No Alteration in Adipose Tissue of Obese and NIDDM Patients. *Diabetes* **1997**, *46*, 1319–1327. [CrossRef]

47. Chinetti, G.; Fruchart, J.-C.; Staels, B. Peroxisome proliferator-activated receptors (PPARs): Nuclear receptors at the crossroads between lipid metabolism and inflammation. *Inflamm. Res.* **2000**, *49*, 497–505. [CrossRef]

48. Yousefnia, S.; Momenzadeh, S.; Seyed Forootan, F.; Ghaedi, K.; Nasr Esfahani, M.H. The influence of peroxisome proliferator-activated receptor γ (PPARγ) ligands on cancer cell tumorigenicity. *Gene* **2018**, *649*, 14–22. [CrossRef]

49. Ryu, S.; Kim, D.S.; Lee, M.W.; Lee, J.W.; Sung, K.W.; Koo, H.H.; Yoo, K.H. Anti-leukemic effects of PPARγ ligands. *Cancer Lett.* **2018**, *418*, 10–19. [CrossRef]

50. Oshima, H.; Oshima, M. The inflammatory network in the gastrointestinal tumor microenvironment: Lessons from mouse models. *J. Gastroenterol.* **2012**, *47*, 97–106. [CrossRef]

51. Mantovani, A.; Allavena, P.; Sica, A.; Balkwill, F. Cancer-related inflammation. *Nature* **2008**, *454*, 436–444. [CrossRef]

52. Grivennikov, S.I.; Greten, F.R.; Karin, M. Immunity, inflammation, and cancer. *Cell* **2010**, *140*, 883–899. [CrossRef]

53. Sheng, K.C.; Wright, M.D.; Apostolopoulos, V. Inflammatory mediators hold the key to dendritic cell suppression and tumor progression. *Curr. Med. Chem.* **2011**, *18*, 5507–5518. [CrossRef]

54. Sombroek, C.C.; Stam, A.G.M.; Masterson, A.J.; Lougheed, S.M.; Schakel, M.J.A.G.; Meijer, C.J.L.M.; Pinedo, H.M.; van den Eertwegh, A.J.M.; Scheper, R.J.; de Gruijl, T.D. Prostanoids play a major role in the primary tumor-induced inhibition of dendritic cell differentiation. *J. Immunol.* **2002**, *168*, 4333–4343. [CrossRef]

55. Shimabukuro-Vornhagen, A.; Liebig, T.M.; Koslowsky, T.; Theurich, S.; von Bergwelt-Baildon, M.S. The ratio between dendritic cells and T cells determines whether prostaglandin E2 has a stimulatory or inhibitory effect. *Cell Immunol.* **2013**, *281*, 62–67. [CrossRef]

56. Lindau, D.; Gielen, P.; Kroesen, M.; Wesseling, P.; Adema, G.J. The immunosuppressive tumour network: Myeloid-derived suppressor cells, regulatory T cells and natural killer T cells. *Immunology* **2013**, *138*, 105–115. [CrossRef]

57. Greene, E.R.; Huang, S.; Serhan, C.N.; Panigrahy, D. Regulation of inflammation in cancer by eicosanoids. *Prostaglandins Other Lipid Mediat.* **2011**, *96*, 27–36. [CrossRef]

58. Obermajer, N.; Wong, J.L.; Edwards, R.P.; Odunsi, K.; Moysich, K.; Kalinski, P. PGE(2)-driven induction and maintenance of cancer-associated myeloid-derived suppressor cells. *Immunol. Invest.* **2012**, *41*, 635–657. [CrossRef]

59. Sinha, P.; Clements, V.K.; Fulton, A.M.; Ostrand-Rosenberg, S. Prostaglandin E2 promotes tumor progression by inducing myeloid-derived suppressor cells. *Cancer Res.* **2007**, *67*, 4507–4513. [CrossRef]

60. Prima, V.; Kaliberova, L.N.; Kaliberov, S.; Curiel, D.T.; Kusmartsev, S. COX2/mPGES1/PGE2 pathway regulates PD-L1 expression in tumor-associated macrophages and myeloid-derived suppressor cells. *Proc. Natl. Acad. Sci. USA* **2017**, *114*, 1117–1122. [CrossRef] [PubMed]

61. Jacobs, J.F.M.; Nierkens, S.; Figdor, C.G.; de Vries, I.J.M.; Adema, G.J. Regulatory T cells in melanoma: The final hurdle towards effective immunotherapy? *Lancet Oncol.* **2012**, *13*, e32–e42. [CrossRef]

62. Yaqub, S.; Taskén, K. Role for the cAMP-protein kinase A signaling pathway in suppression of antitumor immune responses by regulatory T cells. *Crit. Rev. Oncog.* **2008**, *14*, 57–77. [CrossRef] [PubMed]

63. Kalinski, P. Regulation of Immune Responses by Prostaglandin E2. *J. Immunol.* **2012**, *188*, 21–28. [CrossRef]

64. Trinath, J.; Hegde, P.; Sharma, M.; Maddur, M.S.; Rabin, M.; Vallat, J.-M.; Magy, L.; Balaji, K.N.; Kaveri, S.V.; Bayry, J. Intravenous immunoglobulin expands regulatory T cells via induction of cyclooxygenase-2-dependent prostaglandin E2 in human dendritic cells. *Blood* **2013**, *122*, 1419–1427. [CrossRef] [PubMed]

65. Wang, X.; Zheng, J.; Liu, J.; Yao, J.; He, Y.; Li, X.; Yu, J.; Yang, J.; Liu, Z.; Huang, S. Increased population of CD4(+)CD25(high), regulatory T cells with their higher apoptotic and proliferating status in peripheral blood of acute myeloid leukemia patients. *Eur. J. Haematol.* **2005**, *75*, 468–476. [CrossRef]

66. Zhou, Q.; Munger, M.E.; Highfill, S.L.; Tolar, J.; Weigel, B.J.; Riddle, M.; Sharpe, A.H.; Vallera, D.A.; Azuma, M.; Levine, B.L.; et al. Program death-1 signaling and regulatory T cells collaborate to resist the function of adoptively transferred cytotoxic T lymphocytes in advanced acute myeloid leukemia. *Blood* **2010**, *116*, 2484–2493. [CrossRef]

67. Iachininoto, M.G.; Nuzzolo, E.R.; Bonanno, G.; Mariotti, A.; Procoli, A.; Locatelli, F.; de Cristofaro, R.; Rutella, S. Cyclooxygenase-2 (COX-2) inhibition constrains indoleamine 2,3-dioxygenase 1 (IDO1) activity in acute myeloid leukaemia cells. *Molecules* **2013**, *18*, 10132–10145. [CrossRef]

68. Szczepanski, M.J.; Szajnik, M.; Czystowska, M.; Mandapathil, M.; Strauss, L.; Welsh, A.; Foon, K.A.; Whiteside, T.L.; Boyiadzis, M. Increased frequency and suppression by regulatory T cells in patients with acute myelogenous leukemia. *Clin. Cancer Res.* **2009**, *15*, 3325–3332. [CrossRef]

69. Sulciner, M.L.; Serhan, C.N.; Gilligan, M.M.; Mudge, D.K.; Chang, J.; Gartung, A.; Lehner, K.A.; Bielenberg, D.R.; Schmidt, B.; Dalli, J.; et al. Resolvins suppress tumor growth and enhance cancer therapy. *J. Exp. Med.* **2018**, *215*, 115–140. [CrossRef]

70. Secchiero, P.; Gonelli, A.; Ciabattoni, G.; Melloni, E.; Grill, V.; Rocca, B.; Delbello, G.; Zauli, G. TNF-related apoptosis-inducing ligand (TRAIL) up-regulates cyclooxygenase (COX)-1 activity and PGE(2) production in cells of the myeloid lineage. *J. Leukoc. Biol.* **2002**, *72*, 986–994. [PubMed]

71. North, T.E.; Goessling, W.; Walkley, C.R.; Lengerke, C.; Kopani, K.R.; Lord, A.M.; Weber, G.J.; Bowman, T.V.; Jang, I.-H.; Grosser, T.; et al. Prostaglandin E2 regulates vertebrate haematopoietic stem cell homeostasis. *Nature* **2007**, *447*, 1007–1011. [CrossRef]

72. Broxmeyer, H.E.; Pelus, L.M. Inhibition of DPP4/CD26 and dmPGE$_2$ treatment enhances engraftment of mouse bone marrow hematopoietic stem cells. *Blood Cells Mol. Dis.* **2014**, *53*, 34–38. [CrossRef]

73. Hoggatt, J.; Singh, P.; Sampath, J.; Pelus, L.M. Prostaglandin E2 enhances hematopoietic stem cell homing, survival, and proliferation. *Blood* **2009**, *113*, 5444–5455. [CrossRef] [PubMed]

74. Hagedorn, E.J.; Durand, E.M.; Fast, E.M.; Zon, L.I. Getting more for your marrow: Boosting hematopoietic stem cell numbers with PGE2. *Exp. Cell Res.* **2014**, *329*, 220–226. [CrossRef]

75. Cutler, C.; Multani, P.; Robbins, D.; Kim, H.T.; Le, T.; Hoggatt, J.; Pelus, L.M.; Desponts, C.; Chen, Y.-B.; Rezner, B.; et al. Prostaglandin-modulated umbilical cord blood hematopoietic stem cell transplantation. *Blood* **2013**, *122*, 3074–3081. [CrossRef] [PubMed]

76. Desai, A.; Zhang, Y.; Park, Y.; Dawson, D.M.; Larusch, G.A.; Kasturi, L.; Wald, D.; Ready, J.M.; Gerson, S.L.; Markowitz, S.D. A second-generation 15-PGDH inhibitor promotes bone marrow transplant recovery independently of age, transplant dose and granulocyte colony-stimulating factor support. *Haematologica* **2018**, *103*, 1054–1064. [CrossRef] [PubMed]

77. Pelus, L.M.; Broxmeyer, H.E.; Moore, M.A. Regulation of human myelopoiesis by prostaglandin E and lactoferrin. *Cell Tissue Kinet.* **1981**, *14*, 515–526. [CrossRef] [PubMed]

78. Aglietta, M.; Piacibello, W.; Gavosto, F. Insensitivity of chronic myeloid leukemia cells to inhibition of growth by prostaglandin E1. *Cancer Res.* **1980**, *40*, 2507–2511. [PubMed]

79. Li, P.; Lahvic, J.L.; Binder, V.; Pugach, E.K.; Riley, E.B.; Tamplin, O.J.; Panigrahy, D.; Bowman, T.V.; Barrett, F.G.; Heffner, G.C.; et al. Epoxyeicosatrienoic acids enhance embryonic haematopoiesis and adult marrow engraftment. *Nature* **2015**, *523*, 468–471. [CrossRef] [PubMed]

80. Lahvic, J.L.; Ammerman, M.; Li, P.; Blair, M.C.; Stillman, E.R.; Fast, E.M.; Robertson, A.L.; Christodoulou, C.; Perlin, J.R.; Yang, S.; et al. Specific oxylipins enhance vertebrate hematopoiesis via the receptor GPR132. *Proc. Natl. Acad. Sci. USA* **2018**, *115*, 9252–9257. [CrossRef] [PubMed]

81. Weylandt, K.H.; Serini, S.; Chen, Y.Q.; Su, H.-M.; Lim, K.; Cittadini, A.; Calviello, G. Omega-3 Polyunsaturated Fatty Acids: The Way Forward in Times of Mixed Evidence. *Biomed. Res. Int.* **2015**, *2015*, 143109. [CrossRef]

82. Da Silva Borges Betiati, D.; de Oliveira, P.F.; de Quadros Camargo, D.; Nunes, E.A.; Trindade, E.B. Effects of omega-3 fatty acids on regulatory T cells in hematologic neoplasms. *Rev. Bras. Hematol. Hemoter.* **2013**, *35*, 119–125. [CrossRef]

83. Charbonneau, B.; O'Connor, H.M.; Wang, A.H.; Liebow, M.; Thompson, C.A.; Fredericksen, Z.S.; Macon, W.R.; Slager, S.L.; Call, T.G.; Habermann, T.M.; et al. Trans Fatty Acid Intake Is Associated with Increased Risk and n3 Fatty Acid Intake with Reduced Risk of Non-Hodgkin Lymphoma. *J. Nutr.* **2013**, *143*, 672–681. [CrossRef]

84. Cvetković, Z.; Vučić, V.; Cvetković, B.; Karadžić, I.; Ranić, M.; Glibetić, M. Distribution of plasma fatty acids is associated with response to chemotherapy in non-Hodgkin's lymphoma patients. *Med. Oncol.* **2013**, *30*, 741. [CrossRef] [PubMed]

85. Thanarajasingam, G.; Maurer, M.J.; Habermann, T.M.; Nowakowski, G.S.; Bennani, N.N.; Thompson, C.A.; Cerhan, J.R.; Witzig, T.E. Low Plasma Omega-3 Fatty Acid Levels May Predict Inferior Prognosis in Untreated Diffuse Large B-Cell Lymphoma: A New Modifiable Dietary Biomarker? *Nutr. Cancer* **2018**, *70*, 1088–1090. [CrossRef] [PubMed]

86. Saberi Hosnijeh, F.; Peeters, P.; Romieu, I.; Kelly, R.; Riboli, E.; Olsen, A.; Tjønneland, A.; Fagherazzi, G.; Clavel-Chapelon, F.; Dossus, L.; et al. Dietary intakes and risk of lymphoid and myeloid leukemia in the European Prospective Investigation into Cancer and Nutrition (EPIC). *Nutr. Cancer* **2014**, *66*, 14–28. [CrossRef] [PubMed]

87. Salvador, C.; Entenmann, A.; Salvador, R.; Niederwanger, A.; Crazzolara, R.; Kropshofer, G. Combination therapy of omega-3 fatty acids and acipimox for children with hypertriglyceridemia and acute lymphoblastic leukemia. *J. Clin. Lipidol.* **2018**, *12*, 1260–1266. [CrossRef]

88. Chagas, T.R.; Borges, D.S.; de Oliveira, P.F.; Mocellin, M.C.; Barbosa, A.M.; Camargo, C.Q.; Del Moral, J.Â.G.; Poli, A.; Calder, P.C.; Trindade, E.B.S.M.; et al. Oral fish oil positively influences nutritional-inflammatory risk in patients with haematological malignancies during chemotherapy with an impact on long-term survival: A randomised clinical trial. *J. Hum. Nutr. Diet* **2017**, *30*, 681–692. [CrossRef]

89. Slagsvold, J.E.; Pettersen, C.H.H.; Follestad, T.; Krokan, H.E.; Schønberg, S.A. The Antiproliferative Effect of EPA in HL60 Cells is Mediated by Alterations in Calcium Homeostasis. *Lipids* **2008**, *44*, 103–113. [CrossRef] [PubMed]

90. Yamagami, T.; Porada, C.D.; Pardini, R.S.; Zanjani, E.D.; Almeida-Porada, G. Docosahexaenoic acid induces dose dependent cell death in an early undifferentiated subtype of acute myeloid leukemia cell line. *Cancer Biol. Ther.* **2009**, *8*, 331–337. [CrossRef]

91. Jurczyszyn, A.; Czepiel, J.; Gdula-Argasińska, J.; Czapkiewicz, A.; Biesiada, G.; Dróżdż, M.; Perucki, W.; Castillo, J.J. Erythrocyte membrane fatty acids in multiple myeloma patients. *Leuk. Res.* **2014**, *0*, 1260–1265. [CrossRef]

92. Moloudizargari, M.; Mortaz, E.; Asghari, M.H.; Adcock, I.M.; Redegeld, F.A.; Garssen, J. Effects of the polyunsaturated fatty acids, EPA and DHA, on hematological malignancies: A systematic review. *Oncotarget* **2018**, *9*, 11858–11875. [CrossRef] [PubMed]

93. Fiancette, R.; Vincent, C.; Donnard, M.; Bordessoule, D.; Turlure, P.; Trimoreau, F.; Denizot, Y. Genes encoding multiple forms of phospholipase A(2) are expressed in immature forms of human leukemic blasts. *Leukemia* **2009**, *23*, 1196–1199. [CrossRef]

94. Pabst, C.; Krosl, J.; Fares, I.; Boucher, G.; Ruel, R.; Marinier, A.; Lemieux, S.; Hébert, J.; Sauvageau, G. Identification of small molecules that support human leukemia stem cell activity ex vivo. *Nat. Methods* **2014**, *11*, 436–442. [CrossRef]

95. Moore, M.A.; Mertelsmann, R.; Pelus, L.M. Phenotypic evaluation of chronic myeloid leukemia. *Blood Cells* **1981**, *7*, 217–236.

96. Taetle, R.; Guittard, J.P.; Mendelsohn, J.M. Abnormal modulation of granulocyte/macrophage progenitor proliferation by prostaglandin E in chronic myeloproliferative disorders. *Exp. Hematol.* **1980**, *8*, 1190–1201.

97. Gold, E.J.; Conjalka, M.; Pelus, L.M.; Jhanwar, S.C.; Broxmeyer, H.; Middleton, A.B.; Clarkson, B.D.; Moore, M.A. Marrow cytogenetic and cell-culture analyses of the myelodysplastic syndromes: Insights to pathophysiology and prognosis. *J. Clin. Oncol.* **1983**, *1*, 627–634. [CrossRef]

98. Walter, R.B.; Milano, F.; Brasky, T.M.; White, E. Long-Term Use of Acetaminophen, Aspirin, and Other Nonsteroidal Anti-Inflammatory Drugs and Risk of Hematologic Malignancies: Results From the Prospective Vitamins and Lifestyle (VITAL) Study. *J. Clin. Oncol.* **2011**, *29*, 2424–2431. [CrossRef]

99. Dussault, I.; Miller, S.C. Effect on leukemia cell numbers of in vivo administration of immunotherapeutic agents is age-dependent. *Oncology* **1996**, *53*, 241–246. [CrossRef] [PubMed]

100. Wang, Y.; Krivtsov, A.V.; Sinha, A.U.; North, T.E.; Goessling, W.; Feng, Z.; Zon, L.I.; Armstrong, S.A. The Wnt/beta-catenin pathway is required for the development of leukemia stem cells in AML. *Science* **2010**, *327*, 1650–1653. [CrossRef]

101. Truffinet, V.; Donnard, M.; Vincent, C.; Faucher, J.L.; Bordessoule, D.; Turlure, P.; Trimoreau, F.; Denizot, Y. Cyclooxygenase-1, but not -2, in blast cells of patients with acute leukemia. *Int. J. Cancer* **2007**, *121*, 924–927. [CrossRef]

102. Puhlmann, U.; Ziemann, C.; Ruedell, G.; Vorwerk, H.; Schaefer, D.; Langebrake, C.; Schuermann, P.; Creutzig, U.; Reinhardt, D. Impact of the cyclooxygenase system on doxorubicin-induced functional multidrug resistance 1 overexpression and doxorubicin sensitivity in acute myeloid leukemic HL-60 cells. *J. Pharmacol. Exp. Ther.* **2005**, *312*, 346–354. [CrossRef]

103. Carter, B.Z.; Mak, P.Y.; Wang, X.; Tao, W.; Ruvolo, V.; Mak, D.; Mu, H.; Burks, J.K.; Andreeff, M. An ARC-regulated IL1β/Cox-2/PGE2/β-catenin/ARC circuit controls leukemia-microenvironment interactions and confers drug resistance in AML. *Cancer Res.* **2019**, *79*, 1165–1177. [CrossRef]

104. Carter, B.Z.; Mak, P.Y.; Chen, Y.; Mak, D.H.; Mu, H.; Jacamo, R.; Ruvolo, V.; Arold, S.T.; Ladbury, J.E.; Burks, J.K.; et al. Anti-apoptotic ARC protein confers chemoresistance by controlling leukemia-microenvironment interactions through a NFκB/IL1β signaling network. *Oncotarget* **2016**, *7*, 20054–20067. [CrossRef]

105. Mehrotra, S.; Morimiya, A.; Agarwal, B.; Konger, R.; Badve, S. Microsomal prostaglandin E2 synthase-1 in breast cancer: A potential target for therapy. *J. Pathol.* **2006**, *208*, 356–363. [CrossRef]

106. Wang, H.-W.; Hsueh, C.-T.; Lin, C.-F.J.; Chou, T.-Y.; Hsu, W.-H.; Wang, L.-S.; Wu, Y.-C. Clinical implications of microsomal prostaglandin e synthase-1 overexpression in human non-small-cell lung cancer. *Ann. Surg. Oncol.* **2006**, *13*, 1224–1234. [CrossRef]

107. Li, Y.; Yin, S.; Nie, D.; Xie, S.; Ma, L.; Wang, X.; Wu, Y.; Xiao, J. MK886 inhibits the proliferation of HL-60 leukemia cells by suppressing the expression of mPGES-1 and reducing prostaglandin E2 synthesis. *Int. J. Hematol.* **2011**, *94*, 472–478. [CrossRef]

108. Smyth, E.M.; Grosser, T.; Wang, M.; Yu, Y.; FitzGerald, G.A. Prostanoids in health and disease. *J. Lipid Res.* **2009**, *50* (Suppl.), S423–S428. [CrossRef]

109. Watanabe, K.; Kawamori, T.; Nakatsugi, S.; Ohta, T.; Ohuchida, S.; Yamamoto, H.; Maruyama, T.; Kondo, K.; Ushikubi, F.; Narumiya, S.; et al. Role of the prostaglandin E receptor subtype EP1 in colon carcinogenesis. *Cancer Res.* **1999**, *59*, 5093–5096.

110. Mutoh, M.; Watanabe, K.; Kitamura, T.; Shoji, Y.; Takahashi, M.; Kawamori, T.; Tani, K.; Kobayashi, M.; Maruyama, T.; Kobayashi, K.; et al. Involvement of prostaglandin E receptor subtype EP(4) in colon carcinogenesis. *Cancer Res.* **2002**, *62*, 28–32.

111. Shoji, Y.; Takahashi, M.; Kitamura, T.; Watanabe, K.; Kawamori, T.; Maruyama, T.; Sugimoto, Y.; Negishi, M.; Narumiya, S.; Sugimura, T.; et al. Downregulation of prostaglandin E receptor subtype EP3 during colon cancer development. *Gut* **2004**, *53*, 1151–1158. [CrossRef] [PubMed]

112. Ross, M.E.; Mahfouz, R.; Onciu, M.; Liu, H.-C.; Zhou, X.; Song, G.; Shurtleff, S.A.; Pounds, S.; Cheng, C.; Ma, J.; et al. Gene expression profiling of pediatric acute myelogenous leukemia. *Blood* **2004**, *104*, 3679–3687. [CrossRef]

113. Yagi, T.; Morimoto, A.; Eguchi, M.; Hibi, S.; Sako, M.; Ishii, E.; Mizutani, S.; Imashuku, S.; Ohki, M.; Ichikawa, H. Identification of a gene expression signature associated with pediatric AML prognosis. *Blood* **2003**, *102*, 1849–1856. [CrossRef]

114. Denizot, Y.; Donnard, M.; Truffinet, V.; Malissein, E.; Faucher, J.L.; Turlure, P.; Bordessoule, D.; Trimoreau, F. Functional EP2 receptors on blast cells of patients with acute leukemia. *Int. J. Cancer* **2005**, *115*, 499–501. [CrossRef]

115. Hegde, S.; Kaushal, N.; Ravindra, K.C.; Chiaro, C.; Hafer, K.T.; Gandhi, U.H.; Thompson, J.T.; van den Heuvel, J.P.; Kennett, M.J.; Hankey, P.; et al. Δ12-prostaglandin J3, an omega-3 fatty acid–derived metabolite, selectively ablates leukemia stem cells in mice. *Blood* **2011**, *118*, 6909–6919. [CrossRef]

116. Chen, Y.; Hao, H.; He, S.; Cai, L.; Li, Y.; Hu, S.; Ye, D.; Hoidal, J.; Wu, P.; Chen, X. Lipoxin A4 and its analogue suppress the tumor growth of transplanted H22 in mice: The role of antiangiogenesis. *Mol. Cancer Ther.* **2010**, *9*, 2164–2174. [CrossRef]

117. Stenke, L.; Edenius, C.; Samuelsson, J.; Lindgren, J.A. Deficient lipoxin synthesis: A novel platelet dysfunction in myeloproliferative disorders with special reference to blastic crisis of chronic myelogenous leukemia. *Blood* **1991**, *78*, 2989–2995. [PubMed]

118. Kode, A.; Manavalan, J.S.; Mosialou, I.; Bhagat, G.; Rathinam, C.V.; Luo, N.; Khiabanian, H.; Lee, A.; Murty, V.V.; Friedman, R.; et al. Leukaemogenesis induced by an activating [bgr]-catenin mutation in osteoblasts. *Nature* **2014**, *506*, 240–244. [CrossRef]

119. Brennan, E.P.; Nolan, K.A.; Börgeson, E.; Gough, O.S.; McEvoy, C.M.; Docherty, N.G.; Higgins, D.F.; Murphy, M.; Sadlier, D.M.; Ali-Shah, S.T.; et al. Lipoxins attenuate renal fibrosis by inducing let-7c and suppressing TGFβR1. *J. Am. Soc. Nephrol.* **2013**, *24*, 627–637. [CrossRef] [PubMed]
120. Kinder, M.; Wei, C.; Shelat, S.G.; Kundu, M.; Zhao, L.; Blair, I.A.; Puré, E. Hematopoietic stem cell function requires 12/15-lipoxygenase-dependent fatty acid metabolism. *Blood* **2010**, *115*, 5012–5022. [CrossRef] [PubMed]

International Journal of
Molecular Sciences

Review

Curbing Lipids: Impacts ON Cancer and Viral Infection

Anika Dutta [†] and Neelam Sharma-Walia [*,†]

Department of Microbiology and Immunology, Chicago Medical School, Rosalind Franklin University of
Medicine and Science, 3333 Green Bay Road, North Chicago, IL 60064, USA; anika.dutta@my.rfums.org
* Correspondence: neelam.sharma-walia@rosalindfranklin.edu
† These authors contributed equally to this work.

Received: 30 November 2018; Accepted: 22 January 2019; Published: 2 February 2019

Abstract: Lipids play a fundamental role in maintaining normal function in healthy cells. Their
functions include signaling, storing energy, and acting as the central structural component of cell
membranes. Alteration of lipid metabolism is a prominent feature of cancer, as cancer cells must
modify their metabolism to fulfill the demands of their accelerated proliferation rate. This aberrant
lipid metabolism can affect cellular processes such as cell growth, survival, and migration. Besides the
gene mutations, environmental factors, and inheritance, several infectious pathogens are also linked
with human cancers worldwide. Tumor viruses are top on the list of infectious pathogens to cause
human cancers. These viruses insert their own DNA (or RNA) into that of the host cell and affect
host cellular processes such as cell growth, survival, and migration. Several of these cancer-causing
viruses are reported to be reprogramming host cell lipid metabolism. The reliance of cancer cells and
viruses on lipid metabolism suggests enzymes that can be used as therapeutic targets to exploit the
addiction of infected diseased cells on lipids and abrogate tumor growth. This review focuses on
normal lipid metabolism, lipid metabolic pathways and their reprogramming in human cancers and
viral infection linked cancers and the potential anticancer drugs that target specific lipid metabolic
enzymes. Here, we discuss statins and fibrates as drugs to intervene in disordered lipid pathways in
cancer cells. Further insight into the dysregulated pathways in lipid metabolism can help create more
effective anticancer therapies.

Keywords: PPAR; statins; fibrates; cholesterol; viruses; cancer; fatty acids

---

1. Introduction

1.1. Cancers and Infection Related Cancers

Cancer is a leading cause of death worldwide [1]. In 2018, 609,640 cancer deaths and 1,735,350 new
cancer cases were projected to occur in the United States alone [2]. The most deaths are caused by
breast, gastric, liver, lung, and colon cancer [1]. Lung cancer is the leading cause of cancer-related
death worldwide and in the United States. Lung cancer is also the largest contributor to new cancer
diagnoses [3]. Breast cancer is the second most common cancer in women and accounts for 25% of
all cancer diagnoses in American women [4]. Gastric cancer is the second most commonly occurring
cancer worldwide and the fourth and fifth most common cancer in men and women, respectively [1].
Colon cancer is the third most common cancer worldwide and its likelihood of diagnosis increases
progressively from age 40 [5]. Lastly, liver cancer is the fifth most common cancer in the world and has
a poor survival rate due to its aggressive nature [6].

Viruses are estimated to cause about 15% of all human cancers worldwide, and most of these
tumor viruses are hooked on lipid signaling, synthesis, and metabolism [7]. DNA viruses that
contribute to human cancers include human papillomavirus (HPV), Epstein–Barr virus (EBV), Kaposi's

sarcoma-associated herpesvirus (KSHV)—also known as human herpesvirus 8 (HHV-8), Merkel cell polyomavirus—a polyomavirus (MCPyV) associated with the development of Merkel cell carcinoma (MCC) and hepatitis B virus [7]. The two RNA viruses that can cause the development of human cancer are hepatitis C and human T lymphotropic virus (HTLV-1) [7]. EBV and KSHV are both herpesviruses with DNA genomes [7]. EBV is associated with Hodgkin's disease, B and T cell lymphomas, post-transplant lymphoproliferative disease [8], nasopharyngeal carcinomas, and leiomyosarcomas [7]. It has been associated with up to 10% of all gastric cancers, and up to 200,000 new malignancies every year worldwide [9,10]. A vaccine to prevent or treat EBV has not yet been licensed [10]. KSHV is similar to EBV in that the B lymphocyte is the predominant infected cell, and it has been estimated to cause 34,000 new cancer cases globally [7,11]. It is the leading cause of AIDS-related malignancy and cancer mortality [12]. Kaposi's sarcoma (KS) is the most common AIDS-defining cancer [13–16]. KS is a serious clinical problem prevailing in up to 50% of HIV+KS+ patients in the United States and 19–61% in Sub-Saharan Africa, who never regain remission even after combination of anti-retroviral therapy (cART] [17–19]. HPV is a DNA tumor virus that causes warts or benign papilloma, and persistent infection is associated with the development of cervical cancer [7]. It infects epithelial cells, integrates into host DNA, produces E6 and E7 oncoproteins, and disrupts tumor suppressor pathways to encourage the proliferation of cervical cancer cells [7]. It also plays a role in cancers of the skin, head, and neck [7]. The HPV vaccine is effective against HPV 16 and 18, but it does not protect against all high-risk HPV types and may not benefit women who are already infected [7].

Hepatitis C virus (HCV] and hepatitis B virus (HBV) together cause 80% of hepatocellular carcinoma cases [7]. Hepatitis C is an RNA virus that can infect liver cells and cause acute and chronic hepatitis [7]. Infection with hepatitis C virus can result in cirrhosis, which can then lead to primary hepatocellular carcinoma [7]. By contrast, hepatitis B is a DNA virus, but it can also cause acute and chronic hepatitis, which can lead to cirrhosis, liver failure, and hepatocellular carcinoma [7]. Both of these viruses could use new methods of treatment. Hepatitis C is not well suited to vaccines, because its genome mutates at a high rate and therefore it is able to escape elimination and immune recognition [7]. Hepatitis B does have a vaccine, but up to 10% are non-responders, and HBV infection still causes an estimated 1 million deaths annually [7,20]. HTLV-1 is an RNA retrovirus that has been linked to adult T-cell leukemia and a variety of chronic inflammatory diseases including [7]. Once tumor formation begins, progression is very rapid [7]. Chemotherapy for HTLV-1 associated adult T-cell leukemia can be at first beneficial but relapse is common and survival is an average of eight months [7].

There is a need for therapies that can block viral replication, and studies show that viral entry and release, and consequently replication, can potentially be blocked if membrane lipid composition is altered [21]. Additionally, modification of lipid metabolism may offer new possibilities for antiviral therapies [21].

1.2. Diet and Obesity in Cancer

Nutrition, diet, obesity, hyperlipidemia, hyperglycemia, and other modifiable risk factors such as lack of exercise, hypertension, and insulin resistance have been linked to type 2 diabetes, cardiovascular disease, hypertension, and several cancers such as breast, endometrial, pancreatic, kidney, gallbladder, colorectal, and ovarian cancers [22,23]. Obesity is a risk factor for breast cancer [23,24]. Overweight and obese breast cancer patients have an increased risk of lymph node metastasis, large tumors, and mortality [24]. There are several hypotheses as to why obesity is correlated with breast cancer. (A) The first is that the increased amount of adipose tissue in obese women means more peripheral aromatization of androgens, which causes higher levels of circulating estrogens (Figure 1) [24]. (B) Another hypothesis (Figure 1) is that obesity leads to higher levels of circulating insulin and insulin-like growth factor (IGF), which act as mitogens [24]. Extra adipose tissue releases additional non-esterified fatty acids, which leads to the development of insulin resistance [24]. Tissues are not able to efficiently absorb glucose and so the pancreas increases insulin secretion in both fasted and fed states [24]. Insulin is necessary for cell growth and it promotes DNA synthesis [24]. It also increases levels of insulin-like growth factor 1 (IGF-1), which can target breast epithelial cell receptors and induce anti-apoptotic and mitogenic pathways [24]. Half of the breast tumors have been shown to overexpress the IGF-1 receptor, and inactivation of the receptor leads to diminished mammary tumor growth [24]. Insulin and IGF-1 promote angiogenesis, increase cell proliferation, and inhibit apoptosis [25].

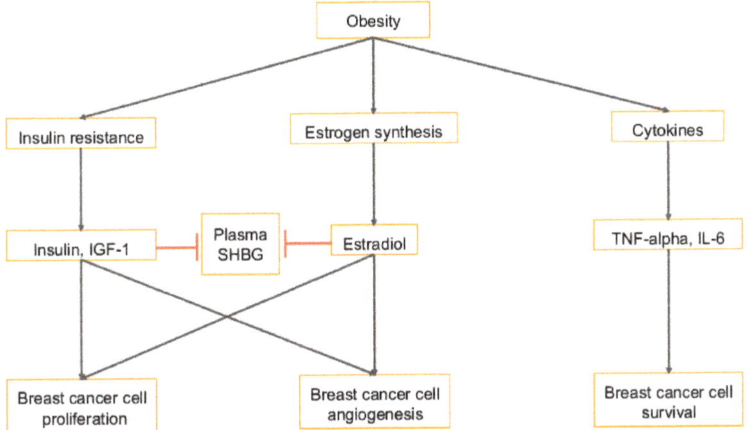

Figure 1. Pathways that link breast cancer with obesity: Several consequences of obesity, such as insulin resistance, higher levels of circulating estrogen, and secreted cytokines play a role in the development of breast cancer. Insulin and IGF-1 promote proliferation and angiogenesis by activating the PI3K/Akt and Ras/Raf/MAPK pathways. Insulin also inhibits sex-hormone-binding globulin (SHBG), which binds testosterone and estradiol so there is increased free estradiol. Circulating estrogens promote growth of breast epithelial cells and lead to more proliferation and angiogenesis as well. Adipocytes can secrete pro-inflammatory cytokines which stimulate more lipolysis and further release of free fatty acids to promote cancer cell survival. T-bars in red denote inhibition.

Estrogen and the insulin/IGF-1 pathway work together (Figure 1) in breast epithelial cells to increase transcriptional activation of estrogen receptor (ER-α) and induce mitogenic responses [24]. Estrogen does this by stimulating resting breast epithelial cells in G0/G1 to re-enter the cell cycle and go through cell division [24]. This is mediated by c-Myc, a transcription factor that is induced with estrogen stimulation [24]. With higher levels of insulin and IGF-1, concentrations of sex-hormone binding globulin (SHBG) are reduced [24]. SHBG binds estradiol and testosterone, so a decrease in its levels leads to an increase in circulating estradiol [24]. SHBG binds to breast cancer cells to inhibit estradiol-induced cell proliferation, and incubation of breast cancer cells with SHBG before treatment with estradiol cancels out the anti-apoptotic effect of estradiol [24]. It has been shown that breast cancer risk is inversely correlated with blood levels of SHBG (Figure 1) [24]. (C) The third hypothesis is that adipocytes are like endocrine cells that secrete hormone-like molecules and cytokines [24]. When invasive tumors penetrate the basement membrane and tissue barriers, the adipocytes and breast cancer cells can simply participate in paracrine interactions [24]. Studies have shown that tumor growth can be directly influenced by adipose tissue; mice injected with adipose tissue and mammary carcinoma cell line SP1 developed tumors, but no tumor growth was observed with injection of SP1 far from any fat [24]. Furthermore, breast cancer cells that were treated with adipocyte-conditioned media upregulated proliferation and metastasis while also downregulating BRCA1-associated RING domain protein 1 (BARD1), a tumor suppressor, and p18, a cell-cycle checkpoint inhibitor [24]. Breast cancer tumors injected with adipocytes grew to be three times as large as the tumors injected with fibroblasts [24].

Adipocytes secrete tumor necrosis factor-alpha (TNFα), an inflammatory cytokine [24]. Its expression is increased in obese rodent models and it inhibits the insulin receptor signaling pathway, thus assisting in the development of insulin resistance [24] (Figure 1). Adipocytes also secrete IL-6, high levels of which are associated with poor prognosis in breast cancer [24]. IL-6 production is also associated with the signals from prostaglandin PGE_2, which induces DNA transcription for IL-6 synthesis [26–28] (Figure 1). IL-6 activates the mitogen-activated protein kinase (MAPK) pathway which promotes cell migration, and it inhibits the activation of proteases that are involved in apoptosis [24]. IL-6 also inhibits cell differentiation and promotes osteoclast formation and therefore promotes metastatic growth [24]. In a previous study from our lab, we imaged three-dimensional (3D) sphere cultures of primary human mammary epithelial cells (HMEC), highly invasive breast cancer (SUM1315MO2) and primary inflammatory breast cancer (SUM149PT and SUM190PT) cells [29]. We found that the SUM1315MO2 and SUM149PT spheres were larger with differences in morphology, composition and their microenvironment [29]. We performed cytokine profiling of HMEC, SUM1315MO2 and SUM149PT conditioned media, which showed an abundance of inflammatory cytokines and chemokines such as interleukins IL-6, IL-8, and IL-17. Levels of survival kinases such as AKT, p44/42 MAPK, p65, and GSK3β were also, higher in breast cancer SUM1315MO2 and SUM149PT spheres when compared to HMEC spheres [29]. Our study [29] for the first time, revealed that osteoprotegerin (OPG) is secreted and expressed at very high levels from the SUM1315MO2 invasive breast cancer cell line, as well as the SUM149PT and SUM190PT inflammatory breast cancer cell lines when compared to healthy HMECs. Our study [29] demonstrated specific OPG staining in inflammatory breast cancer patient tumor sections. Interestingly, immunoprecipation of breast cancer cell extracts by OPG antibody revealed lipid metabolic enzyme, fatty acid synthase (FASN), which is a key enzyme of the fatty acid biosynthetic pathway [30,31]. FASN controls the process of producing de novo fatty acids from carbohydrate and amino acid-derived carbon sources [32]. Adult body mass index (BMI) is a reflection of the accumulation of adipose tissue [33]. Obesity has also been linked with gastric cancer and its complications including gastroesophageal reflux, insulin resistance, high adiponectin, leptin, and an abnormally high blood level of IGF [34] (Figure 1).

2. Lipid Synthesis in Human Cancers and Viral Infection Linked Cancers

2.1. Regulation of Lipids in Membrane Microdomains

The goal of lipid synthesis is to convert carbons derived from nutrients into fatty acids, cholesterol, phosphoglycerides, eicosanoids, and sphingolipids [35,36]. Cancerous cells show an increased rate of lipid synthesis, which has several important functions [35,36]. Lipid (fatty acids) compositional complexity, versatility, repertoire, fluidity, and lipid asymmetry is very essential to determine the characteristics of the membrane, rafts or even cell per se [35–37]. The membranes include the organelles such as the mitochondria, Golgi and the endoplasmic reticulum [35–37]. Therefore, changing lipid properties can drastically affect biomembranes, their topology, spatial organization and overall cellular machinery [35,37]. Higher levels of lipid saturation in the cell membrane protect cancer cells from oxidative damage by reducing oxidative degradation of lipids and may inhibit chemotherapeutic drug uptake [35,36,38]. Breast cancer cells have less membrane fluidity because of the increased levels of lipids, and inhibition of their synthesis is associated with apoptosis and cell cycle inhibition [35–38].

Lipids also function as signaling molecules in cancer [35]. Phosphoinositides are lipid second messengers that relay signals to the cellular machinery from activated growth factor receptors [35,39]. Lysophosphatidic acid (LPA) is another lipid second messenger that binds to G-protein-coupled receptors and activates cell migration, proliferation, and survival [35,40]. Ceramides, which are involved in inducing apoptosis and arresting cell growth, are downregulated in cancer cells [35,41,42]. Conversely, sphingosine-1-phosphate (S1P), which promotes angiogenesis and cell growth, is upregulated in cancer cells [35,43–46]. Eicosanoids regulate inflammation and thus assist in tumor progression [35,47,48]. Lipids function in protein regulation as well [35]. Prenylation facilitates the activity and localization of several signaling proteins [35,49,50]. Glycosylphosphatidylinositol (GPI) targets proteins to the outer layer of the plasma membrane [35]. Protein trafficking and localization requires different kinds of lipid anchors [35]. Association with membrane rafts is promoted by protein modification with saturated acyl chains [35]. In contrast, unsaturated fatty acids keep proteins out of cholesterol-rich membrane rafts. Regulation of growth factors is also associated with protein acylation [35,51]. Lastly, lipids are associated with autophagy—a self-destructive mechanism required under nutrient-poor conditions to remove dysfunctional components [35,52–54]. This allows for cancer cells to conserve their energy during nutrient limitation and therefore promotes cell survival [35]. The overexpression of lipid metabolism-related genes such as ATP-binding cassette transporter (*ABCA1*), acyl-coA synthetase long-chain family member 1 (*ACSL1*), 1-acylglycerol-3-phosphate *O*-acyltransferase 1 (*AGPAT1*) and stearoyl-CoA desaturase (Δ-9-desaturase) (*SCD*) has been proposed as a prognostic marker of stage II colorectal cancer (CRC) and is also called a ColoLipidGene signature [55]. Rectal adenocarcinoma (RAC), a common malignant tumor of the digestive tract is also linked to lipid peroxidation related oxidative stress, and plasmalogen alterations [56]. Pancreatic ductal adenocarcinoma (PDAC), a devastating disease is related to the intake of total fat, but especially of saturated and mono-unsaturated fatty acids (MUFAs) [57]. Given the functions of lipids in membrane structure, cell signaling, and post-translational modification of proteins, it is clear that lipids have vital roles that regulate the survival and proliferation of cancer cells [35,58]. We will focus primarily on aberrant cholesterol and fatty acid synthesis.

Lipids play an important role in viral infection, as they are the structural elements of cellular and viral membranes [21]. Viruses target lipid synthesis and signaling to remodel their host cells and generate lipids for the viral envelope [59]. Lipid interactions such as membrane fusion, envelopment, and remodeling are vital for viral replication, and compounds that affect lipids such as cholesterol and sphingolipids interfere with viral replication [21]. Viruses replicate inside the host cell, so they have to cross the host cell membrane for entry and exit [21]. Lipids have several roles in viral entry. They can function as direct and indirect viral receptors, as entry cofactors, and fusion cofactors [59]. Lipids are involved in viral replication in several ways. They have a role in phosphoinositide signaling to reorganize the membrane or bind viral proteins [59]. Viruses may also generate lipids at sites of replication by promoting lipid biosynthesis [59]. By inducing lipid metabolism, viruses exploit the energy in lipid stores during their replication [59]. Viruses can induce autophagy to degrade lipid droplets and release lipids, which are oxidized in mitochondria to generate ATP [59]. In addition to providing energy, lipid droplets can aid in viral assembly and budding [59]. Lipids may also facilitate viral exit by use of the VLDL secretion machinery [59]. Knockdown of apolipoproteins ApoE and ApoB decreased the amount of secreted infectious virus [59].

Hepatitis C affects lipid metabolism and uses it to its advantage throughout the infectious cycle [60]. An increase in lipid droplets has been found in liver biopsies of infected patients [60]. EBV-encoded latent membrane protein 1 (LMP1) has been shown to promote cell proliferation and progression of nasopharyngeal carcinoma via activation of SREBP1-mediated lipogenesis [61]. Short-chain fatty acids (SCFA) stimulate the two related human gamma-herpesviruses to enter the lytic cycle through different pathways of chromatin remodeling [62]. EBV LMP1 has been shown to reorganize membrane lipid rafts and cytoskeleton microdomains to modulate phosphatidylinositol 3-kinase (PI3K) and its downstream target, Akt signal transduction [63].

In patients infected with human T lymphotropic virus (HTLV-1), significantly higher levels of VLDL and triglycerides were detected [64,65].It has also been found that disruption of lipid rafts can lead to a decrease in infection by HTLV-1 [66]. HTLV-1 encoded Tax1 protein has been associated with the accumulation of cellular alterations that promote leukemia in infected HTLV-1-infected individuals [66]. The cytoplasmic Tax1 protein persistently resides in the Golgi-associated lipid raft microdomains and Tax1 directs lipid raft translocation of IKK through selective interaction with IKKγ [67–69]. Depletion of IKKγ impairs Tax1-directed lipid raft recruitment of IKKα and IKKβ suggesting that Tax1 actively recruits IKK to the lipid raft microdomains for continuous/sustained NF-κB activation and contributes to HTLV-1 infection linked tumorigenesis [67–69].

2.2. Association of Lipid Pathways with Glycolysis, Fatty Acid Synthesis, and Glutaminolysis

High-throughput RNA sequencing demonstrated significant changes in genes involved in overlapping lipid-related functions and/or glucose metabolism disorder in KS lesions [70]. KSHV infection has been shown to induce glycolysis, glutaminolysis, and fatty acid synthesis pathways, for the survival of latently infected endothelial cells [71]. KSHV infection of primary endothelial cells utilizes the host lipid raft-dependent macropinocytosis pathway and endosomal sorting complexes required for transport (ESCRT)-0 proteins for entry [72,73]. KSHV ORF45, a viral protein in the tegument layer, connecting capsid and envelope, associates with lipid rafts of host cellular membrane triggering KSHV budding for final envelopment and virion maturation [74–76]. Lipid metabolism plays a vital part in KSHV infection and therefore may be used as a drug target [12,76]. KSHV can manipulate lipid biosynthesis in a host cell to promote viral infection and tumorigenesis in several ways [12,76]. Lipids play a role in the initial infection, survival and proliferation, reactivation, and angiogenesis of KSHV infected cells [12,76]. Studies have shown that lipids are important for the survival of KSHV-infected cells and that they have higher rates of fatty acid synthesis and aerobic glycolysis than primary B cells [12,76]. Reprogramming of cholesteryl ester metabolism has been demonstrated to be involved in regulating neo-angiogenesis and metastasis in KSHV infected endothelial cells [77]. KSHV stabilized hypoxia-inducible factors (HIFs) has been reported to play a critical role in KSHV latency, reactivation and metabolic reprogramming (carbohydrate, lipid, and amino acids) [78].

3. Cholesterol Synthesis in Human Cancers and Viral Infection Linked Cancers

Cholesterol has vital physiological roles such as controlling membrane fluidity and using cell signaling to regulate cell growth, proliferation, and migration [79]. It is also a precursor for steroid hormones which activate nuclear receptors to control inflammation and immune functions [79]. Cholesterol is transported from the liver to cells through the bloodstream in a low-density lipoprotein (LDL) bound form [80]. Cells take-up the LDL using clathrin mediated endocytosis and the endocytic pathway transports it to lysosomes where it is hydrolyzed to free cholesterol molecules [80]. The cholesterol molecules are then taken to the membrane-bound organelles and the cell membrane [80].

Cholesterol levels are tightly regulated in the body [79]. The key regulators are sterol regulatory element-binding protein transcription factor 2 (SREBF2) and liver x receptors (LXR; LXRα and LXRβ) [80]. Levels of endoplasmic reticulum (ER) cholesterol are used to sense for intracellular cholesterol homeostasis [80]. If there is a decrease in ER cholesterol, SREBF2 is translocated from the ER to Golgi to the nucleus to activate gene transcription for cholesterol synthesis [80]. Conversely, if there is an increase in cholesterol levels, its synthesis is shut down and its export is facilitated by activation of LXR receptors [80]. LXRs are sterol-sensitive transcription factors of the nuclear receptor superfamily. LXRs regulate the expression of several genes involved in the uptake, transport, efflux, and excretion of cholesterol in a tissue-dependent manner. These are also crucial regulators of the reverse cholesterol transport pathway and subsequently whole-body cholesterol content [81].

There are several signaling pathways that activate cholesterol synthesis in cancer cells [80]. Intracellular cholesterol levels are promoted by the activation of PI3K/AKT signaling, which induces cholesterol synthesis by activating the SREBP transcription factor, the regulator of cholesterol synthesis encoding genes [80] (Figure 2). SREBP is activated by inhibiting mTORC1 dependent ABCA1 mediated cholesterol export and activating LDL receptor-mediated cholesterol import pathway [80]. Promotion of cholesterol synthesis by the AKT/mTORC1/SREBP pathway contributes to cell growth, bone metastases, and cancer aggressiveness [80] (Figure 2).

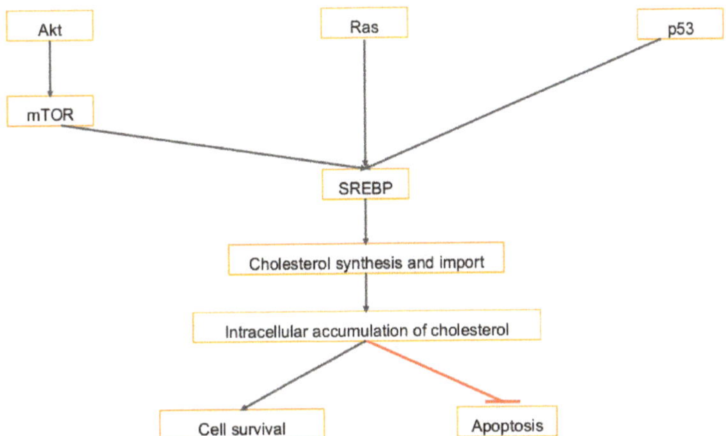

Figure 2. Pathways that lead to an accumulation of cholesterol: Activation of SREBP transcription factor induces cholesterol. This is induced by activation of PI3K/Akt/mTOR signaling, cancer gene RAS, and dysregulation by TP53. The accumulation of cholesterol in the cell promotes survival and inhibits apoptosis. T-bar in red denotes inhibition.

Cholesterol synthesis is also activated through TP53, a frequently mutated gene in cancer [80]. In breast cancer, the cholesterol synthesis pathway is unregulated by the loss of TP53 function [80]. This disrupts the breast tissue architecture and induces proliferation [80]. In studies where the mutant TP53 was knocked down, the morphology of the breast cancer cells changed from the disorganized back to a normal phenotype [80] (Figure 2).

In cancer cells, high levels of mitochondrial cholesterol lead to resistance to apoptotic signals [80]. Cholesterol import into the mitochondria is regulated by the two proteins steroidogenic acute regulatory (STAR) and STAR-related lipid transfer domain containing 3 (STARD3) [80]. In human epidermal growth factor receptor 2 (HER2/neu, c-erbB2) positive breast cancer cells, STARD3 is associated with a poor prognosis, and lower levels of STARD3 increase cell death while reducing cell proliferation [80]. HER2 is a membrane tyrosine kinase and oncogene, when activated it provides the cell with potent proliferative and anti-apoptosis signals, and confers aggressiveness to breast cancers. Higher levels of STARD3 also decrease the adhesiveness of breast cancer cells, which promotes metastases [80]. Another gene that controls cholesterol homeostasis is ABCA1, a cell membrane cholesterol exporter, and it is dysregulated in cancer cells [80]. Lower levels of ABCA1 increase mitochondrial cholesterol levels and promote cancer cell survival [80]. Growing tumors have been found to have 3-fold lower levels of ABCA1 expression as opposed to normal cells [80].

High cholesterol is a risk factor for several pathologies and is associated with the development of cancer [79]. It has been observed that cholesterol promotes cell proliferation and migration [79]. It accelerates the formation of tumors, enhances tumor angiogenesis, and increases their aggressiveness [23]. Cholesterol is also associated with chemotherapy resistance [79]. In breast cancer tumors, higher plasma cholesterol levels are associated with higher expression of cyclin D1, an oncogenic driver [23]. Furthermore, high cholesterol content in lipid rafts is associated with higher rates of cell survival in prostate cancer cells [23]. Lipid rafts are implicated in Akt activation, which then phosphorylates pro-apoptotic proteins and inactivates them [82]. Cholesterol is a major component of cellular and mitochondrial membranes; the inhibition of cholesterol synthesis may inhibit the formation of new membranes demanded by proliferating tumor cells [83]. It was observed that although lipogenesis was upregulated in cancer cells, during tumor development, plasma cholesterol levels were reduced [23]. This suggests that transformed cells may utilize more cholesterol than normal cells; thus regulation of the cholesterol synthesis pathway may limit cellular proliferation [84]. Cholesterol metabolic pathways are required for the replication, secretion, and entry of HCV and drugs targeting cholesterol metabolic pathways have potential in treating HCV infection [85]. Cholesterol has also been identified as a critical factor for EBV latent membrane protein 2A trafficking and protein stability as it regulates LMP2 phosphorylation and ubiquitination [86]. Treatment of cells with methyl-beta-cyclodextrin (MβCD), which depletes cholesterol from the plasma membrane, increased LMP2A levels, its secretion in exosomes and blocked LMP2A endocytosis resulting in LMP2A abundance in the plasma membrane [86].

The mevalonate (MVA) pathway, which leads to the production of cholesterol, can be dysregulated in tumor cells [87] (Figure 3). Many of the downstream products are required for protein synthesis, membrane integrity, signaling, and cell-cycle progression, and are therefore critical in cell proliferation [84]. Higher levels of enzymes in the MVA pathway are associated with rapid progression and poor prognosis in cancer patients, and treatment with mevalonate promotes proliferation of breast cancer cells and tumor growth [82,87] (Figure 3). This pathway is upregulated by mutated p53, a tumor suppressor protein [84]. The increased proliferation rates are associated with a faster entry of cells through the G1 restriction point and into S phase [88]. These cells have more activating phosphorylation of cyclin-dependent kinase-2 (CDK-2), which controls initiation of DNA synthesis and replication, and decreased inhibitory binding of CDK-2 to p21, a regulator of the G1 restriction point [88]. The disruption of this pathway in malignant cells may result in the inhibition of cell-cycle progression and reduce proliferation and metastasis of cancer cells [84].

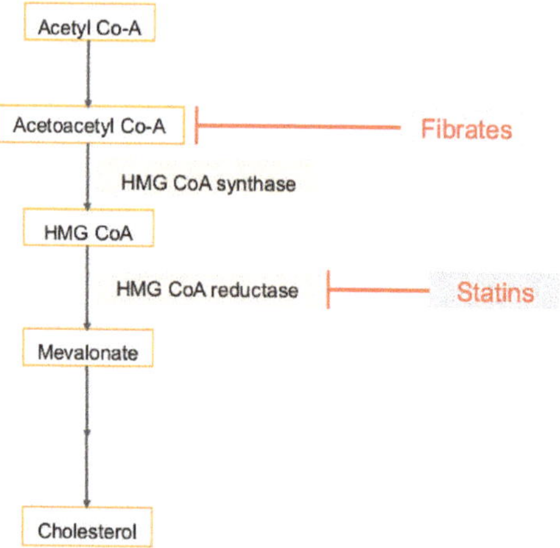

Figure 3. Mevalonate pathway as an important metabolic pathway: The mevalonate pathway is regulated by HMG CoA reductase (HMGCR), and this enzyme is targeted by statins to decrease plasma cholesterol. Fibrates target the mevalonate pathway by inhibiting acetoacetyl coenzyme A. This reverses the effects of cholesterol to inhibit cell proliferation and trigger apoptotic parameters. Downregulating the pathway also suppresses production of farnesyl pyrophosphate and geranylgeranyl phosphate to inhibit the invasive properties of cancer cells. T-bars in red denote inhibition.

Hepatitis C virus requires geranylgeranylation, a metabolite of the mevalonate pathway, to allow binding of viral protein NS5A to viral cofactor FBL2 [89,90]. In the mevalonate pathway, simvastatin interferes with the activity and localization of EBV latent membrane protein 1 LMP-1 to induce apoptosis [91]. KSHV viral microRNAs (miRNAs) have been shown to target 3-Hydroxy-3-methylglutaryl-coenzyme A (CoA) synthase 1 (HMGCS1), 3-hydroxy-3-methylglutaryl-CoA reductase, enzymes in the mevalonate/cholesterol pathway [89]. Addition of 25-hydroxycholesterol to primary cells inhibited KSHV infection suggesting that KSHV miRNAs decrease the level of 25-hydroxycholesterol and promote viral infection [89].

4. Fatty Acid Synthase (FASN) in Cancers and Viral Infection-Associated Cancers

Besides cholesterol, triacylglycerol can also be synthesized from acetyl-CoA [92] (Figure 4). Malonyl-CoA is produced from acetyl-CoA via acetyl-CoA carboxylase (ACC), which is converted to Palmitate. Palmitate is lengthened to form stearate by enzyme fatty acid elongase (Elovl #1–7) [92] (Figure 4). Fatty acid elongation primarily occurs in the endoplasmic reticulum and uses malonyl-CoA and fatty acyl-CoA as substrates. Another endoplasmic reticulum-bound enzyme is stearoyl-CoA desaturase (SCD), which catalyzes the desaturation of saturated palmitoyl- and stearoyl-CoA, those are converted to palmitoleoyl- and oleoyl-CoA, respectively [92] (Figure 4). Multiple drugs targeting lipid pathways are in clinical trials. Antitumor synergy has been observed in vitro and in vivo combining A939572 with an mTOR inhibitor in ccRCC and A939572 is a small molecule that specifically inhibits SCD1 enzymatic activity [93–95]. Similarly, T-3764518, a novel and orally available small molecule inhibitor of SCD1 showed promising antitumor effects in colorectal cancer HCT-116 cells and their growth and mesothelioma [96,97].

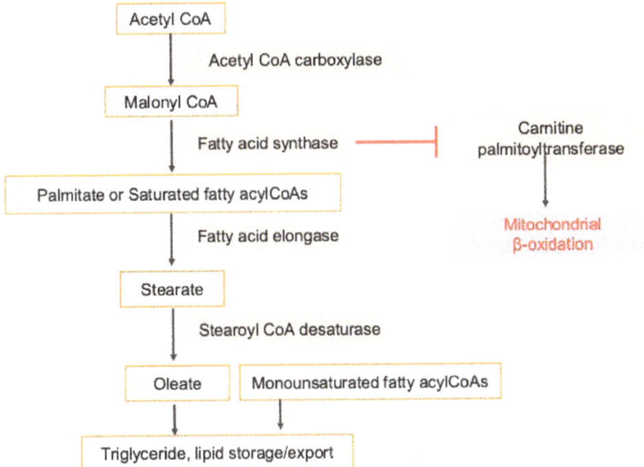

Figure 4. Triacylglycerols are synthesized from acetyl-CoA. Acetyl-coA is metabolized to malonyl-CoA via acetyl-CoA carboxylase, which in turn is converted to Palmitate, the principal product of the fatty acid synthase system in animal cells. Palmitate is lengthened to form stearate by enzyme fatty acid elongase. Stearate, a saturated fatty acid is subsequently metabolized by stearoyl-CoA desaturase enzyme, that forms a double bond in stearoyl-CoA, leading to the monounsaturated fatty acid oleic acid. Carnitine palmitoyltransferase (CPT), the enzyme in the outer mitochondrial membrane, converts long-chain acyl-CoA species to their corresponding long-chain acyl-carnitines for transport into the mitochondria. CPT induces mitochondrial β-oxidation, which is a complex pathway involving energy metabolism. FASN inhibits CPT with resultant inhibition of fatty acid oxidation. T-bar in red denotes inhibition.

FASN, the enzyme responsible for de novo fatty acid synthesis, is expressed at higher levels in breast, prostate, colon, and ovarian cancer cells as opposed to normal human mammary epithelial cells (HMEC) [98]. We analyzed the breast tissue sections of healthy subjects and breast cancer patients for the presence of FASN by immunofluorescence staining using anti-FASN antibody (unpublished results). Abundant FASN expression was detected in breast cancer tissue sections (Figure 5A,B,D,E) compared to the normal healthy control tissue sections (Figure 5C,F). We next evaluated the fold change in FASN expression in all 32 sections by densitometry analysis using ImageJ software. A 0–2, 2–4, 4–6-fold induction in FASN expression was observed in 14, 12 and 6 tumor sections, respectively. Collectively, these results highlight the presence of PPARα expression in human breast cancer tissues.

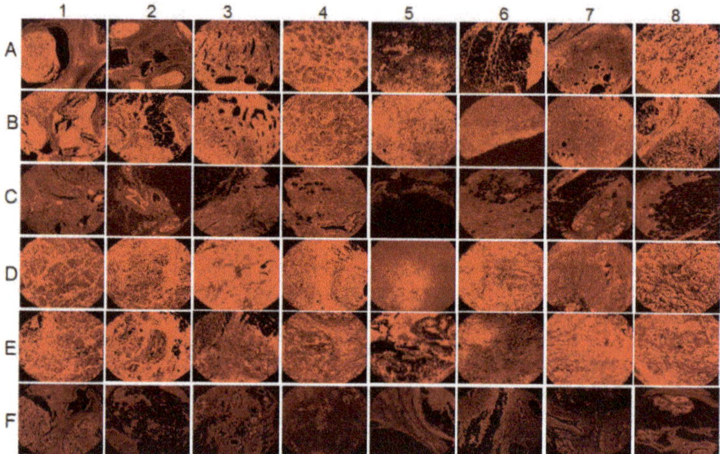

Figure 5. Fatty acid synthase (FASN) levels in human breast cancer tissue samples. 16 breast cancer tissue samples, in duplicates (**A,B,D,E**) along with their controls (**C,F**) were analyzed by IHC staining for FASN. Magnification for the panels is 4×.

Our previous studies show larger numbers of lipid droplets in SUM1315MO2 (invasive breast cancer cell line) and SUM149PT cells (inflammatory breast cancer cell lines) when compared to HMEC using electron microscopy, immunofluorescence, and fluorescence quantitation [30]. SUM1315MO2 and SUM149PT also had higher expression of FASN when compared to HMEC using immunoblotting and immunofluorescence staining [30]. In addition, Western blot analysis showed higher levels of cyclooxygenase-2 (COX2), which is important in cancer progression and inflammation, in SUM1315MO2 and SUM149PT cells [30]. To investigate the importance of FASN and COX2 in cancer cell progression, these two enzymes were blocked with C75 (FASN inhibitor) and/or celecoxib (COX2 inhibitor) and the results showed a reduction in the amount of lipids formed per cell as well as decreased levels of cell survival proteins p-Erk and p-GSK3β [30]. Since high levels of FASN leads to more lipogenesis, which causes more lipids to be integrated into membrane lipid rafts, activating membrane receptor tyrosine kinases and resulting in oncogenic signaling pathways [98]. Therefore, fatty acid synthase is associated with poor prognosis in patients with cancer [99]. It has been associated with clinically more aggressive cancers; in stage I breast cancer, there was a four-fold increase in mortality risk associated with the expression of fatty acid synthase in two studies [100]. FASN is correlated with peritumoral lymphatic vessel invasion and inversely correlated with breast cancer survival [101].

Fatty acid synthase is an attractive therapeutic target because it is expressed at low or undetectable levels in normal tissues because fatty acids are supplied by diet [100]. In contrast, since FASN is overexpressed in malignant cells, a FASN inhibitor would target the cancerous cells while leaving the normal cells unaffected [100]. It is also restricted solely to fatty-acid synthesis, unlike the other lipogenic enzymes. For example, acetyl-CoA carboxylase is the rate-limiting enzyme of fatty-acid synthesis but it is also widely distributed in muscle [100]. In colon cancer cells, inhibition of fatty acid synthase inhibited S-phase progression and DNA replication [100].

Preclinical evaluation of novel FASN inhibitors in primary colorectal cancer cells (CRCs) and a patient-derived xenograft model of colorectal cancer showed that anti-tumor activity was primary due to a significant decrease in the activation of Akt and Erk1/2 oncogenic pathways in CRCs [102]. Treatment with a FASN inhibitor led to the arrest of cell growth and apoptosis of breast tumor cells, further reinforcing the role of FASN in tumorigenesis [101]. FASN inhibition does this by upregulating pro-apoptotic genes BCL2 Interacting Protein 3 (BNIP3) and tumor necrosis factor related apoptosis-inducing ligand (TRAIL) [101]. BNIP3 induces apoptosis by mitochondrial dysfunction [101].

TRAIL has been found to be inversely correlated with FASN expression, which demonstrates that TRAIL is a component of the apoptotic pathway caused by FASN inhibition [101]. FASN can be inhibited using FASN small interfering RNA (siRNA). FASN inhibition by siRNA increases ceramide synthesis, which upregulates BNIP3 and TRAIL [101].

Cerulenin is another inhibitor of fatty acid synthase, and treatment demonstrated a cytotoxic property that was proportional to fatty acid synthase [100]. In other words, Cerulenin was selectively toxic and induced apoptosis in cancer cells but not normal cells in vitro [100]. Cerulenin is, however chemically unstable and therefore a more stable inhibitor of fatty acid synthase is needed as an anticancer agent [100].

Hepatitis B interferes with lipid metabolism as well [103]. Fatty acid synthase is upregulated in HBV infected cells, and many studies have shown that HBV promotes the synthesis of fatty acids and cholesterol [103]. Lipid metabolism is altered during EBV infection as well [8]. Fatty acid synthase expression is increased in infected cells to engage de novo synthesis of palmitate, instead of relying on dietary fatty acids as healthy cells do [8] (Figure 6). This is confirmed by the findings that tumorigenesis in EBV infected cells driven by fatty acid synthase can be blocked by using fatty acid synthase inhibitors [8]. Moreover, treatment of infected cells with a fatty acid synthase inhibitor reduces cell survival [12]. Inhibition of key enzymes such as acetyl-CoA carboxylase also leads to apoptosis in infected cells [12]. EBV immediate-early (IE) protein BRLF1 activates expression of the host fatty acid synthase through a p38 stress mitogen-activated protein kinase and induces a lytic form of EBV replication [104].

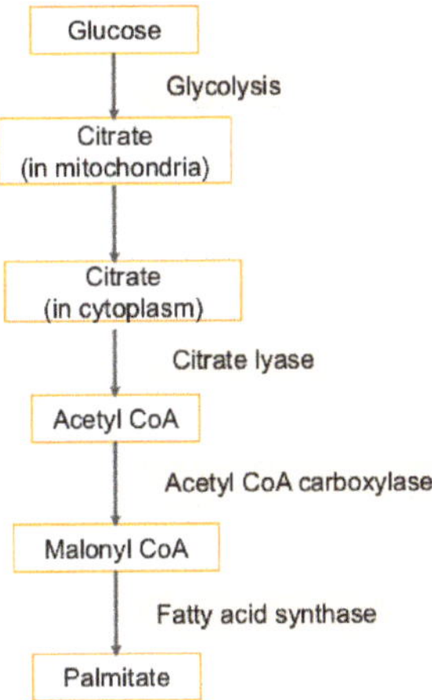

Figure 6. Fatty acid deregulation in pathogenesis of human cancer: A key lipogenic enzyme in fatty acid is fatty acid synthase (FASN), which can be inhibited to downregulate fatty acid synthase to inhibit DNA replication and arrest cell growth.

KSHV infection also increases peroxisome biogenesis, and the proteins involved in the peroxisomal metabolism of very long chain fatty acids are critical for the survival of infected cells [105]. Reactivation is induced by short chain fatty acids, and angiogenesis is promoted by the vitamin D receptor pathway and increased cholesteryl ester synthesis [12].

5. Arachidonic Acid Pathway Metabolites in Cancers and Viral Infection-Associated Cancers

Many cancers have been shown to have aberrant metabolism of arachidonic acid [106–110]. Arachidonic acid is a fatty acid found in the cellular membrane [106] (Figure 7). It is metabolized to eicosanoids through 3 pathways: the cytochrome P450 monooxygenase (ω-hydroxylases and epoxygenases), the cyclooxygenases (COX-1 and COX-2), and the lipoxygenase (5-LO, 12-LO, 15-LOa,

15-LOb) pathways [106]. Metabolism by these pathways generates eicosanoids, such as leukotrienes and prostaglandins, which are pro-inflammatory and promote tumor growth [106–110] (Figure 7). Besides these eicosanoids are the epoxyeicosatrienoic acid (EETs), which are generated via the conversion of arachidonic acid by CYP epoxygenases and are mainly metabolized by soluble epoxide hydrolase (sEH). EETs, lipid signaling molecules are autocrine and paracrine mediators of cell proliferation, migration, inflammation, and angiogenesis in several tissues [111,112]. Pro-inflammatory eicosanoids are highly expressed in cancer cells and can promote tumor progression by inducing the secretion of growth factors by epithelial cells, inducing angiogenesis by binding receptors on stromal cells, regulating apoptosis, cell migration and proliferation by activating receptors on tumor epithelial cells and promoting a microenvironment that supports tumor growth [106].

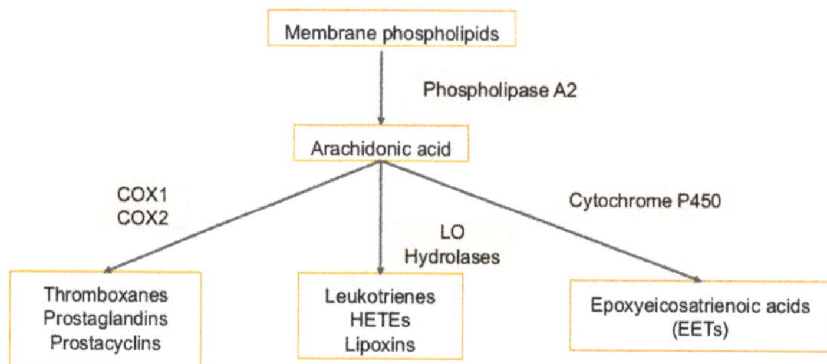

Figure 7. Arachidonic acid cascade: Arachidonic acid is a fatty acid that is freed from cellular membranes by phospholipase A2. It can then be metabolized to prostanoids such as thromboxanes (TXA2), prostacyclin (PGI2) and prostaglandins PGD$_2$, PGE$_2$, and PGF2a through the cyclooxygenase (COX) pathway. The lipoxygenase (LO) pathway along with hydrolases converts arachidonic acid into leukotrienes, hydroxyeicosatetraenoic acids (HETEs), and lipoxins (LXA4 and LXB4). Lastly, the cytochrome P450 monooxygenase pathway converts arachidonic acid into epoxyeicosatrienoic acids (EETs).

The 5-LO expression is promoted by pro-inflammatory stimuli, and it is constitutively expressed in cancers including breast, prostate, lung, colon, and esophagus [106] (Figure 7). Inhibition of this pathway may reduce tumor growth and metastases, and have shown this in mouse models of human breast, esophageal, colon, and skin cancer [106–110]. The COX enzyme has two isoforms: COX1, which is ubiquitously expressed, and COX2, which is inducible [113] (Figure 7). COX2 is highly expressed during tumor formation and in areas of inflammation, and increased expression is associated with decreased rates of survival among patients with mesothelioma, breast, prostate, pancreas, liver, stomach, lung, and esophagus cancers [106–110] (Figure 7). Mammary angiogenesis and tumorigenesis is reduced in rodent models when COX2 is knocked out [113].

Both COX enzymes catalyze the ultimate production of prostaglandin E$_2$ (PGE$_2$), which upregulates the production of aromatase in fat cells and therefore more estrogen production, which promotes tumor cell proliferation [106]. PGE$_2$ also upregulates NFkB activity, which is involved in inhibiting apoptosis and upregulates anti-apoptotic protein Bcl2 via Ras-MAPK signaling [106]. PGE$_2$ has immunosuppressive effects as well: it upregulates immunosuppressive TH2 cytokines while downregulating anti-tumor TH1 cytokines [106]. It suppresses natural killer cells, which have anti-tumor activity, by inhibiting CD8+T activity [106]. A negative correlation has been shown between prostaglandin levels in breast tumors and survival [114].

PGE$_2$ and leukotriene B4 (LTB4) promote angiogenesis by inducing vascular endothelial growth factor (VEGF), fibroblast growth factor 2 (FGF2), and chemokines CCL2 and CXCL1 [106]. They also shift the normal tissue microenvironment to one that supports tumor growth by exacerbating inflammation by recruiting more leukocytes into the tissue from the circulation [106].

Non-steroidal anti-inflammatory drugs (NSAIDs) inhibit COX and therefore have anti-inflammatory effects. These have been reported to reduce the risk of tumors in breast, lung, colon, and prostate cancers [106–110]. However, prolonged use of NSAIDs has cardiovascular and gastrointestinal side effects [106]. One way to avoid these may be to use antagonists of PGE_2 receptors to inhibit the growth of tumors, which has been shown to have an effect in colon, lung, esophageal, and breast cancer animal models [106]. The cyclooxygenase-2-prostaglandin E_2-eicosanoid receptor inflammatory axis has been demonstrated to play a key role in KSHV associated malignancies [115]. KSHV can activate parts of the lipoxygenase pathway such as leukotrienes and 5-lipoxygenase to aid in initial infection [12,116].

Concurrent inhibition of COX-2 and soluble epoxide hydrolase (sEH) using PTUPB, an orally bioavailable COX-2/sEH dual inhibitor results in antitumor, anti-angiogenic activity and has organ-protective effects [117]. PTUPB has the potential for future combination chemotherapy partner for cisplatin [117].

6. Use of Statins in Cancers, Viral Infections and Associated Cancers

Statins are competitive inhibitors of 3-hydroxy-3-methyl-glutaryl-coenzyme A reductase (HMGCR) and are referred to as cholesterol-lowering drugs as these regulate the rate-limiting step of the cholesterol synthesis pathway [118–120]. Statins target the MVA pathway by inhibiting HMGCR and reducing the levels of mevalonate and ultimately reducing plasma cholesterol [121] (Figure 3). HMGCR regulates the mevalonate pathway, and when upregulated, promotes tumor growth in breast cancer cells [122]. High levels of HMGCR are correlated with poor survival outcomes [122]. It has also been shown that HMG CoA reductase can induce growth in breast epithelial cells, independent of Anchorage [35]. Statins, therefore, reduce metastatic potentials and improve the survival rates in metastatic breast cancer mouse models [122,123]. Statins induce apoptosis of many cancer cell types and block pathways driving cell division in several types of leukemias and lymphomas [118–120]. Furthermore, an inverse correlation between cancer risk and statin use has been shown [124]. Studies have shown that statins reduce the incidence of breast cancer, colorectal cancer, prostate cancer, and cholangiocarcinoma [79]. Statins have been associated with lower risk of gastric cancer (due to weakening H. pylori infection), reduced hepatocellular carcinoma risk, and lower risk of melanoma [123,125,126]. By suppressing the mevalonate pathway, statins inhibit cell proliferation, trigger apoptotic parameters, and are powerful anti-inflammatory agents [121]. Another effect of blocking the mevalonate pathway is the regulation of the production of geranylgeranyl and farnesyl pyrophosphates, which modify the Rho and Ras -GTPases [127]. Rho proteins are associated with the invasive and proliferative properties of cancer cells [127]. Statins suppress the expression of matrix metalloproteinases (MMPs) and very late antigens (VLAs), and subsequently inhibit tumor growth and spontaneous metastasis [123]. They also induce cell apoptosis by inhibiting PI3K/Akt signaling [127,128] and produce anti-proliferative effects on lung, liver, colorectal, and prostate tumors [124]. Statins also modulate MAPK and CDK2, which reduce the expression of p21 and p27 cyclin kinase inhibitors, which regulate proliferation and apoptosis of tumor cells [127].

Studies in rodents show that lipophilic statins such as simvastatin, lovastatin, and fluvastatin have a protective effect on the growth of several tumor types by decreasing mevalonate synthesis [88]. Lovastatin arrests the cell cycle at the G1 phase by inducing cell cycle inhibitors p21 and p27, and decreases transition to the S and G2/M phases [83]. Lovastatin treated breast cancer cells show higher levels of caspase activity, which is consistent with apoptosis initiation [83]. Furthermore, cytochrome c was released by a reduction in the mitochondrial membrane potential and Bax translocation into the mitochondria [83]. Fluvastatin increases the rate of apoptosis and reduces breast tumor proliferation [129]. Simvastatin in particular has been shown to reduce the risk of several different cancer types, including breast cancer [125]. In mice that were treated with simvastatin, tumor growth and proliferation were inhibited and reduced [127]. Simvastatin is also associated with a reduced risk of breast cancer recurrence [4]. In bile duct cancer cells, simvastatin suppresses cell proliferation by inducing G1 phase cell cycle arrest [130]. It also activates caspase-3, downregulates

Bcl-2 expression, and enhances Bax expression, which consequently induces apoptosis in bile duct cancer cells [130]. In prostate cancer, simvastatin suppresses metastasis by inhibiting TGF-β1, which promotes tumor progression and is associated with poor outcomes [131]. Statins were reported to prime cancer cells for apoptosis and worked in synergy with an inhibitor of BCL2 called venetoclax, and demonstrated better clinical responses compared to venetoclax alone [118–120]. In breast cancer cells, simvastatin affects several signal transduction pathways and changes expression of Akt, NFkB, BclXL, and PTEN [127]. Simvastatin reduces cell viability through reduction of raft formation, which consequently downregulates survival kinase Akt [82]. Akt promotes cell survival and inactivates pro-apoptotic proteins, and is therefore characteristic of malignant tumors [82]. Simvastatin blocks phosphorylation of Akt, which inhibits Akt activation [127]. Simvastatin inhibits NFkB transcriptional activity, which subsequently decreases expression of anti-apoptotic protein BclXL and increases levels of phosphatase and tensin homolog (PTEN), a tumor suppressor protein [127] (Figure 8). Akt regulates many transcription factors, including NFkB, which is constitutively active in breast cancer cells [127]. NFkB induces expression of several anti-apoptotic proteins including BclXL [127]. Breast cancer cells have higher levels of BclXL, and simvastatin inhibits its transcription by targeting NFkB [127]. In breast cancer cells, a mutation or deletion in PTEN increases levels of PI3 kinase product PIP3 [127], which consequently upregulates Akt activity [127]. Activation of phosphatidylinositol 3 kinase (PI3 kinase) and its target Akt kinase is associated with the anti-apoptotic properties of cancer cells [127]. PI3 kinase is negatively regulated by phosphatase and tensin homolog (PTEN), a tumor suppressor protein [127]. PTEN controls the levels of the product of PI3 kinase: PI 3, 4, 5-triphosphate (PIP3), which activates Akt kinase [127]. Simvastatin increases expression of PTEN to reduce activation of Akt by targeting NF-κB to inhibit its repression of PTEN expression and therefore inhibit breast cancer cell proliferation [127]. Simvastatin thus increases transcription of phosphatase and tensin homolog (PTEN), which suppresses the oncogenic phosphatidylinositol-3-kinase (PI3K) pathway and reduces expression of the anti-apoptotic protein bcl-XL [129].

Two other pathways simvastatin uses to suppress cell growth and induce apoptosis are the MAPK/ERK pathway and the JNK/CHOP/DR5 pathway [132]. Simvastatin suppresses tumor growth by suppressing the MAPK/ERK pathway, which promotes cancer cell survival and metastasis [132]. Simvastatin also activates the JNK/CHOP/DR5 pathway, which induces apoptosis in breast cancer cells [129]. Simvastatin has been reported to inhibit the activity of HBV minichromosome maintenance (MCM) 7 complex, an important host factor aiding virus genome replication in host cells by increasing the phosphorylation of eIF2α, which is mediated by the liver kinase B1 (LKB1)-AMP-activated protein kinase (AMPK) signaling pathway [133]. Statin use may have potential protective effects and reduce the risk for hepatocellular carcinoma (HCC) in HBV-infected patients in a dose-dependent manner [134].

Treatment with statins, which downregulate the mevalonate pathway, has been shown to block hepatitis C virus [89]. Statin treatment affected cell cycle, apoptosis and alternative splicing genes between the native (B-lymphocytes) and EBV transformed cells human lymphoblastoid cells (LCLs) [135]. Simvastatin has been shown to delay the development of EBV-lymphomas in severe combined immunodeficiency mice models and could be considered for the treatment of EBV-lymphomas [136]. Anti-HCV effects or reduced HCV replication via Ceestatin, a drug that binds to 3-hydroxy-3-methylglutaryl-coenzyme A (HMG-CoA) synthase and irreversibly inhibits HMG-CoA synthase in a dose-dependent manner are reversed by addition of HMG-CoA, mevalonic acid, or geranylgeraniol [137]. Side effects of statins include acute renal failure, myopathy, myoglobinuria, and hepatotoxicity [138]. Also, increased risk of diabetes mellitus and muscle pain has been observed [79]. Recent studies have found an increased risk of cancer after use of pravastatin, which caused an increase of mevalonate synthesis in extrahepatic tissues and thus promoted the growth of breast cancer cells. No such association has been found with use of lipophilic statins such as simvastatin, lovastatin or fluvastatin [88].

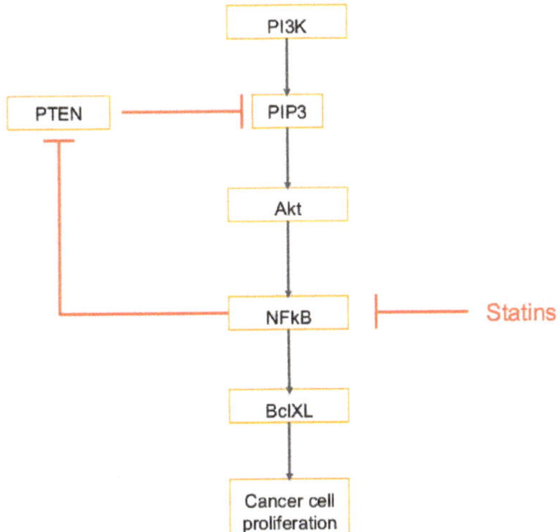

Figure 8. Statin therapy and downstream consequences: Statins block transcription of NFkB. This decreases expression of anti-apoptotic BclXL and inhibits cancer cell proliferation. Inhibition of NFkB also increases expression of pro-apoptotic phosphatase and tensin homolog (PTEN). PTEN inhibits PI3 kinase, and therefore also inhibits the target of PI3 kinase, Akt kinase, which are both associated with the anti-apoptotic properties of cancer cells. T-bars in red denote inhibition.

7. Use of Fibrates (PPARα Agonist) in Cancers, Viral Infections and Associated Cancers

Fibrates are an effective and safe group of hypolipidemic drugs, frequently used to treat patients with atherogenic dyslipidemia and have shown tremendous potential in cardiovascular outcomes [139–141]. Fibrates also have benefits in carbohydrate metabolism, adipokines levels, thrombosis, atherosclerosis (hardening of arteries), heart attack, stroke and inflammation [139–141]. Fibrates such as fenofibrate can inhibit triglyceride (TG) synthesis via reducing the availability of free fatty acids by promoting β-oxidation and also has anticancer agent potential [139–141]. Fibrates are well-known ligands of peroxisome proliferator-activated receptor α (PPARα). PPARs are ligand-activated transcription factors that modulate lipid metabolism and are involved in transcription of genes that regulate cell proliferation [24,98]. This family of nuclear receptors is expressed in tumors cells as well as in tumor endothelium [142]. PPARα is a particularly important transcriptional regulator of inflammatory and metabolic processes, which means its agonists can be used to treat dyslipidemia [98]. Functional activity of PPAR is maintained by histone deacetylase (HDAC), co-repressors/co-activators, retinoid X receptors (RXRs), RNA polymerase II, histone acetyltransferases (HATs), phosphorylation and dephosphorylation [139–141].

PPARα ligands suppress cancer cell growth in several lines, including breast, colon, and skin [142]. They also suppress the metastasis of melanoma cells [142]. PPARα ligands can induce apoptosis of endothelial cells and inhibit proliferation and migration [142]. PPARα ligands produce anti-inflammatory effects by inhibiting NOS, COX-2, and tumor necrosis factor TNFα, and reducing inflammation in tumors may be associated with inhibition of cell growth and improved prognosis [142]. Moreover, PPARα deficiency suppresses angiogenesis by producing excess thrombospondin (TSP-1) and also prevents tumor growth [142]. Given these reasons, PPARα ligands could potentially be tumor-preventing agents. Fibrates in particular target the MVA pathway by inhibiting acetoacetyl coenzyme A [143,144] (Figure 3).

Clofibrate inhibits cell growth by changing the levels of checkpoint kinases, cell cycle inhibitors, and tumor suppressors [98]. In breast cancer cells, higher levels of PPARα are found when compared to HMEC cells [98]. In these cells, Clofibrate reduces the amount of fatty acid synthase, inhibits growth, and reduces survival kinases [98]. In breast cancer cells, COX-2 expression promotes cell adhesion and migration, which in turn accelerates cancer progression [98]. Clofibrate, in turn, downregulates COX-2 as well as 5-LO inflammatory pathway components [98]. In rodents, the use of fibrates for prolonged periods of time can cause peroxisome proliferation, which leads to hepatomegaly and tumor formation. As this has not been demonstrated in humans, a low dose of fibrates for a short duration of time can be used [98]. PPAR agonists have also been associated with weight gain [24].

8. Other Lipid Metabolites and Pathways in Cancer and Viral Infections

Beside the pathways and metabolites mentioned here, there are many other very important lipid pathways mediating enzymes and metabolites. Gangliosides along with other glycosphingolipids, phospholipids, and cholesterol in glycolipid-enriched microdomains present at the outer leaflet of the plasma membrane interact with signaling molecules including receptor tyrosine kinases and signal transducers and contribute to activation of cell signaling, increasing cell proliferation and migration, as well as tumor growth [145].

Phospholipase D (isoenzymes PLD1 and PLD2) catalyzes the hydrolysis of cell membrane phospholipids and plays the role in several cancers and infectious diseases [146]. PLD inhibitors have been proposed as novel compounds for the treatment of cancers, neurodegenerative disorders, and viral infections [147]. PLD activity has been linked with promoting survival and metastatic phenotype of malignant prostate cancer cells [148] and PLD inhibitors reduce human prostate cancer cell proliferation and colony formation [149]. HPV infection, the key risk factor for the development and progression of cervical cancer utilizes PLD and Phosphatidic acid (PA) and regulate the PI3K/AKT/mTOR pathway [150].

ATP citrate lyase (ACLY) is involved in lipid during membrane biogenesis (fatty acid synthesis) linked with aerobic glycolysis. ACLY catalyzes the conversion of citrate to oxaloacetic acid (OAA) and acetyl-CoA [151]. Acetyl-CoA is the acetyl donor for lysine acetylation that links metabolism, signaling, and epigenetics. ACLY overexpression and activation increase metabolic activity in proliferating cells via activation of Akt signaling in glioblastoma, colorectal cancer, breast cancer, non-small cell lung cancer, and hepatocellular carcinoma etc. [152]. There are varying reports on the effect of ACLY depletion on cancer cells. Few suggest that genetic deletion of ACLY does not kill cells but these cells start proliferating at the impaired rate by utilizing exogenous acetate for de novo lipogenesis and histone acetylation [153]. There are studies in breast cancer, which suggest that depletion of ACLY suppressed breast tumor growth and progression [151]. Low molecular weight cleaved form of cyclin E, a powerful independent predictor of survival in women with progressed breast cancer, was shown to interact with ACLY and promote aberrant lipid metabolism pathways in breast cancer tumorigenesis [154]. ACLY has been implicated in integrin signaling in glioblastoma leading to cell adhesion and migration required for its progression and metastasis [155]. ACLY activation during hepatitis B virus (HBV) has been reported to be involved in the disturbed lipid metabolism and proliferation of hepatocytes [156]. ACSS2-mediated acetyl-CoA synthesis from acetate has been shown to take over ACLY functions during HCMV infection, which has been associated with malignancies including colon cancer, malignant glioma, prostate carcinoma, and breast [157–160].

9. Summary

Diet and obesity play an important role in breast, prostate, gastric, lung and skin cancer development, and increased lipogenesis has been observed in tumor cells of virally infected hosts [23,161]. This may be caused by disruptions in the same signaling pathways responsible for the oncogenic transformation of cancer cells [161]. The challenge, therefore, is to discover novel, non-toxic therapies that target these essential steps in the development of lipid/obesity-associated cancers. One such pathway, the mevalonate pathway, can be dysregulated in cancer cells and lead to poor prognosis in cancer patients, as cholesterol has been shown to accelerate tumor formation [23,87]. Statins target this pathway by inhibiting HMGCR, which reduces plasma cholesterol and has to been shown to ultimately reduce cancer risk [121,124]. Another possible therapeutic target is fatty acid synthase, which is expressed at high levels in cancer cells and leads to more lipogenesis and oncogenic signaling pathways [98]. Treatment with a FASN inhibitor increases apoptosis and decreases cell growth in cancer cells [101]. PPARs are another modulator of lipid metabolism and regulate cell proliferation [24,98]. PPARα ligands, fibrates, suppress breast cancer cell growth [98]. Lastly, abnormal metabolism of arachidonic acid in cancer cells leads to high levels of pro-inflammatory eicosanoids, which promote tumor progression [106]. NSAIDs inhibit the inflammatory COX pathway and have been reported to reduce the risk of cancer [106]. Since viral infection associated cancers also heavily utilize lipid metabolism for their growth and progression, therefore there could be common targets which could be tested to combat viral cancers. Numerous studies discuss and show promising results targets for NSAIDs [116,162–168], mTOR pathway regulators [169–172], and FASN inhibitors [75] in viral infection associated cancers but additional in vivo studies need to be done to fully exploit these pathways. These studies support the further development of statins, FASN inhibitors, fibrates, and NSAIDs as potential anticancer agents.

Funding: We are grateful to funding from H. M. Bligh Cancer Research Laboratories and NIH-funded grant R01CA192970 to NSW. The funders had no role in the design, decision to publish, or preparation of the manuscript. All authors have read the journal's policy on conflicts of interest.

Conflicts of Interest: The authors declare no conflicts of interest.

Abbreviations

KSHV	Kaposi's sarcoma-associated herpesvirus
MCPyV	Merkel cell polyomavirus—a polyoma virus
MCC	Merkel cell carcinoma
EBV	Epstein–Barr virus
HPV	human papilloma virus
HTLV	human T lymphotropic virus
cART	combination of anti-retroviral therapy
IGF-1	insulin-like growth factor 1
SHBG	sex-hormone binding globulin
BARD1	BRCA1-associated RING domain protein 1
TNFα	tumor necrosis factor-alpha
MAPK	mitogen activated protein kinase
HMEC	human mammary epithelial cells
OPG	osteoprotegerin
SREBF2	Sterol regulatory element-binding protein transcription factor 2
LXR	Liver x receptors
PPARs	Peroxisome proliferator—activated receptors
HCC	hepatocellular carcinoma
LKB1	liver kinase B1
HMG-CoA synthase	3-hydroxy-3-methylglutaryl-coenzyme A synthase
MCM	minichromosome maintenance
PTEN	phosphatase and tensin homolog
LTB4	leukotriene B4
MMPs	matrix metalloproteinases
PLD	Phospholipase D
ACLY	ATP citrate lyase
VEGF	vascular endothelial growth factor
FGF2	fibroblast growth factor 2
NSAIDs	Non-steroidal anti-inflammatory drugs
EETs	epoxyeicosatrienoic acid

STAR	steroidogenic acute regulatory
FASN	fatty acid synthase
ABCA1	ATP-binding cassette transporter
ACSL1	Acyl-CoA Synthetase Long Chain Family Member 1
AGPAT1	1-Acylglycerol-3-Phosphate O-Acyltransferase 1
SCD	Stearoyl-CoA desaturase (Δ-9-desaturase)
RAC	Rectal adenocarcinoma
PDAC	Pancreatic ductal adenocarcinoma
MUFAs	mono-unsaturated fatty acids
SCFA	Short-chain fatty acids
PI3K	phosphatidylinositol 3-kinase
HIFs	hypoxia inducible factors
ESCRT	endosomal sorting complexes required for transport
ELOVL6	fatty acid elongase
CPT	Carnitine palmitoyltransferase.

References

1. Zali, H.; Rezaei-Tavirani, M.; Azodi, M. Gastric cancer: Prevention, risk factors and treatment. *Gastroenterol. Hepatol. Bed Bench* **2011**, *4*, 175–185. [PubMed]

2. Siegel, R.L.; Miller, K.D.; Jemal, A. Cancer statistics, 2018. *CA Cancer J. Clin.* **2018**, *68*, 7–30. [CrossRef] [PubMed]

3. Dela Cruz, C.S.; Tanoue, L.T.; Matthay, R.A. Lung cancer: Epidemiology, etiology, and prevention. *Clin. Chest Med.* **2011**, *32*, 605–644. [CrossRef] [PubMed]

4. Ahern, T.P.; Pedersen, L.; Tarp, M.; Cronin-Fenton, D.P.; Garne, J.P.; Silliman, R.A.; Sorensen, H.T.; Lash, T.L. Statin prescriptions and breast cancer recurrence risk: A Danish nationwide prospective cohort study. *J. Natl. Cancer Inst.* **2011**, *103*, 1461–1468. [CrossRef] [PubMed]

5. Haggar, F.A.; Boushey, R.P. Colorectal cancer epidemiology: Incidence, mortality, survival, and risk factors. *Clin. Colon Rectal Surg.* **2009**, *22*, 191–197. [CrossRef] [PubMed]

6. Wong, M.C.; Jiang, J.Y.; Goggins, W.B.; Liang, M.; Fang, Y.; Fung, F.D.; Leung, C.; Wang, H.H.; Wong, G.L.; Wong, V.W.; et al. International incidence and mortality trends of liver cancer: A global profile. *Sci. Rep.* **2017**, *7*, 45846. [CrossRef] [PubMed]

7. Liao, J.B. Viruses and human cancer. *Yale J. Biol. Med.* **2006**, *79*, 115–122.

8. Piccaluga, P.P.; Weber, A.; Ambrosio, M.R.; Ahmed, Y.; Leoncini, L. Epstein-Barr Virus-Induced Metabolic Rearrangements in Human B-Cell Lymphomas. *Front. Microbiol.* **2018**, *9*, 1233. [CrossRef]

9. Stanfield, B.A.; Luftig, M.A. Recent advances in understanding Epstein-Barr virus. *F1000Res* **2017**, *6*, 386. [CrossRef]

10. Cohen, J.I. Epstein-barr virus vaccines. *Clin. Transl. Immunol.* **2015**, *4*, e32. [CrossRef]

11. Wu, T.T.; Qian, J.; Ang, J.; Sun, R. Vaccine prospect of Kaposi sarcoma-associated herpesvirus. *Curr. Opin. Virol.* **2012**, *2*, 482–488. [CrossRef] [PubMed]

12. Dai, L.; Lin, Z.; Jiang, W.; Flemington, E.K.; Qin, Z. Lipids, lipid metabolism and Kaposi's sarcoma-associated herpesvirus pathogenesis. *Virol. Sin.* **2017**, *32*, 369–375. [CrossRef] [PubMed]

13. Campbell, J.H.; Hearps, A.C.; Martin, G.E.; Williams, K.C.; Crowe, S.M. The importance of monocytes and macrophages in HIV pathogenesis, treatment, and cure. *AIDS* **2014**, *28*, 2175–2187. [CrossRef] [PubMed]

14. Jemal, A.; Bray, F.; Center, M.M.; Ferlay, J.; Ward, E.; Forman, D. Global cancer statistics. *CA Cancer J. Clin.* **2011**, *61*, 69–90. [CrossRef] [PubMed]

15. Mesri, E.A.; Cesarman, E.; Boshoff, C. Kaposi's sarcoma and its associated herpesvirus. *Nat. Rev. Cancer* **2010**, *10*, 707–719. [CrossRef] [PubMed]

16. Dittmer, D.P.; Damania, B. Kaposi sarcoma associated herpesvirus pathogenesis (KSHV)—An update. *Curr. Opin. Virol.* **2013**, *3*, 238–244. [CrossRef] [PubMed]

17. Mosam, A.; Shaik, F.; Uldrick, T.S.; Esterhuizen, T.; Friedland, G.H.; Scadden, D.T.; Aboobaker, J.; Coovadia, H.M. A randomized controlled trial of highly active antiretroviral therapy versus highly active antiretroviral therapy and chemotherapy in therapy-naive patients with HIV-associated Kaposi sarcoma in South Africa. *J. Acquir. Immune Defic. Syndr.* **2012**, *60*, 150–157. [CrossRef]

18. Letang, E.; Almeida, J.M.; Miro, J.M.; Ayala, E.; White, I.E.; Carrilho, C.; Bastos, R.; Nhampossa, T.; Menendez, C.; Campbell, T.B.; et al. Predictors of immune reconstitution inflammatory syndrome-associated with kaposi sarcoma in mozambique: A prospective study. *J. Acquir. Immune Defic. Syndr.* **2010**, *53*, 589–597. [CrossRef]

19. Borok, M.; Fiorillo, S.; Gudza, I.; Putnam, B.; Ndemera, B.; White, I.E.; Gwanzura, L.; Schooley, R.T.; Campbell, T.B. Evaluation of plasma human herpesvirus 8 DNA as a marker of clinical outcomes during antiretroviral therapy for AIDS-related Kaposi sarcoma in Zimbabwe. *Clin. Infect. Dis.* **2010**, *51*, 342–349. [CrossRef]

20. MacLachlan, J.H.; Cowie, B.C. Hepatitis B virus epidemiology. *Cold Spring Harb. Perspect. Med.* **2015**, *5*, a021410. [CrossRef]

21. Lorizate, M.; Krausslich, H.G. Role of lipids in virus replication. *Cold Spring Harb. Perspect. Biol.* **2011**, *3*, a004820. [CrossRef] [PubMed]

22. Stone, T.W.; McPherson, M.; Gail Darlington, L. Obesity and Cancer: Existing and New Hypotheses for a Causal Connection. *EBioMedicine* **2018**, *30*, 14–28. [CrossRef] [PubMed]

23. Llaverias, G.; Danilo, C.; Mercier, I.; Daumer, K.; Capozza, F.; Williams, T.M.; Sotgia, F.; Lisanti, M.P.; Frank, P.G. Role of cholesterol in the development and progression of breast cancer. *Am. J. Pathol.* **2011**, *178*, 402–412. [CrossRef] [PubMed]

24. Lorincz, A.M.; Sukumar, S. Molecular links between obesity and breast cancer. *Endocr. Relat. Cancer* **2006**, *13*, 279–292. [CrossRef]

25. Bianchini, F.; Kaaks, R.; Vainio, H. Overweight, obesity, and cancer risk. *Lancet Oncol.* **2002**, *3*, 565–574. [CrossRef]

26. Bagga, D.; Wang, L.; Farias-Eisner, R.; Glaspy, J.A.; Reddy, S.T. Differential effects of prostaglandin derived from omega-6 and omega-3 polyunsaturated fatty acids on COX-2 expression and IL-6 secretion. *Proc. Natl. Acad. Sci. USA* **2003**, *100*, 1751–1756. [CrossRef] [PubMed]

27. Calder, P.C. Polyunsaturated fatty acids and inflammation. *Prostaglandins Leukot. Essent. Fat. Acids* **2006**, *75*, 197–202. [CrossRef]

28. Calder, P.C. n-3 polyunsaturated fatty acids, inflammation, and inflammatory diseases. *Am. J. Clin. Nutr.* **2006**, *83*, 1505S–1519S. [CrossRef]

29. Goswami, S.; Sharma-Walia, N. Osteoprotegerin secreted by inflammatory and invasive breast cancer cells induces aneuploidy, cell proliferation and angiogenesis. *BMC Cancer* **2015**, *15*, 935. [CrossRef]

30. Goswami, S.; Sharma-Walia, N. Crosstalk between osteoprotegerin (OPG), fatty acid synthase (FASN) and, cycloxygenase-2 (COX-2) in breast cancer: Implications in carcinogenesis. *Oncotarget* **2016**, *7*, 58953–58974. [CrossRef]

31. Goswami, S.; Sharma-Walia, N. Osteoprotegerin rich tumor microenvironment: Implications in breast cancer. *Oncotarget* **2016**, *7*, 42777–42791. [CrossRef] [PubMed]

32. Suburu, J.; Shi, L.; Wu, J.; Wang, S.; Samuel, M.; Thomas, M.J.; Kock, N.D.; Yang, G.; Kridel, S.; Chen, Y.Q. Fatty acid synthase is required for mammary gland development and milk production during lactation. *Am. J. Physiol. Endocrinol. Metab.* **2014**, *306*, E1132–E1143. [CrossRef] [PubMed]

33. Dal Maso, L.; Zucchetto, A.; Talamini, R.; Serraino, D.; Stocco, C.F.; Vercelli, M.; Falcini, F.; Franceschi, S. Prospective Analysis of Case-control studies on Environmental, f.; health study, g. Effect of obesity and other lifestyle factors on mortality in women with breast cancer. *Int. J. Cancer* **2008**, *123*, 2188–2194. [CrossRef] [PubMed]

34. Li, Q.; Zhang, J.; Zhou, Y.; Qiao, L. Obesity and gastric cancer. *Front. Biosci.* **2012**, *17*, 2383–2390. [CrossRef]

35. Baenke, F.; Peck, B.; Miess, H.; Schulze, A. Hooked on fat: The role of lipid synthesis in cancer metabolism and tumour development. *Dis. Model. Mech.* **2013**, *6*, 1353–1363. [CrossRef] [PubMed]

36. Van Meer, G.; Voelker, D.R.; Feigenson, G.W. Membrane lipids: Where they are and how they behave. *Nat. Rev. Mol. Cell Biol.* **2008**, *9*, 112–124. [CrossRef] [PubMed]

37. Zalba, S.; Ten Hagen, T.L. Cell membrane modulation as adjuvant in cancer therapy. *Cancer Treat. Rev.* **2017**, *52*, 48–57. [CrossRef]

38. Beloribi-Djefaflia, S.; Vasseur, S.; Guillaumond, F. Lipid metabolic reprogramming in cancer cells. *Oncogenesis* **2016**, *5*, e189. [CrossRef]

39. Delmas, P.; Coste, B.; Gamper, N.; Shapiro, M.S. Phosphoinositide lipid second messengers: New paradigms for calcium channel modulation. *Neuron* **2005**, *47*, 179–182. [CrossRef]

40. Moolenaar, W.H. Lysophosphatidic acid, a multifunctional phospholipid messenger. *J. Biol. Chem.* **1995**, *270*, 12949–12952. [CrossRef]

41. Wang, J.; Lv, X.W.; Shi, J.P.; Hu, X.S. Mechanisms involved in ceramide-induced cell cycle arrest in human hepatocarcinoma cells. *World J. Gastroenterol.* **2007**, *13*, 1129–1134. [CrossRef] [PubMed]

42. Guenther, G.G.; Peralta, E.R.; Rosales, K.R.; Wong, S.Y.; Siskind, L.J.; Edinger, A.L. Ceramide starves cells to death by downregulating nutrient transporter proteins. *Proc. Natl. Acad. Sci. USA* **2008**, *105*, 17402–17407. [CrossRef] [PubMed]

43. Watters, R.J.; Wang, H.G.; Sung, S.S.; Loughran, T.P.; Liu, X. Targeting sphingosine-1-phosphate receptors in cancer. *Anticancer Agents Med. Chem.* **2011**, *11*, 810–817. [CrossRef] [PubMed]

44. Pyne, N.J.; Pyne, S. Sphingosine 1-phosphate and cancer. *Nat. Rev. Cancer* **2010**, *10*, 489–503. [CrossRef] [PubMed]

45. Long, J.S.; Fujiwara, Y.; Edwards, J.; Tannahill, C.L.; Tigyi, G.; Pyne, S.; Pyne, N.J. Sphingosine 1-phosphate receptor 4 uses HER2 (ERBB2) to regulate extracellular signal regulated kinase-1/2 in MDA-MB-453 breast cancer cells. *J. Biol. Chem.* **2010**, *285*, 35957–35966. [CrossRef] [PubMed]

46. Watson, C.; Long, J.S.; Orange, C.; Tannahill, C.L.; Mallon, E.; McGlynn, L.M.; Pyne, S.; Pyne, N.J.; Edwards, J. High expression of sphingosine 1-phosphate receptors, S1P1 and S1P3, sphingosine kinase 1, and extracellular signal-regulated kinase-1/2 is associated with development of tamoxifen resistance in estrogen receptor-positive breast cancer patients. *Am. J. Pathol.* **2010**, *177*, 2205–2215. [CrossRef] [PubMed]

47. Wymann, M.P.; Schneiter, R. Lipid signalling in disease. *Nat. Rev. Mol. Cell Biol.* **2008**, *9*, 162–176. [CrossRef]

48. Morad, S.A.; Cabot, M.C. Ceramide-orchestrated signalling in cancer cells. *Nat. Rev. Cancer* **2013**, *13*, 51–65. [CrossRef]

49. Robbins, S.M.; Quintrell, N.A.; Bishop, J.M. Myristoylation and differential palmitoylation of the HCK protein-tyrosine kinases govern their attachment to membranes and association with caveolae. *Mol. Cell. Biol.* **1995**, *15*, 3507–3515. [CrossRef]

50. Resh, M.D. Fatty acylation of proteins: New insights into membrane targeting of myristoylated and palmitoylated proteins. *Biochim. Et Biophys. Acta* **1999**, *1451*, 1–16. [CrossRef]

51. Drazic, A.; Myklebust, L.M.; Ree, R.; Arnesen, T. The world of protein acetylation. *Biochim. Et Biophys. Acta* **2016**, *1864*, 1372–1401. [CrossRef]

52. Zhang, Y.; Sowers, J.R.; Ren, J. Targeting autophagy in obesity: From pathophysiology to management. *Nat. Rev. Endocrinol.* **2018**, *14*, 356–376. [CrossRef] [PubMed]

53. Burke, J.E. Structural Basis for Regulation of Phosphoinositide Kinases and Their Involvement in Human Disease. *Mol. Cell* **2018**, *71*, 653–673. [CrossRef] [PubMed]

54. Yang, M.; Zhang, Y.; Ren, J. Autophagic Regulation of Lipid Homeostasis in Cardiometabolic Syndrome. *Front. Cardiovasc. Med.* **2018**, *5*, 38. [CrossRef]

55. Fernandez, L.P.; Ramos-Ruiz, R.; Herranz, J.; Martin-Hernandez, R.; Vargas, T.; Mendiola, M.; Guerra, L.; Reglero, G.; Feliu, J.; Ramirez de Molina, A. The transcriptional and mutational landscapes of lipid metabolism-related genes in colon cancer. *Oncotarget* **2018**, *9*, 5919–5930. [CrossRef] [PubMed]

56. Fernandes Messias, M.C.; Mecatti, G.C.; Figueiredo Angolini, C.F.; Eberlin, M.N.; Credidio, L.; Real Martinez, C.A.; Rodrigues Coy, C.S.; de Oliveira Carvalho, P. Plasma Lipidomic Signature of Rectal Adenocarcinoma Reveals Potential Biomarkers. *Front. Oncol.* **2017**, *7*, 325. [CrossRef] [PubMed]

57. Sunami, Y.; Rebelo, A.; Kleeff, J. Lipid Metabolism and Lipid Droplets in Pancreatic Cancer and Stellate Cells. *Cancers (Basel)* **2017**, *10*. [CrossRef] [PubMed]

58. Ahearn, I.M.; Haigis, K.; Bar-Sagi, D.; Philips, M.R. Regulating the regulator: Post-translational modification of RAS. *Nat. Rev. Mol. Cell Biol.* **2011**, *13*, 39–51. [CrossRef]

59. Heaton, N.S.; Randall, G. Multifaceted roles for lipids in viral infection. *Trends Microbiol.* **2011**, *19*, 368–375. [CrossRef]

60. Popescu, C.I.; Riva, L.; Vlaicu, O.; Farhat, R.; Rouille, Y.; Dubuisson, J. Hepatitis C virus life cycle and lipid metabolism. *Biology (Basel)* **2014**, *3*, 892–921. [CrossRef]

61. Lo, A.K.; Lung, R.W.; Dawson, C.W.; Young, L.S.; Ko, C.W.; Yeung, W.W.; Kang, W.; To, K.F.; Lo, K.W. Activation of SREBP1-mediated lipogenesis by the Epstein-Barr Virus-encoded LMP1 promotes cell proliferation and progression of nasopharyngeal carcinoma. *J. Pathol.* **2018**. [CrossRef] [PubMed]

62. Gorres, K.L.; Daigle, D.; Mohanram, S.; Miller, G. Activation and repression of Epstein-Barr Virus and Kaposi's sarcoma-associated herpesvirus lytic cycles by short- and medium-chain fatty acids. *J. Virol.* **2014**, *88*, 8028–8044. [CrossRef] [PubMed]

63. Meckes, D.G., Jr.; Menaker, N.F.; Raab-Traub, N. Epstein-Barr virus LMP1 modulates lipid raft microdomains and the vimentin cytoskeleton for signal transduction and transformation. *J. Virol.* **2013**, *87*, 1301–1311. [CrossRef] [PubMed]

64. Carvalho, L.D.; Gadelha, S.R.; Marin, L.J.; Brito-Melo, G.E.; Martins, C.P.; Fonseca, F.G.; Barbosa-Stancioli, E.F. Are lipid disorders involved in the predominance of human T-lymphotropic virus-1 infections in women? *Rev. Soc. Bras. Med. Trop.* **2015**, *48*, 759–761. [CrossRef] [PubMed]

65. Hoshino, H. Cellular Factors Involved in HTLV-1 Entry and Pathogenicit. *Front. Microbiol.* **2012**, *3*, 222. [CrossRef] [PubMed]

66. Currer, R.; Van Duyne, R.; Jaworski, E.; Guendel, I.; Sampey, G.; Das, R.; Narayanan, A.; Kashanchi, F. HTLV tax: A fascinating multifunctional co-regulator of viral and cellular pathways. *Front. Microbiol.* **2012**, *3*, 406. [CrossRef] [PubMed]

67. Zhang, H.; Chen, L.; Cai, S.H.; Cheng, H. Identification of TBK1 and IKKepsilon, the non-canonical IκB kinases, as crucial pro-survival factors in HTLV-1-transformed T lymphocytes. *Leukemia Res.* **2016**, *46*, 37–44. [CrossRef] [PubMed]

68. Ren, T.; Takahashi, Y.; Liu, X.; Loughran, T.P.; Sun, S.C.; Wang, H.G.; Cheng, H. HTLV-1 Tax deregulates autophagy by recruiting autophagic molecules into lipid raft microdomains. *Oncogene* **2015**, *34*, 334–345. [CrossRef] [PubMed]

69. Huang, J.; Ren, T.; Guan, H.; Jiang, Y.; Cheng, H. HTLV-1 Tax is a critical lipid raft modulator that hijacks IκB kinases to the microdomains for persistent activation of NF-κB. *J. Biol. Chem.* **2009**, *284*, 6208–6217. [CrossRef] [PubMed]

70. Tso, F.Y.; Kossenkov, A.V.; Lidenge, S.J.; Ngalamika, O.; Ngowi, J.R.; Mwaiselage, J.; Wickramasinghe, J.; Kwon, E.H.; West, J.T.; Lieberman, P.M.; et al. RNA-Seq of Kaposi's sarcoma reveals alterations in glucose and lipid metabolism. *PLoS Pathog.* **2018**, *14*, e1006844. [CrossRef] [PubMed]

71. Sanchez, E.L.; Pulliam, T.H.; Dimaio, T.A.; Thalhofer, A.B.; Delgado, T.; Lagunoff, M. Glycolysis, Glutaminolysis, and Fatty Acid Synthesis Are Required for Distinct Stages of Kaposi's Sarcoma-Associated Herpesvirus Lytic Replication. *J. Virol.* **2017**, *91*. [CrossRef] [PubMed]

72. Veettil, M.V.; Kumar, B.; Ansari, M.A.; Dutta, D.; Iqbal, J.; Gjyshi, O.; Bottero, V.; Chandran, B. ESCRT-0 Component Hrs Promotes Macropinocytosis of Kaposi's Sarcoma-Associated Herpesvirus in Human Dermal Microvascular Endothelial Cells. *J. Virol.* **2016**, *90*, 3860–3872. [CrossRef] [PubMed]

73. Raghu, H.; Sharma-Walia, N.; Veettil, M.V.; Sadagopan, S.; Caballero, A.; Sivakumar, R.; Varga, L.; Bottero, V.; Chandran, B. Lipid rafts of primary endothelial cells are essential for Kaposi's sarcoma-associated herpesvirus/human herpesvirus 8-induced phosphatidylinositol 3-kinase and RhoA-GTPases critical for microtubule dynamics and nuclear delivery of viral DNA but dispensable for binding and entry. *J. Virol.* **2007**, *81*, 7941–7959. [CrossRef] [PubMed]

74. Wang, X.; Zhu, N.; Li, W.; Zhu, F.; Wang, Y.; Yuan, Y. Mono-ubiquitylated ORF45 Mediates Association of KSHV Particles with Internal Lipid Rafts for Viral Assembly and Egress. *PLoS Pathog.* **2015**, *11*, e1005332. [CrossRef] [PubMed]

75. Bhatt, A.P.; Jacobs, S.R.; Freemerman, A.J.; Makowski, L.; Rathmell, J.C.; Dittmer, D.P.; Damania, B. Dysregulation of fatty acid synthesis and glycolysis in non-Hodgkin lymphoma. *Proc. Natl. Acad. Sci. USA.* **2012**, *109*, 11818–11823. [CrossRef] [PubMed]

76. Bhatt, A.P.; Damania, B. AKTivation of PI3K/AKT/mTOR signaling pathway by KSHV. *Front. Immunol.* **2012**, *3*, 401. [CrossRef] [PubMed]

77. Angius, F.; Uda, S.; Piras, E.; Spolitu, S.; Ingianni, A.; Batetta, B.; Pompei, R. Neutral lipid alterations in human herpesvirus 8-infected HUVEC cells and their possible involvement in neo-angiogenesis. *BMC Microbiol.* **2015**, *15*, 74. [CrossRef] [PubMed]

78. Singh, R.K.; Lang, F.; Pei, Y.; Jha, H.C.; Robertson, E.S. Metabolic reprogramming of Kaposi's sarcoma associated herpes virus infected B-cells in hypoxia. *PLoS Pathog.* **2018**, *14*, e1007062. [CrossRef]

79. Huang, C.; Freter, C. Lipid metabolism, apoptosis and cancer therapy. *Int. J. Mol. Sci.* **2015**, *16*, 924–949. [CrossRef]

80. Kuzu, O.F.; Noory, M.A.; Robertson, G.P. The Role of Cholesterol in Cancer. *Cancer Res.* **2016**, *76*, 2063–2070. [CrossRef]

81. Hong, C.; Tontonoz, P. Liver X receptors in lipid metabolism: Opportunities for drug discovery. *Nat. Rev. Drug Discov.* **2014**, *13*, 433–444. [CrossRef] [PubMed]

82. Li, Y.C.; Park, M.J.; Ye, S.K.; Kim, C.W.; Kim, Y.N. Elevated levels of cholesterol-rich lipid rafts in cancer cells are correlated with apoptosis sensitivity induced by cholesterol-depleting agents. *Am. J. Pathol.* **2006**, *168*, 1107–1118. [CrossRef] [PubMed]

83. Shibata, M.A.; Ito, Y.; Morimoto, J.; Otsuki, Y. Lovastatin inhibits tumor growth and lung metastasis in mouse mammary carcinoma model: A p53-independent mitochondrial-mediated apoptotic mechanism. *Carcinogenesis* **2004**, *25*, 1887–1898. [CrossRef] [PubMed]

84. Nielsen, S.F.; Nordestgaard, B.G.; Bojesen, S.E. Statin use and reduced cancer-related mortality. *N. Engl. J. Med.* **2012**, *367*, 1792–1802. [CrossRef] [PubMed]

85. Ye, J. Reliance of host cholesterol metabolic pathways for the life cycle of hepatitis C virus. *PLoS Pathog.* **2007**, *3*, e108. [CrossRef] [PubMed]

86. Ikeda, M.; Longnecker, R. Cholesterol is critical for Epstein-Barr virus latent membrane protein 2A trafficking and protein stability. *Virology* **2007**, *360*, 461–468. [CrossRef] [PubMed]

87. Pandyra, A.A.; Mullen, P.J.; Goard, C.A.; Ericson, E.; Sharma, P.; Kalkat, M.; Yu, R.; Pong, J.T.; Brown, K.R.; Hart, T.; et al. Genome-wide RNAi analysis reveals that simultaneous inhibition of specific mevalonate pathway genes potentiates tumor cell death. *Oncotarget* **2015**, *6*, 26909–26921. [CrossRef] [PubMed]

88. Duncan, R.E.; El-Sohemy, A.; Archer, M.C. Mevalonate promotes the growth of tumors derived from human cancer cells in vivo and stimulates proliferation in vitro with enhanced cyclin-dependent kinase-2 activity. *J. Biol. Chem.* **2004**, *279*, 33079–33084. [CrossRef] [PubMed]

89. Serquina, A.K.P.; Kambach, D.M.; Sarker, O.; Ziegelbauer, J.M. Viral MicroRNAs Repress the Cholesterol Pathway, and 25-Hydroxycholesterol Inhibits Infection. *mBio* **2017**, *8*. [CrossRef]

90. Alvisi, G.; Madan, V.; Bartenschlager, R. Hepatitis C virus and host cell lipids: an intimate connection. *RNA Biol.* **2011**, *8*, 258–269. [CrossRef] [PubMed]

91. Katano, H.; Pesnicak, L.; Cohen, J.I. Simvastatin induces apoptosis of Epstein-Barr virus (EBV)-transformed lymphoblastoid cell lines and delays development of EBV lymphomas. *Proc. Natl. Acad. Sci. USA* **2004**, *101*, 4960–4965. [CrossRef] [PubMed]

92. Berquin, I.M.; Edwards, I.J.; Kridel, S.J.; Chen, Y.Q. Polyunsaturated fatty acid metabolism in prostate cancer. *Cancer Metast. Rev.* **2011**, *30*, 295–309. [CrossRef] [PubMed]

93. Leung, J.Y.; Kim, W.Y. Stearoyl co-A desaturase 1 as a ccRCC therapeutic target: Death by stress. *Clin. Cancer Res. Off. J. Am. Assoc. Cancer Res.* **2013**, *19*, 3111–3113. [CrossRef] [PubMed]

94. Fenner, A. Kidney cancer: Stearoyl-CoA desaturase: A novel therapeutic target for RCC. *Nat. Rev. Urol.* **2013**, *10*, 370. [CrossRef] [PubMed]

95. Von Roemeling, C.A.; Marlow, L.A.; Wei, J.J.; Cooper, S.J.; Caulfield, T.R.; Wu, K.; Tan, W.W.; Tun, H.W.; Copland, J.A. Stearoyl-CoA desaturase 1 is a novel molecular therapeutic target for clear cell renal cell carcinoma. *Clin. Cancer Res. Off. J. Am. Assoc. Cancer Res.* **2013**, *19*, 2368–2380. [CrossRef] [PubMed]

96. Imamura, K.; Tomita, N.; Kawakita, Y.; Ito, Y.; Ono, K.; Nii, N.; Miyazaki, T.; Yonemori, K.; Tawada, M.; Sumi, H.; et al. Discovery of Novel and Potent Stearoyl Coenzyme A Desaturase 1 (SCD1) Inhibitors as Anticancer Agents. *Bioorg. Med. Chem.* **2017**, *25*, 3768–3779. [CrossRef] [PubMed]

97. Nishizawa, S.; Sumi, H.; Satoh, Y.; Yamamoto, Y.; Kitazawa, S.; Honda, K.; Araki, H.; Kakoi, K.; Imamura, K.; Sasaki, M.; et al. In vitro and in vivo antitumor activities of T-3764518, a novel and orally available small molecule stearoyl-CoA desaturase 1 inhibitor. *Eur. J. Pharm.* **2017**, *807*, 21–31. [CrossRef] [PubMed]

98. Chandran, K.; Goswami, S.; Sharma-Walia, N. Implications of a peroxisome proliferator-activated receptor alpha (PPARalpha) ligand clofibrate in breast cancer. *Oncotarget* **2016**, *7*, 15577–15599. [CrossRef]

99. Kuhajda, F.P. Fatty acid synthase and cancer: New application of an old pathway. *Cancer Res.* **2006**, *66*, 5977–5980. [CrossRef]

100. Kuhajda, F.P. Fatty-acid synthase and human cancer: New perspectives on its role in tumor biology. *Nutrition* **2000**, *16*, 202–208. [CrossRef]

101. Bandyopadhyay, S.; Zhan, R.; Wang, Y.; Pai, S.K.; Hirota, S.; Hosobe, S.; Takano, Y.; Saito, K.; Furuta, E.; Iizumi, M.; et al. Mechanism of apoptosis induced by the inhibition of fatty acid synthase in breast cancer cells. *Cancer Res.* **2006**, *66*, 5934–5940. [CrossRef] [PubMed]

102. Zaytseva, Y.Y.; Rychahou, P.G.; Le, A.T.; Scott, T.L.; Flight, R.M.; Kim, J.T.; Harris, J.; Liu, J.; Wang, C.; Morris, A.J.; et al. Preclinical evaluation of novel fatty acid synthase inhibitors in primary colorectal cancer cells and a patient-derived xenograft model of colorectal cancer. *Oncotarget* **2018**, *9*, 24787–24800. [CrossRef] [PubMed]

103. Shi, Y.X.; Huang, C.J.; Yang, Z.G. Impact of hepatitis B virus infection on hepatic metabolic signaling pathway. *World J. Gastroenterol.* **2016**, *22*, 8161–8167. [CrossRef] [PubMed]

104. Li, Y.; Webster-Cyriaque, J.; Tomlinson, C.C.; Yohe, M.; Kenney, S. Fatty acid synthase expression is induced by the Epstein-Barr virus immediate-early protein BRLF1 and is required for lytic viral gene expression. *J. Virol.* **2004**, *78*, 4197–4206. [CrossRef] [PubMed]

105. Sychev, Z.E.; Hu, A.; DiMaio, T.A.; Gitter, A.; Camp, N.D.; Noble, W.S.; Wolf-Yadlin, A.; Lagunoff, M. Integrated systems biology analysis of KSHV latent infection reveals viral induction and reliance on peroxisome mediated lipid metabolism. *PLoS Pathog.* **2017**, *13*, e1006256. [CrossRef] [PubMed]

106. Wang, D.; Dubois, R.N. Eicosanoids and cancer. *Nat. Rev. Cancer* **2010**, *10*, 181–193. [CrossRef] [PubMed]

107. Jiang, J.; Qiu, J.; Li, Q.; Shi, Z. Prostaglandin E2 Signaling: Alternative Target for Glioblastoma? *Trends Cancer* **2017**, *3*, 75–78. [CrossRef]

108. Larsson, K.; Jakobsson, P.J. Inhibition of microsomal prostaglandin E synthase-1 as targeted therapy in cancer treatment. *Prostaglandins Other Lipid Mediat.* **2015**, *120*, 161–165. [CrossRef]

109. Harris, R.E. Cyclooxygenase-2 (cox-2) blockade in the chemoprevention of cancers of the colon, breast, prostate, and lung. *Inflammopharmacology* **2009**, *17*, 55–67. [CrossRef]

110. Nuvoli, B.; Galati, R. Cyclooxygenase-2, epidermal growth factor receptor, and aromatase signaling in inflammation and mesothelioma. *Mol. Cancer* **2013**, *12*, 844–852. [CrossRef]

111. Fleming, I. Vascular cytochrome p450 enzymes: Physiology and pathophysiology. *Trends Cardiovasc. Med.* **2008**, *18*, 20–25. [CrossRef] [PubMed]

112. Rand, A.A.; Barnych, B.; Morisseau, C.; Cajka, T.; Lee, K.S.S.; Panigrahy, D.; Hammock, B.D. Cyclooxygenase-derived proangiogenic metabolites of epoxyeicosatrienoic acids. *Proc. Natl. Acad. Sci. USA* **2017**, *114*, 4370–4375. [CrossRef] [PubMed]

113. Howe, L.R. Inflammation and breast cancer. Cyclooxygenase/prostaglandin signaling and breast cancer. *Breast Cancer Res.* **2007**, *9*, 210. [CrossRef] [PubMed]

114. Mazhar, D.; Ang, R.; Waxman, J. COX inhibitors and breast cancer. *Br. J. Cancer* **2006**, *94*, 346–350. [CrossRef] [PubMed]

115. Paul, A.G.; Chandran, B.; Sharma-Walia, N. Cyclooxygenase-2-prostaglandin E2-eicosanoid receptor inflammatory axis: A key player in Kaposi's sarcoma-associated herpes virus associated malignancies. *Transl. Res.* **2013**, *162*, 77–92. [CrossRef] [PubMed]

116. Sharma-Walia, N.; Chandran, K.; Patel, K.; Veettil, M.V.; Marginean, A. The Kaposi's sarcoma-associated herpesvirus (KSHV)-induced 5-lipoxygenase-leukotriene B4 cascade plays key roles in KSHV latency, monocyte recruitment, and lipogenesis. *J. Virol.* **2014**, *88*, 2131–2156. [CrossRef] [PubMed]

117. Wang, F.; Zhang, H.; Ma, A.H.; Yu, W.; Zimmermann, M.; Yang, J.; Hwang, S.H.; Zhu, D.; Lin, T.Y.; Malfatti, M.; et al. COX-2/sEH Dual Inhibitor PTUPB Potentiates the Antitumor Efficacy of Cisplatin. *Mol. Cancer* **2018**, *17*, 474–483. [CrossRef] [PubMed]

118. Goulitquer, S.; Croyal, M.; Lalande, J.; Royer, A.L.; Guitton, Y.; Arzur, D.; Durand, S.; Le Jossic-Corcos, C.; Bouchereau, A.; Potin, P.; et al. Consequences of blunting the mevalonate pathway in cancer identified by a pluri-omics approach. *Cell Death Dis.* **2018**, *9*, 745. [CrossRef]

119. Hajar, R. Statins: Past and present. *Heart Views Off. J. Gulf Heart Assoc.* **2011**, *12*, 121–127. [CrossRef]

120. Lee, J.S.; Roberts, A.; Juarez, D.; Vo, T.T.; Bhatt, S.; Herzog, L.O.; Mallya, S.; Bellin, R.J.; Agarwal, S.K.; Salem, A.H.; et al. Statins enhance efficacy of venetoclax in blood cancers. *Sci. Transl. Med.* **2018**, *10*. [CrossRef]

121. Walther, U.; Emmrich, K.; Ramer, R.; Mittag, N.; Hinz, B. Lovastatin lactone elicits human lung cancer cell apoptosis via a COX-2/PPARgamma-dependent pathway. *Oncotarget* **2016**, *7*, 10345–10362. [CrossRef] [PubMed]

122. Brewer, T.M.; Masuda, H.; Liu, D.D.; Shen, Y.; Liu, P.; Iwamoto, T.; Kai, K.; Barnett, C.M.; Woodward, W.A.; Reuben, J.M.; et al. Statin use in primary inflammatory breast cancer: A cohort study. *Br. J. Cancer* **2013**, *109*, 318–324. [CrossRef] [PubMed]

123. Tsubaki, M.; Takeda, T.; Kino, T.; Obata, N.; Itoh, T.; Imano, M.; Mashimo, K.; Fujiwara, D.; Sakaguchi, K.; Satou, T.; et al. Statins improve survival by inhibiting spontaneous metastasis and tumor growth in a mouse melanoma model. *Am. J. Cancer Res.* **2015**, *5*, 3186–3197. [PubMed]

124. Karlic, H.; Thaler, R.; Gerner, C.; Grunt, T.; Proestling, K.; Haider, F.; Varga, F. Inhibition of the mevalonate pathway affects epigenetic regulation in cancer cells. *Cancer Genet.* **2015**, *208*, 241–252. [CrossRef] [PubMed]

125. Lin, C.J.; Liao, W.C.; Lin, H.J.; Hsu, Y.M.; Lin, C.L.; Chen, Y.A.; Feng, C.L.; Chen, C.J.; Kao, M.C.; Lai, C.H.; et al. Statins Attenuate Helicobacter pylori CagA Translocation and Reduce Incidence of Gastric Cancer: In Vitro and Population-Based Case-Control Studies. *PLoS ONE* **2016**, *11*, e0146432. [CrossRef] [PubMed]

126. Zhou, Y.Y.; Zhu, G.Q.; Wang, Y.; Zheng, J.N.; Ruan, L.Y.; Cheng, Z.; Hu, B.; Fu, S.W.; Zheng, M.H. Systematic review with network meta-analysis: Statins and risk of hepatocellular carcinoma. *Oncotarget* **2016**, *7*, 21753–21762. [CrossRef] [PubMed]

127. Ghosh-Choudhury, N.; Mandal, C.C.; Ghosh-Choudhury, N.; Ghosh Choudhury, G. Simvastatin induces derepression of PTEN expression via NFkappaB to inhibit breast cancer cell growth. *Cell Signal.* **2010**, *22*, 749–758. [CrossRef] [PubMed]

128. Liu, Y.; Chen, L.; Gong, Z.; Shen, L.; Kao, C.; Hock, J.M.; Sun, L.; Li, X. Lovastatin enhances adenovirus-mediated TRAIL induced apoptosis by depleting cholesterol of lipid rafts and affecting CAR and death receptor expression of prostate cancer cells. *Oncotarget* **2015**, *6*, 3055–3070. [CrossRef]

129. Ahern, T.P.; Lash, T.L.; Damkier, P.; Christiansen, P.M.; Cronin-Fenton, D.P. Statins and breast cancer prognosis: Evidence and opportunities. *Lancet Oncol.* **2014**, *15*, e461–e468. [CrossRef]

130. Lee, J.; Hong, E.M.; Jang, J.A.; Park, S.W.; Koh, D.H.; Choi, M.H.; Jang, H.J.; Kae, S.H. Simvastatin Induces Apoptosis and Suppresses Insulin-Like Growth Factor 1 Receptor in Bile Duct Cancer Cells. *Gut Liver* **2016**, *10*, 310–317. [CrossRef]

131. Xie, F.; Liu, J.; Li, C.; Zhao, Y. Simvastatin blocks TGF-beta1-induced epithelial-mesenchymal transition in human prostate cancer cells. *Oncol. Lett.* **2016**, *11*, 3377–3383. [CrossRef] [PubMed]

132. Wang, T.; Seah, S.; Loh, X.; Chan, C.W.; Hartman, M.; Goh, B.C.; Lee, S.C. Simvastatin-induced breast cancer cell death and deactivation of PI3K/Akt and MAPK/ERK signalling are reversed by metabolic products of the mevalonate pathway. *Oncotarget* **2016**, *7*, 2532–2544. [CrossRef] [PubMed]

133. Li, W.; Cao, F.; Li, J.; Wang, Z.; Ren, Y.; Liang, Z.; Liu, P. Simvastatin exerts anti-hepatitis B virus activity by inhibiting expression of minichromosome maintenance protein 7 in HepG2.2.15 cells. *Mol. Med. Rep.* **2016**, *14*, 5334–5342. [CrossRef] [PubMed]

134. Tsan, Y.T.; Lee, C.H.; Wang, J.D.; Chen, P.C. Statins and the risk of hepatocellular carcinoma in patients with hepatitis B virus infection. *J. Clin. Oncol.* **2012**, *30*, 623–630. [CrossRef] [PubMed]

135. Bolotin, E.; Armendariz, A.; Kim, K.; Heo, S.J.; Boffelli, D.; Tantisira, K.; Rotter, J.I.; Krauss, R.M.; Medina, M.W. Statin-induced changes in gene expression in EBV-transformed and native B-cells. *Hum. Mol. Genet.* **2014**, *23*, 1202–1210. [CrossRef] [PubMed]

136. Cohen, J.I. HMG CoA reductase inhibitors (statins) to treat Epstein-Barr virus-driven lymphoma. *Br. J. Cancer* **2005**, *92*, 1593–1598. [CrossRef] [PubMed]

137. Peng, L.F.; Schaefer, E.A.; Maloof, N.; Skaff, A.; Berical, A.; Belon, C.A.; Heck, J.A.; Lin, W.; Frick, D.N.; Allen, T.M.; et al. Ceestatin, a novel small molecule inhibitor of hepatitis C virus replication, inhibits 3-hydroxy-3-methylglutaryl-coenzyme A synthase. *J. Infect. Dis.* **2011**, *204*, 609–616. [CrossRef] [PubMed]

138. Zhong, G.C.; Liu, Y.; Ye, Y.Y.; Hao, F.B.; Wang, K.; Gong, J.P. Meta-analysis of studies using statins as a reducer for primary liver cancer risk. *Sci. Rep.* **2016**, *6*, 26256. [CrossRef] [PubMed]

139. Lalloyer, F.; Staels, B. Fibrates, glitazones, and peroxisome proliferator-activated receptors. *Arterioscler. Thromb. Vasc. Biol.* **2010**, *30*, 894–899. [CrossRef]

140. Okopien, B.; Buldak, L.; Boldys, A. Fibrates in the management of atherogenic dyslipidemia. *Expert Rev. Cardiovasc. Ther.* **2017**, *15*, 913–921. [CrossRef]

141. Lian, X.; Wang, G.; Zhou, H.; Zheng, Z.; Fu, Y.; Cai, L. Anticancer Properties of Fenofibrate: A Repurposing Use. *J. Cancer* **2018**, *9*, 1527–1537. [CrossRef] [PubMed]

142. Panigrahy, D.; Kaipainen, A.; Huang, S.; Butterfield, C.E.; Barnes, C.M.; Fannon, M.; Laforme, A.M.; Chaponis, D.M.; Folkman, J.; Kieran, M.W. PPARalpha agonist fenofibrate suppresses tumor growth through direct and indirect angiogenesis inhibition. *Proc. Natl. Acad. Sci. USA* **2008**, *105*, 985–990. [CrossRef]

143. Moosmann, B.; Behl, C. Selenoprotein synthesis and side-effects of statins. *Lancet* **2004**, *363*, 892–894. [CrossRef]

144. Miziorko, H.M. Enzymes of the mevalonate pathway of isoprenoid biosynthesis. *Arch. Biochem. Biophys.* **2011**, *505*, 131–143. [CrossRef] [PubMed]

145. Groux-Degroote, S.; Rodriguez-Walker, M.; Dewald, J.H.; Daniotti, J.L.; Delannoy, P. Gangliosides in Cancer Cell Signaling. *Prog. Mol. Biol. Transl. Sci.* **2018**, *156*, 197–227. [CrossRef] [PubMed]

146. Ramenskaia, G.V.; Melnik, E.V.; Petukhov, A.E. Phospholipase D: Its role in metabolism processes and disease development. *Biomed. Khim.* **2018**, *64*, 84–93. [CrossRef] [PubMed]

147. Brown, H.A.; Thomas, P.G.; Lindsley, C.W. Targeting phospholipase D in cancer, infection and neurodegenerative disorders. *Nat. Rev. Drug Discov.* **2017**, *16*, 351–367. [CrossRef] [PubMed]

148. Utter, M.; Chakraborty, S.; Goren, L.; Feuser, L.; Zhu, Y.S.; Foster, D.A. Elevated phospholipase D activity in androgen-insensitive prostate cancer cells promotes both survival and metastatic phenotypes. *Cancer Lett.* **2018**, *423*, 28–35. [CrossRef]

149. Noble, A.R.; Maitland, N.J.; Berney, D.M.; Rumsby, M.G. Phospholipase D inhibitors reduce human prostate cancer cell proliferation and colony formation. *Br. J. Cancer* **2018**, *118*, 189–199. [CrossRef]

150. Rabachini, T.; Boccardo, E.; Andrade, R.; Perez, K.R.; Nonogaki, S.; Cuccovia, I.M.; Villa, L.L. HPV-16 E7 expression up-regulates phospholipase D activity and promotes rapamycin resistance in a pRB-dependent manner. *BMC Cancer* **2018**, *18*, 485. [CrossRef]

151. Wang, D.; Yin, L.; Wei, J.; Yang, Z.; Jiang, G. ATP citrate lyase is increased in human breast cancer, depletion of which promotes apoptosis. *Tumour. Biol.* **2017**, *39*. [CrossRef] [PubMed]

152. Khwairakpam, A.D.; Shyamananda, M.S.; Sailo, B.L.; Rathnakaram, S.R.; Padmavathi, G.; Kotoky, J.; Kunnumakkara, A.B. ATP citrate lyase (ACLY): A promising target for cancer prevention and treatment. *Curr. Drug Targets* **2015**, *16*, 156–163. [CrossRef] [PubMed]

153. Zhao, S.; Torres, A.; Henry, R.A.; Trefely, S.; Wallace, M.; Lee, J.V.; Carrer, A.; Sengupta, A.; Campbell, S.L.; Kuo, Y.M.; et al. ATP-Citrate Lyase Controls a Glucose-to-Acetate Metabolic Switch. *Cell Rep.* **2016**, *17*, 1037–1052. [CrossRef] [PubMed]

154. Lucenay, K.S.; Doostan, I.; Karakas, C.; Bui, T.; Ding, Z.; Mills, G.B.; Hunt, K.K.; Keyomarsi, K. Cyclin E Associates with the Lipogenic Enzyme ATP-Citrate Lyase to Enable Malignant Growth of Breast Cancer Cells. *Cancer Res.* **2016**, *76*, 2406–2418. [CrossRef] [PubMed]

155. Lee, J.V.; Berry, C.T.; Kim, K.; Sen, P.; Kim, T.; Carrer, A.; Trefely, S.; Zhao, S.; Fernandez, S.; Barney, L.E.; et al. Acetyl-CoA promotes glioblastoma cell adhesion and migration through Ca(2+)-NFAT signaling. *Genes Dev.* **2018**, *32*, 497–511. [CrossRef] [PubMed]

156. Teng, C.F.; Wu, H.C.; Hsieh, W.C.; Tsai, H.W.; Su, I.J. Activation of ATP citrate lyase by mTOR signal induces disturbed lipid metabolism in hepatitis B virus pre-S2 mutant tumorigenesis. *J. Virol.* **2015**, *89*, 605–614. [CrossRef] [PubMed]

157. Vysochan, A.; Sengupta, A.; Weljie, A.M.; Alwine, J.C.; Yu, Y. ACSS2-mediated acetyl-CoA synthesis from acetate is necessary for human cytomegalovirus infection. *Proc. Natl. Acad. Sci. USA* **2017**, *114*, E1528–E1535. [CrossRef]

158. Kamphorst, J.J.; Cross, J.R.; Fan, J.; de Stanchina, E.; Mathew, R.; White, E.P.; Thompson, C.B.; Rabinowitz, J.D. Hypoxic and Ras-transformed cells support growth by scavenging unsaturated fatty acids from lysophospholipids. *Proc. Natl. Acad. Sci. USA* **2013**, *110*, 8882–8887. [CrossRef]

159. Sansone, A.; Tolika, E.; Louka, M.; Sunda, V.; Deplano, S.; Melchiorre, M.; Anagnostopoulos, D.; Chatgilialoglu, C.; Formisano, C.; Di Micco, R.; et al. Hexadecenoic Fatty Acid Isomers in Human Blood Lipids and Their Relevance for the Interpretation of Lipidomic Profiles. *PLoS ONE* **2016**, *11*, e0152378. [CrossRef]

160. Wang, X.; Lin, H.; Gu, Y. Multiple roles of dihomo-gamma-linolenic acid against proliferation diseases. *Lipids Health Dis.* **2012**, *11*, 25. [CrossRef]

161. Swinnen, J.V.; Brusselmans, K.; Verhoeven, G. Increased lipogenesis in cancer cells: New players, novel targets. *Curr. Opin. Clin. Nutr. Metab. Care* **2006**, *9*, 358–365. [CrossRef] [PubMed]

162. Paul, A.G.; Sharma-Walia, N.; Chandran, B. Targeting KSHV/HHV-8 latency with COX-2 selective inhibitor nimesulide: A potential chemotherapeutic modality for primary effusion lymphoma. *PLoS ONE* **2011**, *6*, e24379. [CrossRef]

163. Paul, A.G.; Chandran, B.; Sharma-Walia, N. Concurrent targeting of eicosanoid receptor 1/eicosanoid receptor 4 receptors and COX-2 induces synergistic apoptosis in Kaposi's sarcoma-associated herpesvirus and Epstein-Barr virus associated non-Hodgkin lymphoma cell lines. *Transl. Res.* **2013**, *161*, 447–468. [CrossRef] [PubMed]

164. Gandhi, J.; Gaur, N.; Khera, L.; Kaul, R.; Robertson, E.S. COX-2 induces lytic reactivation of EBV through PGE2 by modulating the EP receptor signaling pathway. *Virology* **2015**, *484*, 1–14. [CrossRef] [PubMed]

165. Rader, J.S.; Sill, M.W.; Beumer, J.H.; Lankes, H.A.; Benbrook, D.M.; Garcia, F.; Trimble, C.; Tate Thigpen, J.; Lieberman, R.; Zuna, R.E.; et al. A stratified randomized double-blind phase II trial of celecoxib for treating patients with cervical intraepithelial neoplasia: The potential predictive value of VEGF serum levels: An NRG Oncology/Gynecologic Oncology Group study. *Gynecol. Oncol.* **2017**, *145*, 291–297. [CrossRef] [PubMed]

166. Waris, G.; Siddiqui, A. Hepatitis C virus stimulates the expression of cyclooxygenase-2 via oxidative stress: Role of prostaglandin E2 in RNA replication. *J. Virol.* **2005**, *79*, 9725–9734. [CrossRef] [PubMed]

167. Bassiouny, A.R.; Zaky, A.; Neenaa, H.M. Synergistic effect of celecoxib on 5-fluorouracil-induced apoptosis in hepatocellular carcinoma patients. *Ann. Hepatol.* **2010**, *9*, 410–418.

168. Chandrasekharan, J.A.; Huang, X.M.; Hwang, A.C.; Sharma-Walia, N. Altering the Anti-inflammatory Lipoxin Microenvironment: A New Insight into Kaposi's Sarcoma-Associated Herpesvirus Pathogenesis. *J. Virol.* **2016**, *90*, 11020–11031. [CrossRef]

169. Roy, D.; Sin, S.H.; Lucas, A.; Venkataramanan, R.; Wang, L.; Eason, A.; Chavakula, V.; Hilton, I.B.; Tamburro, K.M.; Damania, B.; et al. mTOR inhibitors block Kaposi sarcoma growth by inhibiting essential autocrine growth factors and tumor angiogenesis. *Cancer Res.* **2013**, *73*, 2235–2246. [CrossRef]

170. Dittmer, D.P.; Krown, S.E. Targeted therapy for Kaposi's sarcoma and Kaposi's sarcoma-associated herpesvirus. *Curr. Opin. Oncol.* **2007**, *19*, 452–457. [CrossRef]

171. Sin, S.H.; Roy, D.; Wang, L.; Staudt, M.R.; Fakhari, F.D.; Patel, D.D.; Henry, D.; Harrington, W.J., Jr.; Damania, B.A.; Dittmer, D.P. Rapamycin is efficacious against primary effusion lymphoma (PEL) cell lines in vivo by inhibiting autocrine signaling. *Blood* **2007**, *109*, 2165–2173. [CrossRef] [PubMed]

172. Molinolo, A.A.; Marsh, C.; El Dinali, M.; Gangane, N.; Jennison, K.; Hewitt, S.; Patel, V.; Seiwert, T.Y.; Gutkind, J.S. mTOR as a molecular target in HPV-associated oral and cervical squamous carcinomas. *Clin. Cancer Res. Off. J. Am. Assoc. Cancer Res.* **2012**, *18*, 2558–2568. [CrossRef] [PubMed]

International Journal of
Molecular Sciences

Article

Structural Identification of Antibacterial Lipids from Amazonian Palm Tree Endophytes through the Molecular Network Approach

Morgane Barthélemy [1], Nicolas Elie [1], Léonie Pellissier [2], Jean-Luc Wolfender [2], Didier Stien [3], David Touboul [1,*] and Véronique Eparvier [1,*]

1   CNRS-Institut de Chimie des Substances Naturelles, UPR2301, Université Paris-Saclay, 91198 Gif-sur-Yvette CEDEX, France; morgane.barthelemy@cnrs.fr (M.B.); nicolas.elie@cnrs.fr (N.E.)
2   School of Pharmaceutical Sciences, EPGL, University of Geneva, University of Lausanne, Rue Michel Servet 1, CH-1211 Geneva, Switzerland; leonie.pellissier@unige.ch (L.P.); jean-luc.wolfender@unige.ch (J.-L.W.)
3   Sorbonne Université, CNRS, Laboratoire de Biodiversité et Biotechnologie Microbienne, USR3579, Observatoire Océanologique, 66650 Banyuls-sur-mer, France; didier.stien@cnrs.fr
*   Correspondence: david.touboul@cnrs.fr (D.T.); veronique.eparvier@cnrs.fr (V.E.);
    Tel.: +33-(0)16-982-3032 (D.T.); +33-(0)16-982-3679 (V.E.)

Received: 31 March 2019; Accepted: 17 April 2019; Published: 24 April 2019

Abstract: A library of 197 endophytic fungi and bacteria isolated from the Amazonian palm tree *Astrocaryum sciophilum* was extracted and screened for antibacterial activity against methicillin-resistant *Staphylococcus aureus* (MRSA). Four out of five antibacterial ethyl acetate extracts were also cytotoxic for the MRC-5 cells line. Liquid chromatography coupled to tandem mass spectrometry (UPHLC-HRMS/MS) analyses combined with molecular networking data processing were carried out to allow the identification of depsipeptides and cyclopeptides responsible for the cytotoxicity in the dataset. Specific ion clusters from the active *Luteibacter* sp. extract were also highlighted using an MRSA activity filter. A chemical study of *Luteibacter* sp. was conducted leading to the structural characterization of eight fatty acid exhibiting antimicrobial activity against MRSA in the tens of µg/mL range.

Keywords: *Astrocaryum sciophilum*; endophytes; molecular networking; antibacterial; cytotoxicity; fatty acids

1. Introduction

Endophytes are microorganisms that live inside the tissues of a host plant without causing apparent symptoms either for all or a part of their life cycle [1]. This colonization is ubiquitous for all plants in various environments. Indeed, endophytes have been isolated not only in environments ranging from tropical to temperate climates but also in extreme habitats [2,3].

Plants are widely explored for new chemical entities with therapeutic purposes. Plant endophytes are increasingly regarded as an important and a relatively underrated source of natural products for drug discovery. Recent surveys suggested that tropical forests harbour several million hyperdiverse endophyte species [4,5] that are expected to lead to the discovery of unique bioactive secondary metabolites [6]. Additionally, because endophytes contribute to the defence of their host, their metabolites are expected to exert a defensive role, for example acting as antimicrobials or pathogen growth-inhibitors [1].

We have pursued the study of the palm tree *Astrocaryum sciophilum* endophytes. *A. sciophilum* grows as an understorey palm throughout the North-East Amazon (Brazil [Pará, Amapá], French Guiana, Surinam, Guyana) [7]. Its development reflects a long-life cycle with a maturation age reaching approximately 170 years. It has periodical leaf production of 16 months on average, allowing the dating of each leaf [8]. Leaves can reach 20 years of age. As a result, endophytes of this model plant are thought to persist for a long time in leaves and thus are able to resist environmental threats. For all of these reasons, *A. sciophilum* endophytes may have developed antimicrobial weaponry to a greater extent than other microbes.

A collection of 197 strains of cultivable endophytes from young (<16 months) and old (2–21 years) healthy leaves collected from 6 specimens of *A. sciophilum* was cultured, extracted and screened towards methicillin-resistant *Staphylococcus aureus* (MRSA). Additionally, the full strain collection was analysed by high-resolution tandem mass spectrometry profiling (HRMS/MS), and the resulting fragmentation data were organized as molecular networks (MNs). Molecular networking approaches allow the organization of untargeted tandem MS datasets according to their spectral similarity, and generate clusters of structurally related metabolites [9]. These approaches have become powerful tools for navigating the chemical space of complex biological systems and can be used to view the chemical constituents of a wide variety of extracts in a single map.

2. Results

2.1. Biological Activities of Endophyte Extracts

The 197 endophyte isolates (66 bacteria and 131 fungi) were identified, cultivated on solid media and then extracted with ethyl acetate. All of the extracts were tested for antibacterial activity against methicillin-resistant *S. aureus*. Five extracts, four from fungi (BSNB-0575, -0651, -0303 and -0732) and one from bacteria (BSNB -0721) were active (Minimal Inhibitory Concentration (MIC) ≤ 64 µg/mL). The most active extracts arise from endophytic fungi with MIC of 32 (BSNB-0575 and BSNB-0651), 16 (BSNB-0303) and ≤ 8 µg/mL (BSNB-0732) (Table S1). Three of these four fungal extracts come from the *Fusarium* genus (BSNB-0575, -0651 and -0303), and the fourth was identified as *Akanthomyces attenuatus*. Extract obtained from bacteria *Luteibacter* sp. (BSNB-0721) have a MIC value of 64 µg/mL.

The cytotoxicity of all 5 antimicrobial extracts was evaluated using a normal human lung fibroblast MRC-5 cell line. The bacterial extracts of *Luteibacter* sp. (BSNB-0721) was not cytotoxic, with an antibacterial selectivity index above 1 (Table S1).

2.2. Molecular Networking

All active and inactive ethyl acetate (EtOAc) extracts were analysed by Ultra-High Performance Liquid Chromatography-High Resolution Mass Spectrometry (UHPLC-HRMS/MS) using the data-dependent acquisition mode. Data were first processed by MZmine 2.33 [10], generated by MetGem [11] and molecular networks (MNs) were visualized using Cytoscape 3.7.0.

Relative quantification of the ions was represented by pie chart diagrams, with their proportions based on the respective areas of the corresponding extracted ion chromatograph (XIC) areas [12]. Targeted analysis of the MN was made possible by colouring the nodes based on the extract's antibacterial activity, highlighting node clusters among the antibacterial extracts.

A first MN was built with the data from the 131 fungal extracts (Figure S1) composed of 28,719 nodes in 13,060 clusters. Eleven clusters contained nodes from the four active fungal extracts. MS^2 spectra of the ions at *m/z* 640.42, 654.43, 668.45 (Cluster 2), 600.41, 552.41, 586.39 (Cluster 1) and 730.42 (Cluster 9) matched respectively with three depsipeptides, three pentacyclopeptide and Exumolide A (another depsipeptide) referenced in the available libraries (See Figure 1 and Figure S2). Examination of the MS^2 spectra of other active clusters showed that these compounds share MS^2 fragmentation patterns that are similar to that of a peptide because of the typical loss of fragment at *m/z* 113.08 (Leucine/Isoleucine), *m/z* 99.07 (Valine), *m/z* 71.04 (Alanine) and *m/z* 101.05 (Threonine) (See Figure S2). According to the MN, these compounds are also found in extracts with MIC of 128 µg/mL from other *Fusarium* strains. Differences of activity for these extracts can be explained by the relative quantities of metabolites in the extracts (Figure S3).

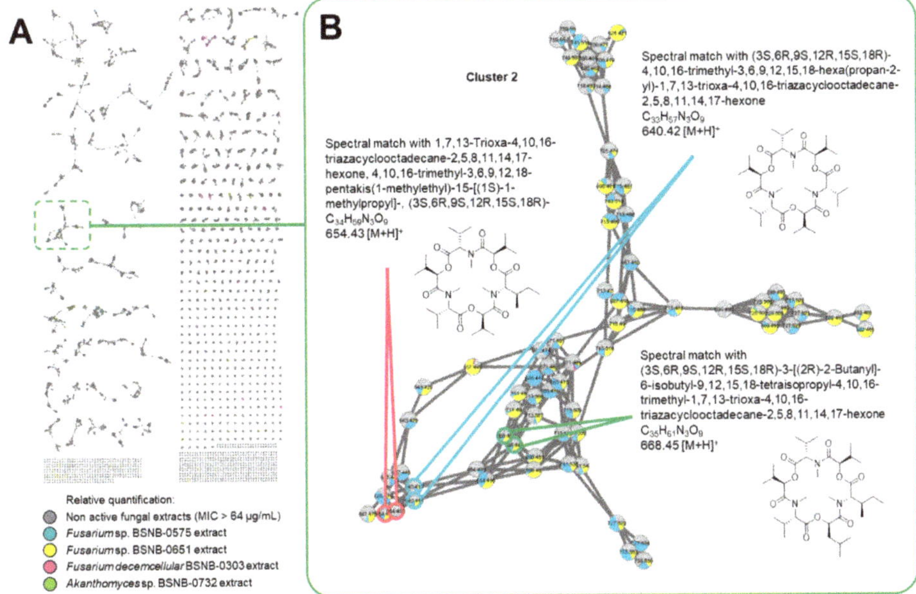

Figure 1. (**A**) Molecular network (MN) of the 131 extracts from fungal endophytes. Relative quantification values of each ion within the extracts are represented as an extracted ion chromatograph (XIC) area-dependent pie chart. (**B**) Dereplication process nodes belongings to cluster 2 composed by ions from the extracts of BSNB-0575 and BSNB-0651. Cluster 2 appears to group the ions corresponding to depsipeptides metabolites. MIC, Minimal Inhibitory Concentration.

The MN generated from the data of the 66 bacterial endophytes extract is composed of 13,763 nodes in 4485 clusters (Figures S4 and S5). This MN showed clusters containing nodes from the one active extract of *Luteibacter* sp. BSNB-0721 (Figure 2A). Comparison of experimental spectral MS^2 data obtained with databases supported the dereplication of a cluster with fatty acids analogues (Figure 2B). Consequently, the dereplication process could highlight the presence of fatty acid derivatives in *Luteibacter* extract. Other clusters contained nodes specific to the extract of the *Luteibacter* sp. strain and indicated a valuable chemical diversity in this particular strain.

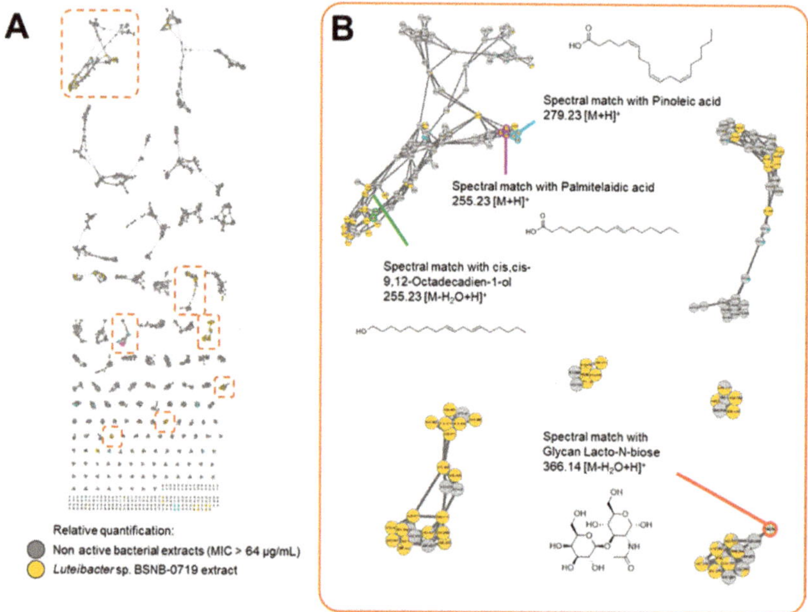

Figure 2. (**A**) Molecular network of the 66 extracts from bacterial endophytes. Relative quantification values of each ion within the extracts are represented as a XIC area-dependent pie chart. (**B**) Clusters of ions from the active extracts of *Luteibacter* sp. BSNB-0721.

2.3. Isolated Compounds from Luteibacter sp. BSNB-0721

The ethyl acetate extract of *Luteibacter* sp. BSNB-0721 was subjected to bioguided fractionation and eight pure compounds were isolated from the three active fractions (See Table S2). All isolated compounds have been determined by spectroscopic data (Figures S6–S29). These compounds were identified as fatty acids with an aliphatic chain of between 15 and 17 carbons that may be terminated with an isopropyl group. Five bear a hydroxyl group on position 2: (*R*)-2-hydroxy-13-methyltetradecanoic acid (1) or on position 3 (*R*)-3-hydroxy-14-methylpentadecanoic acid (2), (*S*)-β-hydroxypalmitic acid (3), (*R*)-3-hydroxy-15-methylhexadecanoic acid (4) and (*R*)-3-hydroxy-13-methyltetradecanoic acid (5). We also isolated and identified 13-methyltetradecanoic acid (6) that does not exhibit a hydroxyl group. The configurations of the compounds were determined by measuring the optical rotation and comparing with the corresponding literature values.

Compounds (7) and (8) presented an unsaturated chain. Based on the coupling constant of the multiplet (δ_H 5.35), the *Z* configuration was determined. The localization of the unsaturation was determined by MS^2 analysis using the method described by Vrkoslav and Cvacka [13]. After esterification, the two molecules were subjected to atmospheric pressure chemical ionization (APCI) leading to the formation of $[M+C_3H_5N]^+$ adducts. Careful analysis of the MS^2 data allowed the determination of the double bonds' position (Figures S30 and S31). These two compounds were finally identified as (9*Z*)-hexadecenoic acid (7) and 15-Methyl-(9*Z*)-hexadecenoic acid (8).

These fatty acids were tested for their antimicrobial activities against MRSA. They showed MIC values ranging from 128 to 32 µg/mL. The most active compound (7) is a C17 fatty acid with a Z double bond at C-9. Fatty acids with hydroxyl at C-3 (2 – 5) or gem-dimethyl at the final position (8) appear to have lower antibacterial activities (Table 1).

Table 1. Minimal Inhibitory Concentration (MIC) of compounds 1–8 against methicillin-resistant *S. aureus*.

	Fatty Acid	MIC on MRSA [1] (µg/mL)
1	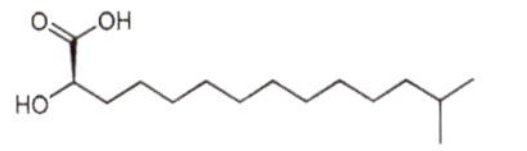	64
2	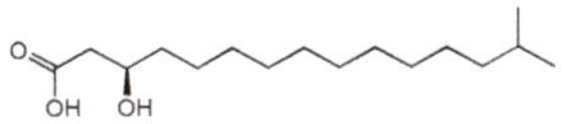	128
3	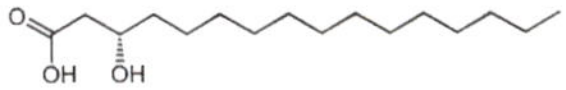	128
4	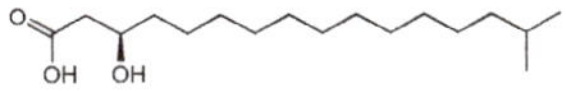	128
5	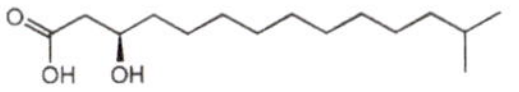	ND
6	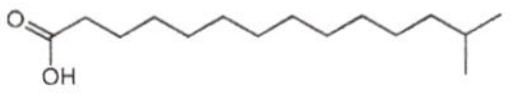	128
7	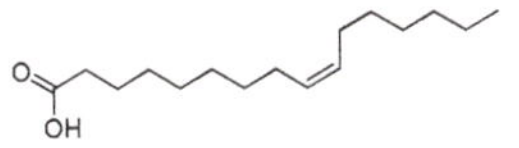	32
8	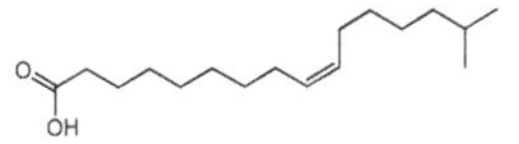	64

[1] Positive Control: Vancomycin (MIC = 1 µg/mL). ND = Not Determined. MRSA = methicillin-resistant *Staphylococcus aureus*.

3. Discussion

Our study, related to the secondary metabolites of the plant host model *Astrocaryum sciophilum*'s associated microorganisms, demonstrated that isolated endophytic strains could produce antibacterial compounds. Indeed, six out of 197 endophytic extracts displayed significant antimicrobial activity.

The four antibacterial extracts derived from fungi (three *Fusarium* strains and one *Akanthomyces* strain) also exhibited cytotoxic activity against MRC-5 cells. Thus, there is a high probability that the antibacterial activity of these extracts depends on the presence of cytotoxic secondary metabolites. Using a molecular networking-based approach, several structurally related analogues of depsipeptide and cyclic peptide - molecules known to be cytotoxic - were highlighted from four active fungal extracts. Depsipeptides have been frequently isolated from fungi belonging to the *Fusarium* genus [14]. More precisely, three cyclic pentapeptides and a cyclic lipopeptide Fusaristatin A were isolated from a strain of *Fusarium decemcellular* [15]. The isolation of a strain from the *Akanthomyces* genus is quite surprising because it is known as a spider parasitic genus. Previous chemical studies of species from this genus have led to the isolation of glycosylated derivatives [16] and pyrone derivatives [17]. Nevertheless, the prediction of numerous peptides in the four active fungal extracts could explain their biological activities. To the best of our knowledge, the present work represents the first time that this class of compounds was detected in the *Akanthomyces* genus.

Of the 66 isolated and extracted bacteria, two shown an antibacterial activity with no cytotoxicity: one *Luteibacter* strain (BSNB-0721) and one *Bacillus* strain (BSNB-0730).

Bacteria from the genus *Luteibacter* were first isolated from the rhizosphere of Barley [18]. To date, three species of this genus have been described: *L. anthropi*, *L. yeojuensis* [19] and *L. rhizovicinus*. No chemical study has ever been performed on this genus, and the analysis of the UHPLC-HRMS/MS data of its EtOAc extract presumed the presence of specific unknown metabolites and fatty acid derivatives. Thus, the extract from the *Luteibacter* sp. (BSNB-0721) was selected for further investigation in order to identify bioactive compounds against MRSA. Eight fatty acid derivatives were isolated in the present study as the major compounds of the EtOAc extract. They appeared to be responsible for the antibacterial activity observed in the crude extract. Nevertheless, ions from the MN cluster unique to the *Luteibacter* extract were not recovered with a fractionation by reverse-phase chromatography because they may be produced in minute amounts and should be further investigated.

The antifungal and bactericidal properties of fatty acids (FA) have been widely studied. Their mechanisms of action are not completely understood; however, it is accepted that FA disrupt cell membranes [20,21]. Many studies have analysed the relationships between FA structure and antibacterial activity, and some general trends have been elucidated. Saturated fatty acids composed of 12 carbons in the chain appear to be the most active, and their activity decreases as the chain becomes either longer or shorter [20,22]. The presence of a Z double bond was found to increase the antibacterial activity of a fatty acid, whereas a *trans* configuration leads to an inactive compound [22]. Moreover, a free carboxyl group allows better antibacterial activity than that of esterified fatty acids [23]. These molecules may be an alternative to the common antimicrobial agents for applications in agriculture, food preservation or cosmetics [22,24]. Such literature data may explain, in part, the highest activity observed in this study for compounds 1, 7 and 8. Indeed, it is observed in our study that there is no difference in the activity between unbranched and isobranched FA. The presence of a hydroxyl group at position 3 does not appear to increase the antibacterial activity but may increase the solubility of these compounds. However, in agreement with the literature data, we noticed that the presence of a Z double bond enhances antibacterial activity. *Luteibacter*'s ability to produce such a panel of FA must be correlated with its ability for defence against aggressors and must improve its survival. The origin of the FA remains uncertain. In fact, they could be sequestered in the membranes until quick release by some lipases during bacterial attack. Further investigation to identify their biosynthetic origin should be performed later.

4. Materials and Methods

4.1. General

Optical rotations were measured using an Anton Paar MCP 200 polarimeter (Anton Paar Graz, Austria) in a 350 µL cell with a length of 100 mm. NMR spectra were recorded in CD_3OD using a Bruker 500 MHz spectrometer or a Bruker 600 MHz spectrometer equipped with a 2 mm invers detection probe (Bruker, Rheinstetten, Germany). Chemical shifts (δ) are reported in ppm relative to the TMS (tetramethylsilane) signal. Coupling constants (J) are in Hertz. High-resolution ESITOFMS measurements were performed using a Waters Acquity UPLC system (Waters, Manchester, England) with a column bypass coupled to a Waters Micromass LCT (Low Chromatography Times-of-flight) Premier time-of-flight mass spectrometer equipped with an electrospray interface (ESI). Flash chromatography was performed using a Grace Reveleris system equipped with a 120 g C_{18} column. The flow rate was 80 mL/min, and detection was performed with dual UV at 210 and 270 nm and ELSD (Evaporative Light Scattering Detector). Analytical and preparative HPLC experiments were conducted using a Gilson system equipped with a 322 pumping device, GX-271 fraction collector, 171 diode array detector and prepELSII detector electrospray nebulizer. The columns used for analytical experiments included a Phenomenex Luna C_{18} 5 µm 4.6 × 250 mm and a Phenomenex Luna C_8 5 µm 4.6 × 250 mm. The columns used for preparative experiments included a Phenomenex Luna C_{18} 5 µm 21.2 × 250 mm and a Phenomenex Luna C_8 5 µm 21.2 × 250 mm (Phenomenex, Le Pecq, France). The flow rates were 1 mL/min and 21 mL/min, respectively, for analytical and preparative experiments and were carried out using a linear gradient of H_2O mixed with an increasing proportion of acetonitrile (CH_3CN). Both solvents were modified with 0.1% formic acid. All of the solvents used for analysis were of HPLC grade.

4.2. Endophyte Material

Astrocaryum sciophilum palm trees were sampled in French Guiana in Piste de Saint-Elie, Sinnamary, in July 2014. The general procedures adopted for the isolation of the microorganisms followed the methodology described by Casella et al. [25]. After collection, the plant material was washed with sterile water and its surface was sterilized by immersion in 70% aqueous ethanol (3 min), followed by immersion in a 5% aqueous sodium hypochlorite (5 min) and finally in 70% aqueous ethanol (1 min). The leaves were cut into small pieces (1–0.5 cm^2) that were placed in a potato dextrose agar medium (potato dextrose agar (PDA), Fluka Analytical, Steinheim, Germany) in Petri dishes at 28 °C (4–5 parts per Petri dishes). Each individual hyphal tip of emerging fungi was removed and placed in a sterile PDA culture medium in 10 cm Petri dishes. The leaf fragments were cultured for a maximum of 1 month. All of the isolated endophytic strains were deposited in the "ICSN/CNRS Strain Library France". The strains are maintained in triplicate at −80 °C in a 2 mL cryotube containing 1 mL of a solution of glycerol and water (1:1).

4.3. Identification of Endophytic Strains

Fungal and bacterial strains were identified using nucleotides sequencing of rDNA ITS (ITS1-5, 8S-ITS2) and rDNA 16S regions, respectively. The obtained sequences were then submitted to BLAST on NCBI to identify the strain. The sequence data were submitted to GenBank with an accession number for each strain (Table S1).

4.4. Cultures and Extraction

Each strain was cultivated at 28 °C in 10 Petri dishes (10 cm diameter) of the PDA culture media. Then, the culture was extracted with ethyl acetate (EtOAc) at room temperature for 24 h. The organic phase was removed via filtration, washed three times with H_2O, dried with anhydrous solid Na_2SO_4 and evaporated using a rotary evaporator under reduced pressure to yield a crude mixture. An extract of the culture media without microorganisms was also conducted.

4.5. UPLC-HRMS Analysis

Chromatographic separation was performed using an Acquity UHPLC system interfaced to a Q-Exactive Plus mass spectrometer (Thermo Scientific, Bremen, Germany), using a heated electrospray ionization (HESI-II) source. Thermo Scientific Xcalibur 2.1 software was used for instrument control and data analysis. The LC conditions were as follows: column, Waters BEH (Ethylene Bridget Hybrid) C18 50 × 2.1 mm, 1.7 μm; mobile phase, (A) water with 0.1% formic acid; (B) acetonitrile with 0.1% formic acid; flow rate, 600 μL/min; injection volume, 1 μL; gradient, linear gradient of 5–100% B over 7 min and isocratic at 100% B for 1 min. An Acquity UPLC photodiode array detector was used to acquire the PDA spectra which were collected in the 200–500 nm range. In the positive ion mode, the di-isooctyl phthalate $C_{24}H_{38}O_4$ [M + H]$^+$ ion (*m/z* 391.28429) was used as the internal lock mass. The optimized HESI-II parameters were as follows: source voltage, 3.5 kV (pos); sheath gas flow rate (N_2), 55 units; auxiliary gas flow rate, 15 units; spare gas flow rate, 3.0; capillary temperature, 275.00 °C (pos), S-Lens RF Level, 45. The mass analyser was calibrated using a mixture of caffeine, methionine–arginine– phenylalanine–alanine–acetate (MRFA), sodium dodecyl sulphate, sodium taurocholate, and Ultramark 1621 in an acetonitrile/ methanol/water solution containing 1% formic acid by direct injection. The data-dependent MS/MS events were performed on the four most intense ions detected in full scan MS (Top3 experiment). The MS/MS isolation window width was 1 Da, and the normalized collision energy (NCE) was set to 35 units. In the data-dependent MS/MS experiments, full scans were acquired at a resolution of 35,000 FWHM (at *m/z* 200) and MS/MS scans were acquired at 17,500 FWHM both with a maximum injection time of 50 ms. After being acquired in a MS/MS scan, the parent ions were placed on the dynamic exclusion list for 2.0 s.

4.6. MZmine 2.33 Data-Preprocessing Parameters

Raw files were converted into MzXML (mass spectrometry data format) files using the MSConvert software. Then, MzXML files were processed using Mzmine 2.37 [10]. Mass detection was carried out with a centroid mass detector with the noise level set to 5.0E5 for the MS level set to all. The ADAP (Automated Data Analysis Pipeline) chromatogram builder [26] was achieved using a minimum group size of scans of 5, a minimum group intensity threshold of 5.0E5, a minimum highest intensity of 5.0E5 and a m/z tolerance of 0.002 or 5 ppm. The wavelets (ADAP) algorithm was used for the chromatogram deconvolution with the following settings: S/N threshold of 10, an intensity window SN, a minimum feature height of 1000, a coefficient area threshold of 100, a peak duration range between 0.01 and 0.5 and an RT wavelet range between 0.001 and 0.05. The m/z and RT range for MS^2 scan pairing were set to 0.001 Da and 0.3 min, respectively. Chromatograms were deisotoped using the isotopic peaks grouper algorithm with an m/z tolerance of 0.003 (5 ppm), RT tolerance of 0.1 (absolute), maximum charge of 2 and the representative isotope used was the most intense. Peak alignment was performed using the join aligner method: m/z tolerance of 0.001 or 5.0 ppm, weight for m/z of 0.001, RT tolerance of 0.3 min, weight for RT of 0.1. Adduct search (Na^+, K^+, NH_4^+, ACN^+) was conducted on the peak list with an RT tolerance set to 1.0 min and the maximum relative peak height at 50%. The found adducts were then removed from the peak list. The peak list was gap-filled with the peak finder module: intensity tolerance of 90%, m/z tolerance of 0.001 or 5.0 ppm and RT tolerance of 0.1 min. For further analysis, the peak list was reduced to the ions with the m/z values between 200 and 900, in order to decrease the number of data.

4.7. Molecular Network Analysis

After preprocessing the UHPLC-HRMS/MS data with MZmine 2.33, the output mgf file was processed with the MetGem software [11] to give a network containing nodes distributed in clusters. To decrease the size of the peak list, ions with the m/z values of 200-900 and/or selfloop nodes can be removed. Networks were generated using the following parameters: m/z tolerance set to 0.02, Minimum Matched Peaks set to 6, topK set to 10, Minimal Cosine score Value of 0.7 and Max. Connected Component Size of 100. Then, the associated CSV file was loaded. For the mapping process, the relative quantification of each ion was represented by pie chart-diagrams for which the proportions were based on the respective areas of the corresponding extracted ion chromatograph areas (XIC). Then, the analogues of the spectra in the network were searched in the available spectral libraries. The library spectra were filtered in the same manner as the input data. All of the matches between network spectra and library spectra were required to have a score above 0.7 and at least 6 matched peaks. The m/z tolerance for the analogues' search was set to 100. For a more advanced retreatment of the network, the MN was also exported to the Cytoscape 3.7.0 software (https://cytoscape.org/).

4.8. Large-Scale Cultivation of Luteibacter sp. and Isolation

Strains were cultivated 15 days at 28 °C in 14 cm Petri dishes of PDA (potato dextrose agar) media. Culture media was extracted three times consecutively with ethyl acetate (EtOAc) at room temperature (the organic phase after 24 h and replacing the remaining agar in EtOAc). The combined organic solution was washed as described above.

Large-scale cultivation of *Luteibacter* sp. was conducted on 210 14 cm Petri dishes to yield 1.96 g of a brown-yellow crude extract. This crude extract (1.8 g) was fractionated by reverse flash chromatography on a C_{18} column with a 5-min-step gradient of water mixed with an increasing proportion of acetonitrile (v/v, 95:5, 75:25, 50:50, 20:80 and 0:100). Six fractions were generated based on the UV and ELSD detection: F1 (22.2 mg, 1.2%), F2 (165.5 mg, 9.2%), F3 (93.6 mg, 5.2%), F4 (41.0 mg, 2.3%), F5 (46.3 mg, 2.6%) and F6 (563.8 mg, 31.3%). A step gradient of acetonitrile—methylene chloride (v/v, 50:50–0:100) was conducted to generate 2 additional fractions: F7 (334.4 mg, 18.6%), F8 (91.8 mg, 5.1%).

F4 (35.0 mg) was purified by preparative HPLC (Luna C_{18}, mobile phase H_2O + 0.1% FA/CH_3CN + 0.1% FA, isocratic elution 60:40 during 6 min then linear gradient from 60:40 to 0:100 over 13 min, flow rate 21 mL/min) to obtain *(R)*-3-hydroxy-13-methyltetradecanoic acid (5) (0.2 mg, t_R = 26.5 min). F5 (41.4 mg) was purified by preparative HPLC (Luna C_8, mobile phase H_2O + 0.1% FA / CH_3CN + 0.1% FA, isocratic elution 40:60 during 6 min then linear gradient from 40:60 to 0:100 over 13 min, flow rate 21 m/min) to obtain *(R)*-2-hydroxy-13-methyltetradecanoic acid (1) (3.7 mg, t_R = 14.0 min), *(R)*-3-hydroxy-14-methylpentadecanoic acid (2) (1.5 mg, t_R = 15.3 min), *(S)*-β-hydroxypalmitic acid (3) (1.9 mg, t_R = 15.6 min) and 9Z-hexadecenoic acid (7) (0.4 mg, t_R = 21.5 min). F6 (258.0 mg) was purified by preparative HPLC (Luna C_8, mobile phase H_2O + 0.1% FA / CH_3CN + 0.1% FA, isocratic elution 35:65 during 30 min, flow rate 21 mL/min) to obtain *(R)*-3-hydroxy-15-methylhexadecanoic acid (4) (7.2 mg, t_R = 15.0 min), 9Z-hexadecenoic acid (7) (30.7 mg, t_R = 21.0 min), 13-methyltetradecanoic acid (6) (62.2 mg, t_R = 24.0 min) and 15-methyl-9Z-hexadecenoic acid (8) (63.8 mg, t_R = 27.0 min).

(R)-2-hydroxy-13-methyltetradecanoic acid (1): White powder. $[\alpha]_D^{20}$ = −5.6 (*c* = 0.5, chloroform). ¹H-NMR (500 MHz, CD_3OD): 4.07 (1H, *dd*, *J* = 7.4, 4.4, H-2), 1.75 (1H, *m*, H-3), 1.63 (1H, *m*, H-3), 1.52 (1H, *non*, *J* = 6.7, H-3), 1.44 (2H, *m*, H-4), 1.30 (14H, *s*, H-5 to H-11), 1.17 (2H, *m*, H-11), 0.88 (6H, *d*, *J* = 6.8, H-14, H-15). ¹³C-NMR (500 MHz, CD_3OD): 178.6 (C-1), 71.9 (C-2), 40.4 (C-12), 35.7 (C-3), 31.2–30.8 (5C, C-6 to C-10), 30.7 (C-5), 29.3 (C-13), 28.7 (C-11), 26.3 (C-4), 23.2 (2C, C-14, C-15). HR-ESI-MS: 257.2109 ([M − H]⁻, $C_{15}H_{29}O_3^-$; calc. 257.2122), 515.4294 ([2M − H]⁻, $C_{30}H_{59}O_6^-$: calc. 515.4317).

(R)-3-hydroxy-14-methylpentadecanoic acid (2): White powder. $[\alpha]_D^{20}$ = −5.6 (*c* = 0.5, chloroform). ¹H-NMR (600 MHz, CD_3OD): 3.96 (1H, *m*, H-3), 2.42 (1H, *dd*, *J* = 15.1, 4.7, H-2), 2.34 (1H, *dd*, *J* = 15.1, 8.2, H-2), 1.52 (1H, *non*, *J* = 6.7, H-14), 1.47 (4H, *m*, H-4, H-5), 1.30 (14H, *s*, H-5 to H-12), 1.18 (2H, *m*, H-13), 0.88 (6H, *d*, *J* = 6.8, H-15, H-16). ¹³C-NMR (600 MHz, CD_3OD): 176.7 (C-1), 69.6 (C-3), 43.7 (C-2), 40.4 (C-13), 38.2 (C-4), 31.2-30.9 (6C, C-6 to C-11), 29.3 (C-14), 28.7 (C-12), 26.8 (C-5), 23.2 (2C, C-15, C-16). HR-ESI-MS: 271.2274 ([M − H]⁻, $C_{16}H_{31}O_3^-$; calc. 271.2279), 543.4611 ([2M − H]⁻, $C_{32}H_{63}O_6^-$; calc. 543.4630).

(S)-β-hydroxypalmitic acid (3): White powder. $[\alpha]_D^{20}$ = +18 (c = 0.1, chloroform). ¹H-NMR (600 MHz, CD_3OD): 3.97 (1H, *m*, H-3), 2.43 (1H, *dd*, *J* = 15.1, 4.7, H-2), 2.36 (1H, *dd*, *J* = 15.1, 8.2, H-2), 1.47 (4H, *m*, H-4, H-5), 1.30 (20H, *s*, H-6 to H-15), 0.90 (3H, *t*, *J* = 7.2, H-16). ¹³C-NMR (600 MHz, CD_3OD): 176.1 (C-1), 69.5 (C-3), 43.5 (C-2), 38.3 (C-4), 33.2 (C-14), 30.9-30.6 (8C, C-6 to C-13), 26.8 (C-5), 23.9 (C-15), 14.6 (C-16). HR-ESI-MS: 271.2274 ([M − H]⁻, $C_{16}H_{31}O_3^-$; calc. 271.2279), 543.4611 ([2M − H]⁻, $C_{32}H_{63}O_6^-$; calc. 543.4630).

(R)-3-hydroxy-15-methylhexadecanoic acid (4): White powder. $[\alpha]_D^{20}$ = −20 (c = 0.1, chloroform). ¹H-NMR (500 MHz, CD_3OD): 3.96 (1H, *m*, H-3), 2.43 (1H, *dd*, *J* = 15.1, 4.7, H-2), 2.35 (1H, *dd*, *J* = 15.2, 8.2, H-2), 1.53 (1H, *non*, *J* = 6.7, H-15), 1.47 (2H, *m*, H-4), 1.30 (18H, *s*, H-5 to H-12), 1.18 (2H, *m*, H-14), 0.88 (6H, *d*, *J* = 6.6, H-16, H-17). ¹³C-NMR (500 MHz, CD_3OD): 176.5 (C-1), 69.7 (C-3), 43.7 (C-2), 40.4 (C-14), 38.3 (C-4), 31.2-30.9 (7C, C-6 to C-12), 29.3 (C-15), 28.7 (C-12), 26.8 (C-5), 23.2 (2C, C-15, C-16). HR-ESI-MS: 285.2431 ([M − H]⁻, $C_{17}H_{33}O_3^-$; calc. 285.2435), 571.4926 ([2M − H]−′ $C_{34}H_{67}O_6^-$; calc. 571.4943).

(R)-3-hydroxy-15-methylhexadecanoic acid (4): White powder. $[\alpha]_D^{20}$ = −20 (c = 0.1, chloroform). ¹H-NMR (500 MHz, CD_3OD): 3.96 (1H, *m*, H-3), 2.43 (1H, *dd*, *J* = 15.1, 4.7, H-2), 2.35 (1H, *dd*, *J* = 15.2, 8.2, H-2), 1.53 (1H, *non*, *J* = 6.7, H-15), 1.47 (2H, *m*, H-4), 1.30 (18H, *s*, H-5 to H-12), 1.18 (2H, *m*, H-14), 0.88 (6H, *d*, *J* = 6.6, H-16, H-17). ¹³C-NMR (500 MHz, CD_3OD): 176.5 (C-1), 69.7 (C-3), 43.7 (C-2), 40.4 (C-14), 38.3 (C-4), 31.2-30.9 (7C, C-6 to C-12), 29.3 (C-15), 28.7 (C-12), 26.8 (C-5), 23.2 (2C, C-15, C-16). HR-ESI-MS: 285.2431 ([M − H]⁻, $C_{17}H_{33}O_3^-$; calc. 285.2435), 571.4926 ([2M − H]−′ $C_{34}H_{67}O_6^-$; calc. 571.4943).

(R)-3-hydroxy-13-methyltetradecanoic acid (5): White powder. $[\alpha]_D^{20}$ = −15 (c = 0.1, chloroform). ¹H-NMR (600 MHz, CD_3OD): 3.89 (1H, *s*, H-3), 2.33 (1H, *m*, H-2), 2.24 (1H, *m*, H-2), 1.52 (1H, *m*, H-13), 1.45 (4H, *s*, H-4, H-5), 1.30 (12H, *s*), 1.18 (2H, *s*, H-12), 0.88 (6H, *d*, *J* = 6.6, H-14, H-15). ¹³C-NMR (600

MHz, CD$_3$OD): 176.5 (C-1), 70.4 (C-3), 45.3 (C-2), 40.4 (C-12), 38.2 (C-4), 31.2-30.9 (5C, C-6 to C-10), 29.3 (C-13), 28.7 (C-11), 26.8 (C-5), 23.1 (2C, C-14, C-15). HR-ESI-MS: 257.2101 ([M − H]$^-$, C$_{15}$H$_{29}$O$_3^-$; calc. 257.2122), 515.4285 ([2M − H]$^-$, C$_{30}$H$_{59}$O$_6^-$; calc. 515.4317).

13-methyltetradecanoic acid (6): White powder. ^{1}H-NMR (500 MHz, CD$_3$OD): 2.27 (1H, t, *J* = 7.4, H-2), 1.60 (1H, m, H-3), 1.52 (1H, *non*, *J* = 6.6, H-13), 1.30 (16H, s, H-4 to H-11), 1.18 (1H, m, H-12), 0.88 (6H, d, *J* = 6.7, H-14, H-15). ^{13}C-NMR (500 MHz, CD$_3$OD): 177.9 (C-1), 40.4 (C-11), 35.2 (C-2), 31.2-30.4 and 28.7 (8C, C-4 to C-10), 29.3 (C-13), 26.3 (C-3), 23.2 (2C, C-14, C-15). HR-ESI-MS: 241.2168 ([M − H]$^-$, C$_{15}$H$_{29}$O$_2^-$; calc. 241.2173), 483.4432 ([2M − H]$^-$, C$_{30}$H$_{59}$O$_4^-$; calc. 483.4419).

9Z-hexadecenoic acid (7): yellow oily liquid. ^{1}H-NMR (500 MHz, CD$_3$OD): 5.35 (2H, dt, *J* = 11.3, 6.2, H-9, H-10), 2.27 (2H, t, *J* = 7.5, C-2), 2.03 (4H, m, H-8, H-11), 1.60 (2H, m, H-3), 1.33 (16H, s), 0.90 (3H, t, *J* = 6.9, H-16). ^{13}C-NMR (500 MHz, CD$_3$OD): 178.0 (C-1), 131.1 (C-10), 130.9 (C-11), 35.3 (C-2), 33.1 (C-14), 31.0 (s, C-13), 31.0-30.2 (5C, C-4 to C-8), 28.3 (2C, C-9, C-12), 26.3 (C-3), 23.9 (C-15), 14.6 (C-16). HR-ESI-MS: 255.2326 ([M + H]$^+$, C$_{16}$H$_{31}$O$_2^+$; calc. 255.2319), 253.2164 ([M − H]$^-$, C$_{16}$H$_{29}$O$_2^-$; calc. 253.2162), 507.4426 ([2M − H]$^-$, C$_{32}$H$_{59}$O$_4^-$; calc. 507.4408).

15-methyl-9Z-hexadecenoic acid (8): yellow oily liquid. ^{1}H-NMR (500 MHz, CD$_3$OD): 5.35 (2H, dt, *J* = 11.3, 6.2, H-10, H-11), 2.27 (2H, t, *J* = 7.4, H-2), 2.04 (4H, m, H-9, H-12), 1.60 (2H, m, H-3), 1.53 (1H, *non*, *J* = 6.7, H-15), 1.33 (12H, s), 1.19 (2H, m, H-14), 0.88 (6H, d, *J* = 6.6, H-16, H-17). ^{13}C-NMR (500 MHz, CD$_3$OD): 177.9 (C-1), 131.0 (2C, C-10, C-11), 40.3 (C-14), 35.2 (C-2), 31.3-30.3 (5C, C-4 to C-8), 29.3 (C-15), 28.3 (2C, C-9, C-11), 28.3 (C-13), 26.3 (C-3), 23.2 (2C, C-16, C-17). HR-ESI-MS: 269.2484 ([M + H]$^+$, C$_{17}$H$_{33}$O$_2^+$; calc. 269.2475), 267.2334 ([M − H]$^-$, C$_{17}$H$_{31}$O$_2^-$; calc. 267.2330), 535.4756 ([2M − H]$^-$, C$_{16}$H$_{31}$O$_2^-$; calc. 535.4721).

4.9. Preparation of Fatty Acid Methyl Esters (FAMEs)

Preparation of FAMEs was carried out based on methanolysis/methylation using conc. Hydrochloric acid (HCl) as described by Ichihara and Fukubayashi [27]. A solution of 8.0% (*w/v*) HCl was obtained by diluting conc. HCl (37%, *w/w*; 9.1 mL) in methanol (40.9 mL). Each fatty acid was dissolved in toluene to reach a concentration of 0.005 g/mL. Then, methanol (7.5-fold) and 8.0% HCl solution (1.5-fold) were added sequentially to this solution. The solution was stirred at 45°C overnight. After cooling at room temperature, hexane (5-fold) and water (5-fold) were added for the extraction of FAMEs. HCl (37%, *w/w*) was purchased from Carlo Erba©.

FAMEs were then diluted in CH$_2$Cl$_2$ to a concentration of 2 mg/mL and then were diluted in CH$_3$CN to the concentration 200 µg/mL. These solutions were analysed by MS/MS on a Q-ToF 6540 mass spectrometer (Agilent, Les Ulis, France) by direct introduction at a flow rate of 10 µL/min and using an APCI (Atmospheric Pressure Chemical Ionization) ion source in the positive mode. The corona current was set to 2 µA, the nebulizer pressure was 60 psig and 8 L/min nitrogen flow heated at 300 °C was used for desolvation. Capillary, fragmentor and skimmer voltages were set to 3000 V, 100 V, and 45 V, respectively. CH$_2$Cl$_2$ and CH$_3$CN were purchased from J.T. Baker©. The MS/MS collision energy was 15 (arbitrary units).

4.10. Determination of Minimal Inhibitory Concentration

The ATCC strains were purchased from the Pasteur Institute. The strain used in this study was methicillin-resistant *S. aureus* ATCC33591. Extracts, fractions and pure compounds were tested according to the reference protocol of the European Committee on Antimicrobial Susceptibility Testing [28]. The standard microdilution test as described by the Clinical and Laboratory Standards Institute guidelines (M7-A8) was used to determine minimal inhibition concentrations (MIC) against bacteria [29]. Crude extracts and pure compounds were tested at concentrations ranging from 256 to 0.5 µg/mL. The microplates were incubated at 35 °C, and MIC values were calculated after 24 h.

The MIC values reported in Table S2 refer to the lowest concentration preventing visible growth in the wells. All assays were conducted in triplicate.

4.11. Cytotoxicity Evaluation

The crude extract for each strain was tested to determine its cytotoxicity using the MRC-5 (ATCC CCL-171) normal human lung fibroblast cells (LGC standards, Molsheim, France). MRC-5 cells were seeded into 96-well microplates at 2000 cells per well. The assay was conducted according to the procedure described by Tempête et al. [30]. After 24 h in wells, solubilized extracts in DMSO are deposited in triplicate as well as the solvent and docetaxel (Taxotère) controls. Cell viability is evaluated in comparison with untreated control cultures after 3 days. Extracts are tested at the concentration of 10 µg/mL.

Supplementary Materials: Supplementary materials can be found at http://www.mdpi.com/1422-0067/20/8/2006/s1.

Author Contributions: M.B. performed the experiments, analysed the data and wrote the article. N.E., L.P. contributed to analysis tools and analysed the data. J.-L.W., D.S., D.T. and V.E. designed the study. All authors read and approved the manuscript.

Funding: This work has benefited from a joint Agence Nationale de la Recherche - Swiss National Science Foundation (ANR-SNF) grant (SECIL, ref ANR-15-CE21-0016 and SNF N° 310030E-164289) and from an "Investissement d'Avenir" grant (CEBA, ref ANR-10-LABX-0025) managed by the ANR.

Conflicts of Interest: The authors declare no conflict of interest.

Abbreviations

MRSA	Methicillin-resistant *Staphylococcus aureus*
UHPLC-HRMS/MS	Ultra-High Performance Liquid Chromatography-High Resolution Mass Spectrometry
MIC	Minimal Inhibitory Concentration
EtOAc	Ethyl acetate
MN	Molecular Network
XIC	Extracted Ion Chromatogram
PDA	Potato Dextrose Agar
FAME	Fatty Acid Methyl Ester
FA	Formic Acid
HCl	Hydrochloric acid

References

1. Arnold, A.E.; Mejía, L.C.; Kyllo, D.; Rojas, E.I.; Maynard, Z.; Robbins, N.; Herre, E.A. Fungal endophytes limit pathogen damage in a tropical tree. *Proc. Natl. Acad. Sci. USA* **2003**, *100*, 15649–15654. [CrossRef] [PubMed]
2. Rosa, L.H.; Vaz, A.B.M.; Caligiorne, R.B.; Campolina, S.; Rosa, C.A. Endophytic fungi associated with the Antarctic grass *Deschampsia antarctica* Desv. (Poaceae). *Polar. Biol.* **2009**, *32*, 161–167. [CrossRef]
3. Ali, A.H.; Abdelrahman, M.; Radwan, U.; El-Zayat, S.; El-Sayed, M.A. Effect of *Thermomyces* fungal endophyte isolated from extreme hot desert-adapted plant on heat stress tolerance of cucumber. *Appl. Soil Ecol.* **2018**, *124*, 155–162. [CrossRef]
4. Arnold, A.E.; Lutzoni, F. Diversity and host range of foliar fungal endophytes: Are tropical leaves biodiversity hotspots? *Ecology* **2007**, *88*, 541–549. [CrossRef]
5. Higgins, K.L.; Arnold, A.E.; Coley, P.D.; Kursar, T.A. Communities of fungal endophytes in tropical forest grasses: Highly diverse host- and habitat generalists characterized by strong spatial structure. *Fungal Ecol.* **2014**, *8*, 1–11. [CrossRef]
6. Rosa, L.H.; Vieira, M.L.A.; Cota, B.B.; Johann, S.; Alves, T.M.A.; Zani, C.L.L.; Rosa, C.A. Endophytic Fungi of Tropical Forests: A Promising Source of Bioactive Prototype Molecules for the Treatment of Neglected Diseases. In *Drug Development—A Case Study Based Insight into Modern Strategies*; Rundfeldt, C., Ed.; In-Tech: Rijeka, Croatia, 2011; pp. 469–486.

7. Kahn, F. El género *Astrocaryum* (Arecaceae). *Rev. Peru. Biol.* **2014**, *15*, 31–48. [CrossRef]

8. Charles-dominique, P.; Chave, J.; Dubois, M.A.; De Granville, J.J.; Riera, B.; Vezzoli, C. Colonization front of the understorey palm *Astrocaryum sciophilum* in a pristine rain forest of French Guiana. *Glob. Ecol. Biogeogr.* **2003**, *12*, 237–248. [CrossRef]

9. Wang, M.; Carver, J.J.; Phelan, V.V.; Sanchez, L.M.; Garg, N.; Peng, Y.; Nguyen, D.D.; Watrous, J.; Kapono, C.A.; Luzzatto-Knaan, T.; et al. Sharing and community curation of mass spectrometry data with Global Natural Products Social Molecular Networking. *Nat. Biotechnol.* **2016**, *34*, 828–837. [CrossRef]

10. MZmine 2: Modular Framework for Processing, Visualizing, and Analyzing Mass Spectrometry-Based Molecular Profile Data. Available online: https://bmcbioinformatics.biomedcentral.com/articles/10.1186/1471-2105-11-395 (accessed on 30 August 2018).

11. Olivon, F.; Elie, N.; Grelier, G.; Roussi, F.; Litaudon, M.; Touboul, D. MetGem Software for the Generation of Molecular Networks Based on the t-SNE Algorithm. *Anal. Chem.* **2018**, *90*, 13900–13908. [CrossRef]

12. Olivon, F.; Grelier, G.; Roussi, F.; Litaudon, M.; Touboul, D. MZmine 2 Data-Preprocessing To Enhance Molecular Networking Reliability. *Anal. Chem.* **2017**, *89*, 7836–7840. [CrossRef]

13. Vrkoslav, V.; Cvačka, J. Identification of the double-bond position in fatty acid methyl esters by liquid chromatography/atmospheric pressure chemical ionisation mass spectrometry. *J. Chromatogr. A* **2012**, *1259*, 244–250. [CrossRef]

14. Wang, X.; Gong, X.; Li, P.; Lai, D.; Zhou, L. Structural Diversity and Biological Activities of Cyclic Depsipeptides from Fungi. *Molecules* **2018**, *23*, 169. [CrossRef]

15. Li, G.; Kusari, S.; Golz, C.; Strohmann, C.; Spiteller, M. Three cyclic pentapeptides and a cyclic lipopeptide produced by endophytic *Fusarium decemcellulare* LG53. *RSC Adv.* **2016**, *6*, 54092–54098. [CrossRef]

16. Helaly, S.E.; Kuephadungphan, W.; Phongpaichit, S.; Luangsa-Ard, J.J.; Rukachaisirikul, V.; Stadler, M. Five Unprecedented Secondary Metabolites from the Spider Parasitic Fungus *Akanthomyces novoguineensis*. *Molecules* **2017**, *22*, 991. [CrossRef]

17. Kuephadungphan, W.; Helaly, S.E.; Daengrot, C.; Phongpaichit, S.; Luangsa-ard, J.J.; Rukachaisirikul, V.; Stadler, M. Akanthopyrones A–D, α-Pyrones Bearing a 4-O-Methyl-β-D-glucopyranose Moiety from the Spider-Associated Ascomycete *Akanthomyces novoguineensis*. *Molecules* **2017**, *22*, 1202. [CrossRef]

18. Johansen, J.E.; Binnerup, S.J.; Kroer, N.; Mølbak, L. *Luteibacter rhizovicinus* gen. nov., sp. nov., a yellow-pigmented gammaproteobacterium isolated from the rhizosphere of barley (*Hordeum vulgare* L.). *Int. J. Syst. Evol. Microbiol.* **2005**, *55*, 2285–2291. [CrossRef]

19. Kämpfer, P.; Lodders, N.; Falsen, E. *Luteibacter anthropi* sp. nov., isolated from human blood, and reclassification of *Dyella yeojuensis* Kim et al. 2006 as *Luteibacter yeojuensis* comb. nov. *Int. J. Syst. Evol. Microbiol.* **2009**, *59*, 2884–2887. [CrossRef] [PubMed]

20. Bergsson, G.; Arnfinnsson, J.; Steingrímsson, O.; Thormar, H. Killing of Gram-positive cocci by fatty acids and monoglycerides. *Acta Pathol. Microbiol. Immunol. Scand.* **2001**, *109*, 670–678. [CrossRef]

21. Parsons, J.B.; Yao, J.; Frank, M.W.; Jackson, P.; Rock, C.O. Membrane Disruption by Antimicrobial Fatty Acids Releases Low-Molecular-Weight Proteins from *Staphylococcus aureus*. *J. Bacteriol.* **2012**, *194*, 5294–5304. [CrossRef] [PubMed]

22. Kabara, J.J.; Swieczkowski, D.M.; Conley, A.J.; Truant, J.P. Fatty Acids and Derivatives as Antimicrobial Agents. *Antimicrob. Agents Chemother.* **1972**, *2*, 23–28. [CrossRef] [PubMed]

23. Kodicek, E.; Worden, A.N. The effect of unsaturated fatty acids on *Lactobacillus helveticus* and other Gram-positive micro-organisms. *Biochem. J.* **1945**, *39*, 78–85. [CrossRef]

24. Desbois, A.P.; Smith, V. Antibacterial free fatty acids: Activities, mechanisms of action and biotechnological potential. *Appl. Microbiol. Biotechnol.* **2009**, *85*, 1629–1642. [CrossRef]

25. Casella, T.M.; Eparvier, V.; Mandavid, H.; Bendelac, A.; Odonne, G.; Dayan, L.; Duplais, C.; Espindola, L.S.; Stien, D. Antimicrobial and cytotoxic secondary metabolites from tropical leaf endophytes: Isolation of antibacterial agent pyrrocidine C from *Lewia infectoria* SNB-GTC2402. *Phytochemistry* **2013**, *96*, 370–377. [CrossRef]

26. Myers, O.D.; Sumner, S.J.; Li, S.; Barnes, S.; Du, X. One Step Forward for Reducing False Positive and False Negative Compound Identifications from Mass Spectrometry Metabolomics Data: New Algorithms for Constructing Extracted Ion Chromatograms and Detecting Chromatographic Peaks. *Anal. Chem.* **2017**, *89*, 8696–8703. [CrossRef]

27. Ichihara, K.; Fukubayashi, Y. Preparation of fatty acid methyl esters for gas-liquid chromatography. *J. Lipid Res.* **2010**, *51*, 635–640. [CrossRef]
28. Rodriguez, A.M.S.; Theodoro, P.; Eparvier, V.; Basset, C.; Silva, M.R.R.; Beauchêne, J.; Espindola, L.; Stien, D. Search for Antifungal Compounds from the Wood of Durable Tropical Trees. *J. Nat. Prod.* **2010**, *73*, 1706–1707. [CrossRef]
29. Clinical and Laboratory Standards. *Methods for Dilution Antimicrobial Susceptibility Tests for Bacteria That Grow Aerobically*, 8th ed.; Approved Standard M7-A8; Clinical and Laboratory Standards Institute: Wayne, PA, USA, 2009.
30. Tempête, C.; Werner, G.; Favre, F.; Rojas, A.; Langlois, N. In vitro cytostatic activity of 9-demethoxyporothramycin B. *Eur. J. Med. Chem.* **1995**, *30*, 647–650. [CrossRef]

International Journal of
Molecular Sciences

Article

Characterization, Diversity, and Structure-Activity Relationship Study of Lipoamino Acids from *Pantoea* sp. and Synthetic Analogues

Seindé Touré [1], Sandy Desrat [1], Léonie Pellissier [2], Pierre-Marie Allard [2], Jean-Luc Wolfender [2], Isabelle Dusfour [3], Didier Stien [4] and Véronique Eparvier [1,*]

[1] CNRS, Institut de Chimie des Substances Naturelles, UPR 2301, Université Paris-Saclay,
     1 avenue de la Terrasse, 91198 Gif-sur-Yvette, France; seindet@gmail.com (S.T.); sandy.desrat@cnrs.fr (S.D.)
[2] School of Pharmaceutical Sciences, EPGL, University of Geneva, University of Lausanne,
     Rue Michel Servet 1, CH-1211 Geneva 4, Switzerland; leonie.pellissier@unige.ch (L.P.);
     pierre-marie.allard@unige.ch (P.-M.A.); jean-luc.wolfender@unige.ch (J.-L.W.)
[3] Institut Pasteur de la Guyane (IPG), Unité de Contrôle et Adaptation des
     Vecteurs, 97306 Cayenne, Guyane, France; isabelle.dusfour@pasteur.fr
[4] Sorbonne Université, CNRS, Laboratoire de Biodiversité et Biotechnologie Microbienne, LBBM,
     Observatoire Océanologique, 66650 Banyuls-sur-mer, France; didier.stien@cnrs.fr
*   Correspondence: veronique.eparvier@cnrs.fr; Tel.: +33-(0)16-982-3679

Received: 30 January 2019; Accepted: 26 February 2019; Published: 2 March 2019

Abstract: A biological evaluation of a library of extracts from entomopathogen strains showed that *Pantoea* sp. extract has significant antimicrobial and insecticidal activities. Three hydroxyacyl-phenylalanine derivatives were isolated from this strain. Their structures were elucidated by a comprehensive analysis of their NMR and MS spectroscopic data. The antimicrobial and insecticidal potencies of these compounds were evaluated, and compound **3** showed 67% mortality against *Aedes aegypti* larvae at a concentration of 100 ppm, and a minimum inhibitory concentration (MIC) of 16 µg/mL against methicillin-resistant *Staphylococcus aureus*. Subsequently, hydroxyacyl-phenylalanine analogues were synthesized to better understand the structure-activity relationships within this class of compounds. Bioassays highlighted the antimicrobial potential of analogues containing saturated medium-chain fatty acids (12 or 14 carbons), whereas an unsaturated long-chain fatty acid (16 carbons) imparted larvicidal activity. Finally, using a molecular networking-based approach, several close analogues of the isolated and newly synthesized lipoamino acids were discovered in the *Pantoea* sp. extract.

Keywords: hydroxyacyl-phenylalanine; lipoamino acid; entomopathogen; bacteria; antimicrobial; larvicidal; *Aedes aegypti*

---

1. Introduction

The development of novel antimicrobial and insecticidal agents is urgently needed to improve public health worldwide. Indeed, drug-resistant pathogenic microorganisms such as methicillin-resistant *Staphylococcus aureus* (MRSA), vancomycin-resistant *S. aureus* (VRSA), and vancomycin-resistant *Enterococci* (VRE) have emerged [1]. Other Gram-negative pathogens are particularly worrisome because they are becoming resistant to nearly all of the antibiotic drugs currently available, including carbapenems [2]. These pathogens have become a major clinical problem, causing significant mortality in both healthy hosts and in those with underlying comorbidities [3]. Thus, it is essential to investigate new drugs to address the decreasing efficiency of the currently available antibiotics [2]. Another major public health problem concerns mosquitoes and their ability to spread pathogens such as malaria parasites, dengue, chikungunya, and more recently the Zika virus. The World Health Organization (WHO) estimated that more than 80% of the world's population is at risk of contracting vector-borne diseases, and each year more than 700,000 people die as a result [4]. *Aedes* sp. mosquitoes are the primary vector for transmitting arboviruses worldwide [5]. Today, the two main approaches to control them involve genetic modification or the application of chemicals. Synthetic chemicals including pyrethroids or organophosphates such as temephos and malathion have been sprayed into the environment for decades. However, most of these synthetic chemicals have adverse effects, leading to the development of resistance, environmental pollution, and the introduction of toxic hazards to humans and other nontarget organisms [6]. Furthermore, the abuse and/or misuse of these compounds has resulted in the loss of vector control efficacy. Thus, specific safer pesticides are urgently needed [7].

Natural products have proven to be an immeasurable source of bioactive compounds [8]. Entomopathogenic microorganisms, the natural enemies of insects, are known to produce bioactive metabolites that have been implicated in complex defense and self-protection mechanisms. Indeed, to achieve ecological success, they produce several chemical entities including insecticides and/or antimicrobial metabolites [9].

As part of our investigation into the secondary metabolites produced by entomopathogenic microorganisms, a collection of 53 strains was extracted and screened against *Aedes aegypti* mosquito larvae and human pathogenic microorganisms (*Staphylococcus aureus*, MRSA, *Candida albicans*, and *Trichophyton rubrum*). The extract from the bacterium *Pantoea* sp. SNB-VECD14B exhibited a mortality rate of 97.2% against *Ae. aegypti.* larvae at a concentration of 100 ppm and a minimum inhibitory concentration (MIC) of 16 µg/mL against *T. rubrum*. In addition, the full strain collection was profiled by high-resolution mass spectrometry and the resulting fragmentation data were organized as a single molecular network. This molecular networking approach allowed us to organize untargeted tandem MS datasets according to their spectral similarity and thus to group analytes by structural similarity [10]. Similar approaches have served as powerful tools to navigate the chemical space of complex biological matrices and can be used to view the chemical constituents of a wide variety of extracts in a single map [11,12]. Appropriate taxonomical color mapping allowed us to highlight a structurally related series of compounds also produced by *Pantoea* sp. that was selected for further investigation.

2. Results and Discussion

2.1. Isolation and Structure Elucidation

The EtOAc extract of *Pantoea* sp. SNB-VECD14B was subjected to bioguided preparative high performance liquid chromatography (HPLC) using a C_{18} silica gel column to yield three pure compounds (**1–3**) (Figure 1).

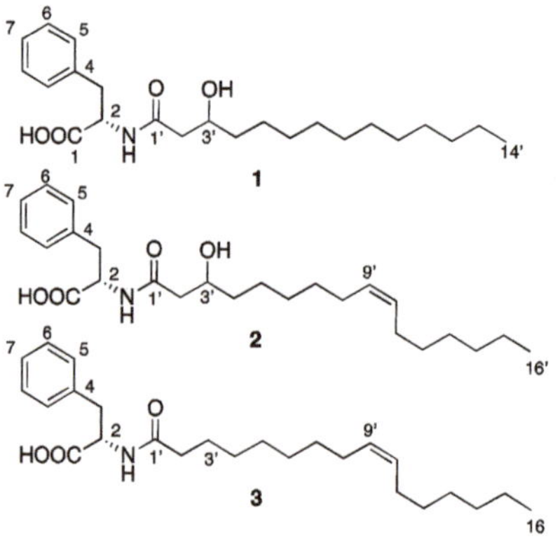

Figure 1. Compounds (**1–3**) isolated from *Pantoea* sp.

The molecular formula of compound **1** was determined to be $C_{23}H_{37}NO_4$ via HRESI-TOFMS analysis which gave a pseudomolecular ion at *m/z* 392.2769 [M+H]$^+$ (calculated for $C_{23}H_{38}NO_4{}^+$, 392.2795), indicating six degrees of unsaturation. The structure of this compound was deduced from NMR spectral data (Table 1). The ^1H and ^{13}C NMR spectra of **1** suggested the presence of one methyl group (δ_H 0.90) that appeared as a triplet in the ^1H NMR spectrum, fatty acid methylenes integrating for 20 protons (δ_C 30.9, δ_H 1.29) that appeared as a characteristic broad signal in the ^1H NMR spectrum, two other distinct methylenes at δ_H 1.40 and 2.29/2.34, and six methines, one of which was hydroxylated (δ_H 3.85) with the other five methines corresponding to aromatic protons with chemical shifts in the ^1H NMR spectrum between 7.18 and 7.24 ppm. The sequence of ^1H-^1H COSY signals from H2' to H5' and from H12' to H14' allowed us to determine the presence of a hydroxylated fatty acid moiety. ^1H-^{13}C correlations observed in the HMBC spectrum between H2'/C1', C3' and C4' confirmed that compound **1** has a tetradecanoate moiety that is hydroxylated at C3' (Figure 2). Another ^1H-^1H correlation was observed between H2 and H3 and ^1H-^{13}C correlations between H2/C1, C3 and H3/C1, C2, C4, and C5 along with the C1 chemical shift at δ 176.0 demonstrated the presence of a phenylalanine unit. Although no correlation was observed (COSY, HMBC, NOESY) between the phenylalanine and fatty acid moieties, the only possible linkage between these two units is via an amide bond. This is consistent with the chemical shifts at C1 (δ_C 173.8), C2 (δ_C 55.9), and H2 (δ_H 4.60). Thus, compound **1** was identified as *N*-(3-hydroxytetradecanoyl)phenylalanine. At this stage, the asymmetric carbons configurations could not be determined. The stereochemistry of this molecule and the following compounds will be demonstrated by synthesis.

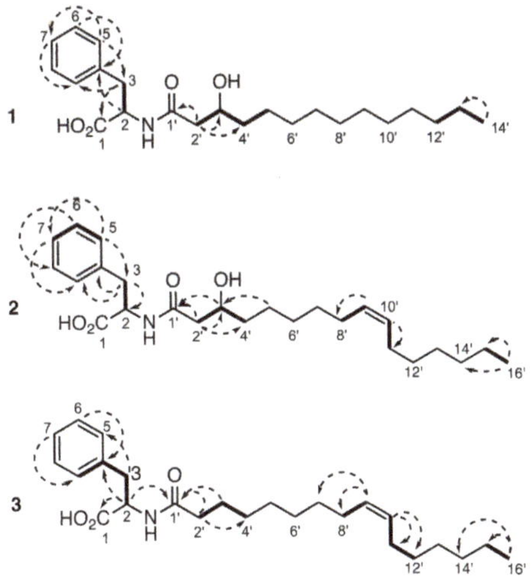

Figure 2. Key ^1H-^1H COSY (bold lines) and HMBC (dashed arrows) correlations.

The molecular formula of compound **2** was determined to be $C_{25}H_{39}NO_4$ based on the ion observed at *m/z* 418.2926 [M+H]$^+$ in the HRESI-TOFMS experiment (calculated for $C_{25}H_{40}NO_4{}^+$, 418.2952), which corresponded to seven degrees of unsaturation. The ^1H and ^{13}C NMR spectra revealed several similarities to **1**. Only minor differences were observed, including the presence of a broad methylene signal integrating for just fourteen protons, along with two de-shielded protons at δ_H 5.35 and two methylenes at δ_H 2.04 corresponding to the methines and adjacent methylenes of a double bond. The COSY and HMBC data of **2** allowed us to determine that the fatty acid moiety is 3-hydroxyhexadec-9-enoate and that it is hydroxylated at C3. The (Z) configuration of the double bond was determined by comparison with the literature data based on the distinctive splitting of the olefinic protons [13]. The smaller coupling constant in the (Z) double bond impacts on the spacing of the olefinic protons multiplet peaks and the overall aspect of the multiplet in ^{1}H NMR. The double bond configuration was confirmed by comparison with analytical data of synthetic compounds described below. Thus, compound **2** was finally identified as (Z)-N-(3-hydroxyhexadec-9-enoyl)phenylalanine.

Table 1. Nuclear magnetic resonance (NMR) Spectroscopic Data (CD$_3$OD) for compounds **1**, **2**, and **3**.

Position	Compound 1		Compound 2		Compound 3	
	δ_C [a], type	δ_H [b], m (*J* in Hz)	δ_C [c], type	δ_H [d], m (*J* in Hz)	δ_C [a], type	δ_H [b], m (*J* in Hz)
1	176.0, C		176.3, C		176.5, C	
2	55.9, CH	4.60, m	57.6, CH	4.53, m	56.0, CH	4.62, dd (8.0, 4.7)
3	38.7, CH$_2$	2.98, dd (13.9, 8.2) 3.21, dd (13.9, 4.5)	39.4, CH$_2$	2.97, dd (14.0, 7.7) 3.21, dd (14.0, 4.5)	39.0, CH$_2$	2.94, dd (13.8, 8.7) 3.22, dd (13.8, 4.7)
4	139.0, C		139.8, C		139.2, C	
5/9	130.5, CH	7.24, m	130.7, CH	7.24, m	130.5, CH	7.23, m
6/8	129.3, CH	7.25, m	129.3, CH	7.22, m	129.4, CH	7.24, m
7	127.6, CH	7.18, m	127.3, CH	7.15, m	127.6, CH	7.17, m
1$'$	173.8, C		173.5, C		175.8, C	
2$'$	44.6, CH$_2$	2.29, dd (14.4, 5.2) 2.34, dd (14.4, 7.4)	44.9, CH$_2$	2.25, dd (14.4, 7.8) 2.30, dd (14.5, 4.9)	37.2, CH$_2$	2.13, t (7.4) 2.23, t (7.5)
3$'$	69.6, CH	3.85, m	67.9, CH	3.85, m	27.1, CH$_2$	1.48, m
4$'$	38.0, CH$_2$	1.40, m	38.2, CH$_2$	1.39, m	30.3, CH$_2$	1.29, br s
5$'$	30.9, CH$_2$	1.29, br s	31.0, CH$_2$	1.30, m	23.8, CH$_2$	1.31, br s
6$'$	30.9, CH$_2$	1.29, br s	31.0, CH$_2$	1.30, m	30.3, CH$_2$	1.29, br s
7$'$	30.9, CH$_2$	1.29, br s	31.0, CH$_2$	1.30, m	30.3, CH$_2$	1.29, br s
8$'$	30.9, CH$_2$	1.29, br s	28.3, CH$_2$	2.04, m	30.3, CH$_2$	2.03, m
9$'$	30.9, CH$_2$	1.29, br s	130.9, CH	5.40, m	131.0, CH	5.35, m
10$'$	30.9, CH$_2$	1.29, br s	130.9, CH	5.40, m	131.0, CH	5.35, m
11$'$	30.9, CH$_2$	1.29, br s	28.3, CH$_2$	2.04, m	30.3, CH$_2$	2.03, m
12$'$	30.9, CH$_2$	1.29, br s	31.0, CH$_2$	1.30, m	30.5, CH$_2$	1.33, br s
13$'$	23.6, CH$_2$	1.31, m	31.0, CH$_2$	1.30, m	30.5, CH$_2$	1.33, br s
14$'$	14.4, CH$_3$	0.90, t (6.8)	31.0, CH$_2$	1.30, m	30.5, CH$_2$	1.33, br s
15$'$			23.9, CH$_2$	1.30, m	23.8, CH$_2$	1.31, br s
16$'$			14.6, CH$_3$	0.90, t (6.8)	14.6, CH$_3$	0.90, t (6.8)

[a] Data recorded at 125 MHz. [b] Data recorded at 500 MHz. [c] Data recorded at 600 MHz. [d] Data recorded at 200 MHz.

The HRESI-TOFMS experiment for compound **3** indicated a molecular formula of C$_{25}$H$_{39}$NO$_3$ with a *m/z* of 402.3004 [M+H]$^+$ (calculated for C$_{25}$H$_{40}$NO$_3{}^+$, 402.3003), which corresponded to seven degrees of unsaturation. After examining the ^{1}H and ^{13}C NMR spectra, compound **3** was obviously similar to **2**. The only differences observed include signals pertaining to the hydroxy group that are absent in compound **3** as well as differences in the signals of the neighboring protons and carbons (Table 1). Ultimately, the fatty acid moiety was determined to be hexadec-9-enoate. The COSY and HMBC spectra showed the same correlations as in **2**, along with a correlation between H2/C1$'$ that further confirmed the amide linkage between the two moieties. Compound **3** was therefore identified as (Z)-N-hexadec-9-enoylphenylalanine (Figure 2). The configuration of the amino acid subunit was determined to be L (*S*) after synthesis of **3** and ***ent*-3** (described below) and comparison of their optical rotations. All synthetic (2*S*)-lipoamino acids had a positive optical rotation, as did compounds **1–3**. Altogether, the analytical data along with the obvious biosynthetic resemblance between **1**, **2**, and **3** led to the conclusion that the absolute stereochemistry of the phenylalanine moiety was L in **1** and **2** as well. The configuration of C-3$'$ could not be determined. These three compounds were isolated for the first time from natural sources and the complete NMR data for molecules **1–3** is described for the first time in the literature [14,15].

2.2. Synthesis of Lipoaminoacid Analogues

Our structure activity relationship (SAR) investigation began with the synthesis of various acylated phenylalanine analogues of compound **3**. We modified the carbon chain length of the fatty acid, the configuration of the amino acid (L vs. D), and the number of double bonds in the fatty acid alkyl chain (Scheme 1). The synthetic method used classical conditions for peptides coupling [16]. Amide coupling between L- or D- amino acid methyl esters and fatty acids in the presence of HATU and Hunig's base, followed by saponification of the ester moiety with lithium hydroxide gave a series of 32 acylated phenylalanine derivatives. Thus, starting from the methyl ester of either L- or

D-phenylalanine, 16 ester derivatives (compounds **4** to **19**) and 16 free acids (compounds **20** to **35**) were prepared with varying lengths of the fatty acid chain (from 12 to 20 carbons) and different degrees of unsaturation. Compounds **3**, **3-OMe**, *ent*-**3**, *ent*-**3-OMe**, **8**, **15**, **24**, and **31** were *Z*-alkenes, whereas compounds **9**, **16**, **25**, and **32** had an *E*-alkene configuration. Four additional lipoamino acids (**18**, **19**, **34**, **35**) were prepared from palmitoleic acid and the methyl ester of either L-alanine or L-tyrosine following the same synthetic procedures. The general synthetic process is shown in Scheme 1. The structures of the newly synthesized compounds were confirmed by NMR and HRESI-TOFMS analysis.

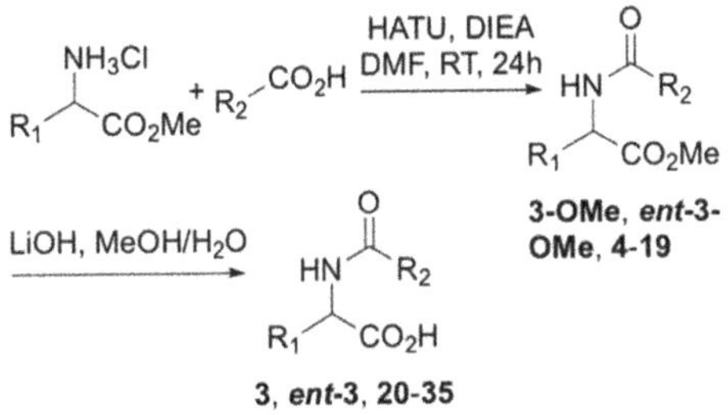

Cmpd #	Amino acid	R_2
3-OMe, 3	L-Phe	(Z)-Δ_9-$C_{15}H_{29}$
4, 20	L-Phe	n-$C_{15}H_{31}$
5, 21	L-Phe	n-$C_{19}H_{39}$
6, 22	L-Phe	n-$C_{11}H_{23}$
7, 23	L-Phe	n-$C_{13}H_{25}$
8, 24	L-Phe	(Z)-Δ_9-$C_{17}H_{33}$
9, 25	L-Phe	(E)-Δ_9-$C_{17}H_{33}$
10, 26	L-Phe	n-$C_{17}H_{35}$
ent-**3-OMe**, *ent*-**3**	D-Phe	(Z)-Δ_9-$C_{15}H_{29}$
11, 27	D-Phe	n-$C_{15}H_{31}$
12, 28	D-Phe	n-$C_{19}H_{39}$
13, 29	D-Phe	n-$C_{11}H_{23}$
14, 30	D-Phe	n-$C_{13}H_{25}$
15, 31	D-Phe	(Z)-Δ_9-$C_{17}H_{33}$
16, 32	D-Phe	(E)-Δ_9-$C_{17}H_{33}$
17, 33	D-Phe	n-$C_{17}H_{35}$
18, 34	L-Ala	(Z)-Δ_9-$C_{15}H_{29}$
19, 35	L-Tyr	(Z)-Δ_9-$C_{15}H_{29}$

Scheme 1. Synthetic compounds **3**, **3-OMe**, *ent*-**3**, *ent*-**3-OMe**, and **4–35**.

2.3. Biological Activities and SAR Investigation

Compounds **1–3**, isolated from the bacteria *Pantoea* sp., were assayed for their antimicrobial and insecticidal activities against human pathogenic microorganisms and *Ae. aegypti* larvae (Table 2). The bioassays showed that compound **1** exhibited minimal larvicidal and antibacterial activities. In contrast, compound **3** demonstrated significant larvicidal activity with a mortality rate of 67.3% and antibacterial activity with an MIC of 16 µg/mL against methicillin-resistant *S. aureus*. Based on the observed differences of these compounds in the bioassays, we evaluated analogues of **1–3** to gain more insight into the structure-activity relationship (SAR) of this class of compounds.

Table 2. Antimicrobial and insecticide activities of compounds **1–3** and synthetic analogues **4–39**.

Compound	Larvicidal Activity (% Mortality at 10 ppm)	MIC (µg/mL)			
		C. albicans ATCC10231	*S. aureus* ATCC29213	**MRSA** ATCC33591	*T. rubrum* SNB-TR1
1 *	32.1	nd	−	64	nd
2 *	nd	nd	−	128	nd
3 */**3**	67.3/42.9	nd/−	−/−	16/64	128/−
ent-**3**	71.1	nd	−	−	−
3-OMe	8.2	−	−	−	−
ent-**3-OMe**	20.4	−	−	−	−
4	0	−	−	−	nd
5	0	>64	−	>64	nd
6	2.2	−	−	−	−
7	10.6	−	−	8	−
8	0	−	−	−	−
9	10.2	−	−	−	−
10	6.2	−	−	−	−
11	6.0	−	−	−	−
12	2.1	>64	>64	>64	>256
13	4	−	−	−	−
14	1.9	−	−	−	−
15	0	−	−	−	−
16	1.8	−	−	−	−
17	5.5	−	−	−	−
18	3.8	−	−	nd	256
19	29.4	−	−	−	64
20	2.3	−	−	−	−
21	4.5	−	−	−	−
22	2.1	−	64	32	32
23	26.5	−	−	−	32
24	31.2	−	−	−	−
25	19.1	−	−	−	−
26	4.0	−	−	−	−
27	23.3	−	−	−	nd
28	4.5	−	−	−	−
29	6.1	64	64	64	32
30	18.7	−	−	<8	<8
31	53.3	−	−	−	−
32	49.0	nd	−	−	−
33	2.3	−	−	−	−
34	4.5	nd	32	32	32
35	4.5	−	−	32	32
Pos. Ctrl.[a]	40	4	0.2	0.6	4

[a] Postitive control: Oxacillin for *S. aureus*; fluconazole for *C. albicans* and *T. rubrum*; vancomycin for MRSA and rotenone at 1 µg/mL for larvicidal activity. −: MIC >256µg/mL, considered not active. nd: Not determined. * natural product.

The synthetic compounds were evaluated for their larvicidal and antimicrobial activities (Table 2). First, the results showed that the best antimicrobial activities were obtained using compounds **1**, **3**, **7**, **19**, **22**, **23**, **29**, **30**, **34**, and **35** (MIC ≤64 µg/mL against MRSA and/or *T. rubrum*) with the highest antimicrobial activity being observed with compounds **30** (MIC <8 µg/mL against MRSA and *T. rubrum*) and **7** (MIC = 8 µg/mL against MRSA). Compounds **19** and **23** only showed activity against *T. rubrum*, whereas compounds **1**, **3**, and **7** only showed antibacterial activity against MRSA. Finally, compound **29** was the only compound to demonstrate MICs ≤64 µg/mL on all tested pathogenic strains. Interestingly, compounds **1**, **3**, **7**, **23**, and **35** showed antibacterial activity against MRSA only and not *S. aureus*. Most of these compounds contain a carboxyl group and a saturated alkyl chain with 12 to 14 carbons. Similar biological activities were observed irrespective of the stereochemical configuration of the free acids (**22–35**). On the other hand, given the antibacterial activities of compounds **8** and **16** against MRSA, the D configuration appears deleterious for lipoaminoesters.

Regarding larvicide potential, compounds **1**, **3**, *ent*-**3**, **19**, **23**, **24**, **27**, **31**, and **32** were the most active with mortality rates ranging from 26.5% to 71.1% at a concentration of 10 ppm. These compounds all possess long-chain acyl groups (mostly C16 and C18). Moreover, the compounds prepared from D-phenylalanine (*ent*-**3**, **27**, **31**, and **32**) were more active than their L-phenylalanine counterparts (**3**, **20**, **24**, and **25**). We next evaluated the effect of alkyl chain unsaturation on larvicide potential and found that unsaturated molecules were more active than saturated ones. For example, the mortality rate of compound **27** was 23.3%, while it was 71.1% for *ent*-**3**. Likewise, the mortality rate of compound **33** was 2.3% while it was 53.3% for **31**. Finally, unsaturated compounds with the Z-configuration were found to be more active than those with the E-configuration (i.e., compounds **24** and **31** compared to **25** and **32**).

We also investigated the effect of amino acid replacement on compound bioactivity. In general, the methyl esters of tyrosine alanine derivatives (**18** and **19**, respectively) were not antimicrobial. In contrast, the corresponding acids (**34** and **35**) appeared to be more active than the phenylalanine analogue **3**. The ester compounds were less soluble than the acids, which might explain their weaker activity. Regarding larvicidal activity, only the methyl ester of the acylated tyrosine derivative (**19**) was active (mortality rate of approximately 30%).

In summary, compound **3** was the only compound that demonstrated larvicidal and slight antibacterial activity with mortality rates of 42.9% and 67.3% and MIC values of 16 and 64 µg/mL against *Ae. aegypti* larvae and MRSA, respectively.

2.4. Molecular Networking

A molecular network (MN) was generated to organize the tandem MS data acquired from the entire collection of entomopathogenic strains that were biologically tested in this work, including *Pantoea* sp. SNB-VECD14B. After applying appropriate data treatment to align the common features among the collection, the resulting spectral data were submitted to the Global Natural Products Social molecular networking (GNPS) platform for molecular network generation [10–12]. The resulting molecular network grouped 20,859 spectra into 2619 clusters (Figure 3). Taxonomical mapping was applied by attributing a given color code to each strain in the collection, allowing us to visually navigate through the MN and check for the presence and distribution of specific compound classes within the full collection. A cluster was identified that was primarily related to the *Pantoea* genus. Together with the results from biological screening of the 53 strains collection, we identified a series of structurally related compounds that could be responsible for the observed larvicidal activity. This was confirmed by injecting the isolated compounds **1–3**, which were effectively found to be present in this particular cluster. Interestingly, this cluster of lipoamino acid-related structures indicated the following: (i) There is high structural diversity in the compounds generated by the *Pantoea* genus, and (ii) some of the members of this structural family were also found to be present in other entomopathogenic strains of the collection (Figure 3).

To study in more depth the chemistry of *Pantoea* lipoamino acids, a molecular network was generated organizing the tandem MS data acquired from the *Pantoea* sp. extracts together with the MS data generated from the isolated and synthetic lipoamino acid derivatives (Figure 4). In addition to the three isolated lipoamino acids (**1–3**), it was possible to align six of the synthetic lipoamino acids (**4**, **8**, **9**, **14**, **20** and **22**) with features present in the crude *Pantoea* sp. extracts, thus indicating that these compounds were also naturally present in the bacterium extract (represented as squares in Figure 4). Two of the synthesized derivatives (**5** and **10**) were also found to be present in the cluster but were not detected in the natural extracts (arrow-like node in Figure 4).

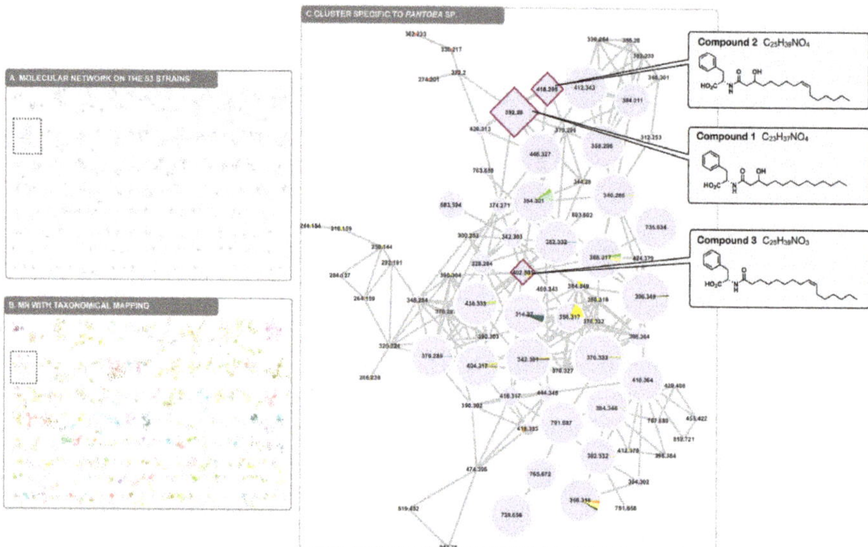

Figure 3. (**A**) Generation of a massive molecular network from the UHPLC-HRMS^2 analysis of 53 entomopathogenic microorganism extracts. (**B**) Taxonomical mapping showing a cluster containing ions mainly found in the *Pantoea* sp. extract. (**C**) Cluster of lipoaminoacids with compounds **1**, **2**, and **3** corresponding to compounds isolated from the *Pantoea* sp. extract.

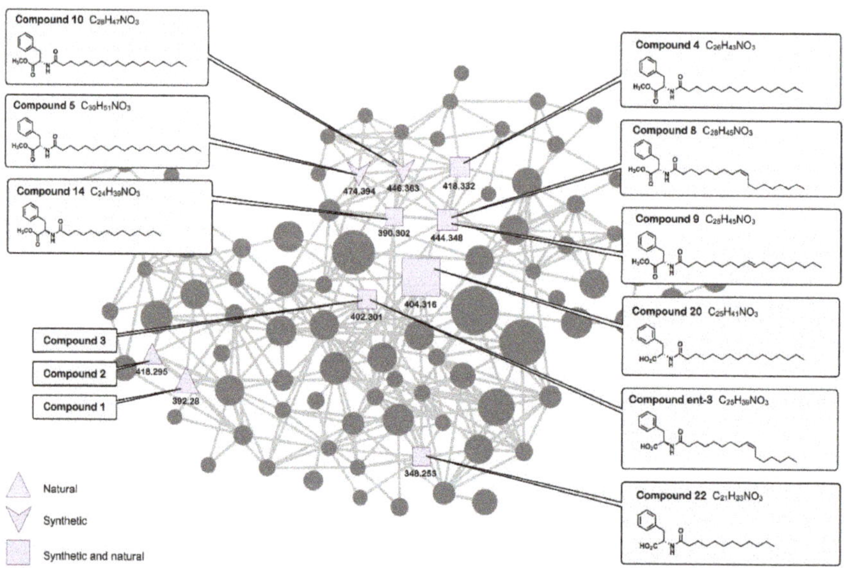

Figure 4. Cluster corresponding to natural lipoamino acids in the ethyl acetate extract of the *Pantoea* species together with synthetic lipoamino acid derivatives. The size of the nodes is proportional to the peak height of each feature within the *Pantoea* sp. extract.

The co-injection of synthetic analogues indicated that the studied extract contained additional lipoamino acid derivatives and allowed us to propagate annotations to neighboring nodes. To further annotate such compounds in the cluster, the mass spectral signature of each individual node was manually curated (Figure 5). For each feature, the molecular formula was established, annotations were made regarding adducts or complexes, and each fragmentation spectrum was treated using Sirius GUI software [17]. The results are summarized in Table S4 and allowed us to putatively annotate 16 additional compounds. Inspection of the fragmentation pattern revealed that five diagnostic fragments corresponding to the amino acid portion of the compounds could be observed within this cluster. The phenylalanine (Phe) and Phe-methyl ester [M+H]$^+$ moieties gave a typical fragment at *m/z* 166.09 and 188.10, respectively. Fragments at *m/z* 132 and 146 indicated potential leucine/isoleucine (Leu/Ile) and Leu/Ile-methyl ester [M+H]$^+$ moieties, respectively. Finally, a fragment at *m/z* 118 indicated a potential valine (Val) [M+H]$^+$ moiety. It is to be noted that an additional lipoamino acid derivative bearing a tyrosine-methyl ester could also be identified in another cluster of the molecular network by co-injection with a synthetic analogue (**19**). The presence of double bonds was inferred from the calculated round double bond equivalent (RDBE). The position of the OH groups was inferred from the isolated compounds. Together with the molecular formula determination, these signature indications allowed us to propose putative partial structures for the other nodes of the cluster and highlights the significant diversity of the lipoamino acids produced by this *Pantoea* sp. strain (Figure 5).

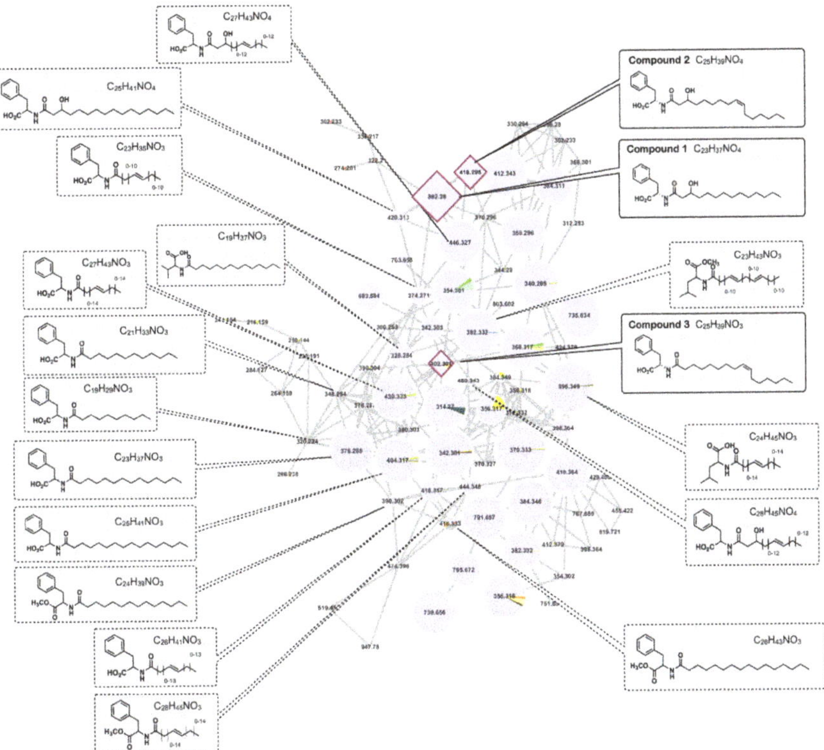

Figure 5. Cluster corresponding to hypothetic analogues of lipoamino acids putatively annotated through manual curation of their MSMS spectra.

3. Materials and Methods

3.1. General Procedures

Optical rotations were obtained on an Anton Paar MCP 200 polarimeter (Anton Paar Graz, Austria) in a 100 mm-long 350 µL cell using MeOH as the solvent at 20 °C. High resolution ESITOFMS measurements were performed using a Waters ACQUITY UHPLC system coupled to a Waters Micromass LCT-Premier Time-of-Flight mass spectrometer equipped with electrospray interface (ESI) (Waters, Manchester, England). Nuclear magnetic resonance (NMR) data were recorded in CD_3OD on Bruker 500 MHz and 600 MHz spectrometers equipped with a 1 mm inverse detection probe. Chemical shifts (δ) are reported in ppm based on the TMS signal, and coupling constant (J) is reported in Hertz (Bruker, Rheinstetten, Germany). Flash chromatography was performed on a Grace Reveleris system with UV and ELSD detectors using 120 g C18 or 80 g silica gel columns (Grace, Maryland, USA). HPLC analysis was conducted on a Gilson system equipped with a 322 pumping device, a GX-271 fraction collector, a 171 diode array detector, and a prepELSII detector electrospray nebulizer (Gilson Middelton, USA). Analytical analysis was conducted using Phenomenex Luna C18 (5 μm, 4.6 × 250 mm) and Phenomenex Kinetex C8 (5 μm, 4.6 × 250 mm) columns. A preparation analysis was conducted using Phenomenex Luna C18 (5 μm, 21.2 × 250 mm) and Kinetex C8 (5 μm, 21.2 × 250 mm) columns (Phenomenex, Le Pecq, France). All the chemicals were purchased from Sigma-Aldrich (Sigma-Aldrich chimie, Saint-Quentin-Fallavier, France).

3.2. Collection and Identification of Pantoea sp. SNB-VECD14B

An individual insect from the Diaspididae family that was infected by entomopathogenic microorganisms was collected in Montsinéry, French Guiana. The cuticle of the insect was scraped with a handle and transplanted onto a Petri dish containing a solid potato dextrose agar (PDA) medium and then stored at 28 °C. After one day, growing bacteria were removed and transferred onto another Petri dish. The strain SNB-VECD14B was stored in triplicate at −80 °C in a H_2O–glycerol mixture (50:50). A sample was submitted for amplification of the nuclear ribosomal internal transcribed spacer region 16S, which allowed for identification after comparison to the NCBI sequence. The sequence was registered in the NCBI GenBank database with the accession number KX858894 and identified as *Pantoea* sp. A molecular analysis was performed externally by BACTUP, France.

3.3. Culture, Extraction, and Isolation

The bacteria strain was initially cultivated on a small scale using Petri dishes with Mueller Hinton (MH) solid medium and then on a large scale using 176 14 cm Petri dishes at 28 °C for 8 days. The culture medium containing the mycelium was cut into small pieces and macerated with ethyl acetate (EtOAc) at room temperature on a rotary shaker (70 rpm) for 48 h. The contents were extracted three times with 5 L of EtOAc using a separatory funnel. The combined organic layers were washed with water. The organic solvent was evaporated to dryness under reduced pressure to yield the crude extract (2.5 g). A portion of the crude extract (2.35 g) was fractionated by reversed-phase flash chromatography on a C18 column using a linear gradient of water—acetonitrile (v/v, 1:0 to 0:1 over 20 min, flow rate = 80 mL/min) followed by another gradient of acetonitrile—methylene chloride (v/v, 1:1 to 0:1 over 10 min, flow rate = 80 mL/min) to generate 8 fractions that were labeled A to H. The larvicidal and antimicrobial activities were concentrated in fractions E (60.5 mg), F (445.7 mg), G (898.7 mg) and H (108.4 mg).

Fraction E was fractionated by preparative HPLC using H_2O–ACN (37:63 to 26:74 over 30 min, flow rate = 21 mL/min) to afford compound **1** (1.9 mg, t_R = 20.40 min) and compound **2** (1.2 mg, t_R = 24.10 min).

Fraction F was fractionated by preparative HPLC using H_2O–ACN (33:67 to 17:83 over 30 min, flow rate = 21 mL/min) to afford compound **3** (10.3 mg, t_R = 17.10 min).

Compound **1**: White powder; $[\alpha]_D^{20}$+ 29 (*c* 0.2, MeOH); ^1H and ^{13}C NMR spectroscopic data, see Table S1; HRESITOFMS *m/z* 392.2769 $[M+H]^+$ (calculated for $C_{23}H_{38}NO_4^+$, 392.2795). MSMS spectra were deposited at CCMSLIB00004684200.

Compound **2**: White powder; $[\alpha]_D^{20}$ + 6 (*c* 0.2, MeOH); ^1H and ^{13}C NMR spectroscopic data, see Table S2; HRESITOFMS *m/z* 418.2926 $[M+H]^+$ (calculated for $C_{25}H_{40}NO_4^+$, 418.2952). MSMS spectra were deposited at CCMSLIB00004684202.

Compound **3**: White powder; $[\alpha]_D^{20}$+ 12 (*c* 0.2, MeOH); ^{1}H and ^{13}C NMR spectroscopic data, see Table S3; HRESITOFMS *m/z* 402.3004 [M+H]$^+$ (calculated for $C_{25}H_{40}NO_3^+$, 402.3003). MSMS spectra were deposited at CCMSLIB00004684204.

3.4. General Synthetic Procedure for Lipoamino Acid Methyl Ester Synthesis

To a mixture of the fatty acid (100 mg), amino acid methyl ester (1 equiv.) and HATU (1.5 equiv.) was added Dimethylformamide (DMF) (5 mL) followed by *N,N*-Diisopropylethylamine (DIEA) (3 equiv.) under a N_2 atmosphere. The reaction mixture was stirred at room temperature (RT) overnight. A saturated solution of ammonium chloride (NH_4Cl) was added and the reaction mixture was extracted with *tert*-butylmethylether (3 times). The combined organic phases were dried with Na_2SO_4 and concentrated in vacuo. The crude mixture was purified by flash chromatography on silica gel to provide compounds **4–21**. Analytical data including optical rotations are reported in the Supplementary Materials.

3.5. General Synthetic Procedure for Lipoamino Acid Synthesis

To a solution of the lipoamino acid methyl ester (50 mg) in methanol (5 mL) and water (0.5 mL) was added $LiOH \cdot H_2O$ (5 equiv.). The reaction mixture was stirred at RT for 4 h and then aqueous HCl (1 M) was added to adjust the pH to 2–3. The resulting mixture was extracted twice with ethyl acetate, dried with Na_2SO_4, and concentrated in vacuo to provide the pure compounds **22–39**. Analytical data, including optical rotations, are reported in the Supplementary Materials.

3.6. Synthesis of Compound 3

To a mixture of the palmitoleic acid (50 mg), L-phenylalanine methyl ester (80 mg) and HATU (143 mg) was added DMF (5 mL) followed by DIEA (97 mg) under a N_2 atmosphere. The reaction mixture was stirred at RT overnight. A saturated solution of ammonium chloride (NH_4Cl) was added and the reaction mixture was extracted with *tert*-butylmethylether (3 times). The combined organic phases were dried with Na_2SO_4 and concentrated in vacuo. The crude mixture was purified by flash chromatography using C18 4g silica gel to give 52.5 mg of **3-OMe**.

To a solution of **3-OMe** (15 mg) in MeOH (5ml) and water (0.5 mL) was added $LiOH \cdot H_2O$ (5 equiv.). The reaction mixture was stirred at RT for 4 h and aqueous HCl (1M) was added to adjust the pH to 2–3. The resulting mixture was extracted twice with ethyl acetate, dried with Na_2SO_4 and concentrated in vacuo to provide 14 mg of **3**. $[\alpha]_D^{20}$ +49 (*c* 0.5, CHCl$_3$); HRESITOFMS *m/z* 402.2976 [M+H]$^+$ (calculated for $C_{25}H_{40}NO_3^+$, 402.3003); ^{1}H and ^{13}C NMR spectroscopic data, see S76-S77.

The same synthesis was done with D-phenylalanine methyl ester to give *ent*-**3-OMe** (65.6 mg) and *ent*-**3** (13.5 mg) with $[\alpha]_D^{20}$-49 (*c* 0.5, CHCl$_3$); HRESITOFMS *m/z* 402.2977 [M+H]$^+$ (calculated for $C_{25}H_{40}NO_3^+$, 402.3003); ^{1}H and ^{13}C NMR spectroscopic data, see S100-S101.

3.7. Molecular Networking

3.7.1. Mass Spectrometry Analysis Parameters

Chromatographic separation was performed on a Waters ACQUITY UPLC system interfaced to a Q-Exactive Focus mass spectrometer (Thermo Scientific, Bremen, Germany), with a heated electrospray ionization (HESI-II) source. Thermo Scientific Xcalibur 3.1 software was used for instrument control. The LC (Liquid Chromatography) conditions were as follows: Column, Waters BEH C_{18} 50 × 2.1 mm, 1.7 µm; mobile phase, (A) water with 0.1% formic acid, (B) acetonitrile with 0.1% formic acid; flow rate, 600 µL·min^{-1}; injection volume, 1 µL; gradient, linear gradient of 2−100% B over 6 min and then isocratic at 100% B for 0.6 min. An ACQUITY UPLC photodiode array detector was used to acquire PDA spectra, which were collected from 210 to 450 nm. In positive ion mode, the diisooctyl phthalate $C_{24}H_{38}O_4$ [M+H]$^+$ ion (*m/z* 391.28429) was used as the internal lock mass. The optimized HESI-II parameters were as follows: Source voltage, 4.0 kV (pos); sheath gas flow rate (N$_2$), 55 units; auxiliary gas flow rate, 15 units; spare gas flow rate, 3.0; capillary temperature, 275.00 °C (pos), S-lens RF level, 45. The mass analyzer was calibrated using a mixture of caffeine, methionine−arginine−phenylalanine−alanine−acetate (MRFA), sodium dodecyl sulfate, sodium taurocholate, and Ultramark 1621 in an acetonitrile/methanol/water solution containing 1% formic acid by direct injection. The data-dependent MS/MS events were performed on the three most intense ions detected in the full scan MS (Top3 experiment). The MS/MS isolation window width was 1 Da, and the stepped normalized collision energy (NCE) was set to 15, 30, and 45 units. In data-dependent MS/MS experiments, full scans were acquired at a resolution of 35,000 FWHM (at *m/z* 200) and MS/MS scans at 17,500 FWHM both with an automatically determined maximum injection time. After being acquired in a MS/MS scan, parent ions were placed in a dynamic exclusion list for 2.0 s.

3.7.2. MS Data Pretreatment

The MS data were converted from the .RAW (Thermo) data format to the .mzXML format using MSConvert software, which is part of the ProteoWizard package [15]. The converted files were treated using the MzMine software suite [17].

The parameters were adjusted as follows: The centroid mass detector was used for mass detection with the noise level set to 1.0×10^6 when the MS level was set to 1, and to 0 when the MS level was set to 2. The ADAP chromatogram builder was used and set to a minimum group size of 5 scans, minimum group intensity threshold of 1.0×10^5, minimum highest intensity of 1.0×10^5, and *m/z* tolerance of 8.0 ppm. For chromatogram deconvolution, the wavelets (ADAP) algorithm was used. The intensity window S/N was used as the S/N estimator with a signal-to-noise ratio set at 25, a minimum feature height at 10,000, a coefficient area threshold at 100, a peak duration range from 0.02 to 0.9 min and the RT wavelet range from 0.02 to 0.05 min. Isotopes were detected using the isotope peak grouper with a m/z tolerance of 5.0 ppm, an RT tolerance of 0.02 min (absolute), and the maximum charge set at 2. The representative isotope used was the isotope that was the most intense. An adduct (Na$^+$, K$^+$, NH$_4^+$, ACN$^+$, CH$_3$OH$^+$, Isoprop$^+$) search was performed with the RT tolerance set at 0.1 min and the maximum relative peak height at 500 %. A complex search was also performed using [M+H]$^+$ in ESI positive mode, with the RT tolerance set at 0.1 min and the maximum relative peak height at 500%. Finally, a custom database search was performed using the Dictionary of Natural Products 2018 (v. 26.2) database (http://dnp.chemnetbase.com) where the search was restricted to fungal or bacterial metabolites. Peak alignment was performed using the join aligner method where the *m/z* tolerance was set at 8 ppm, absolute RT tolerance at 0.065 min, weight for *m/z* at 10 and weight for RT at 10. The peak list was gap-filled with the same RT and *m/z* range gap filler (*m/z* tolerance at 8 ppm). Eventually the resulting aligned peak list was filtered using the peak-list rows filter option to keep only those features associated with MS2 scans. Full parameters are available as a .xml file in the Supplementary Materials (Entomopathogens_MzMine_parameters.mzmine).

3.7.3. Molecular Networks Generation

To maintain the retention time and exact mass information and to allow the separation of isomers, the molecular networks were created using the .mgf file resulting from the MzMine pretreatment step detailed above: https://bix-lab.ucsd.edu/display/Public/Feature+Based+Molecular+Networking. Spectral data were uploaded on the Global Natural Products Social molecular networking platform [10]. A network was then created where edges were filtered to have a cosine score above 0.65 and more than 6 matched peaks. Further edges between two nodes were kept in the network if and only if each of the nodes appeared in each other's respective top 10 most similar nodes. The spectra in the network were then searched against GNPS spectral libraries. The library spectra were filtered in the same manner as the input data. All matches kept between network spectra and library spectra were required to have a score above 0.7 and at least 6 matched peaks. The output was visualized using Cytoscape 3.6 software [18]. The full MS data set was uploaded and is accessible on the GNPS servers as Massive Data sets N° MSV000082940. The molecular network analysis and clustered spectra are accessible at the following link: https://gnps.ucsd.edu/ProteoSAFe/status.jsp?task=b036dd9fcb964ca49fe8dc4345944b86.

3.8. Evaluation of Larvicidal Activity

3.8.1. Insect Collection and Rearing

The *Aedes aegypti* strain was used for extract and compound Testing. The laboratory strain Paea, came from French Polynesia and had been maintained at the Institut Pasteur de la Guyane in French Guiana for over a decade. This strain is susceptible to all insecticides. *Ae. aegypti* eggs were conserved on dried paper strips. Hatching was induced by dropping these strips in water and placing them under vacuum pressure for at least 1 h. The larvae were fed with yeast pellets. Larval rearing was performed under natural conditions at 28 °C ± 2 °C, 80 % ± 20 % RH and with a photoperiod of 14 h dark and 10 h light. Late third or early fourth-instar larvae were used in the tests. All crude extracts and fractions were investigated using the WHO procedure for testing mosquito larvicides [19,20]. The larvicidal activity of pure compounds was evaluated using a tube assay.

3.8.2. Cup Assays

A total of 100 larvae were exposed in each bioassay. Four replicates of 25 larvae were prepared in cups containing 99 mL of distillated water. A 1 mL aliquot of the extract/fraction in absolute ethanol was added to each cup. The crude extract and the fractions were all tested at a concentration of 100 ppm. Absolute ethanol (1 mL) served as a negative control. Larval mortality was recorded 24 h after exposure.

3.8.3. Tube Assays

Fifty larvae were tested at each concentration with ten replicates per concentration (10 test tubes × 5 larvae). Each compound was tested at a concentration of 10 µg/mL. A stock solution was prepared at 1 mg/mL in absolute ethanol, and then, 30 µL of this solution was added to each test tube (75 × 12 × 0.8 − 1.0 soda rimLess, catalog #212-0013, VWR, International) containing 2.97 mL of distilled water. Larval mortality was recorded 24 h after exposure.

3.8.4. Statistical Analysis

Abbott's formula for mortality was applied if the mortality rate in the control was between 5 and 20%. The test was invalidated if the mortality rate in the control was greater than 20% [21].

3.9. Antimicrobial Assays

The ATCC strains were purchased from the Pasteur Institute and clinical isolates were provided by Phillipe Loiseau (University Paris Sud, Châtenay-Malabry, France). The strains used in this study were: *Candida albicans* ATCC10231, *Staphylococcus aureus* ATCC29213, methicillin-resistant *S. aureus* ATCC33591, and *Trichophyton rubrum* SNB-TR1. Extracts, fractions, and pure compounds were tested according to the reference protocol of the European Committee on Antimicrobial Susceptibility Testing (EUCAST, http://www.eucast.org, 11 April 2016). The standard microdilution test as described by the Clinical and Laboratory Standards Institute guidelines (M27-A2, M7-A8 and M38-A) was used to determine minimal inhibition concentrations (MIC) against dermatophyte fungi, bacteria, and yeasts [22–24]. Crude extracts and pure compounds were tested at concentrations ranging from 256 to 0.5 µg/mL. The microplates were incubated at 32 °C, and MIC values were calculated after 24 h for bacteria, 48 h for yeast and 5 days for *T. rubrum*. The MIC values reported in Table 2 refer to the lowest concentration preventing visible growth in the wells. All assays were conducted in triplicate.

4. Conclusions

In conclusion, our study highlights that *Pantoea* sp. SNB-VECD14B can biosynthesize diverse lipoamino acids with antimicrobial and/or insecticidal activities. Three hydroxyacyl-phenylalanine derivatives were isolated and characterized from EtOAc extracts, and different analogs were synthesized to demonstrate structure-activity relationships. In short, we observed that the free carboxylic acid group was important for biological activity. Compounds with a long carbon chain (16 to 18 carbons) and a *Z* double bond, exhibited the highest larvicidal activity, while compounds with a saturated C12 or C14 chain demonstrated the best antimicrobial activity. The studied lipoamino acids appear to have similar biological and structural properties compared to free fatty acids [25,26]. Their amphipathic nature provides a wide range of activities. Given the observed biological activities, these molecules could provide an alternative to common antimicrobial agents for application in agriculture, food preservation, or cosmetics.

Finally, the molecular networking-based approach revealed several close analogues of lipoamino acids in the *Pantoea* sp. extract. The global molecular network established over a wide collection of entomopathogenic strains indicated that this structural family was indeed typical of the *Pantoea* genus. These lipoamino acids have diverse biological activities, and it can be hypothesized that the strain has acquired the ability to protect itself from various environmental aggressions (fungi, bacteria, and insects) by producing a truly diverse mix of lipoamino acids that exhibit complementary activities.

Supplementary Materials: Supplementary materials can be found at http://www.mdpi.com/1422-0067/20/5/1083/s1.

Author Contributions: Conceptualization, D.S. and V.E.; formal analysis, S.T., L.P., P.-M.A. and I.D.; funding acquisition, D.S. and V.E.; investigation, S.T.; methodology, S.D., D.S. and V.E.; project administration, V.E.; resources, I.D.; supervision, J.-L.W., I.D., D.S., and V.E.; validation, P.-M.A. and I.D.; writing—original draft, S.T. and D.S.; writing—review and editing, V.E.

Funding: This work has benefited from an "Investissement d'Avenir" grant managed by Agence Nationale de la Recherche (CEBA, ref. ANR-10-LABX-0025).

Conflicts of Interest: The authors declare no conflict of interest.

Abbreviations

ACN	Acetonitrile
DMF	Dimethylformamide
DIEA	*N*,*N*-Diisopropylethylamine
EtOAc	Ethyl acetate
HATU	*N*-[(Dimethylamino)-1*H*-1,2,3-triazolo-[4,5-b]pyridin-1-ylmethylene]-*N*-methylmethanaminium hexafluorophosphate *N*-oxide
HCl	Hydrochloric acid
HPLC	High performance liquid chromatography
MH	Mueller hinton
MIC	Minimum inhibitory concentration
MN	Molecular networking
MRSA	Methicillin-resistant *Staphylococcus aureus*
MS	Mass spectrometry
NMR	Nuclear magnetic resonance
PDA	Potato dextrose agar
RDBE	Round double bond equivalent
SAR	Structure activity relationship
UPLC	Ultra-high performance liquid chromatography
VRE	Vancomycin-resistant *Enterococci*
VRSA	Vancomycin-resistant *Staphylococcus aureus*
WHO	World Health Organization

Int. J. Mol. Sci. **2019**, *20*, 1083

References

1. Gould, I.M. The epidemiology of antibiotic resistance. *Int. J. Antimicrob. Agents* **2008**, *32*, S2–S9. [CrossRef] [PubMed]
2. Pilacik, N.; Kaminska, T.; Augustynowicz-Kopec, E.; Krasowski, G. Etiology of bacterial infections and incidence of comorbidities in patients with tuberculosis, treated in Mazovian Treatment Centre of Tuberculosis and Lung Diseases during years 2012-2014. *Wiad. Lek.* **2015**, *68*, 347–353. [PubMed]
3. Ventola, C.L. The Antibiotic Resistance Crisis. Part 1: Causes and Threats. *Pharm. Ther* **2015**, *40*, 277–283.
4. Word Health Organization. Available online: http://www.who.int/news-room/fact-sheets/detail/antibiotic-resistance (accessed on 18 September 2018).
5. Kraemer, M.U.; Sinka, M.E.; Duda, K.A.; Mylne, A.Q.; Shearer, F.M.; Barker, C.M.; Moore, C.G.; Carvalho, R.G.; Coelho, G.E.; Van Bortel, W.; et al. The global distribution of the arbovirus vectors *Aedes aegypti* and *Ae. albopictus. Ecol. Epidemiol. Glob. Health.* **2015**, *4*, e08437. [CrossRef] [PubMed]
6. Aktar, M.W.; Sengupta, D.; Chowdhury, A. Impact of pesticides use in agriculture: Their benefits and hazards. *Interdiscip. Toxicol.* **2009**, *2*, 1–12. [CrossRef] [PubMed]
7. Sokhna, C.; Ndiath, M.O.; Rogier, C. The changes in mosquito vector behaviour and the emerging resistance to insecticides will challenge the decline of malaria. *Clin. Microbiol. Infect.* **2013**, *19*, 902–907. [CrossRef] [PubMed]
8. Newman, D.J.; Cragg, G.M. Natural Products as Sources of New Drugs from 1981 to 2014. *J. Nat. Prod.* **2016**, *79*, 629–661. [CrossRef] [PubMed]
9. Bode, H.B. Entomopathogenic bacteria as a source of secondary metabolites. *Curr. Opin. Chem. Biol.* **2009**, *13*, 224–230. [CrossRef] [PubMed]
10. Wang, M.; Carver, J.J.; Phelan, V.V.; Sanchez, L.M.; Garg, N.; Peng, Y.; Nguyen, D.D.; Watrous, J.; Kapono, C.A.; Luzzatto-Knaan, T.; et al. Sharing and community curation of mass spectrometry data with Global Natural Products Social Molecular Networking. *Nat. Biotechnol.* **2016**, *34*, 828–837. [CrossRef] [PubMed]
11. Dührkop, K.; Shen, H.; Meusel, M.; Rousu, J.; Böcker, S. Searching molecular structure databases with tandem mass spectra using CSI:FingerID. *Proc. Natl. Acad. Sci. USA* **2015**, *112*, 12580–12585. [CrossRef] [PubMed]
12. Pluskal, T.; Castillo, S.; Villar-Briones, A.; Orešič, M. MZmine 2: Modular framework for processing, visualizing, and analyzing mass spectrometry-based molecular profile data. *BMC Bioinf.* **2010**, *11*, 395. [CrossRef] [PubMed]
13. Frost, D.J.; Gunstone, F.D. The PMR analysis of non-conjugated alkenoic and alkynoic acids and esters. *Chem. Phys. Lipids* **1975**, *15*, 53–85. [CrossRef]
14. Vast, A.; Singh, A.K.; Mukherjee, R.; Chopra, T.; Ravindran, M.S.; Mohanty, D.; Chatterji, D.; Reyrat, J.-M.; Gokhale, R.J. Retrobiosynthetic approach delineates the biosynthetic pathway and the structure of the acyl chain of mycobacterial glycopeptidolipids. *J. Biol. Chem.* **2012**, *287*, 30677–30687.
15. Givaudin, S.A.; Staghouwer, H.; Thoen, C.; Van Buel, M. Improvements in or Relating to Organic Compounds, U.S. Patent 20160214928, 28 July 2016.
16. Sheppard, G.S.; Wang, J.; Kawei, M.; BaMaung, N.Y.; Craig, R.A.; Erickson, S.A.; Lynch, L.; Patel, J.; Yang, F.; Searle, X.B.; et al. 3-Amino-2-hydroxyamides and related compounds as inhibitors of methionine aminopeptidase-2. *Biorg. Med. Chem.* **2004**, *14*, 865–868. [CrossRef] [PubMed]
17. Chambers, M.C.; Maclean, B.; Burke, R.; Amodei, D.; Ruderman, D.L.; Neumann, S.; Gatto, L.; Fischer, B.; Pratt, B.; Egertson, J.; et al. A cross-platform toolkit for mass spectrometry and proteomics. *Nat. Biotechnol.* **2012**, *30*, 918–920. [CrossRef] [PubMed]
18. ©Cytoscape Consortium 2001-2017. Available online: https://cytoscape.org/ (accessed on 2 March 2018).
19. World Health Organization. Guidelines for Laboratory and Field Testing of Mosquito Larvicides. Available online: http://apps.who.int/iris/bitstream/10665/69101/1/WHO_CDS_WHOPES_GCDPP_2005.13.pdf (accessed on 18 September 2018).
20. Falkowski, M.; Jahn-Oyac, A.; Ferrero, E.; Issaly, J.; Eparvier, V.; Girod, R.; Rodrigues, A.M.; Stien, D.; Houël, E.; Dusfour, I. Assessment of A Simple Compound-Saving Method To Study Insecticidal Activity of Natural Extracts and Pure Compounds Against Mosquito Larvae. *J. Am. Mosq. Control Assoc.* **2016**, *32*, 337–340. [CrossRef] [PubMed]

21. Abott, W.S. A Method of Computing the Effectiveness of an Insecticide. *J. Econ Entomol.* **1925**, *18*, 265–267. [CrossRef]

22. Clinical and Laboratory Standards. *Reference Method for Broth Dilution Antifungal Susceptibility Testing of Filamentous Fungi*, 2nd ed.; CLSI: Wayne, PA, USA, 2008.

23. Clinical and Laboratory Standards. *Reference Method for Broth Dilution Antifungal Susceptibility Testing of Yeasts*, 3rd ed.; CLSI: Wayne, PA, USA, 2008.

24. Clinical and Laboratory Standards. *Methods for Dilution Antimicrobial Susceptibility Tests for Bacteria That Grow Aerobically*, 8th ed.; CLSI: Wayne, PA, USA, 2009.

25. Desbois, A.P.; Lebl, T.; Yan, L.; Smith, V.J. Isolation and structural characterisation of two antibacterial free fatty acids from the marine diatom, *Phaeodactylum tricornutum*. *Appl. Microbiol. Biotechnol.* **2008**, *81*, 755–764. [CrossRef] [PubMed]

26. Desbois, A.P.; Smith, V.J. Antibacterial free fatty acids: Activities, mechanisms of action and biotechnological potential. *Appl. Microbiol. Biotechnol.* **2010**, *85*, 1629–1642. [CrossRef] [PubMed]

International Journal of
Molecular Sciences

Article

Lipidomic Analysis of the Outer Membrane Vesicles from Paired Polymyxin-Susceptible and -Resistant *Klebsiella pneumoniae* Clinical Isolates

Raad Jasim [1], Mei-Ling Han [2], Yan Zhu [2], Xiaohan Hu [3], Maytham H. Hussein [3], Yu-Wei Lin [2], Qi (Tony) Zhou [4], Charlie Yao Da Dong [1], Jian Li [2,*] and Tony Velkov [3,*]

[1] Drug Delivery, Disposition and Dynamics, Monash Institute of Pharmaceutical Sciences, Monash University, Parkville, Victoria 3052, Australia; raad.jasim@monash.edu (R.J.); charlie.dong@monash.edu (C.Y.D.D.)

[2] Monash Biomedicine Discovery Institute, Immunity and Infection Program and Department of Microbiology, Monash University, VIC 3800, Australia; meiling.han@monash.edu (M.-L.H.); yan.zhu@monash.edu (Y.Z.); yu-wei.lin@monash.edu (Y.-W.L.)

[3] Department of Pharmacology and Therapeutics, University of Melbourne, Parkville, Victoria 3010, Australia; xiaohanh2@student.unimelb.edu.au (X.H.); maytham.hussein@unimelb.edu.au (M.H.H.)

[4] Department of Industrial and Physical Pharmacy, College of Pharmacy, Purdue University, 575 Stadium Mall Drive, West Lafayette, IN 47907, USA; tonyzhou@purdue.edu

* Correspondence: Jian.Li@monash.edu (J.L.); Tony.Velkov@unimelb.edu.au (T.V.)

Received: 29 July 2018; Accepted: 7 August 2018; Published: 10 August 2018

Abstract: Gram-negative bacteria produce outer membrane vesicles (OMVs) as delivery vehicles for nefarious bacterial cargo such as virulence factors, which are antibiotic resistance determinants. This study aimed to investigate the impact of polymyxin B treatment on the OMV lipidome from paired polymyxin-susceptible and -resistant *Klebsiella pneumoniae* isolates. *K. pneumoniae* ATCC 700721 was employed as a reference strain in addition to two clinical strains, *K. pneumoniae* FADDI-KP069 and *K. pneumoniae* BM3. Polymyxin B treatment of the polymyxin-susceptible strains resulted in a marked reduction in the glycerophospholipid, fatty acid, lysoglycerophosphate and sphingolipid content of their OMVs. Conversely, the polymyxin-resistant strains expressed OMVs richer in all of these lipid species, both intrinsically and increasingly under polymyxin treatment. The average diameter of the OMVs derived from the *K. pneumoniae* ATCC 700721 polymyxin-susceptible isolate, measured by dynamic light scattering measurements, was ~90.6 nm, whereas the average diameter of the OMVs isolated from the paired polymyxin-resistant isolate was ~141 nm. Polymyxin B treatment (2 mg/L) of the *K. pneumoniae* ATCC 700721 cells resulted in the production of OMVs with a larger average particle size in both the susceptible (average diameter ~124 nm) and resistant (average diameter ~154 nm) strains. In light of the above, we hypothesize that outer membrane remodelling associated with polymyxin resistance in *K. pneumoniae* may involve fortifying the membrane structure with increased glycerophospholipids, fatty acids, lysoglycerophosphates and sphingolipids. Putatively, these changes serve to make the outer membrane and OMVs more impervious to polymyxin attack.

Keywords: outer membrane vesicles; lipidomics; Gram-negative; polymyxin; extremely drug resistant

1. Introduction

Over the last decade, extremely drug-resistant (XDR) *Klebsiella pneumoniae* has emerged as one of the most deadly Gram-negative 'superbugs' [1–3]. *K. pneumoniae* is responsible for numerous lethal nosocomial outbreaks [4]; more worryingly, the mortality of nosocomial *K. pneumoniae* infections can be up to 50% [5]. Carbapenem resistance in *K. pneumoniae* mediated by carbapenemase was firstly reported in 1996 in New York City and has spread to most global centres [5,6]. In 2008, bla_{NDM-1}, which encodes the class B New Delhi Metallo-β-lactamase-1 (NDM-1) that inactivates carbapenems, was first detected in a Swedish patient who had contracted an infection in India [7]. Polymyxins (i.e., colistin and polymyxin B) are increasingly used as the last-line therapy against XDR *K. pneumoniae* [8]. Indeed, considerable in vitro activity against *K. pneumoniae* strains has been demonstrated [9]; 98.2% of general clinical strains of *K. pneumoniae* are susceptible to polymyxin B and colistin [10–15]. Ominously, XDR strains that are resistant to polymyxins have recently emerged [16,17], which highlights the need for a greater appreciation of the mechanism(s) of polymyxin resistance in *K. pneumoniae* to assist targeted drug discovery strategies.

The Gram-negative outer membrane (OM) constitutes a formidable barrier limiting the permeability of various noxious substances such as antimicrobial drugs [18,19]. This complex asymmetrical structure comprises an inner phospholipid leaflet, as well as an outer leaflet that predominantly contains lipopolysaccharide (LPS), proteins and phospholipids. Additionally, *K. pneumoniae* commonly expresses a capsular polysaccharide that coats the OM, the expression levels of which have been related to polymyxin susceptibility [20–23]. The antimicrobial action of polymyxins is mediated through a direct and very specific interaction with the lipid A component of the LPS, which leads to a disruption of the OM barrier [8]. The cationic L-α,γ-diaminobutyric acid residues of the polymyxin molecule produce an electrostatic attraction to the negatively charged lipid A phosphate groups, displacing the divalent cations (Mg^{2+} and Ca^{2+}) [8]. The displacement leads to the disorganization of the LPS leaflet, enabling the insertion of the hydrophobic tail and the hydrophobic side chains of amino acids 6 and 7 of the polymyxin molecule into the OM [24]. Polymyxin resistance in *K. pneumoniae* primarily involves the multi-tier upregulation of capsular polysaccharide expression, and the systems required for the modification of lipid A with 4-amino-4-deoxy-L-arabinose and palmitoyl addition [20,23,25–32]. In *K. pneumoniae* the expression of 4-amino-4-deoxy-L-arabinose modifications to the lipid A phosphates is under control of the two component regulatory systems [PhoPQ–PmrD]–PmrAB that are activated in response to low pH, low magnesium, high iron and in response to cationic antimicrobial peptides [23]. More specifically, PhoP–PhoQ regulates the magnesium regulon, which activates polymyxin resistance under low magnesium conditions. This PhoP–PhoQ system is connected by the small basic protein PmrD. PhoP regulates the activation of PmrD, which can then bind to PmrA and prolong its phosphorylation state, eventually activating the expression of the PmrA–PrmB system to promote lipid A modifications that confer polymyxin resistance. The under-acylation of lipid A increases the polymyxin susceptibility of *K. pneumoniae*, which highlights that the decoration of lipid A with additional fatty acyl chains is important for polymyxin resistance [33,34].

Gram-negative bacteria naturally shed their OM via outer membrane vesicles (OMVs), which are spherical bilayer structures of approximately 20–200 nm in diameter [35]. OMVs are believed to serve as delivery vehicles for nefarious bacterial cargo such as virulence factors, antibiotic resistance determinants, toxins and factors that modulate the host immune response to facilitate pathogen evasion [35–39]. This underscores the need to understand the compositional differences between OMVs of MDR *K. pneumoniae* clinical isolates and how this relates to their pathogenicity. In the present study, we aimed to perform a comparative analysis of the lipidome of OMVs isolated from of polymyxin-susceptible and -resistant *K. pneumoniae* clinical isolates and to identify key lipid species that are selectively packaged from the OM into the OMV sub-lipidome of the resistant isolates. The obtained data sheds new light on the OMV lipidomes associated with high-level polymyxin resistance in the problematic Gram-negative pathogen *K. pneumoniae*.

2. Results and Discussion

2.1. Lipidomics Analysis of OMVs from Polymyxin-Susceptible and -Resistant K. pneumoniae Isolates

The OMV lipidome from paired polymyxin-susceptible and -resistant strains from two clinical isolates (*K. pneumoniae* BM3 and FADDI-KP069) and a laboratory type strain (*K. pneumoniae* ATCC 700721) were characterised following lipid extraction using LC-MS analysis. Compositional analysis revealed that the OMV lipid composition of all the *K. pneumoniae* strains mostly consisted of glycerophospholipids (~35%), fatty acids (~33%) and sphingolipids (~20%). Similarly, across all three strains the OMV minor lipid components consisted of lipids from the following classes, glycerolipids (~4%), sterol lipids (~3%) and prenol lipids (~4%).

Principle component analysis (PCA) score plots and the heat map revealed significant global lipidomic differences between the OMVs of the polymyxin-susceptible and -resistant *K. pneumoniae* strains (Figures 1 and 2). Notably, following treatment with a clinically relevant concentration of polymyxin B (2 mg/L) we observed marked global lipidome perturbations in the OMVs of the polymyxin-susceptible *K. pneumoniae* strains; whereas the OMVs of the resistant strains showed moderate global lipidome perturbations in response to polymyxin B treatment of the cells. For univariate analyses, all of the putatively identified lipids were further analysed to reveal those showing at least 2-fold differences ($p < 0.05$, FDR < 0.05, one-way ANOVA test) in relative abundance (Figure 3). The cluster algorithm and fold-change analysis highlighted that, compared to the untreated controls, polymyxin B treatment (2 mg/L) of the polymyxin-susceptible *K. pneumoniae* ATCC 700721 significantly reduced the phosphatidylcholine, phosphatidylethanolamine and 1-acyl-glycerophosphocholine content of its OMVs. Additionally, the sphingolipids namely, sphingosine, N-acyl-sphingosine (ceramide), N-acyl-sphinganine(dihydro-ceramide), sphingomyelin, glucosyl-ceramide and lactosyl-ceramide were significantly reduced following polymyxin B treatment (Figure 3Ai). Moreover, certain saturated fatty acids (e.g., hexadecanoic acid and octadecanoic acid), and polyunsaturated fatty acids (α-linolenic acid and arachidonic acid) were also reduced in the OMVs of the polymyxin B treated susceptible isolate. Polymyxin B treatment of the its paired polymyxin-resistant *K. pneumoniae* ATCC 700721 laboratory isolate significantly increased the content of lysoglycerophosphates, phosphatidylcholines and phosphatidylethanolamines in its OMVs (Figure 3Aii). Similarly, to the polymyxin-susceptible ATCC 700721 isolate, most of the glycerophospholipid and fatty acid content of the OMVs isolated from the polymyxin B treated polymyxin-susceptible clinical isolates (*K. pneumoniae* BM3 and FADDI-KP069) were significantly reduced compared to untreated controls (Figure 3Bi,Ci). In particular, glycerophospholipids (e.g., phosphatidylethanolamines, phosphatidylcholines, lysophosphatidylcholines, and lysoglycerophosphates) were remarkably reduced in response to polymyxin B treatment. In addition, fatty acids (e.g., docosanoic acid, octadecenoic acid and hexadecanoic acid); and sphingolipids (mainly dihydro-ceramides) were also significantly reduced in response to polymyxin B be treatment. In contrast, the majority of glycerophospholipids, fatty acids and sphingolipids content of OMVs isolated from their paired polymyxin-resistant *K. pneumoniae* BM3 a FADDI-KP069 isolates were significantly increased in response to polymyxin B treatment (Figure 3Bii,Cii). Notably, all of the polymyxin B-resistant strains secreted OMVs significantly are richer in glycerophospholipids, fatty acids, lysoglycerophosphates and sphingolipids compared to polymyxin B-susceptible isolates even when grown in the absence of polymyxin B (Figure 4). Glycerophospholipids, fatty acids, glycerolipids and sphingolipids play a crucial role in maintain outer membrane integrity, bacterial survival and pathogenesis [40]. Phospholipids (including glycerophospholipids) are essential components of bacterial membranes and they are responsible for maintaining membrane integrity and the selective permeability of the outer membrane [41]; they contribute to cationic peptide resistance, protect bacteria from osmotic stress and regulate flagellum-mediated motility [42]. In addition, sphingolipids are involved in maintaining normal bacterial growth and membrane integrity; and trigger bacterial pathogenesis via induction of the host immune system [43].

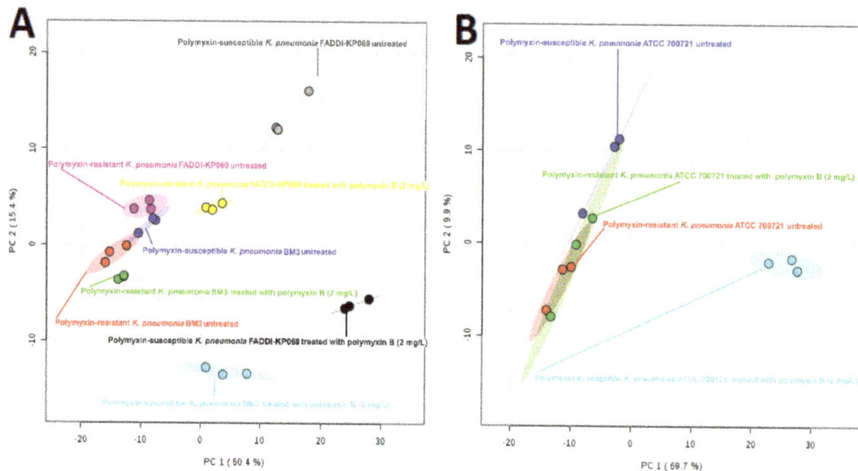

Figure 1. Principal component analysis (PCA) score plot for OMVs isolated from polymyxin-susceptible and -resistant *K. pneumoniae* isolates. (**A**) PCA score plot for the two clinical isolates. Polymyxin-resistant *K. pneumoniae* BM3 untreated (red); polymyxin-resistant *K. pneumoniae* BM3 treated with polymyxin B (2 mg/L) (green); polymyxin-susceptible *K. pneumoniae* BM3 untreated (blue); polymyxin-susceptible *K. pneumoniae* BM3 treated with polymyxin B (2 mg/L) (cyan); polymyxin-resistant *K. pneumoniae* FADDI-KP069 untreated (purple); polymyxin-resistant *K. pneumoniae* FADDI-KP069 treated with polymyxin B (2 mg/L) (yellow); polymyxin-susceptible *K. pneumoniae* FADDI-KP069 untreated (grey); polymyxin-susceptible *K. pneumoniae* FADDI-KP069 treated with polymyxin B (2 mg/L) (black). (**B**) PCA score plot for the paired *K. pneumoniae* ATCC 700721 laboratory type isolates. Polymyxin-resistant *K. pneumoniae* ATCC 700721 untreated (red); polymyxin-resistant *K. pneumoniae* ATCC 700721 treated with polymyxin B (2 mg/L) (green); polymyxin-susceptible *K. pneumoniae* ATCC 700721 untreated (blue); polymyxin-susceptible *K. pneumoniae* ATCC 700721 treated with polymyxin B (2 mg/L) (cyan). Each data point represents three biological replicates.

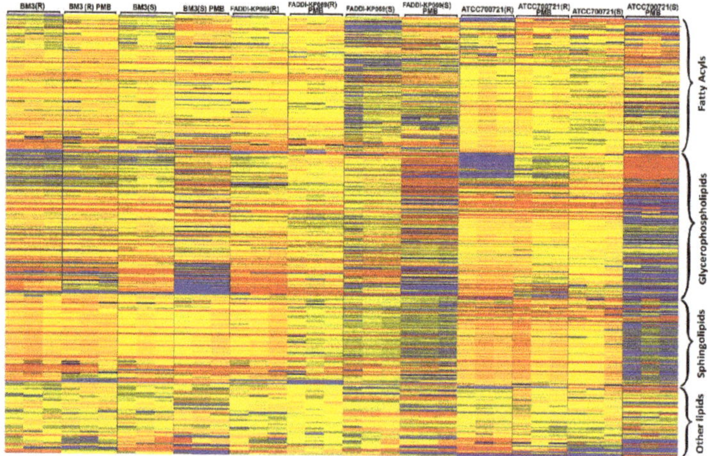

Figure 2. The heat map illustrates the relative peak intensity of lipids within each class in the OMVs of the paired polymyxin-susceptible and -resistant *K. pneumoniae* isolates. (R) = polymyxin-resistant; (S) = Polymyxin-susceptible. Colours indicate relative abundance of lipidomes based on the relative peak intensity (red = high, yellow = no change, blue = undetectable).

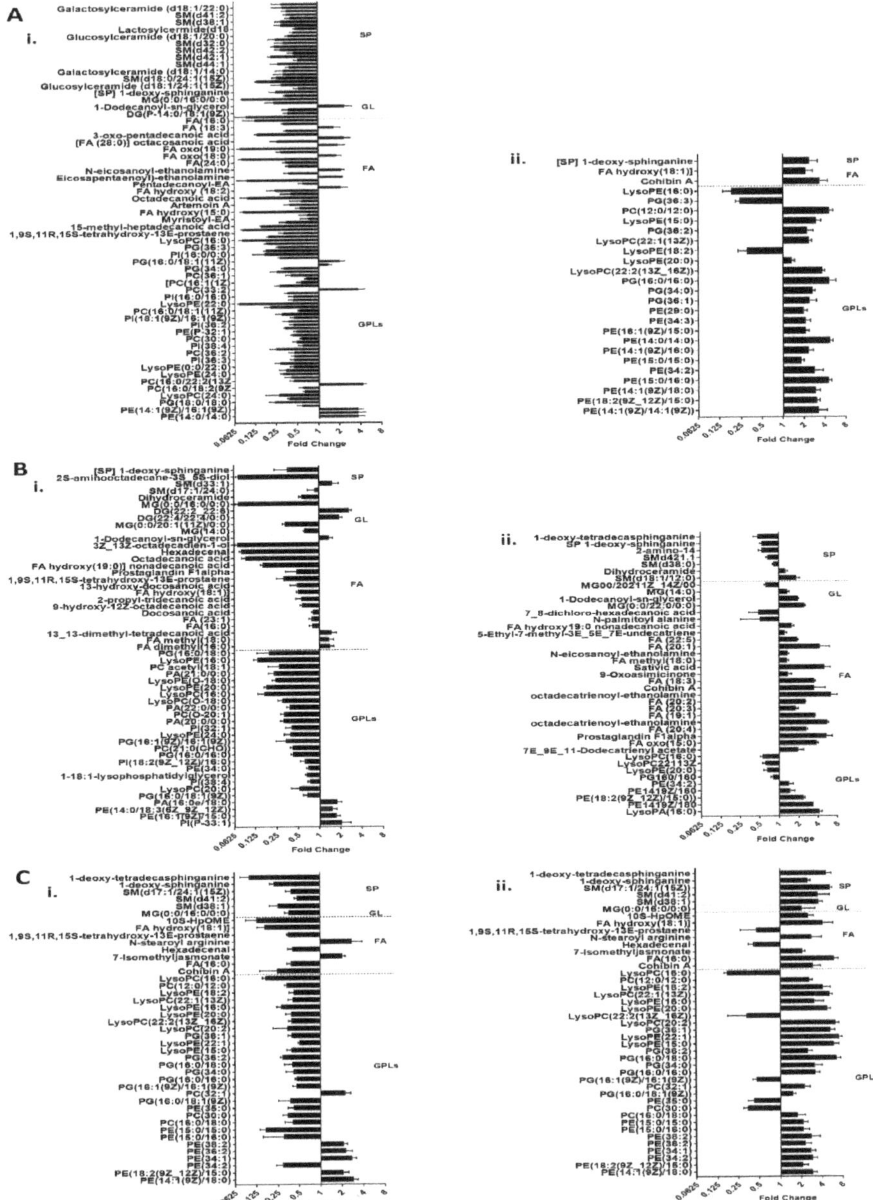

Figure 3. Lipidomic perturbations of OMVs isolated from polymyxin-susceptible and -resistant *K. pneumoniae* isolates. Fold-change of lipids relative to the untreated control cells, in OMVs of the polymyxin-susceptible (**i**) and -resistant (**ii**) strains of paired *K. pneumoniae* isolates in response to polymyxin B treatment (2 mg/L). (**A**) *K. pneumoniae* ATCC 700721. (**B**) *K. pneumoniae* BM3 and (**C**) *K. pneumoniae* FADDI-KP069. GPLs = glycerophospholipids; FA = fatty acids; GL = glycerolipids; SP = sphingolipids.

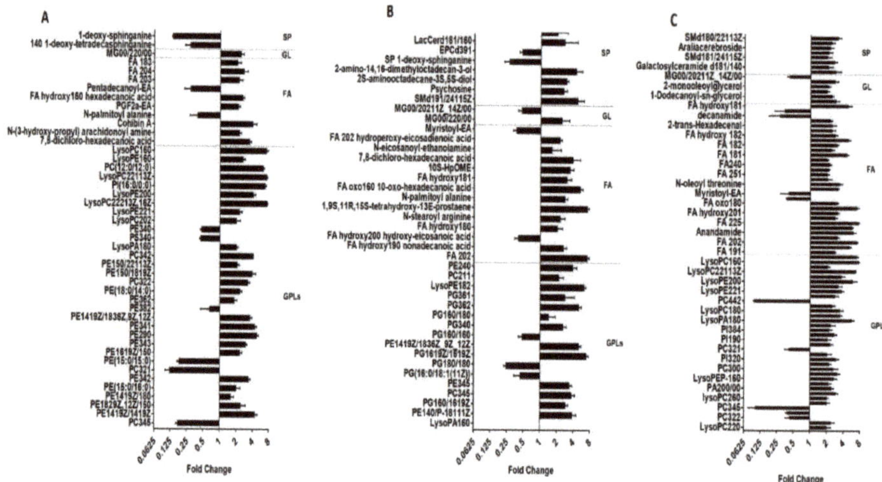

Figure 4. Major differences in the lipid abundance between the OMVs of paired polymyxin-susceptible and -resistant *K. pneumoniae* isolates. The differences are expressed as the fold-change in the OMV lipids of the paired susceptible vs. resistant *K. pneumoniae* isolates. All cultures were grown in the absence of polymyxins. (**A**) *K. pneumoniae* ATCC 700721. (**B**) *K. pneumoniae* BM3 and (**C**) *K. pneumoniae* FADDI-KP069. GPLs = glycerophospholipids; FA = fatty acids; GL = glycerolipids; SP = sphingolipids.

2.2. Transmission Electron Microscopy Imaging and Dynamic Light Scattering Size Estimation of K. pneumoniae OMVs

Dynamic light-scattering (DLS) analysis revealed that the average hydrodynamic radius of the OMVs derived from the *K. pneumoniae* ATCC 700721 polymyxin-susceptible isolate is ~90.6 nm; the profile was symmetrical and the OMV scatter ranged from ~30–500 nm (Figure 5A). The average hydrodynamic radius of the OMVs isolated from the paired *K. pneumoniae* ATCC 700721 polymyxin-resistant isolate was ~141 nm and the OMV scatter ranged from ~30 to 1000 nm (Figure 5C), which indicates that the resistant isolae sheds larger OMVs than the susceptible one. Polymyxin B treatment (2 mg/L) of the *K. pneumoniae* cells resulted in the production of OMVs with slightly larger average particle size in both the susceptible (average diameter ~124 nm, OMV scatter ~30–900 nm; Figure 5B) and resistant (average hydrodynamic radius ~154 nm, OMV scatter ~30–1500 nm; Figure 5D) strains. Notably, the OMV scatter profile in the resistant strain is asymmetrical, with and without polymyxin B treatment. In line with the DLS data [44,45], transmission electron microscopy imaging of *K. pneumoniae* OMVs revealed a similar size distribution wherein the polymyxin-resistant *K. pneumoniae* ATCC 700721 strain produced larger OMVs than the susceptible strain (Figure 6). Moreover, the OMVs isolated from the polymyxin-resistant isolate stained darker with the TEM contrast reagent uranyl acetate, which enhances the contrast by interaction with lipids; in line with the lipidomics findings, this would suggest that the OMVs of the resistant strains contain more lipids. Similarly, in *Salmonella enterica*, LPS remodelling in the outer membrane in response to polymyxins or other environmental PhoP/Q–PmrA/B activating conditions, has been shown to stimulate the biogenesis of larger-diameter OMVs [36–39].

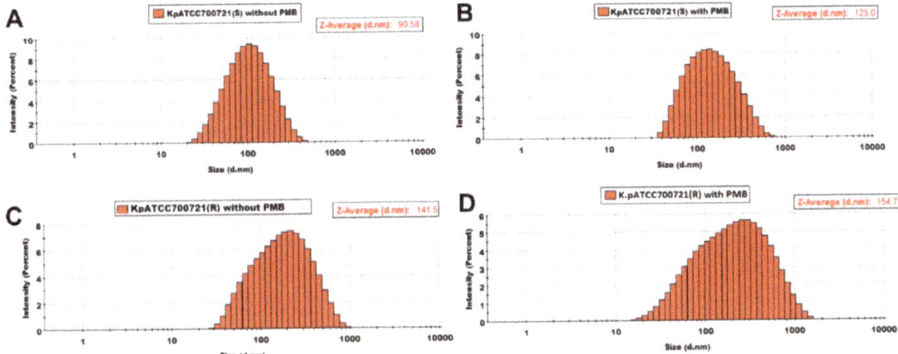

Figure 5. Size distribution measured by dynamic light scattering of OMVs isolated from paired polymyxin-susceptible and -resistant strains of *K. pneumoniae* ATCC 700721. OMVs isolated from the polymyxin-susceptible *K. pneumoniae* ATCC 700721 (**A**) without polymyxin B treatment and (**B**) with polymyxin B (2 mg/L) treatment. OMVs isolated from the polymyxin-resistant *K. pneumoniae* ATCC 700721 (**C**) without polymyxin B treatment and (**D**) with polymyxin B (2 mg/L) treatment.

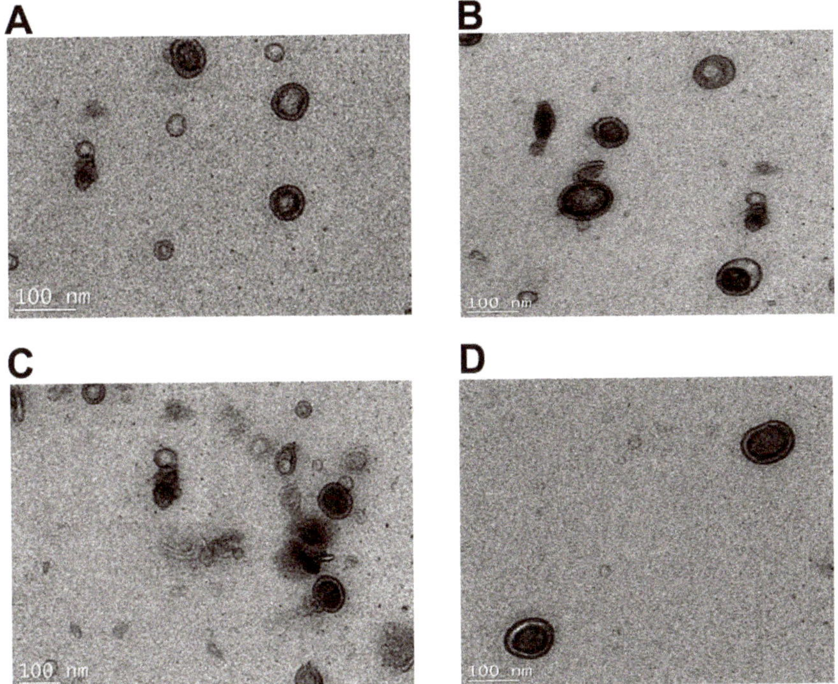

Figure 6. Transmission electron microscopy images of OMVs isolated from paired polymyxin-susceptible and -resistant strains of *K. pneumoniae* ATCC 700721. (**A**) OMVs from untreated *K. pneumoniae* ATCC 700721 (susceptible). (**B**) OMVs from polymyxin B (2 mg/L) treated *K. pneumoniae* ATCC 700721 (susceptible). (**C**) OMVs from untreated *K. pneumoniae* ATCC 700721 (resistant). (**D**) OMVs from polymyxin B (2 mg/L) treated *K. pneumoniae* ATCC 700721 (resistant).

3. Materials and Methods

3.1. Materials

Polymyxin B was supplied by Betapharma (Shanghai, China). All chemicals were purchased from Sigma-Aldrich (Melbourne, VIC, Australia) at the highest research grade; ultrapure water was from Fluka (Castle Hill, New South Wales, Australia). Stock solutions of polymyxin B (10 mg/L) were freshly prepared in ultrapure water and filtered through 0.22 μm syringe filters (Sartorius, Melbourne, Victoria, Australia).

3.2. Bacterial Isolates and Growth Conditions

All bacterial strains used in this study are described in Table S1. Resistance to polymyxin B was defined as MICs of ≥ 8 mg/L [46]. A total of six different *K. pneumoniae* isolates were studied: The clinical isolates *K. pneumoniae* FADDI-KP069 (polymyxin-susceptible strain polymyxin B MIC = 0.5 mg/L; polymyxin-resistant strain polymyxin B MIC > 32 mg/L; Both positive for ESBL and KPC carbapenemase) and *K. pneumoniae* BM3 (polymyxin-susceptible strain polymyxin B MIC = 0.5 mg/L; polymyxin-resistant strain polymyxin B MIC $\geq$ 32 mg/L; Both positive for NDM, CTX-M, SHV, TEM, AAC-6'-1B); and a reference strain *K. pneumoniae* ATCC 700721 (polymyxin-susceptible strain polymyxin B MIC = 0.5 mg/L; polymyxin-resistant strain polymyxin B MIC > 32 mg/L). The antibiograms of the two clinical isolates are documented in Table S1. All bacteria were stored at $-80\ ^{\circ}$C in tryptone soya broth (TSB, Oxoid, Melbourne, Australia). Prior to experiments, parent strains were subcultured onto nutrient agar plates (Medium Preparation Unit, University of Melbourne, Victoria, Australia). Overnight broth cultures were subsequently grown in 5 mL of cation-adjusted Mueller–Hinton broth (CaMHB, Oxoid, West Heidelberg, Victoria, Australia), from which a 1 in 100 dilution was performed in fresh broth to prepare mid-logarithmic cultures according to the optical density at 500 nm (OD_{500nm} = 0.4 to 0.6). All broth cultures were incubated at 37 $^{\circ}$C in a shaking water bath (180 rpm).

3.3. Minimum Inhibitory Concentration (MIC) Microbiological Assay

MICs were performed according to the Clinical and Laboratory Standards Institute (CLSI) guidelines [47]. MICs were determined for all isolates in three replicates on separate days using broth microdilution method in cation-adjusted Mueller–Hinton broth (CAMHB) in 96-well polypropylene microtitre plates. Wells were inoculated with 100 μL of bacterial suspension prepared in CaMHB (containing 10^6 colony-forming units (cfu) per mL) and 100 μL of CaMHB containing increasing concentrations of polymyxin B (0.25–256 mg/L). The MICs were defined as the lowest concentration at which visible growth was inhibited following 18 h incubation at 37 $^{\circ}$C. Cell viability was determined by sampling wells at polymyxin B concentrations greater than the MIC. These samples were diluted in normal saline and spread plated onto nutrient agar. After incubation at 37 $^{\circ}$C for 20 h, viable colonies were counted on these plates. The limit of detection was 10 cfu/mL.

3.4. Isolation of Outer Membrane Vesicles (OMVs)

Mid-logarithmic cultures (6 L) of each isolate were grown at 37 $^{\circ}$C with shaking (1800 rpm) and cell-free supernatants were collected through centrifugation (15 min at 10,000× g, 4 $^{\circ}$C). Where indicated, polymyxin B was added to the culture volume at a final concentration of 2 mg/L. The OMV containing supernatants were filtered through 0.22-μm membrane (Sigma-Aldrich) to remove any remaining cell debris, then concentrated through a tangential filtration concentrator unit (Pall Life Science, Ann Arbor, MI, USA) and collected using 100 kDa Pellicon filtration cassettes (Millipore, Melbourne, Australia). Also, a portion of the supernatant was plated for growth on agar plates overnight at 37 $^{\circ}$C to make sure that the supernatant is free of bacterial cells. OMVs in the cell-free supernatants were then pelleted down by ultracentrifugation at 150,000× g for 2 h at 4 $^{\circ}$C in a

Beckman Ultracentrifuge (SW28 rotor). Purified OMVs were concentrated re-suspended in 1 mL sterile PBS and the concentration was determined by Bio-Rad (Gladesville, NSW, Australia) protein assay.

3.5. Lipidomics Analysis

OMV lipids were extracted with the single-phase Bligh–Dyer method ($CHCl_3$/MeOH/H_2O, 1:3:1, *v*/*v*) [48]. For further analysis, samples were reconstituted in 100 µL of $CHCl_3$ and 200 µL of MeOH, centrifuged at $14,000 \times g$ for 10 min at 4 °C to obtain particle-free supernatants. LC-MS for lipidomic analysis was conducted on a Dionex U3000 high-performance liquid chromatography system (HPLC) in tandem with a Q-Exactive Orbitrap mass spectrometer (Thermo Fisher, Melbourne, Australia) in both positive and negative mode with a resolution at 35,000. The mass scanning range was from 167 to 2000 *m*/*z*. The electrospray voltage was set as 3.50 kV and nitrogen was used as collision gas. The Ascentis Express C_8 column (5 cm × 2.1 mm, 2.7 µm, Sigma-Aldrich, 53831-U) was maintained at 40 °C, and the samples were controlled at 4 °C. The flow rate was 0.2 mL/min at first 24 min, but increased to 0.5 mL/min from 25 min to 30 min. The multi-step gradient started from 100% to 80% mobile phase A over the first 1.5 min, then to 72% mobile phase A at 7 min, over the next 1 min, the gradient changed to 65% mobile phase A, from 8 min to 24 min, the gradient reached a final composition of 35% mobile phase A and 65% mobile phase B. This was followed by a washing step from 65% to 100% mobile phase B over the next 1 min, and maintained for 2 min. A 2-min re-equilibration of the column with 100% A was performed between injections. Untargeted lipidomic analyses were performed through mzMatch [49]; and IDEOM [50] (http://mzmatch.sourceforge.net/ideom.php). Raw LC-MS data files were converted to mzXML format through a proteowizard tool, Msconvert. Automated chromatography peaks were picked by XCMS [51], and then converted to peakML files, which were combined and filtered by mzMatch based on the intensity (1000), reproducibility (RSD for all replicates < 0.8), and peak shape (codadw > 0.8). The mzMatch program was used for retrieving intensities for missing peaks and the annotation of related peaks. Unmatched peaks and noises were rejected through IDEOM. The database used in IDEOM included KEGG, MetaCyc and Lipidmaps [52]. Univariate statistics analysis was performed using a Welch's T-test ($p < 0.05$), while multivariate analysis was conducted using the metabolomics R package.

3.6. Transmission Electron Microscopy (TEM)

Carbon-coated Formvar copper grids were placed on a drop of OMV suspension (1 mg/mL protein) for 5 min then washed three times with PBS and fixed in 1% glutaraldehyde for 4 min. Grids were then washed three times with PBS, two times with Milli-Q water and stained for 20 s with 4% uranyl acetate. Grids were finally washed with Milli-Q water and incubated on ice for 10 min in methyl–cellulose with 4% uranyl acetate (9:1). Grids were then air-dried and viewed with a Tecnai Spirit (T12) transmission electron microscope, and the images were acquired using TIA software (FEI, Melbourne, Australia).

3.7. Dynamic Light Scattering

The particle size of the OMVs was measured using dynamic light scattering (DLS). OMVs were diluted with PBS to a protein concentration of 0.05 mg/L and the scatter was recorded using a Zetasizer NanoS (Malvern, PA, USA) at 173° with a laser of wavelength 632 nm. Data were analysed with Zetasizer Software (V7.11, Malvern, UK) to obtain the average hydrodynamic radius.

4. Conclusions

In this study, we show that polymyxin B treatment of the susceptible *K. pneumoniae* strains significantly reduced the glycerophospholipid, fatty acid, lysoglycerophosphate and sphingolipid content of their OMVs, compare to the untreated control. On the other hand, in the OMVs of their paired polymyxin-resistant strains these lipids were increased both intrinsically and in response to polymyxin B treatment. In view of these findings, it is reasonable to hypothesize that the outer membrane remodelling associated with polymyxin-resistance in *K. pneumoniae* entails fortifying the membrane with increased glycerophospholipids, fatty acids, lysoglycerophosphates and sphingolipids, which are lipids to which polymyxins cannot avidly bind. It is important to mention that polymyxins primarily target the lipid A in the Gram-negative outer membrane—hence their narrow spectrum of activity against Gram-negative bacteria that do not express LPS. These outer membrane changes may be accompanied by the modification of the lipid A with cationic moieties and/or a reduction in the lipid A content, which, together with the increased content of the aforementioned lipids, serve to make the *K. pneumoniae* outer membrane and OMVs more impervious to polymyxin attack.

Supplementary Materials: Supplementary materials can be found at: http://www.mdpi.com/1422-0067/19/8/2356/s1.

Author Contributions: R.J., M.-L.H., Y.Z., X.H., M.H.H., Y.-W.L., Q.Z., and C.Y.D.-D. All contributed to the experimental data collection, reporting of the results and write-up of the manuscript. J.L. and T.V. developed the experimental design and concepts and helped write the manuscript.

Acknowledgments: J.L. and T.V. are supported by research grants from the National Institute of Allergy and Infectious Diseases of the National Institutes of Health (R01 AI132681). J.L. and T.V. are also supported by the Australian National Health and Medical Research Council (NHMRC) as Senior Research and Career Development Level 2 Fellows. The content is solely the responsibility of the authors and does not necessarily represent the official views of the National Institute of Allergy and Infectious Diseases or the National Institutes of Health.

Conflicts of Interest: The authors declare no conflict of interest.

References

1. Walsh, T.R.; Weeks, J.; Livermore, D.M.; Toleman, M.A. Dissemination of NDM-1 positive bacteria in the New Delhi environment and its implications for human health: An environmental point prevalence study. *Lancet Infect. Dis.* **2011**, *11*, 355–362. [CrossRef]

2. Sidjabat, H.; Nimmo, G.R.; Walsh, T.R.; Binotto, E.; Htin, A.; Hayashi, Y.; Li, J.; Nation, R.L.; George, N.; Paterson, D.L. Carbapenem resistance in *Klebsiella pneumoniae* due to the New Delhi Metallo-beta-lactamase. *Clin. Infect. Dis.* **2011**, *52*, 481–484. [CrossRef] [PubMed]

3. Kumarasamy, K.K.; Toleman, M.A.; Walsh, T.R.; Bagaria, J.; Butt, F.; Balakrishnan, R.; Chaudhary, U.; Doumith, M.; Giske, C.G.; Irfan, S.; et al. Emergence of a new antibiotic resistance mechanism in India, Pakistan, and the UK: A molecular, biological, and epidemiological study. *Lancet Infect. Dis.* **2011**, *10*, 597–602. [CrossRef]

4. Holt, K.E.; Wertheim, H.; Zadoks, R.N.; Baker, S.; Whitehouse, C.A.; Dance, D.; Jenney, A.; Connor, T.R.; Hsu, L.Y.; Severin, J.; et al. Genomic analysis of diversity, population structure, virulence, and antimicrobial resistance in *Klebsiella pneumoniae*, an urgent threat to public health. *Proc. Natl. Acad. Sci. USA* **2015**, *112*, E3574–E3581. [CrossRef] [PubMed]

5. Nordmann, P.; Cuzon, G.; Naas, T. The real threat of *Klebsiella pneumoniae* carbapenemase-producing bacteria. *Lancet Infect. Dis.* **2009**, *9*, 228–236. [CrossRef]

6. Yigit, H.; Queenan, A.M.; Anderson, G.J.; Domenech-Sanchez, A.; Biddle, J.W.; Steward, C.D.; Alberti, S.; Bush, K.; Tenover, F.C. Novel carbapenem-hydrolyzing beta-lactamase, KPC-1, from a carbapenem-resistant strain of *Klebsiella pneumoniae*. *Antimicrob. Agents Chemother.* **2001**, *45*, 1151–1161. [CrossRef] [PubMed]

7. Yong, D.; Toleman, M.A.; Giske, C.G.; Cho, H.S.; Sundman, K.; Lee, K.; Walsh, T.R. Characterization of a new metallo-beta-lactamase gene, bla(NDM-1), and a novel erythromycin esterase gene carried on a unique genetic structure in *Klebsiella pneumoniae* sequence type 14 from India. *Antimicrob. Agents Chemother.* **2009**, *53*, 5046–5054. [CrossRef] [PubMed]

8. Velkov, T.; Roberts, K.D.; Nation, R.L.; Thompson, P.E.; Li, J. Pharmacology of polymyxins: New insights into an old class of antibiotics. *Future Microbiol.* **2013**, *8*, 711–724. [CrossRef] [PubMed]

9. Gales, A.C.; Jones, R.N.; Sader, H.S. Global assessment of the antimicrobial activity of polymyxin B against 54 731 clinical isolates of Gram-negative *bacilli*: Report from the SENTRY antimicrobial surveillance programme (2001–2004). *Clin. Microbiol.* **2006**, *12*, 315–321. [CrossRef] [PubMed]

10. Weterings, V.; Zhou, K.; Rossen, J.W.; van Stenis, D.; Thewessen, E.; Kluytmans, J.; Veenemans, J. An outbreak of colistin-resistant *Klebsiella pneumoniae* carbapenemase-producing *Klebsiella pneumoniae* in the Netherlands (July to December 2013), with inter-institutional spread. *Eur. J. Clin. Microbiol. Infect. Dis.* **2015**, *34*, 1647–1655. [CrossRef] [PubMed]

11. van Duin, D.; Doi, Y. Outbreak of Colistin-Resistant, Carbapenemase-Producing *Klebsiella pneumoniae*: Are We at the End of the Road? *J. Clin. Microbiol.* **2015**, *53*, 3116–3117. [CrossRef] [PubMed]

12. Mezzatesta, M.L.; Gona, F.; Caio, C.; Petrolito, V.; Sciortino, D.; Sciacca, A.; Santangelo, C.; Stefani, S. Outbreak of KPC-3-producing, and colistin-resistant, *Klebsiella pneumoniae* infections in two Sicilian hospitals. *Clin. Microbiol.* **2011**, *17*, 1444–1447. [CrossRef] [PubMed]

13. Marchaim, D.; Chopra, T.; Pogue, J.M.; Perez, F.; Hujer, A.M.; Rudin, S.; Endimiani, A.; Navon-Venezia, S.; Hothi, J.; Slim, J.; et al. Outbreak of colistin-resistant, carbapenem-resistant *Klebsiella pneumoniae* in metropolitan Detroit, Michigan. *Antimicrob. Agents Chemother.* **2011**, *55*, 593–599. [CrossRef] [PubMed]

14. Giani, T.; Arena, F.; Vaggelli, G.; Conte, V.; Chiarelli, A.; Henrici De Angelis, L.; Fornaini, R.; Grazzini, M.; Niccolini, F.; Pecile, P.; et al. Large Nosocomial Outbreak of Colistin-Resistant, Carbapenemase-Producing *Klebsiella pneumoniae* Traced to Clonal Expansion of an mgrB Deletion Mutant. *J. Clin. Microbiol.* **2015**, *53*, 3341–3344. [CrossRef] [PubMed]

15. Jin, Y.; Shao, C.; Li, J.; Fan, H.; Bai, Y.; Wang, Y. Outbreak of multidrug resistant NDM-1-producing *Klebsiella pneumoniae* from a neonatal unit in Shandong Province, China. *PloS ONE* **2015**, *10*, e0119571. [CrossRef] [PubMed]

16. Bratu, S.; Tolaney, P.; Karumudi, U.; Quale, J.; Mooty, M.; Nichani, S.; Landman, D. Carbapenemase-producing *Klebsiella pneumoniae* in Brooklyn, NY: Molecular epidemiology and in vitro activity of polymyxin B and other agents. *J. Antimicrob. Chemother.* **2005**, *56*, 128–132. [CrossRef] [PubMed]

17. Elemam, A.; Rahimian, J.; Mandell, W. Infection with panresistant *Klebsiella pneumoniae*: A report of 2 cases and a brief review of the literature. *Clin. Infect. Dis.* **2009**, *49*, 271–274. [CrossRef] [PubMed]

18. Nikaido, H. Molecular basis of bacterial outer membrane permeability revisited. *Microbiol. Mol. Bio. Rev.* **2003**, *67*, 593–656. [CrossRef]

19. Nikaido, H. Outer membrane barrier as a mechanism of antimicrobial resistance. *Antimicrob. Agents Chemother.* **1989**, *33*, 1831–1836. [CrossRef] [PubMed]

20. Llobet, E.; Campos, M.A.; Gimenez, P.; Moranta, D.; Bengoechea, J.A. Analysis of the networks controlling the antimicrobial-peptide-dependent induction of *Klebsiella pneumoniae* virulence factors. *Infect. Immun.* **2011**, *79*, 3718–3732. [CrossRef] [PubMed]

21. Llobet, E.; March, C.; Gimenez, P.; Bengoechea, J.A. *Klebsiella pneumoniae* OmpA confers resistance to antimicrobial peptides. *Antimicrob. Agents Chemother.* **2009**, *53*, 298–302. [CrossRef] [PubMed]

22. Campos, M.A.; Vargas, M.A.; Regueiro, V.; Llompart, C.M.; Alberti, S.; Bengoechea, J.A. Capsule polysaccharide mediates bacterial resistance to antimicrobial peptides. *Infect. Immun.* **2004**, *72*, 7107–7114. [CrossRef] [PubMed]

23. Cheng, H.Y.; Chen, Y.F.; Peng, H.L. Molecular characterization of the PhoPQ-PmrD-PmrAB mediated pathway regulating polymyxin B resistance in *Klebsiella pneumoniae* CG43. *J. Biomed. Sci.* **2010**, *17*, 60. [CrossRef] [PubMed]

24. Rida, S.M.; Soad, F.A.A.; El-Hawash, A.M.; et al. Synthesis of Some Novel Substituted Purine Derivatives As Potential Anticancer, Anti-HIV-1 and Antimicrobial Agents. *Archiv der Pharmazie* **2007**, *340*, 185–194. [CrossRef] [PubMed]

25. Vinogradov, E.; Lindner, B.; Seltmann, G.; Radziejewska-Lebrecht, J.; Holst, O. Lipopolysaccharides from Serratia marcescens possess one or two 4-amino-4-deoxy-L-arabinopyranose 1-phosphate residues in the lipid A and D-glycero-D-talo-oct-2-ulopyranosonic acid in the inner core region. *Chemistry* **2006**, *12*, 6692–6700. [CrossRef] [PubMed]

26. Raetz, C.R.; Reynolds, C.M.; Trent, M.S.; Bishop, R.E. Lipid A modification systems in gram-negative bacteria. *Annu. Rev. Biochem.* **2007**, *76*, 295–329. [CrossRef] [PubMed]

27. Falagas, M.E.; Rafailidis, P.I.; Matthaiou, D.K. Resistance to polymyxins: Mechanisms, frequency and treatment options. *Drug Resist Updat.* **2010**, *13*, 132–138. [CrossRef] [PubMed]

28. Raetz, C.R.; Whitfield, C. Lipopolysaccharide endotoxins. *Annu. Rev. Biochem.* **2002**, *71*, 635–700. [CrossRef] [PubMed]

29. Loutet, S.A.; Flannagan, R.S.; Kooi, C.; Sokol, P.A.; Valvano, M.A. A complete lipopolysaccharide inner core oligosaccharide is required for resistance of Burkholderia cenocepacia to antimicrobial peptides and bacterial survival in vivo. *J. Bacteriol.* **2006**, *188*, 2073–2080. [CrossRef] [PubMed]

30. Mitrophanov, A.Y.; Jewett, M.W.; Hadley, T.J.; Groisman, E.A. Evolution and dynamics of regulatory architectures controlling polymyxin B resistance in enteric bacteria. *PLoS Genet.* **2008**, *4*, e1000233. [CrossRef] [PubMed]

31. Helander, I.M.; Kilpelainen, I.; Vaara, M. Increased substitution of phosphate groups in lipopolysaccharides and lipid A of the polymyxin-resistant pmrA mutants of *Salmonella typhimurium*: A 31P-NMR study. *Mol. Microbiol.* **1994**, *11*, 481–487. [CrossRef] [PubMed]

32. Helander, I.M.; Kato, Y.; Kilpelainen, I.; Kostiainen, R.; Lindner, B.; Nummila, K.; Sugiyama, T.; Yokochi, T. Characterization of lipopolysaccharides of polymyxin-resistant and polymyxin-sensitive *Klebsiella pneumoniae* O3. *Eur. J. Biochem.* **1996**, *237*, 272–278. [CrossRef] [PubMed]

33. Clements, A.; Tull, D.; Jenney, A.W.; Farn, J.L.; Kim, S.H.; Bishop, R.E.; McPhee, J.B.; Hancock, R.E.; Hartland, E.L.; Pearse, M.J.; et al. Secondary acylation of *Klebsiella pneumoniae* lipopolysaccharide contributes to sensitivity to antibacterial peptides. *J. Biol. Chem.* **2007**, *282*, 15569–15577. [CrossRef] [PubMed]

34. Velkov, T.; Soon, R.L.; Chong, P.L.; Huang, J.X.; Cooper, M.A.; Azad, M.A.K.; Baker, M.A.; Thompson, P.E.; Roberts, K.; Nation, R.L.; et al. Molecular basis for the increased polymyxin susceptibility of *Klebsiella pneumoniae* strains with under-acylated lipid A. *Innate Immun.* **2012**, *19*, 265–277. [CrossRef] [PubMed]

35. Schwechheimer, C.; Kuehn, M.J. Outer-membrane vesicles from Gram-negative bacteria: Biogenesis and functions. *Nat. Rev. Microbiol.* **2015**, *13*, 605–619. [CrossRef] [PubMed]

36. Bonnington, K.E.; Kuehn, M.J. Outer Membrane Vesicle Production Facilitates LPS Remodeling and Outer Membrane Maintenance in *Salmonella* during Environmental Transitions. *mBio* **2016**, *7*, e01532-16. [CrossRef] [PubMed]

37. Elhenawy, W.; Bording-Jorgensen, M.; Valguarnera, E.; Haurat, M.F.; Wine, E.; Feldman, M.F. LPS Remodeling Triggers Formation of Outer Membrane Vesicles in *Salmonella*. *mBio* **2016**, *7*, e00940-16. [CrossRef] [PubMed]

38. Roier, S.; Zingl, F.G.; Cakar, F.; Durakovic, S.; Kohl, P.; Eichmann, T.O.; Klug, L.; Gadermaier, B.; Weinzerl, K.; Prassl, R.; et al. A novel mechanism for the biogenesis of outer membrane vesicles in Gram-negative bacteria. *Nat. Commun.* **2016**, *7*, 10515. [CrossRef] [PubMed]

39. Roier, S.; Zingl, F.G.; Cakar, F.; Schild, S. Bacterial outer membrane vesicle biogenesis: A new mechanism and its implications. *Microb. Cell* **2016**, *3*, 257–259. [CrossRef] [PubMed]

40. Sohlenkamp, C.; Geiger, O. Bacterial membrane lipids: Diversity in structures and pathways. *FEMS Microbiol. Rev.* **2016**, *40*, 133–159. [CrossRef] [PubMed]

41. van Dalen, A.; de Kruijff, B. The role of lipids in membrane insertion and translocation of bacterial proteins. *Biochim. Biophys. Acta Mol. Cell. Res.* **2004**, *1694*, 97–109. [CrossRef] [PubMed]

42. Dare, K.; Shepherd, J.; Roy, H.; Seveau, S.; Ibba, M. LysPGS formation in *Listeria monocytogenes* has broad roles in maintaining membrane integrity beyond antimicrobial peptide resistance. *Virulence* **2014**, *5*, 534–546. [CrossRef] [PubMed]

43. Heung, L.J.; Luberto, C.; Del Poeta, M. Role of sphingolipids in microbial pathogenesis. *Infect. Immun.* **2006**, *74*, 28–39. [CrossRef] [PubMed]

44. Bootz, A.; Vogel, V.; Schubert, D.; Kreuter, J. Comparison of scanning electron microscopy, dynamic light scattering and analytical ultracentrifugation for the sizing of poly(butyl cyanoacrylate) nanoparticles. *Eur. J. Pharm. Biopharm.* **2004**, *57*, 369–375. [CrossRef]

45. Hallett, F.R.; Watton, J.; Krygsman, P. Vesicle sizing: Number distributions by dynamic light scattering. *Biophys. J.* **1991**, *59*, 357–362. [CrossRef]

46. Clinical Breakpoints (Bacterial v6.0). The European Committee on Antimicrobial Susceptibility Testing. Published 20 January 2016 (bacteria). Available online: http://www.eucast.org/clinical_breakpoints (accessed on 15 July 2016).

47. Hsueh, P.R.; Ko, W.C.; Wu, J.J.; Lu, J.J.; Wang, F.D.; Wu, H.Y.; Wu, T.L.; Teng, L.J. Consensus statement on the adherence to Clinical and Laboratory Standards Institute (CLSI) Antimicrobial Susceptibility Testing Guidelines (CLSI-2010 and CLSI-2010-update) for Enterobacteriaceae in clinical microbiology laboratories in Taiwan. *J. Microbiol. Immunol. Infect.* **2010**, *43*, 452–455. [CrossRef]

48. Bligh, E.G.; Dyer, W.J. A rapid method of total lipid extraction and purification. *Can. J. Biochem. Physiol.* **1959**, *37*, 911–917. [CrossRef] [PubMed]

49. Scheltema, R.A.; Jankevics, A.; Jansen, R.C.; Swertz, M.A.; Breitling, R. PeakML/mzMatch: A file format, Java library, R library, and tool-chain for mass spectrometry data analysis. *Anal. Chem.* **2011**, *83*, 2786–2793. [CrossRef] [PubMed]

50. Creek, D.J.; Jankevics, A.; Burgess, K.E.; Breitling, R.; Barrett, M.P. IDEOM: An Excel interface for analysis of LC-MS-based metabolomics data. *Bioinformatics* **2012**, *28*, 1048–1049. [CrossRef] [PubMed]
51. Smith, C.A.; Want, E.J.; O'Maille, G.; Abagyan, R.; Siuzdak, G. XCMS: Processing mass spectrometry data for metabolite profiling using nonlinear peak alignment, matching, and identification. *Anal. Chem.* **2006**, *78*, 779–787. [CrossRef] [PubMed]
52. Creek, D.J.; Jankevics, A.; Breitling, R.; Watson, D.G.; Barrett, M.P.; Burgess, K.E. Toward global metabolomics analysis with hydrophilic interaction liquid chromatography-mass spectrometry: Improved metabolite identification by retention time prediction. *Anal. Chem.* **2011**, *83*, 8703–8710. [CrossRef] [PubMed]

International Journal of
Molecular Sciences

Article

Lipomatrix: A Novel Ascorbyl Palmitate-Based Lipid Matrix to Enhancing Enteric Absorption of Serenoa Repens Oil

Andrea Fratter [1,2,*], Vera Mason [1], Marzia Pellizzato [1], Stefano Valier [3], Arrigo Francesco Giuseppe Cicero [4], Erik Tedesco [5], Elisa Meneghetti [5] and Federico Benetti [5]

[1]   Nutraceutical Research and Innovation Technology Department, Labomar Research, Via Fabio Filzi 55, 31036 Istrana (TV), Italy; vera.mason@labomar.com (V.M.); marzia.pellizzato@labomar.com (M.P.)
[2]   Head, Nutraceutical Research and Innovation Technology Department, Labomar Research, 31036 Istrana (TV), Italy
[3]   PHF SA, 6900 Lugano, Switzerland; stefano.valier@phfsa.com
[4]   Faculty of Medicine and Surgery, Department of Cardiovascular Disease Prevention Research Unit, Internal Medicine and Lipidology, University Alma Mater, Bologna, Via Massarenti 9, Pavilion 2, 40138 Bologna, Italy; afgcicero@gmail.com
[5]   ECSIN-European Center for the Sustainable Impact of Nanotechnology, ECAMRICERT SRL, I-45100 Rovigo, Italy; e.tedesco@ecamricert.com (E.T.); e.meneghetti@ecamricert.com (E.M.); f.benetti@ecamricert.com (F.B.)
*   Correspondence: andrea.fratter@labomar.com or andrea.fratter74@alice.it

Received: 6 January 2019; Accepted: 31 January 2019; Published: 4 February 2019

Abstract: The class of lipophilic compounds coming from vegetal source represents a perspective in the adjuvant treatment of several human diseases, despite their poor bioavailability in humans. These compounds are generally soluble in fats and poorly soluble in water. The major reason for the poor bioavailability of lipophilic natural compounds after oral uptake in humans is related to their reduced solubility in enteric water-based fluids, leading to an ineffective contact with absorbing epithelium. The main goal to ensure efficacy of such compounds is then creating technological conditions to deliver them into the first enteric tract as hydro-dispersible forms to maximize epithelial absorption. The present work describes and characterizes a new technological matrix (Lipomatrix, Labomar Research, Istrana, TV, Italy) based on a molten fats core in which Ascorbyl Palmitate is embedded, able to deliver lipophilic compounds in a well-dispersed and emulsified form once exposed to duodenal fluids. Authors describe and quantify Lipomatrix delivery of *Serenoa repens* oil through an innovative in vitro model of human gastro-enteric digestion, reporting results of its improved bioaccessibility, enteric absorption and efficacy compared with not formulated *Serenoa repens* oil-containing commercial products using in vitro models of human intestine and prostatic tissue.

Keywords: ascorbyl palmitate; mono and diglycerides of fatty acids; natural lipophilic compounds; nutraceutical products; *Serenoa repens* oil; enteric bioaccessibility

1. Introduction

Natural compounds with specific reference to plant-derived active ingredients are raising interest as coadjuvant treatment of human diseases. In recent years, this trend has consolidated and many natural products have been introduced into the market under the label of nutritional supplements (NS) or foods with special medical purposes (SMPF). Despite the effective and charming marketing efforts of companies in selling these products following medical routes, the real efficacy of nutraceuticals is still a challenging issue because of the inadequate technological framework for enhancing and testing their bioavailability.

Lipophilic active ingredients belonging to the families of terpens, saponins, and vegetal oils are poorly absorbed by intestine mainly because of their low water solubility [1,2]. The interaction of active compounds with intestinal epithelium depends on the dissolution of molecules in water-based intestinal fluids [3]. A viscous mucopolysaccharides rich layer strictly adhering to the enteric epithelium and synthesized by specialized Goblet cells, called unstirred water phase (UWP), (Figure 1) plays a pivotal role in regulating the absorption rate of active ingredients. Mucins, indeed, can interact with poorly-soluble molecules preventing their penetration [4,5].

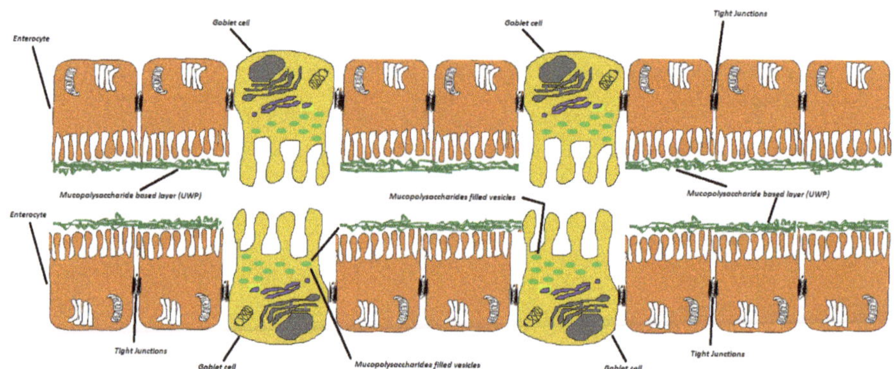

Figure 1. Representation of UWP in the intestine. It is possible to recognize tight junctions and Goblet cells.

Only water-soluble molecules or intimately water-dispersed molecules can readily diffuse through this layer, favoring an effective intestinal absorption [6]. According to this premise, lipophilic terpens, such as boswellic acids (BA) extracted from the resin of *boswellia* genus trees, vegetal oils such as ginger oil (GO), *Serenoa repens* (Bartram) Small oil (SRO) and many others structurally similar compounds, are not effectively absorbed through the intestine even in the presence of bile salts [7,8]. Advanced technological matrices and delivery systems able of emulsifying lipophilic compounds or making them hydro-dispersible in the duodenal tract represent promising and effective strategies to improve their bioavailability.

Over the last decades, many companies and research institutes all over the world, tried to discover new performing technologies for enhancing natural compounds bioavailability, in some cases with great success as the patented technology Phytosome™. Phytosome™ consists in a physical complexation of the active ingredient with a phospholipid enabling it to overcoming the enteric barrier, through an enterocyte membrane partitioning mechanism [9–11].

Many lipophilic plant-derived active principles have shown several biological relevant actions such as anti-inflammatory activity, enteric physiology restoring activity, and osteoarticular health promoting activity. However, their positive impact on human health is often limited by their poor solubility in aqueous systems, such as digestive fluids, so leading to low enteric bio-accessibility and bioavailability. Among them, SRO is a vegetal oil largely used in clinical practice to reduce symptoms of human prostatic hyperplasia (HPH). This oil that mainly works by inhibiting 5-α-reductase enzyme in prostatic cells and other androgen-sensible cells (sebaceous glands, hair follicles etc.) and by modulating expression of PSA and inflammatory cytokines [12,13] is mainly composed of free fatty acids (80–90% *w/w*) and several phytosterols of which β-Sitosterol is the most represented. SRO has shown to be effective in reducing lower urinary tract symptoms (LUTS) such as normalizing urine flow and reducing gland volume [14,15] at dosage of 320 mg of oil titrated at 80–90% in fatty acids per day. Despite this efficacy and the ascertained prostatic tropism [16], not many data are available about its overall bioavailability in humans [16,17] and in particular no data are available about the role of peculiar delivery systems in improving it in humans.

Several technological forms such as nanoemulsions (NE), solid lipid nanoparticles (SLN), and self-emulsifying drug delivery systems (SEDDS) have been developed over the years to deliver insoluble or poorly bio-accessible molecules. On the other side, it must be pointed out, that the most part of natural compounds in form of oils and lipids-based formulations (LF) are mainly composed of triglycerides (TG) in form of emulsion, NE and SLN and for this reason they are partially digested by gastric lipase (GL) that cleaves TG generating free fatty acids (FFA) and diacylglycerols (DAG) [18–20]. This underestimated physiological phenomenon can dramatically affect the strategy of delivery of the molecules entrapped into these technological systems [21,22], partially or totally compromising their overall bioavailability.

To design a platform for increasing enteric absorption of lipophilic molecules, some requirements should be considered pivotal to achieve a good pharmaceutical delivery tool: (i) delivery systems able of retaining active ingredients in the stomach and releasing them in the duodenum; (ii) systems able of enforcing bile salts effect in forming fine emulsions or micellar dispersions of lipophilic molecules in the first duodenum; (iii) systems stable to enzymes and peculiar environment of the gastro-enteric tract (iii) systems able of entrapping lipophilic compounds, even in oily form, in a hydro-dispersible solid matrix powder to be easily transformed in capsules or tablets. Taking into consideration the aforementioned requirements, a new technological matrix has been designed and developed to enhance lipophilic compounds enteric absorption. This technological platform, named Lipomatrix (Labomar Research, Istrana, TV, Italy), contains a molten fats core aimed to entrapping lipophilic compounds into a gastric-refractory environment and to emulsify them once exposed to duodenal fluids. This technology is based on a peculiar association of low melting point fats such as mono and diglycerides of fatty acids (MDGFA) commonly used in food industry (E471), and ascorbyl palmitate (ASP), an ester of L-ascorbic acid (AA) and palmitic acid (PA) widely employed in food, pharmaceutical and cosmetic industry as antioxidant agent (E304) [23,24]. The precise chemical features of ASP and its peculiar mechanism of action as emulsifying agent and drugs absorption enhancer are fully described in the Supplementary file section.

In this work, we present for the first time, the technology Lipomatrix and its efficacy in favoring apparent absorption and improving biological activities of SRO through human cell lines models. Beside this main goal, the apparent enteric permeability rate of an extract of Cranberry (Flowens^{TM}) and Lycopene vehiculated in the mentioned technological matrix with SRO was assessed.

2. Results and Discussion

2.1. Differential Scanning Calorimetry (DSC) Analysis

DSC analysis was performed on MDGFA, mannitol, ASP, SRO, and Lipomatrix as described in Section 3.2.1.2, in order to evaluate possible changes in the crystallinity structure of the single raw material once embedded into the matrix and thereby speculate on its prevalent amorphous or crystalline state. In Supplementary file DSC thermograms for the single substances are reported.

From the analysis of the Lipomatrix thermogram achieved in the range temperature of 25–125 °C (Supplementary file) it is possible to recognize a melting pick between 47 and 55 °C coherent with the melting point of MDGFA even though there is a significant decrease in the melting temperature pick of about 10 °C in respect to MDGFA alone. This could mean that some interaction phenomena of MDGFA with the other matrix components occurs during Lipomatrix preparation, with an overall increase of the amorphous structure respect to the pure raw material [25,26]. Interestingly, comparing the overlapping thermograms of single components and Lipomatrix one in the range 30–300 °C (Figure 2, Table 1), it is possible to ascertain that both endothermic and exothermic picks of ASP completely disappear, while the only distinctive endothermic pick seems to coincide to Mannitol one. This fact could be explained considering the effective dissolution of ASP into the warmed lipidic matrix, according to the method of production, taking place to a prevalent amorphous or disordered crystalline phase molecular dispersion in the lipid matrix after cooling [27]. Finally, the appearing of a well-shaped exothermic pick in the Lipomatrix sample around 220 °C, not recognized in none of the single components thermograms, could means that a sort of crystallization phenomenon occurs during the warming of the mix of lipophilic components and its further rapid cooling onto the Mannitol to get the final matrix. This phenomenon could be a further proof of an occurring structural change.

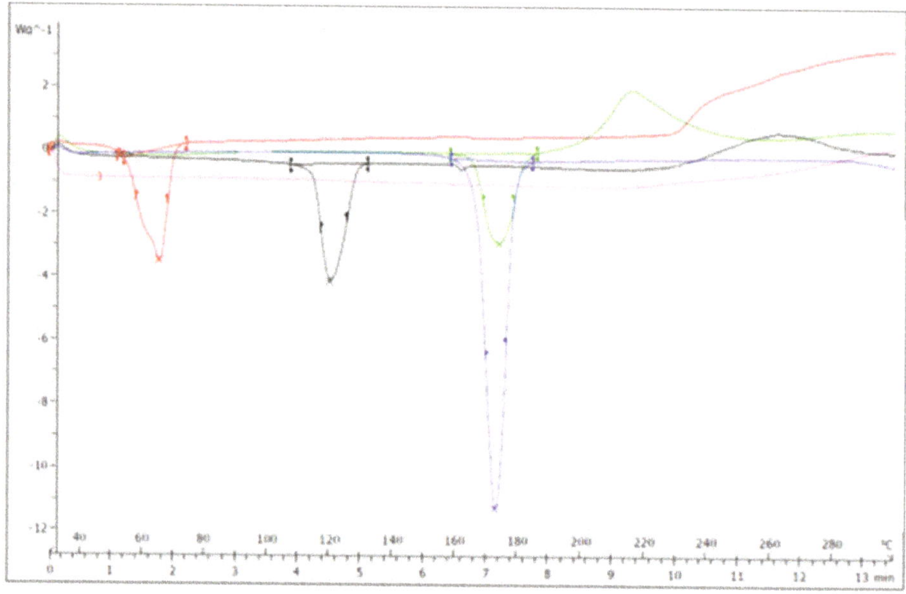

Figure 2. Overlapping thermograms of LIPOMATRIX containing SRO and single components. Black line represents ASP. Red line represents MDGFA. Green line represents Lipomatrix with SRO. Blue line represents Mannitol. Pink line represents SRO.

Table 1. Cristallinity index, entropy, and extrapolation pick time of Lipomatrix and any of its constituents.

Cryst. Ind./Entropy/Extr. P	MDGFA	Mannitol	ASP	SRO	LIPOMATRIX with SRO
Crystallinity Index (CI) (%)	15.75	12.50	11.49	n.r.	12.99
Entropy (mJ)	−529.19	−2122.40	−837.90	n.r.	−705.02
Exprapol. Peak (min)	1.74	7.17	4.52	n.r.	7.20

2.2. In Vitro Assessment of Lipomatrix Containing SRO Gastric Resistance

One of the three aliquots (triplicate) coming from the gastric-resistance experiment with Lipomatrix powder containing SRO described in Materials and Methods (Section 3.2.4) and the one coming from the same experiment assessed on equivalent amount of not formulated SRO contained in two common gelatin soft-gel capsules have been collected and visually compared each other. An oily layer on the GSF surface is appreciable only in the beaker containing SRO from the soft-gel capsules, while no oily separation is appreciable in the beaker containing Lipomatrix powder. This provides a significant evidence that no SRO release occurred throughout the GSF during the test (Figure 3).

Figure 3. Visual comparison of the liquids coming from disaggregation test in GSF between Lipomatrix containing SRO (left beaker) and soft-gel SRO contained in two soft-gel capsules (right beaker). It is clearly appreciable the presence of an oily layer of SRO in the right beaker relating to the soft-gel capsules and the absence of the oily layer in the surface of the GSF liquid relating to Lipomatrix powder.

GC-MS analysis confirmed the absence of SRO derived fatty acids in GSF with Lipomatrix, suggesting that the technology guarantees a complete gastric resistance of SRO at full therapeutic dosage (Figure 4). The major explanation of this observed behavior is that ASP remains unionized in the stomach since low pH does not permit its enediolic ionization (pH < pKa) and this phenomenon contributes, in synergy with MDGFA, to the overall gastric-resistance of the powder (see the supplemental file section for more detailed explanation).

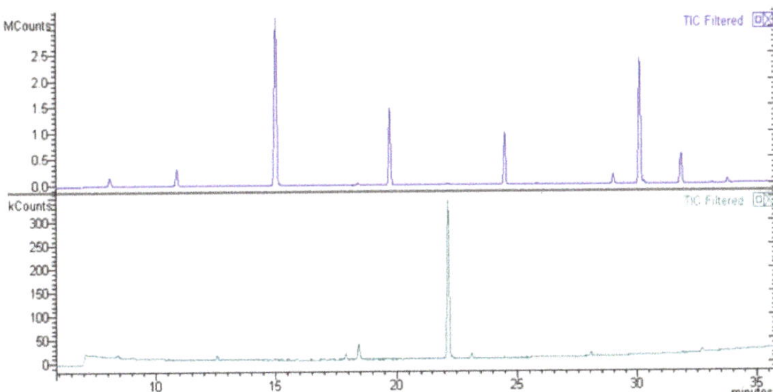

Figure 4. GC-MS overlapping chromatograms of SRO raw material and one of the aliquots of GSF tested (in triplicate, see the supplementary file). The upper graph shows the peculiar picks of fatty acids present in SRO (raw material) that completely disappeared in the lower graph (Lipomatrix specimen), clearly indicating that no SRO release from Lipomatrix has occurred.

2.3. In Vitro Emulsification

2.3.1. Emulsifying Behavior in FaSSIF-V2

The amount of Lipomatrix powder containing 640 mg of SRO has been dispersed in FaSSIF-V2 at 37 °C under moderate stirring conditions for 1 h, according to the method described in Materials and Methods section (Section 3.2.1.2 and Section 3.2.3). The same experiment has been conducted testing two soft-gel capsules containing the same amount of unformulated SRO. Lipomatrix, in force of the auto-emulsifying profile awarded by the association of MDGFA and ASP at the duodenal simulated conditions, allows a much better emulsification of SRO in FaSSIF-V2, creating the postulated synergy with bile salts that prelude to a better bio-accessibility of the oil to the absorbing epithelium. On the contrary, the soft-gel containing unformulated SRO does not take place to any significant emulsion or micellar dispersion that could be visually recognizable (Figure 5). This experimental evidence confirms the hypothesis of reduced and incomplete emulsifying property of sole bile towards oily active ingredients.

Figure 5. Differential behavior in FaSSIF-V2 at 37 °C, 60 min. of 2 gelatin based soft-gel containing not formulated SRO, 640 mg (left) and Lipomatrix powder containing the same amount of SRO (right). It is possible to recognize the emulsifying capability of Lipomatrix in comparison with soft-gel in which no apparent emulsification occurs. Some insoluble particles suspended in FaSSIF-V2 and ascribable to insoluble excipients such as Magnesium Stearate and amorphous silica are appreciable.

2.3.2. DLS Analysis of Emulsified Dispersion of Lipomatrix in FaSSIF-V2

The DLS analysis of the sample containing SRO entrapped in Lipomatrix and dispersed in FaSSIF-V2 produced the size distribution curve reported in Figure 6 (red line) whose descriptive parameters are reported in Table 2.

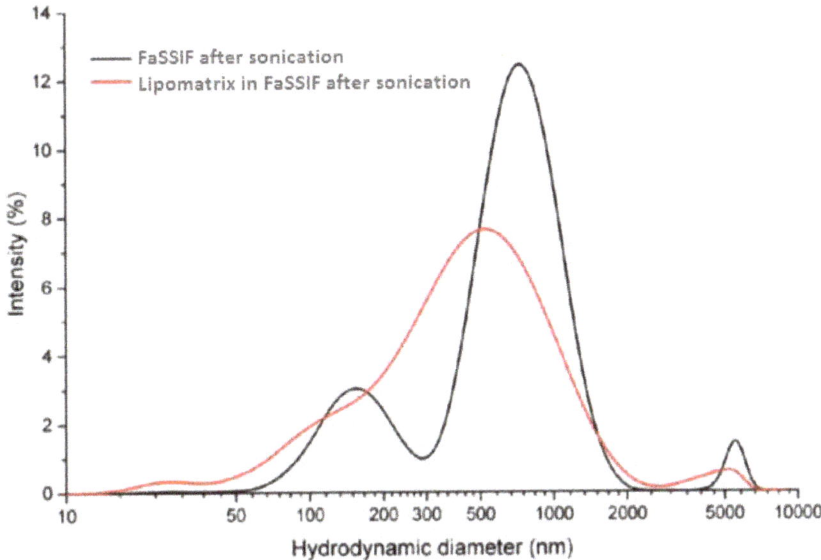

Figure 6. Particle size distribution of FaSSIF-V2 (black Line) and LIPOMATRIX in FaSSIF-V2 (red line) after sonication (*n* = 5).

Table 2. Principal parameters of particle size distribution of the LIPOMATRIX in FaSSIF-V2 sample after sonication.

Sample Name	Hydrodynamic Diameter (Average ± Standard Deviation, nm)	Polydispersity Index
FaSSIF-V2	769 ± 274 161 ± 55 5468 ± 248	0.703
LIPOMATRIX in FaSSIF-V2	528 ± 378 27 ± 6 4511 ± 881	0.529

The FaSSIF-V2 sample shows a high polydispersity index (0.703) with three well-defined particles populations (769, 161 and 5468 nm) of which the first one is the most abundant and the third the least one. The sample of Lipomatrix dispersed in FaSSIF-V2 shows an elevated polydispersity index (0.529) as well, even though lower than FaSSIF-V2 alone and is composed of three groups of particles (528, 27, and 4511 nm), of which the first one is largely the most represented. The DLS evaluation seems to consolidate the visual hypothesis that Lipomatrix can effectively disperse SRO in FaSSIF-V2, giving rise to an even more finely dispersed system than FaSSIF-V2 alone (lower polydispersion index and presence of a prevalent population around 500 nm). Lipomatrix seems therefore to create a more uniform micellar dispersion in force of the presence of both ionized ASP and MDGFA. ASP becomes partially ionized in the presence of enteric fluids (pH > pKa) and in force of this ionization it behaves as hydro-soluble surfacting agent, capable of emulsifying SRO in synergy with MDGFA and bile salts, generating mixed micellar structures [28,29] (Figure 7). This data gives an analytical confirmation of the visual macroscopic difference of emulsification capability of SRO entrapped in Lipomatrix compared to the same unformulated oil dispersed in FaSSIF-V2 as described in Section 2.3.1. Finally, the comparison of the two specimens shows that a new population ascribable to Lipomatrix, with average hydrodynamic diameter of 27 ± 6 nm, was generated.

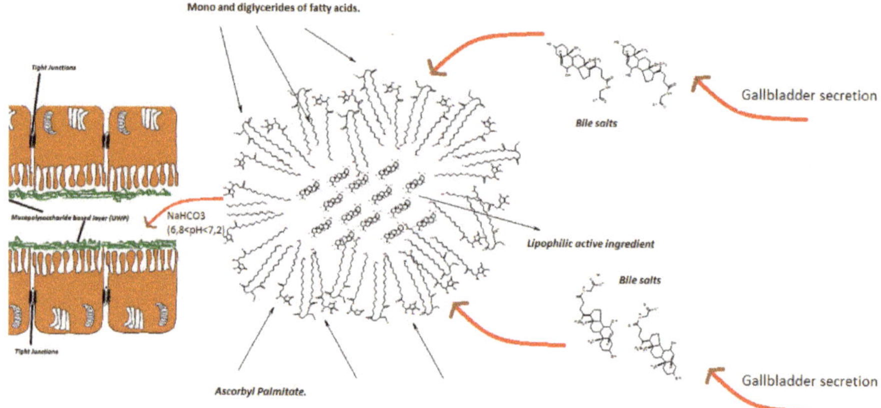

Figure 7. Representation of Lipomatrix mechanism of delivery: lipophilic molecule entrapped in a mixed micellar structure composed of MDGFA, ASP, and bile salts.

2.4. SRO, PCA, and LYC Bioaccessibility

As previously mentioned, therapeutic applications of active principles from plants, such as SRO oil or curcumine, are hindered by their poor solubility in aqueous medium like the digestive fluids. Although their solubility during the digestive process is slightly improved by the emulsifying activity of bile salts, new technologies aimed at improving lipophilic active principle bioaccessibility are needed. To investigate Lipomatrix-based formulation (LBF) performance compared to the commercial formulations, we exposed a single dose of each formulation to in vitro digestion procedure mimicking human adulthood and evaluated the total amount of active principles and the apparent bioaccessible fraction released from its matrix. The apparent bioaccessible fraction includes the portion available to be absorbed. LBF has similar SRO apparent bioaccessibility of CF1 and CF2 formulations, with a good emulsifying effect of Lipomatrix technology (Table 3 and Figure 8). Lipomatrix emulsifying efficiency is similar to CF2, which contains soy lecithin emulsifier. CF1 has a lower emulsifying efficiency than the others do since it does not contain any specific emulsifier in the formulation. CF1 emulsion is only due to the presence of bile salts in the intestinal compartment. Bile acts, to some extent, as a surfactant, helping to emulsify lipids and lipophilic molecules and increasing their absorption.

Table 3. Total amount and apparent bioaccessible fraction of SRO for the three tested formulations, expressed as a percentage of the total SRO in a single dose ($n = 3$).

	Serenoa Repens (Fatty Acids)		
Formulation	Total Amount Post-Digestion (%)	Apparent Bioaccessible Fraction (%)	Emulsifying Efficiency (%)
LBF	54.0	27.7 ± 3.7	51.3
CF1	62.4	30.1 ± 4.6	48.2
CF2	57.2	29.8 ± 7.4	52.0

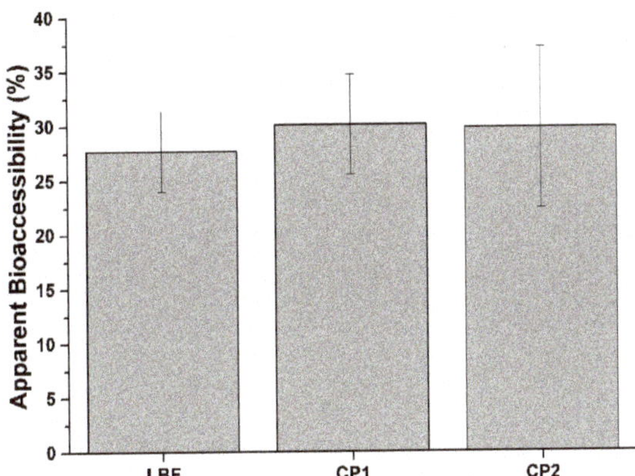

Figure 8. Apparent bioaccessibility of SRO contained in LBF and the two commercial formulations, CF1 and CF2. (*n* = 3).

In addition to SRO, LBF contains PCA and LYC as active principles. As expected from the gastric resistance test, Lipomatrix technology preserves PCA from degradation, as indicated from their total amount and apparent bioaccessible fraction after digestion (Table 4). Protective effect was also observed on LYC, which is preserved from digestion (Table 4).

Table 4. Total amount and bioaccessible fraction of PCA and LYC, expressed as a percentage of the total Flowens^{TM} and total lycopene in a single dose of LBF. (*n* = 3).

	Total Amount Post-Digestion (%)	Apparent Bioaccessibility (%)
Proanthocyanidins	100	98.6
Lycopene	100	62.5

2.5. Impact of Digested Formulations on Intestinal Epithelium Viability

Therapeutic formulations must respond to safety criteria. Taking into consideration their dose and posology, formulations should not negatively affect patients. In particular, damages to the intestinal epithelium must be carefully avoided. Before measuring apparent permeability of the three formulations, the impact of digested formulations on intestinal epithelium viability and integrity was assessed. To this aim, intestinal monolayers were exposed to increasing concentrations of the three formulations, and dose-response curves were obtained (Figure 9). From these curves, half maximal effective concentrations (EC50) were calculated (Table 5). As emerged from dose-responses curves and EC50 values, LBF is the safest formulation.

Table 5. Half maximal effective concentrations (EC$_{50}$) from dose-response curves.

Formulation	EC$_{50}$ (mg/mL)
LBF	8.8 ± 2.1
CF1	2.1 ± 0.1
CF2	7.8 ± 0.7

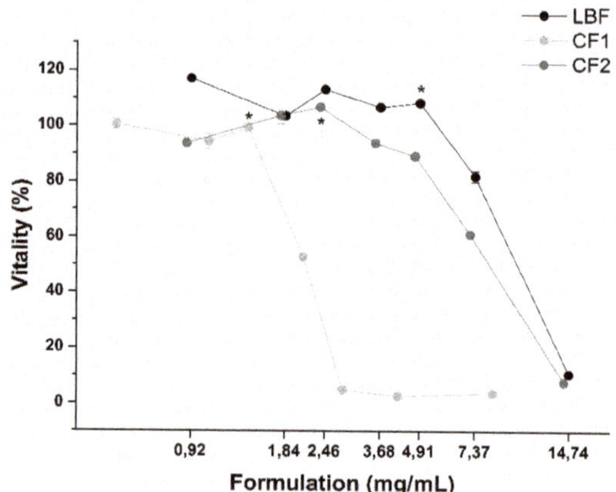

Figure 9. Impact of formulations on intestinal mucosa viability evaluated by determining dose-response curves on formulation concentrations. * Highest non-toxic concentration. ($n = 3$).

2.6. SRO, PCA, and LYC Absorption Rate

Based on the impact of digested formulations on intestinal epithelium viability and posology (2 capsule/day LBF, 1 capsule/day CF1 and CF2), we set experiments for determining SRO, PCA, and LYC absorption rate. In vitro intestinal epithelia were exposed to the digested formulations for 3 h, and SRO (as fatty acids), PCA and LYC were measured in both apical (lumen) and basolateral (serosal) chambers. Absorption rate was then calculated and expressed as percentage of absorption.

PCA and LYC values in the basolateral compartment were below the detection limits (<0.005 mg/mL and <0.001 mg/mL respectively), as expected considering the low amount and titer of LYC (1.33%) and PCA (2.45%) in LBF. Concerning SRO, its absorption rate is higher in LBF than CF1 and CF2 (Table 6), though this difference is considered to be not statistically significant with respect to CF1.

Table 6. SRO absorption rate expressed as a percentage of absorption ± standard error (SE) ($n = 3$).

Formulation	*S. repens* Absorption Rate (% ± SE)
LBF	22.0 ± 12.7
CF1	18.8 ± 12.2
CF2	3.2 ± 3.2

Despite apparent bioaccessibility values, the highest absorption rate of LBF suggests that the Lipomatrix technology supports the bioaccessible form of SRO more than the other two formulations (Figure 10).

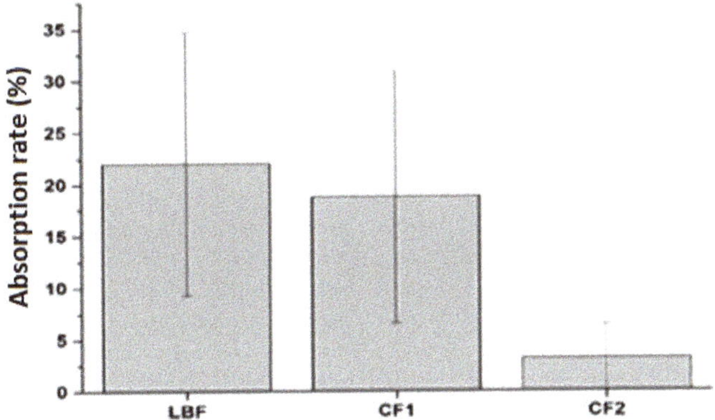

Figure 10. Absorption rate of the three formulations: LBF, CF1, and CF2. ($n = 3$).

2.7. Impact of Digested Formulations on Intestinal Mucosa Viability and Integrity

After exposure of intestinal epithelia to digested formulations, Caco-2 monolayer viability and barrier integrity were analyzed. As expected, no viability reduction was observed during absorption rate experiments, while a slight but significant increase in apparent permeability (P_{app}) was observed for all the three digested formulations (Figure 11).

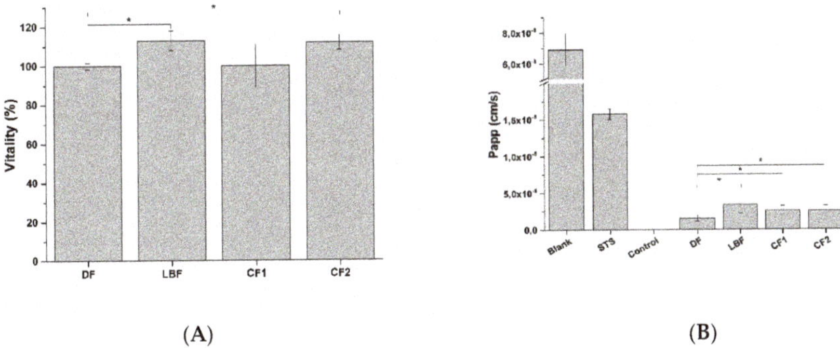

(A) (B)

Figure 11. Cell vitality (**A**) and apparent permeability (P_{app}) (**B**) of intestinal epithelium exposed to digestive fluids (DF; control) and digested formulations ($n = 3$). LBF is composed of SRO, PAC, and lycopene bioaccessible fractions as reported in Tables 5 and 6. CF1 and CF2 represent the bioaccessible fractions as reported in Table 3.

Increase of the absorption rate parallels to a reduction of intestinal epithelia trans-epithelial electrical resistance (TEER) after exposure (Figure 12). Both digested formulations and digestive fluids reduced TEER transiently, since they fully recovered to pre-treatment values.

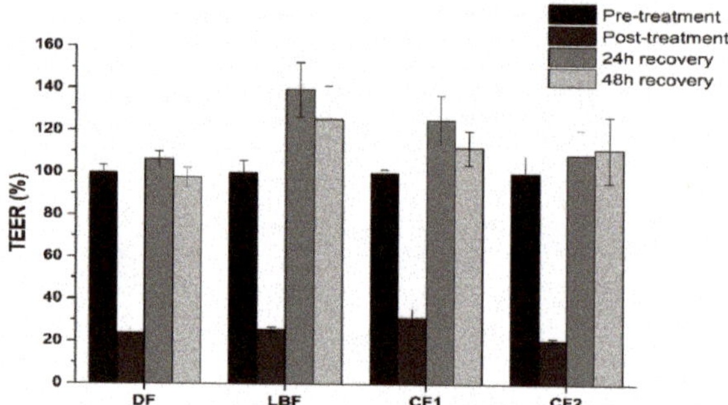

Figure 12. TEER values recorded before the treatment (Pre-treatment), after exposure to digested fluids (DF) or formulations (Post-treatment) at the bioaccessible concentrations, and upon 24 and 48 h recovery. Values are expressed as percentage of the pre-treatment TEER value ($n = 3$).

2.8. Cytotoxic Effect of Diclofenac and Permeable Fractions on Prostatic Epithelium Model

SRO is one of the most popular natural treatments for treating HPH as mentioned. The permeable fractions were tested on prostatic epithelium in vitro model to evaluate their efficacy against inflammation process. Before performing efficacy tests, cytotoxicity of the anti-inflammatory positive control Diclofenac and bioaccessible fractions on LNCaP was measured.

As shown in Figure 13, diclofenac significantly decreases prostatic cell viability at a concentration of 80 µg/mL after 6 h exposure, with a highest non-toxic concentration of 32 µg/mL. Conversely, no effect on cell vitality was observed on LNCaP cells after treatment with bioaccessible fractions (Figure 14).

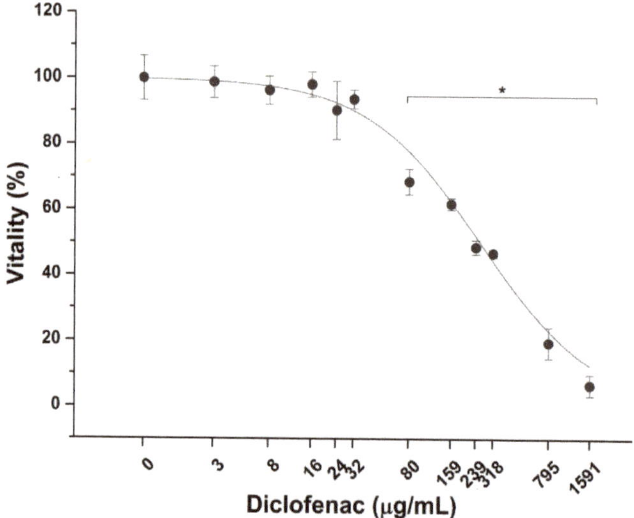

Figure 13. Effect of diclofenac on LNCaP prostatic cells vitality. * $p < 0.05$ ($n = 3$).

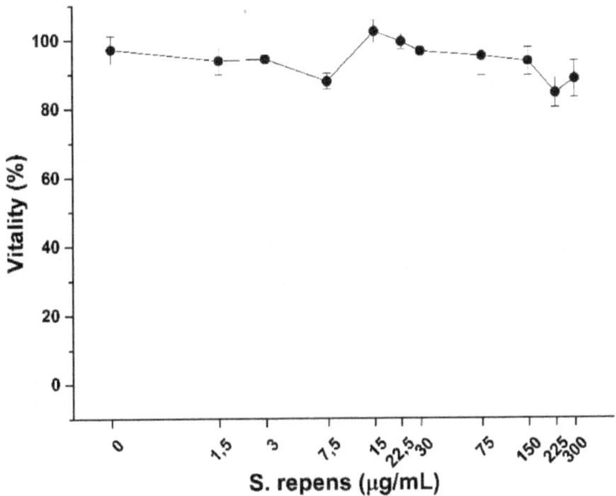

Figure 14. Effect on LNCaP prostatic cells vitality after 6 h exposure to SRO. ($n = 3$).

2.9. Prostate-Specific Anti-Inflammatory Activity of Permeable Fractions

In recent years, inflammation has been recognized as the main phenomenon responsible for the onset of HPH, a noncancerous increase in size of the prostate, leading to the appearance of bothersome symptoms, such as frequent urination, difficult urination, weak stream, inability to urinate, and loss of bladder control. Prostatic-specific anti-inflammatory activity of bioaccessible fractions was evaluated on the in vitro prostatic epithelium model based on LNCaP cells, by measuring expression levels of pro-inflammatory cytokines IL-1β and TNF-α.

As shown in Figure 15, bioaccessible fractions significantly reduce the production of IL-1β compared to the control. In particular, the higher amount of SRO of LBF compared to the two commercially available formulations, presents the strongest effect, with an 85% reduction in IL-1β production.

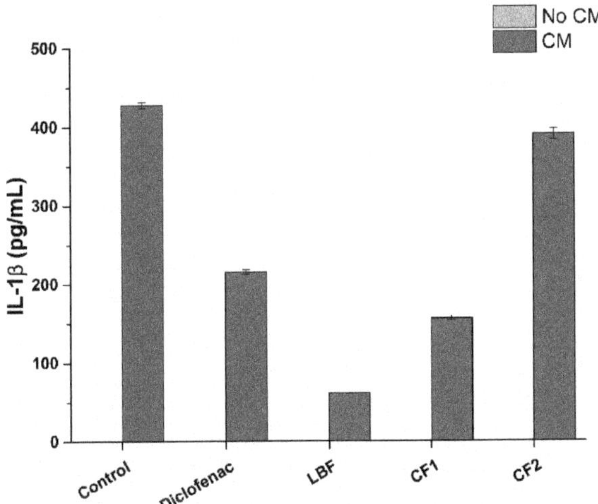

Figure 15. Production profile of IL-1β following treatment of LNCaP prostatic cells with bioavailable fractions: LBF: 22% SRO; CF1: 18.8% SRO; CF2: 3.2% SRO (Table 6). Diclofenac was used at 32 µg/mL. CM: Conditioned medium. * $p < 0.05$ ($n = 3$).

Interestingly, no TNF-α reduction was observed for the other considered pro-inflammatory cytokine, TNF-α (Figure 16).

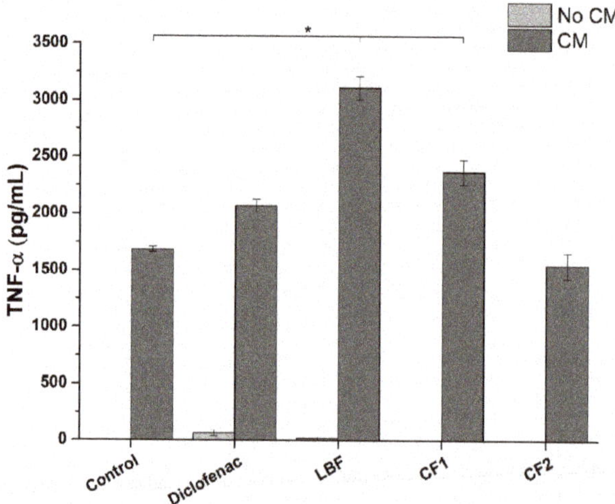

Figure 16. Production profile of TNF-α following treatment of LNCaP prostatic cells with bioavailable fractions: LBF: 22% SRO; CF1: 18.8% SRO; CF2: 3.2% SRO (Table 6). Diclofenac was used at 32 μg/mL. CM: Conditioned medium. * $p < 0.05$ ($n = 3$).

Conversely, to IL-1β, the production of TNF-α significantly increases upon exposure to LBF and CF1 bioaccessible fractions. No effect was observed for CF2. This peculiar trend could be explained considering the pro-apoptotic activity of SRO. Indeed, TNF-α is a cytokine known to be involved in the apoptotic process. Silvestri and colleagues [30] demonstrated that SRO extract induces apoptosis in LNCaP cells. To test this hypothesis, we evaluated the pro-apoptotic effect of SRO absorbable fractions on the prostatic epithelium model in both uninflamed and inflamed conditions (Figure 17).

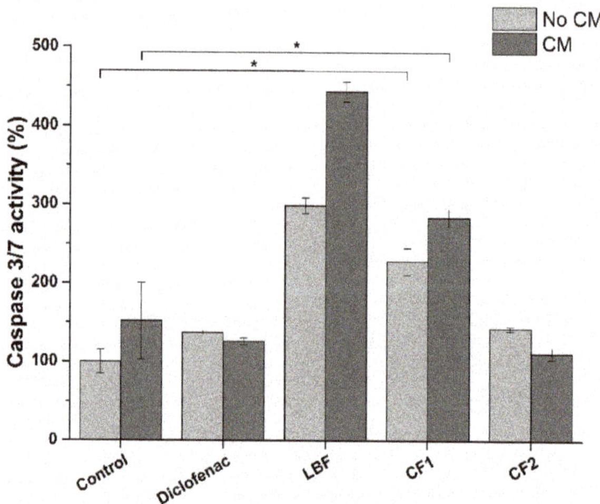

Figure 17. Pro-apoptotic activity of bioavailable fractions on LNCaP prostatic cells: LBF: 22% SRO; CF1: 18.8% SRO; CF2: 3.2% SRO (Table 6). Diclofenac was used at 32 μg/mL. CM: Conditioned medium. Samples are normalized on no CM control. * $p < 0.05$ ($n = 3$).

As expected, pro-apoptotic activity induced by bioaccessible fractions well correlates with TNF-α expression, meaning a role of SRO in inducing apoptosis against inflamed and tumor prostatic cells. The apparent contrast between LNCaP prostatic cells vitality (Figure 14) and apoptosis (Figure 17) results could be explained by an early-apoptosis phenomenon. During early apoptosis phenomenon, indeed, cells retain their vitality. Activation of early apoptosis cascade leads to cell dead at later times, suggesting a decrease in LNCaP prostatic cells viability following prolonged exposure to SRO. While no effects were observed after 6 h incubation (Figure 18A), cell viability significantly decreases after 24 h exposure (Figure 18B).

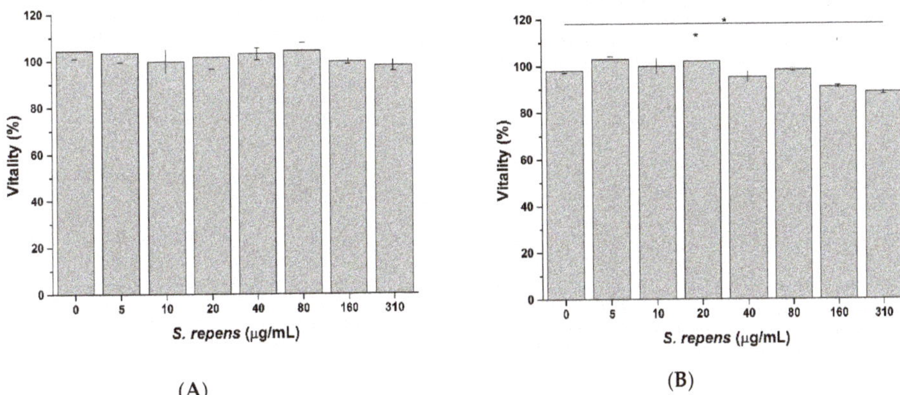

(A)

(B)

Figure 18. Effect on LNCaP prostatic cells vitality after 6 h (**A**) and 24 h (**B**) exposure to SRO. * $p < 0.05$ ($n = 3$).

2.10. Activity of Bioaccessible Fractions on PSA Secretion

Prostate-Specific Antigen (PSA) is considered the main serum marker for the progression of prostate cancer [31]. A decrease in PSA secretion, following treatment with bioaccessible fractions, indicates a potential therapeutic effect of the formulations. To verify the effect of the bioaccessible formulations on PSA secretion, LNCaP hormone-sensitive cell line was used. Compared to control conditions, PSA secretion decreased in cells treated with LBF and CF1 bioaccessible fractions, while no differences were observed in cells treated with CF2 formulation (Figure 19).

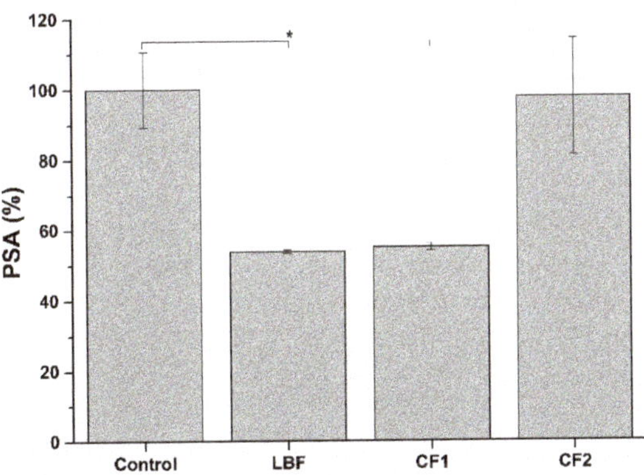

Figure 19. PSA secretion in LNCaP prostatic cells treated with bio-accessible fraction of the different formulations. * $p < 0.05$ ($n = 3$).

2.11. Smooth Muscle Myorelaxing Activity

The decrease in epithelial-to-stromatic tissue ratio is a well-known marker of benign prostatic hypeplasia (BPH) development. In the prostate, the stromal component is mainly composed of smooth muscle tissue, which is normally contracted in response to adrenergic stimulation. As a result, urethra lumen is reduced and urination made difficult. To explore the potential effect of tested formulations on this BPH symptom, the myorelaxing activity of the bioaccessible fractions on smooth muscles was evaluated using WPMY-1 myofibroblast in vitro model.

As shown in Figure 20, no muscle relaxation was induced by CF2 bioaccessible fraction. Conversely, the bioaccessible fractions of LBF and CF1 showed a significant myorelaxing activity on smooth muscles, with LBF presenting the highest myorelaxing activity.

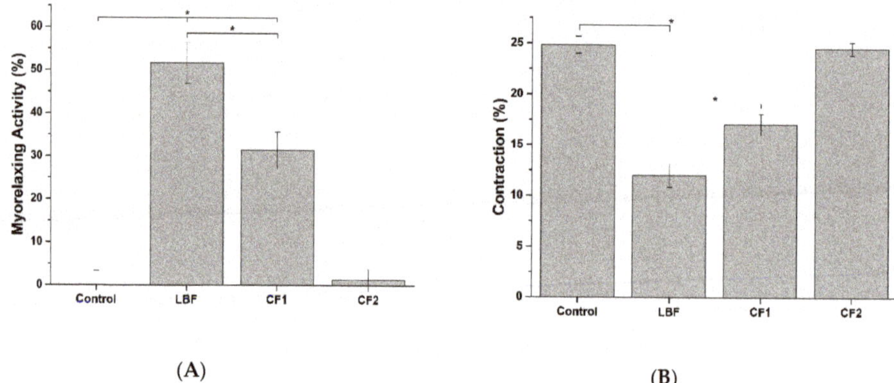

(A)   **(B)**

Figure 20. Myorelaxing (**A**) and contraction inhibitory (**B**) activity of the bioaccessible fraction of the different formulations on an in vitro model of smooth muscle. * $p < 0.05$ ($n = 3$).

3. Materials and Methods

3.1. Materials

Serenoa repens oil 85% fatty acids GC was purchased from Naturex S.p.a. (Caronno Pertusella, VA, Italy). Ascorbyl palmitate and lecithin were purchased from A.C.E.F. (Fiorenzuola D'Arda, PC, Italy). Mono- and diglycerides of fatty acids were from BASF Italia S.p.a. (Cesano Maderno, MB, Italy). Mannitol, synthetic amorphous silica, magnesium stearate and sodium bicarbonate were purchased from Giusto Faravelli S.p.a. (Milano, Italy). Sodium chloride was purchased from Fagron Italia S.r.l. (Quarto Inferiore, BO, Italy). Thaurocholic acid was purchased from Shangai T and W Pharmaceuticals Co. (Shangai, China), maleic acid and lipase from porcine pancreas were purchased from Sigma-Aldrich (St Louis, MO, USA), Flowens™ and lycopene (powder titrated at min. 6% w/w) were purchased from Naturex.

Caco-2 human colon adenocarcinoma cell line (ATCC® HTB-37™), LNCaP androgen-sensitive human prostate adenocarcinoma cell line (ATCC® CRL-1740™), WPMY-1 human myofibroblast stromal cell line (ATCC® CRL-2854™) and THP-1 (ATCC® TIB-202™) were purchased from ATCC (Manassas, VA, USA). High glucose Dulbecco's Modified Eagle Medium (DMEM), Roswell Park Memorial Institute (RPMI) 1640 Medium, Hanks' Balanced Salt Saline (HBSS), non-essential amino acids (NEAA), L-glutamine, penicillin-streptomycin mix, lipolysaccharide (LPS), diclofenac, dihydrotestosterone (DHT), phorbol 12-myristate 13-acetate (PMA), proantocyanidine A, B1, and B2 standards and Lucifer Yellow (LY) were purchased from Sigma-Aldrich (St Louis, MO, USA). Foetal bovine serum (FBS) was purchased from Euroclone (Milan, IT). Interleukin 1β (IL-1β), Tumor Necrosis Factor α (TNF-α) and prostate-specific antigen (PSA) ELISA kit were purchased was purchased from R and D Systems, PeproTech (London, UK) and Abcam (Cambridge, UK), respectively. Cell contraction assay was purchased from Cell Biolabs (San Diego, CA, USA). Transwell® insert were purchased from Millipore (Burlington, MA, USA). CellTiter 96® AQueous One Solution Cell Proliferation Assay (MTS) and Apo-ONE® Homogeneous Caspase-3/7 Assay were purchased from Promega (Madison, WI, USA). C18 cromatographic columns were purchased from Agilent (Santa Clara, CA, USA). Cyclo-oxygenase (COX) activity assay kit was purchased from Cayman Chemicals (Ann Arbor, MI, USA).

3.2. Methods

3.2.1. Preparation of Lipomatrix Powder

3.2.1.1. Preparation of Lipomatrix Powder with SRO, Flowens, and Lycopene

MDGFA and ASP were dry mixed in a beaker and the mixture was melted at temperature of 80 °C under mechanical stirring (IKA® RTC Basic Staufen, Germany). The resulting oily liquid phase was added of SRO 85% fatty acids GC and Lycopene (powder titrated at min 6% w/w), maintaining the temperature at 75 °C. Mannitol and Flowens™, previously mixed and cooled down in fridge at 15 °C, were put in a planetary mixer (Kenwood KMX750RD, De Longhi Treviso, Italia, Kenwood group) and the hot oily phase was added on it in small continuous additions under mixing so to induce an instant and uniform solidification of the lipid phase onto the cold powder. The formed composite powder was cooled down at room temperature for 24 h. Finally, the granules were passed through a 1.5 mm mesh net connected to the mechanical sieve (Vasquali, Marchesini group, Cerro Maggiore (MI), ITALY). The resulting powder was added of synthetic amorphous silica and magnesium stearate to improve powder flowability. Type 0 animal gelatin capsules were filled with the powder using a manual encapsulator (MultiGel) so that the content of any capsule corresponded to 160 mg of SRO, 125 mg of Flowens and 5 mg of lycopene (powder titrated at 6% w/w).

3.2.1.2. Preparation of Lipomatrix Powder with SRO Alone

The same method of Section 3.2.1.1. in which no Flowens™ and lycopene powder have been introduced.

3.2.2. Flow Property

Flow through an orifice was measured according to Ph. Eur. Chapter 2.9.36. The test was performed using a metal truncated cone Flowability Tester (Flowability Tester BEP Auto Copley Scientific, Colwick, Nottingham) with different size of the orifice. The time it took for 100 g of powder to pass through 25–15–10 mm diameter orifices was measured in triplicate.

3.2.3. In Vitro Emulsification Test

In vitro emulsification in FaSSIF-V2 was performed as follows: a 100 mL glass beaker was filled with 50 mL of simulated gastric fluid (GSF, according to Eur. Pharmcopea) and placed at 37 °C ± 0.5 °C (IKA® RTC Basic). FaSSIF-V2 was chosen as model of enteric fluid to assess the real capability of Lipomatrix to create soluble micellar forms in the duodenum without any interfering molecules coming from ingested foods such as higher Lecithin concentration than FaSSIF-V2 and oleic acid (OA) esters and salts (FeSSIF-V2) [32]. GSF was realized from demineralized water (inverse osmosis process) added of dilute solution of HCl up to pH = 1. The quantity of Lipomatrix powder corresponding to 4 capsules as described in Section 3.2.1.2. containing a total amount of 640 mg of SRO, has been dispersed in the afore described GSF at 37 °C under moderate magnetic stirring (200 rpm, IKA® RTC Basic) for 90 min simulating gastric transit time. After that, the suspended Lipomatrix powder was decanted, the GSF was removed and replaced with 50 mL of FaSSIF-V2. FaSSIF-V2, composition was the following: sodium chloride 68.62 mM, thaurocholic acid 3 mM, lecithin 0.2 mM, maleic acid 19.12 mM, lipase from porcine pancreas 100 units/mL, pH 7.20 adjusted with sodium bicarbonate.

We preferred to use sodium bicarbonate instead of phosphate buffer, as described in the consolidated FaSSIF-V2 model [33], to simulate the secretion of sodium bicarbonate by pancreas [34]. This model showed furthermore to be useful to verify the pH lowering behavior of Lipomatrix ascribable to ASP ionization, so confirming the postulated mechanism of action. The temperature of the media was set at 37 ± 0.5 °C by means of a thermostatic probe. The powder was maintained in the dissolution medium for 60 min under magnetic stirring (200 rpm, IKA® RTC Basic). During the test, pH probe was introduced in the FaSSIF-V2 dispersion and pH was monitored continuously (Sension + PH3, Hach). Since ASP embedded in Lipoamtrix ionizes in FaSSIF-V2 (pKa < pH), ionization leads to a pH reduction so each 15 min pH was adjusted with sodium bicarbonate to maintain pH not less than 7.00 and simulate the continuous pancreatic secretion of sodium bicarbonate. The test has been assessed with 50 mL of both GSF and FaSSIF-V2 to set the models close to the average volumes, of physiologic fasted condition of gastro-enteric tract in humans [35,36].

3.2.4. In Vitro Assessment of Lipomatrix Containing SRO Gastric Resistance (the Test on Lipomatrix Specimens was Performed in Triplicate)

A Semi-Automatic Disintegration Tester (Charles Ischi AG DISI-M) was used to simulate gastric transit of Lipomatix powder as described in Section 3.2.1.2. The dissolution medium was 800 mL of GSF at 37 °C (according to Eur. Ph). Three aliquots of Lipomatrix powder as described in Section 3.2.1.2 containing a total amount of 640 mg of SRO was maintained for 90 min in GSF anyone placed in a single chamber of the mentioned device.

3.2.5. Analytical Characterization

3.2.5.1. Differential Scanning Calorimetry (DSC) Analysis

DSC scans were performed by a Mettler-Toledo DSC calorimeter using the following analysis conditions: 25.00–300.00 °C, 20K/min and 25.00–125.00 °C, 10K/min. The Lipomatrix powder with SRO alone (as described in Section 3.2.1.2) has been processed and each single excipient of which it is composed as well. The main scope of this analysis is to understand the thermal behavior of Lipomatrix powder and if any change in the structure occurs during fusion of the fatty phase and its further rapid cooling during absorption onto the polyol-based powder. The thermal exchanges profile of any component and of the final powder have been determined.

3.2.5.2. Gas Chromatography-Mass Spectrometry (GC-MS) Analysis on Gastric-Resistance Test Samples

To proceed with GC-MS assay, 5 mL of liquid coming from gastric resistance test chamber were added of 100 µL of metyl pentadecanoate solution (115 µg/mL) as internal standard. Extraction with ethyl acetate was performed three times. The obtained organic phases were pooled together, anhydrified with sodium sulfate and led to dryness. The extracts were added of 3 mL of methanol and 1 mL of dichloromethane and were acidified with concentrated sulfuric acid. Samples were heated under reflux for 40 min, then the solution were cooled down and diluted with water and diethyl ether. The etheric phase was investigated.

The standard was prepared mixing 100 mg of SRO fatty acids GC with 1 mL of metyl pentadecanoate solution (115 µg/mL) as internal standard, and adding 2 mL of dichloromethane, 10 mL of methanol and 0.2 mL of concentrated sulfuric acid. Standard sample was heated under reflux, then cooled down and diluted with water and diethyl ether. The etheric phase was investigated.

The GC-MS system was comprised of a Varian 3800 equipped with autosampler and coupled with a Varian Saturn 2100 MS/MS ion trap mass spectrometer. A HP-88 column (60 m × 0.25 mm) was used for separation (J and W Scientific, Agilent technologies Inc., Santa Clara, CA 95051, USA)

3.2.5.3. Dynamic Light Scattering (DLS) Analysis of Lipomatrix Dispersion in FaSSIF-V2

DLS was performed on emulsified dispersion of Lipomatrix in *FaSSIF-V2* as described in Section 3.2.1.2, corresponding to 640 mg of SRO resulting from in vitro emulsification test described in Section 3.2.3. The size distribution of the particles was evaluated as a signal intensity function only. The conversion of the signal intensity distribution into particles volume or number distribution can cause error propagation since it requires some unavailable parameters (e.g., the particles refractive index). The liquid samples (FaSSIF-V2 and Lipomatrix dispersion in FaSSIF-V2) were analyzed by DLS (Zetasizer Nano S, Malvern).

3.2.6. Cell Cultures

3.2.6.1. Caco-2 Cell Culture

The Caco-2 human colon adenocarcinoma cells (passage 32 to 42) were seeded in adhesion flask at a density of 2×10^3 cell/cm^2 in Caco-2 Complete Medium (CCM) (high glucose DMEM, 10% heat inactivated FBS (FBS), 1% non-essential amino acids, 4 mM L-glutamine, and 1% penicillin-streptomycin mix) and cultured at 37 °C and 5% CO$_2$ in a humidified incubator. Cell were seeded at 2000 cells/cm^2 and subcultivated by tryspinization every 7 d when 80–90% confluent. The medium was refreshed every other day.

3.2.6.2. LNCaP Cell Culture

The human prostate cancer cell line LNCaP cells (passage 25 to 40) were maintained in RPMI-1640 medium supplemented with 10% FBS and 1% penicillin-streptomycin mix. The cells were grown at 37 °C in a humidified atmosphere with 5% CO$_2$. Cell were seeded at 10,000 cell/cm^2 and medium changes every other day. As for Caco-2, cells were subcultivated by tryspinization every 7 d when 80–90% confluent. For vitality and anti-inflammatory experiments, LNCaP were seeded 96-well plates and six-well plates, respectively, at a density of 1×10^5 cell/cm^2 and allowed to adhere for two days prior to experiments, while for prostate-specific antigen (PSA) experiment cells were seeded at a density of 50,000 cells/cm^2 in 24-well plates.

3.2.6.3. THP-1 Cell Culture

Human THP-1 monocytes (passage was maintained in RPMI-1640 medium with glutamate supplemented with 10% FBS, 100 U/mL penicillin and 100 µg/mL streptomycin (GIBCO, Winsford CW7 3GA, UK). Cells were cultured at a density of 5×10^5 cells/mL in 5% CO$_2$ humidified atmosphere at 37 °C and subcultured twice a week. Macrophage differentiation was induced by incubation with 500 nM phorbol myristate acetate (PMA; Sigma-Aldrich, MO, USA) for 24 h. Culture medium was then replaced and cells cultured for an additional 24 h. For medium conditioning, 6×10^6 cell were seeded in 75 cm^2 flask, differentiate into macrophages as described before and treated with 1 ng/mL LPS for 6 h. At the end of the LPS treatment, medium was recovered and stored at −80 °C until use.

3.2.6.4. WPMY-1 Cell Culture

Human prostate stromal (myofibroblast) (WPMY-1) (passage 40 to 50) were cultured in DMEM, supplemented with 10 % fetal bovine serum (FBS) and 1 % penicillin-streptomycin mix. As for Caco-2, cells were seeded at 2000 cells/cm^2 and subcultivated by tryspinization every 7 d when 80–90% confluent. The medium was refreshed every other day. For cell contraction experiments, 1×10^6 cells were loaded into each collagen gel.

3.2.7. Determination of Active Principle Absorption Rate

To evaluate the effectiveness of Lipomatrix technology in increasing the enteric absorption rate of lipophilic active principles, we compared the performance of a Lipomatrix-based formulation (LBF) as described in *3.2.1.1* against two commercial formulations, as indicated in Table 7.

Table 7. List and composition of analyzed formulations.

Formulation	Active Composition	mg/capsule
Lipomatrix-based formulation (LBF)	*S. repens* (Bartram) small oil (85% fatty acids, GC)	160
	Flowens™	125
	Lycopene (6% minimum)	5
Commercial formulation 1 (CF1)	*S. repens* (Bartram) small oil extract from fruit	320
Commercial formulation 2 (CF2)	*S. repens* (Bartram) small oil extract from fruit (88% fatty acids)	320
	Pinus massoniana L. dry extract from bark (95% proanthocyanidin)	120
	Crocus sativus L. dry extract from stigmas (0.3% safranal)	100

3.2.7.1. Digestion Process

A single dose of each formulation listed in Table 7 was exposed to in vitro digestion process simulating the physiological human digestion in the oral, gastric and intestinal compartments. Briefly, the formulations were incubated for 5 min in saliva at 37 $\pm$ 1 °C, rotating head-over-heels at 55 rpm, simulating peristaltic movements. Subsequently, gastric juice (pH 1.3 $\pm$ 0.1) was added to the mixture and the pH of the sample was checked and, if necessary, adjusted to 2.5 $\pm$ 0.5 with NaOH (1 M) or HCl (37% w/w). The sample was further incubated rotating at 37 °C for 2 h. Subsequently, duodenal juice (pH 8.1 $\pm$ 0.1), bile (pH 8.2 $\pm$ 0.1) and sodium bicarbonate were added. The pH of this mixture was set at 6.5 $\pm$ 0.5 with NaOH (1 M) or HCl (37%) and it was rotated head-over-heels for another 2 h. For simulated digestive fluids composition refer to Walczak et al., 2013. Once completed the digestion process, SRO bioaccessibility was determined by measuring fatty acids with GC/MS, while proanthocyanidin (PAC) and lycopene loaded in Lipomatrix formulation were determined by high pressure liquid chromatography (HPLC).

3.2.7.2. HPLC Analysis of LYC and PCA

Bioaccessible fractions of LYC and PAC, released during the digestive process and absorbed at the intestinal epithelium level, were determined by HPLC. Reversed phase HPLC with absorbance detection (at 475 nm) based on the modified method of Thadikamala et al. (2009) was used for analyzing LYC. HPLC separation was carried out using a Varian Prostar 210 pump system (Agilent) and a UV–VIS detector (Variant Prostar, Agilent), operated at 25 °C. A C18 column (4.6 × 150 mm, Agilent) was used with methanol-acetonitrile-methanol-tetrahydrofuran (THF) (70:25:5, v/v) as an isocratic eluent. Each sample was dissolved first in methanol-THF (50:50, $\%v/v$) and diluted as needed in the same solvent. Quantification of the eluted LYC was accomplished by the peak area method using the calibration range of 15 to 150 ppm of LYC (Sigma-Aldrich, Milan, Italy) as external standards. For PACs, HPLC analysis was performed with Variant Prostar HPLC (Agilent) with the same pump system and UV–VIS detector described above. The phenolic compounds were detected at 280 nm with a flow rate of 1 mL/min. The column was operated at a temperature of 25 °C. Separations were carried out with a C18 column (Agilent) in a dual pumping system by varying the proportion of 2.5% (v/v) acetic acid in water (mobile phase A) and 70% methanol in water (mobile phase B). The solvent gradient elution program was as follows: 10% to 26% B (*v/v*) in 10 min, to 70% B at 20 min and finally to 90% B at 25 to 31 min. The injection volume for all samples was 100 µL. The phenolic compounds were analyzed by matching the retention time and their spectral characteristics against those of standards PAC A, B1, or B2 (20 to 400 ppm).

3.2.8. In Vitro Model of Human Intestinal Epithelium

Permeability and absorption rate of SRO (i.e., fatty acids), PAC and LYC were determined using an in vitro model of human intestinal epithelium based on Caco-2 cells. Briefly, Caco-2 cells are seeded on Transwell® polytetrafluoroethylene inserts (1 µm pore size) at an initial density of 1.5×10^5 cells/cm^2 and allowed to mature and differentiate for 21 days. Indeed, thanks to the compartmentalized nature of the Transwell® system (apical (or lumen) and basolateral (or serosal) compartment), Caco-2 cells differentiate, acquiring morphological and functional feature typical of enterocytes, as the presence of microvilli, tight junctions and P-glycoprotein. Absorption experiment were performed between 21 and 28 days post seeding.

3.2.8.1. Evaluation of the Impact of Digested Formulations on Intestinal Epithelium Viability

To evaluate the impact of digested formulations on intestinal epithelium viability, digested formulations were serially diluted in digestive fluids (from 1:2 up to 1:32 dilution) and added to the apical side of the in vitro intestinal epithelia, while HBSS buffer was placed in the basolateral compartment. Digestive fluids (without formulations) were added to the apical side of the in vitro intestinal epithelia as a negative control. After 3 h incubation, monolayers were washed twice with pre-warmed HBSS and viability of intestinal epithelia was evaluated with MTS assay, according to manufacturer's instructions. This assay is based on MTS tetrazolium compound reduction by viable cells to generate a colored formazan product that can be quantified by measuring the absorbance at 490 nm. The color intensity at 490 nm was determined with a microplate reader (Synergy4, Biotek, Colonia Santo Domingo Azcapotzalco Distrito Federal, México). Cell viability (%) was expressed as the ratio of the color intensity in the treated groups to that in the control (untreated) group. Absorption rate experiments were performed using non-toxic concentrations determined by dose-response curves.

3.2.8.2. Evaluation of SRO (Fatty Acids), PCA, and LYC Enteric Absorption Rate

Based on dose-response curve information and their posology, digested formulations were added to the apical side of the in vitro intestinal epithelium, while HBSS buffer supplemented with 1% BSA was placed in the basolateral compartment. Due to the lipophilicity of formulation active components, 1% BSA was added to the basolateral compartment for improving their absorption rate. According to the literature (Fossati et al., 2008) [37] the addition of BSA improves the correlation between absorption occurring in Caco-2 cell monolayer and humans. After 3 h incubation, apical and basolateral solutions were collected and fatty acids, proanthocyanidin and lycopene content was determined by GC/MS and HPLC respectively. Absorption rate of SRO fatty acids, PCA and LYC is expressed as percentage of absorption, derived from three independent experiments.

3.2.8.3. Barrier Integrity and Cell Viability

After exposure to digested formulations, cell viability and barrier integrity of the intestinal epithelium model were evaluated. Briefly, at the end of the incubation with the digested formulations, the in vitro intestinal epithelia were washed twice with pre-warmed HBSS and equilibrated in the same buffer for 30 min. Once equilibrated, epithelia barrier integrity was evaluated by measuring the trans-epithelial electrical resistance (TEER) of the cell monolayer with an ERS2 Voltohmmeter (Millipore), equipped with a chopstick electrode. Intestinal epithelium model paracellular permeability was determined with Lucifer Yellow (LY), a fluorescent polar tracer unable to pass through intact tight junctions. Paracellular permeability was measured by adding 0.5 mL of 100 µg/mL LY in HBSS in the apical compartment and 1.5 mL of HBSS in the basolateral compartment. After 1-h incubation, the basolateral fractions were collected and their fluorescence measured with a spectrofluorometer (Synergy 4, Biotek). Apparent permeability coefficient (P_{app}, cm/s) was calculated with the following formula:

$$P_{app} = (\Delta C \cdot V)/(\Delta t \cdot A \cdot C_0) \tag{1}$$

where $\Delta C/\Delta t$ is the flow of the molecule being transported across the monolayer during the incubation time (mM/s), V is the volume of the basolateral compartment (cm^3), A is the area of the membrane (cm^2), C_0 is the initial concentration of the molecule in the apical compartment. Finally, cell viability was evaluated by using MTS assay as described above.

3.2.9. Prostate-Specific Anti-Inflammatory Activity

The prostate-specific anti-inflammatory activity of the bioaccessible fraction of tested formulations was evaluated in a prostatic epithelium in vitro model, based on tumoral prostatic cells (LNCaP). The prostate-specific anti-inflammatory activity was evaluated pre-treating the in vitro model for 2 h with the bioaccessible fractions corresponding to SRO concentrations reported in Table 8, and then exposing the model to inflamed conditions for 4 h. In particular, prostatic epithelium in vitro model was exposed either to normal monocytic/macrophage cell culture medium (uninflamed condition) or THP-1 cell culture conditioned medium (CM, inflamed condition).

Table 8. Absorption rate expressed as a percentage of SRO absorption $\pm$ standard error (SE) and concentration (μg/mL) at the basolateral compartment (serosal).

Formulation	*S. Repens* Absorption Rate (% $\pm$ SE)	*S. Repens* Concentration in the Serosal Compartment (μg/mL)
Lipomatrix	22.0 $\pm$ 12.7	300
CF1	18.8 $\pm$ 12.2	270
CF2	3.2 $\pm$ 3.2	50

CM was obtained by stimulating overnight PMA-differentiated THP-1 cells with the pro-inflammatory compound LPS (1 ng/mL). LNCaP cells in normal monocytic/macrophage cell culture medium and CM were used as negative and positive controls of inflammation respectively. Diclofenac, a well-known anti-inflammatory drug, was used as a positive control of anti-inflammatory activity. A dose-response curve was performed with MTS assay to determine Diclofenac highest non-lethal concentration on prostatic epithelium in vitro model after 6 h exposure. Similarly, the impact of bioaccessible fractions on LNCaP viability was checked by using MTS assay and morphology. At the end of the treatment, in vitro prostate epithelia were washed with pre-warmed DPBS and detached in ice-cold PBS with a cell scraper. Following centrifugation (1000$\times$ *g* for 5 min), LNCaP cells were resuspendend in lysis buffer (protease inhibitor cocktail and 0.1% Triton X-100 in deionized water) and sonicated (5 s pulse-on at 10% amplitude and 25 sec pulse-off, total time 1.5 min) (Sonicator Q700, QSonica, Newtown, CT, USA). Finally, cell lysates were centrifuged for 15 min at 10,000$\times$ *g* and the surnatants recovered. Anti-inflammatory activity was determined by measuring the amount of pro-inflammatory cytokines interleukin-1beta (IL-1β) and tumor necrosis factor-alpha (TNF-α) present in cell lysates. IL-1β and TNF-α were quantified by commercial ELISA (Enzyme-Linked Immunosorbent Assay) kits, following the manufacturer's instructions.

Anti-inflammatory effect of bioaccessible fractions was also evaluated in cell lysate by measuring cyclooxygenase (COX) enzyme 1 and 2 activity. Total COX and COX-2 activities were assessed with a commercial COX activity assay kit, following the manufacturer's instructions.

3.2.9.1. SRO Pro-Apoptotic Activity

To evaluate the pro-apoptotic activity of the bioaccessible fractions of SRO, a fluorimetric assay, based on caspases 3/7 activation, was performed (Apo-ONE® Homogeneous Caspase-3/7 Assay).

This assay is based on the ability of activated caspases 3 and 7 to selectively cleave a specific substrate, making it fluorescent (excitation wavelenght 499 nm, emission wavelength 521 nm). Consequently, the produced fluorescence intensity is linked to the activation of the apoptotic process by cells. To correlate anti-inflammatory with pro-apoptotic activity, the same experimental setup described above for the anti-inflammatory activity was applied. Experiments were performed in triplicate and the assay conducted following manufacturer's instructions.

3.2.9.2. Smooth Muscle Myorelaxing Activity

The myorelaxing activity of the bioaccessible fraction of tested formulations was evaluated in an in vitro model of smooth muscles, based on WPMY-1 human myofibroblast stromal cell. The myorelaxing effect was analyzed by means of a two-step gel-contraction assay, following the manufacturer's instructions. The gel contraction assay is characterized by two phases: mechanical stress-generating phase and floating phase. During the first phase, a mix of myofibroblast and collagen is seeded and left to develop mechanical stress for two days while, during the floating phase, gels are released and allowed to freely contract, dissipating the mechanical stress generated during the first phase. Fibroblast-containing gels were exposed to the bioaccessible fraction of the formulations for the duration of the experiment (24 h). The myorelaxation activity was calculated with the following formula:

$$\text{Myorelaxing Activity } (\%) = 100 - ((C_{exposed}/C_{control}) \times 100) \tag{2}$$

where $C_{fibroblast}$ is the contraction of the fibroblast-containing gels exposed to the bioaccessible fraction of the different formulations and $C_{control}$ the contraction of the untreated fibroblast-containing gel (positive control of contraction).

Gel contraction is calculated as follows:

$$\text{Contraction } (\%) = 100 - ((D_{fibroblast}/D_{collagen}) \times 100) \tag{3}$$

where $D_{fibroblast}$ represents the diameter of the fibroblast-containing gel and $D_{collagen}$ the diameter of the collagen gel in which fibroblast were not seeded (negative control of contraction). Experimental and control cells were plated in triplicate and the diameter of each collagen gel was photographed and measurement via images analysis, performed with ImageJ (University of Wisconsin-Madison, Madison, WI, USA).

3.2.9.3. Measurement of Prostate Specific Antigen (PSA) Secretion by LNCaP Prostatic Cells

The effect of the different formulation bioaccessible fraction on the secretion of the androgen-induced prostate-specific antigen (PSA), commonly used as a marker for prostate tumors, was evaluated in LNCaP prostatic cells. After seeding and adhesion (48 h), normal cell culture medium was replaced with a medium containing low hormone level for 24 h (10% charcoal-stripped FBS and 1% Penicillin-Streptomycin mix in RPMI without phenol red). The pre-treatment medium was then removed and cells were incubated with the bioaccessible fraction of each formulation. A control was included in the assay by treating cells with low-hormones cell culture medium. Additionally, control and cells treated with LBF bioaccessible fraction were stimulated with dihydrotestosterone (DHT) (10 nM), known to stimulate the release of PSA [38]. After 24 h exposure, media were collected for measurement of secreted PSA using a commercial ELISA kit (Abcam), following the manufacturer's instructions. Results were expressed as percentage of secreted PSA in cells treated with different formulations compared to control.

3.3. Statistical Analysis

Results from performed experiments were statistically analyzed using OriginLab (OriginLab Corporation, Northampton, MA, USA) software. Experiments were performed in triplicate, results presented as average ± standard deviation. A *p* value of ≤0.05 was considered significant.

4. Conclusions

In the present paper, we demonstrated the successful application of a new technological lipid-based delivery matrix (Lipomatrix) able to protect lipophilic plant-derived active principles from the gastric compartment and favoring the formation of hydro-dispersible forms in the duodenum, thanks to the pH-dependent ionization of ASP. Lipomatrix-based formulation, despite the not improved apparent bioaccessibility data, seems to enhance overall enteric absorption of SRO in a CACO-2 cells model and this data well correlates with an overall improvement of its biologic effects on in vitro prostatic tissue model, compared to not formulated commercial SRO. Indeed, we showed that the Lipomatrix-associated fraction of SRO reduced inflammation and modulated cytokines pattern expression, enhanced pro-apoptotic caspases-mediated activity and improved myo-relaxation significantly better than not formulated SRO contained in the mentioned commercial products. Lipomatrix-associated SRO shoved furthermore to significantly reduce PSA expression in *LNCaP* prostatic cells but not better than one of the two SRO-containing commercial formulations tested. Moreover, even though as preliminary evidence, SRO formulated in Lipomatrix seems to ensure a higher safety profile on in vitro intestinal model in terms of overall cell vitality, compared to the tested commercial formulations.

To conclude, taken together the data collected in this work, it is possible to consider Lipomatrix as a promising, next generation technological platform for improving enteric delivery of SRO and as consequence its efficacy on human prostatic gland, so confirming the potential significant role of delivery systems in bioavailability of natural lipophilic compounds.

Supplementary Materials: Supplementary materials can be found at http://www.mdpi.com/1422-0067/20/3/669/s1.

Author Contributions: Conceptualization, A.F.; methodology, A.F., V.M., F.B., E.T., E.M.; software, F.B., E.T.; validation, V.M., F.B.; formal analysis, M.P.; investigation, A.F., V.M., F.B., E.T., E.M.; resources, S.V.; data curation V.M., F.B., E.T.; writing—original draft preparation, A.F., V.M., F.B., E.T.; writing—review and editing A.F., V.M., F.B., E.T., M.P.; visualization, A.F., M.P.; supervision A.F.G.C., F.B.; project administration, S.V.

Funding: This research received no external funding.

Acknowledgments: This work was mainly funded and supported by PHF SA, Lugano, Swiss. We are grateful to Prof. Stefano Dall'Acqua, Prof. Alessandra Semenzato, Dr. Marta Faggian, Dr. Elisa Barbieri, and Dr. Alessia Costantini, UNIRED, Department of Pharmaceutical and Pharmacological Sciences of Padova University, for the valuable and competent analytical support provided during the described experiments.

Conflicts of Interest: A.F., V.M., and M.P. are employees of LABOMAR SRL, the company owner of the patented Lipomatrix technology.

References

1. Köhler, A.; Sarkkinen, E.; Tapola, N.; Niskanen, T.; Bruheim, I. Bioavailability of fatty acids from krill oil, krill meal and fish oil in healthy subjects—A randomized, single-dose, cross-over trial. *Lipids Health Dis.* **2015**, *14*, 19. [CrossRef] [PubMed]

2. Raatz, S.K.; Johnson, L.K.; Bukowski, M.R. Enhanced Bioavailability of EPA From Emulsified Fish Oil Preparations Versus Capsular Triacylglycerol. *Lipids* **2016**, *51*, 643–651. [PubMed]

3. Murgia, X.; Loretz, B.; Hartwig, O.; Hittinger, M.; Lehr, C.M. The role of mucus on drug transport and its potential to affect therapeutic outcomes. *Adv. Drug Deliv. Rev.* **2018**, *124*, 82–97. [CrossRef] [PubMed]

4. Boegh, M.; Nielsen, H.M. Mucus as a barrier to drug delivery—Understanding and mimicking the barrier properties. *Basic Clin. Pharmacol. Toxicol.* **2015**, *116*, 179–186. [CrossRef] [PubMed]

5. Sigurdsson, H.H.; Kirch, J.; Lehr, C.M. Mucus as a barrier to lipophilic drugs. *Int. J. Pharm.* **2013**, *453*, 56–64. [CrossRef] [PubMed]

6. Rein, M.J.; Renouf, M.; Cruz-Hernandez, C.; Actis-Goretta, L.; Thakkar, S.K.; da Silva Pinto, M. Bioavailability of bioactive food compounds: A challenging journey to bioefficacy. *Br. J. Clin. Pharmacol.* **2013**, *75*, 588–602. [CrossRef] [PubMed]

7. Wang, Q.; Wei, Q.; Yang, Q.; Cao, X.; Li, Q.; Shi, F.; Tong, S.S.; Feng, C.; Yu, Q.; Yu, J.; et al. A novel formulation of [5]-gingerol: Proliposomes with enhanced oral bioavailability and antitumor effect. *Int. J. Pharm.* **2017**, *535*, 308–315. [CrossRef]

8. Abdel-Tawab, M.; Werz, O.; Schubert-Zsilavecz, M. Boswellia serrata: An overall assessment of *in vitro*, preclinical, pharmacokinetic and clinical data. *Clin. Pharmacokinet.* **2011**, *50*, 349–369. [CrossRef]

9. Hüsch, J.; Bohnet, J.; Fricker, G.; Skarke, C.; Artaria, C.; Appendino, G.; Schubert-Zsilavecz, M.; Abdel-Tawab, M. Enhanced absorption of boswellic acids by a lecithin delivery form (Phytosome®) of Boswellia extract. *Fitoterapia* **2013**, *84*, 89–98. [CrossRef] [PubMed]

10. Mirzaei, H.; Shakeri, A.; Rashidi, B.; Jalili, A.; Banikazemi, Z.; Sahebkar, A. Phytosomal curcumin: A review of pharmacokinetic, experimental and clinical studies. *Biomed. Pharmacother.* **2017**, *85*, 102–112. [CrossRef]

11. Khan, M.S.; Krishnaraj, K. Phospholipids: A novel adjuvant in herbal drug delivery systems. *Crit. Rev. Ther. Drug Carrier Syst.* **2014**, *31*, 407–428. [CrossRef] [PubMed]

12. Latil, A.; Pétrissans, M.T.; Rouquet, J.; Robert, G.; de la Taille, A. Effects of hexanic extract of *serenoa repens* (permixon® 160 mg) on inflammation biomarkers in the treatment of lower urinary tract symptoms related to benign prostatic hyperplasia. *Prostate* **2015**, *75*, 1857–1867. [CrossRef] [PubMed]

13. De Monte, C.; Carradori, S.; Granese, A.; Di Pierro, G.B.; Leonardo, C.; De Nunzio, C. Modern extraction techniques and their impact on the pharmacological profile of *Serenoa repens* extracts for the treatment of lower urinary tract symptoms. *BMC Urol.* **2014**, *14*, 63. [CrossRef] [PubMed]

14. Novara, G.; Giannarini, G.; Alcaraz, A.; Cózar-Olmo, J.M.; Descazeaud, A.; Montorsi, F.; Ficarra, V. Efficacy and Safety of Hexanic Lipidosterolic Extract of *Serenoa repens* (Permixon) in the Treatment of Lower Urinary Tract Symptoms Due to Benign Prostatic Hyperplasia: Systematic Review and Meta-analysis of Randomized Controlled Trials. *Eur. Urol. Focus* **2016**, *2*, 553–561. [CrossRef] [PubMed]

15. Vela-Navarrete, R.; Alcaraz, A.; Rodríguez-Antolín, A.; Miñana López, B.; Fernández-Gómez, J.M.; Angulo, J.C.; Díaz, D.C.; Romero-Otero, J.; Brenes, F.J.; Carballido, J.; et al. Efficacy and safety of a hexanic extract of Serenoa repens (Permixon®) for the treatment of lower urinary tract symptoms associated with benign prostatic hyperplasia (LUTS/BPH): Systematic review and meta-analysis of randomised controlled trials and observational studies. *BJU Int.* **2018**, *122*, 1049–1065. [PubMed]

16. Chevalier, G.; Bernard, P.; Cousse, H.; Bengone, T. Distribution study of radioactivity in rats after oral administration of the lipido/sterolic extract of *Serenoa repens* (PermixonTM) supplemented with [I-14C]lauric acid, [I-14C]oleic acid or [4-14C]b-sitosterol. *Eur. J Drug Metab. Pharamacokinet.* **1997**, *22*, 73–83. [CrossRef] [PubMed]

17. De Bernardi di Valserra, M.; Tripodi, A.S.; Contos, S.; Germogli, R. *Serenoa repens* capsules: A bioequivalence study. *Acta Toxicol. Ther.* **1994**, *15*, 21–39.

18. Patton, J.S.; Rigler, M.W.; Liao, T.H.; Hamosh, P.; Hamosh, M. Hydrolysis of triacylglycerol emulsions by lingual lipase. A microscopic study. *Biochim. Biophys. Acta* **1982**, *712*, 400–407. [CrossRef]

19. Rogalska, E.; Ransac, S.; Verger, R. Stereoselectivity of lipases. II. Stereoselective hydrolysis of triglycerides by gastric and pancreatic lipases. *J. Biol. Chem.* **1990**, *265*, 20271–20276.

20. Sams, L.; Paume, J.; Giallo, J.; Carrière, F. Relevant pH and lipase for in vitro models of gastric digestion. *Food Funct.* **2016**, *7*, 30–45. [CrossRef]

21. Bakala-N'Goma, J.C.; Williams, H.D.; Sassene, P.J.; Kleberg, K.; Calderone, M.; Jannin, V.; Igonin, A.; Partheil, A.; Marchaud, D.; Jule, E.; et al. Toward the establishment of standardized in vitro tests for lipid-based formulations. 5. Lipolysis of representative formulations by gastric lipase. *Pharm Res.* **2015**, *32*, 279–287. [CrossRef] [PubMed]

22. Dahan, A.; Hoffman, A. Rationalizing the selection of oral lipid based drug delivery systems by an in vitro dynamic lipolysis model for improved oral bioavailability of poorly water soluble drugs. *J. Control Release* **2008**, *129*, 1. [CrossRef] [PubMed]

23. Pop, F. Chemical stabilization of oils rich in long-chain polyunsaturated fatty acids during storage. *Food Sci. Technol. Int.* **2011**, *17*, 111–117. [CrossRef] [PubMed]

24. Klein, E.; Weber, N. In vitro test for the effectiveness of antioxidants as inhibitors of thiyl radical-induced reactions with unsaturated fatty acids. *J. Agric. Food Chem.* **2001**, *49*, 1224–1227. [CrossRef] [PubMed]

25. Emara, L.H.; Badr, R.M.; Elbary, A.A. Improving the dissolution and bioavailability of nifedipine using solid dispersions and solubilizers. *Drug Dev. Ind. Pharm.* **2002**, *28*, 795–807. [CrossRef] [PubMed]

26. Samy, A.M.; Marzouk, M.A.; Ammar, A.A.; Ahmed, M.K. Enhancement of the dissolution profile of allopurinol by a solid dispersion technique. *Drug Discov. Ther.* **2010**, *4*, 77–84. [PubMed]

27. Zhou, M.; Li, X.; Li, Y.; Yao, Q.E.; Ming, Y.; Li, Z.; Lu, L.; Shi, S. Ascorbyl palmitate-incorporated paclitaxel-loaded composite nanoparticles for synergistic anti-tumoral therapy. *Drug Deliv.* **2017**, *24*, 1230–1242. [CrossRef]

28. Christensen, J.Ø.; Schultz, K.; Mollgaard, B.; Kristensen, H.G.; Mullertz, A. Solubilisation of poorly water-soluble drugs during in vitro lipolysis of medium- and long-chain triacylglycerols. *Eur. J. Pharm. Sci.* **2004**, *23*, 287–296. [CrossRef]

29. Borgström, B. The micellar hypothesis of fat absorption: Must it be revisited? *Scand. J. Gastroenterol.* **1985**, *20*, 389–394. [CrossRef]

30. Silvestri, I.; Cattarino, S.; Aglianò, A.; Nicolazzo, C.; Scarpa, S.; Salciccia, S.; Frati, L.; Gentile, V.; Sciarra, A. Effect of S. repens (Permixon®) on the expression of inflammation-related genes: Analysis in primary cell cultures of human prostate carcinoma. *J. Inflamm.* **2013**, *10*, 11. [CrossRef]

31. Habib, F.K.; Ross, M.; Ho, C.K.; Lyons, V.; Chapman, K.S. Repens (Permixon) inhibits the 5alpha-reductase activity of human prostate cancer cell lines without interfering with PSA expression. *Int. J. Cancer* **2005**, *114*, 190–194. [CrossRef] [PubMed]

32. Gumpula Swapna, C.; Manasa, M.; Subramanian, N.S.; Sandesh, G.; Suprabath, J. A review on composition and preparation of biorelevant media. *J. Pharm. Pharm. Sci.* **2017**, *6*, 437–449.

33. Klein, S. The use of biorelevant dissolution media to forecast the in vivo performance of a drug. *AAPS J.* **2010**, *12*, 397–406. [CrossRef] [PubMed]

34. Sheng, J.J.; McNamara, D.P.; Amidon, G.L. Toward an In Vivo Dissolution Methodology: A Comparison of Phosphate and Bicarbonate Buffers. *Mol Pharm.* **2009**, *6*, 29–39. [CrossRef] [PubMed]

35. Mudie, D.M.; Amidon, G.L.; Amidon, G.E. Physiological parameters for oral delivery and in vitro testing. *Mol Pharm.* **2010**, *7*, 1388–1405. [CrossRef] [PubMed]

36. Mudie, D.M.; Murray, K.; Hoad, C.L.; Pritchard, S.E.; Garnett, M.C.; Amidon, G.L.; Gowland, P.A.; Spiller, R.C.; Amidon, G.E.; Marciani, L. Quantification of gastrointestinal liquid volumes and distribution following a 240 mL dose of water in the fasted state. *Mol Pharm.* **2014**, *11*, 3039–3047. [CrossRef] [PubMed]

37. Fossati, L.; Dechaume, R.; Hardillier, E.; Chevillon, D.; Prevost, C.; Bolze, S.; Maubon, N. Use of simulated intestinal fluid for Caco-2 permeability assay of lipophilic drugs. *Int. J. Pharm.* **2008**, *360*, 148–155. [CrossRef]

38. Kampa, M.; Papakonstanti, E.A.; Hatzoglou, A.; Stathopoulos, E.N.; Stournaras, C.; Castanas, E. The human prostate cancer cell line LNCaP bears functional membrane testosterone receptors that increase PSA secretion and modify actin cytoskeleton. *FASEB J.* **2002**, *16*, 1429–1431. [CrossRef]

MDPI
St. Alban-Anlage 66
4052 Basel
Switzerland
Tel. +41 61 683 77 34
Fax +41 61 302 89 18
www.mdpi.com

International Journal of Molecular Sciences Editorial Office
E-mail: ijms@mdpi.com
www.mdpi.com/journal/ijms